Progress in Mathematics
Volume 158

Geometry and Representation Theory of Real and *p*-adic groups

Juan Tirao
David A. Vogan, Jr.
Joseph A. Wolf
Editors

Birkhäuser
Boston • Basel • Berlin

Editors:
Juan Tirao
Facultad de Mathemática,
Astronomia y Fisica
Universidad Nacional de Córdoba
Ciudad Universitaria
Córdoba, Argentina 5000

David A. Vogan, Jr.
Department of Mathematics
Massachusetts Institute of Technology
Cambridge, MA 02139

Joseph A. Wolf
Department of Mathematics
University of California
Berkeley, CA 94720

Library of Congress Cataloging-in-Publication Data

Geometry and representation theory of real and p-adic groups / Juan Tirao, David A. Vogan, Jr., Joseph A. Wolf, editors.
p. cm. -- (Progress in mathematics ; v. 158)
Papers from the Fifth Workshop on Representation Theory of Lie Groups and Its Applications.
Includes bibliographical references
ISBN-13: 978-1-4612-8681-3

1. Lie groups. 2. Representations of groups. I. Tirao, Juan, 1942- . II. Vogan, David, A., 1954- . III. Wolf, Joseph Albert, 1936- . IV. Workshop on Representation Theory of Lie Groups and Its Applications (5th : 1995 : Universidad Nacional de Córdoba) V. Series: Progress in mathematics (Boston, Mass.) ; vol. 158.
QA387.G46 1997 97-36150
512'.55-dc21 CIP

AMS Classifications: 22E25, 22E45, 22E46, 11F70, 53C30, 17B30, 32M10

Printed on acid-free paper

 Birkhäuser

Softcover reprint of the hardcover 1st edition 1998

ISBN-13: 978-1-4612-8681-3 e-ISBN-13: 978-1-4612-4162-1
DOI: 10.1007/978-1-4612-4162-1

Reformatted from disk and typeset by TEXniques, Boston, MA

9 8 7 6 5 4 3 2 1

CONTENTS

PREFACE

The representation theory of Lie groups plays a central rôle in both classical and recent developments in many parts of mathematics and physics. In August, 1995, the Fifth Workshop on Representation Theory of Lie Groups and its Applications took place at the Universidad Nacional de Córdoba in Argentina. Organized by Joseph Wolf, Nolan Wallach, Roberto Miatello, Juan Tirao, and Jorge Vargas, the workshop offered expository courses on current research, and individual lectures on more specialized topics. The present volume reflects the dual character of the workshop. Many of the articles will be accessible to graduate students and others entering the field. Here is a rough outline of the mathematical content. (The editors beg the indulgence of the readers for any lapses in this preface in the high standards of historical and mathematical accuracy that were imposed on the authors of the articles.)

Connections between flag varieties and representation theory for real reductive groups have been studied for almost fifty years, from the work of Gelfand and Naimark on principal series representations to that of Beilinson and Bernstein on localization. The article of Wolf provides a detailed introduction to the analytic side of these developments. He describes the construction of standard tempered representations in terms of square-integrable partially harmonic forms (on certain real group orbits on a flag variety), and outlines the ingredients in the Plancherel formula. Finally, he describes recent work on the complex geometry of real group orbits on partial flag varieties. This is related to very active research on geometric realization of singular unitary representations.

The "singular" tempered representations neglected in Wolf's article take center stage in Garnica's. The basic problem he considers is a very old one: to understand the reducibility of representations unitarily induced from discrete series representations. Knapp and Stein showed that for connected G, the irreducible constituents of such an induced representation all have multiplicity one, and are parametrized by the characters of an "R-group," which is a product of copies of $\mathbb{Z}/2\mathbb{Z}$. Garnica extends the work of Knapp and Stein to disconnected groups. He finds that the irreducible constituents all have the same multiplicity (now a power of two), and that the number of constituents is also a power of two. Both of these numbers are described in terms of an R-group.

Galina's article explores further the connection between the geometric constructions that Wolf describes and formulas for the characters of representations. Essentially she finds the geometric content in an old formula of Harish-Chandra, Hirai, and Wolf for characters of induced representations.

The coefficients in character formulas can often be expressed as dimensions of certain Lie algebra homology groups. (This idea goes back at least to Kostant's work on the Bott-Borel-Weil theorem in 1961. One recent incarnation is a conjecture of Osborne proved by Hecht and Schmid.) Because the character of a representation is independent of the particular globalization chosen, the Lie algebra homology should not depend strongly on the globablization. Hecht and Taylor prove an unpublished theorem of Casselman of this nature: that the Lie algebra homology groups of a maximal nilpotent subalgebra with coefficients in a Harish-Chandra module or its smooth globalization coincide.

The problem of geometric realization of representations on flag varieties is taken up again in Bratten's article. A fundamental problem in the field is that representations on Dolbeault cohomology are almost always too large to carry G-invariant Hilbert space structures. (Wong has shown that they always carry the "maximal globalization" of the underlying Harish-Chandra module.) One would like to have an equally compelling geometric realization of the minimal globalization, since that space can carry a G-invariant inner product. Bratten proves that the compactly supported cohomology groups (with coefficients still in sheaves of germs of sections of holomorphic vector bundles) provide such realizations.

The connection between flag varieties and representations has an algebraic side as well. Two cornerstones are Zuckerman's "cohomological induction" construction of representations, and the Beilinson-Bernstein algebraic localization theory. The article of Miličić and Pandžić addresses the foundations of both of these constructions. Their definitions clarify and simplify a number of basic results, including the "duality theorem" of Hecht, Miličić, Schmid, and Wolf.

Another very active research area is the theory of reductive dual pairs. Weil found in the 1960s a group-theoretic framework for the theory of theta functions, and made it into a powerful construction of automorphic forms. Howe abstracted from Weil's work the following setting. One has a large reductive group H, a very small representation ρ of H, and a pair of reductive subgroups G_1 and G_2, each of which is the centralizer of the other in H. Essentially one studies the restriction of ρ to $G_1 \times G_2$. Howe has made this setting into a tool for studying group representations. His paper with Kraft is a beautiful introduction to these ideas. The goal is to understand the restriction to a maximal compact subgroup of a holomorphic discrete series representation. The authors explain why this is problem is interesting, why it is a problem in invariant theory, and then why the theory of reductive dual pairs has something to say about it. Finally they prove some powerful new results.

The article of Graham and Vogan is first of all an introduction to the Kirillov-Kostant philosophy of coadjoint orbits. This philosophy provides a geometric "classical analogue" of unitary representation theory, in the same sense that classical mechanics is an analogue of quantum mechanics. The simplest technique for constructing representations from coadjoint orbits relies on the existence of a G-invariant Lagrangian foliation of the orbit. Graham and Vogan consider nilpotent coadjoint orbits, for which such foliations usually do not exist. They find instead G-invariant Lagrangian coverings: larger families of Lagrangian subspaces, with a compact family passing through each point. They use such coverings to describe representations of G, but are not able to find unitary structures.

Of course there is much more to Lie theory than the theory of reductive groups. The article of Levstein and Tiraboschi concerns a class of two-step nilpotent (or "metabelian") real Lie groups G. The "regular" condition is that every non-trivial irreducible unitary representation of G is square-integrable modulo the center. This condition has a direct algebraic description, but is not easy to check; roughly speaking, one needs to verify that every non-zero element in a linear space of skew-symmetric bilinear forms is non-degenerate. The authors analyze regular metabelian Lie algebras, and classify those with two-dimensional center. They find some unexpected new examples.

A connection between Lie theory and differential geometry is explored in Burde's article. An affine structure on a manifold M is a maximal collection of charts with the property that any change of coordinate map between two charts is locally an affine map $x \mapsto Ax + b$ on $\mathbb{R}^n$. Burde characterizes left-invariant affine structures on a Lie group G in terms of what he calls étale affine representations: homomorphisms α of G into the affine group $\mathrm{Aff}(E)$ of a vector space E, with the property that some $v \in E$ has a discrete stabilizer in G and an open orbit in E. Semisimple groups never admit such representations, and nilpotent groups often do; indeed Milnor conjectured in 1977 that every nilpotent Lie group admits an étale affine representation. Burde makes a thorough study of such representations. He finds very general conditions for their existence, but is also able to construct nilpotent groups for which they do not exist.

In the theory of p-adic reductive groups, one of the fundamental results is Borel and Casselman's description of the representations with an Iwahori-fixed vector. Briefly, suppose G is a split group over a p-adic field, and J is an Iwahori subgroup. Borel showed that a smooth irreducible representation π of G admits a J-fixed vector if and only if π is a subquotient of an unramified principal series representation of G. It follows that such representations may be studied by means of the Iwahori Hecke algebra of compactly supported J-bi-invariant functions on G. This Hecke algebra has a relatively simple structure (found by Matsumoto) and its representations were explicitly parametrized by Kazhdan and Lusztig. In the case of $GL(n)$ over fields of large residual characteristic, Howe and Moy found a way to study arbitrary representations, replacing J and the trival representation of J by smaller compact open subgroups and representations of them. They constructed isomorphisms between the corresponding Hecke algebras and Iwahori Hecke algebras for smaller groups over extension fields. Finally Bushnell and Kutzko removed the restrictions on the residual characteristic. Kutzko's article here provides a careful and elementary introduction to his work with Bushnell, as well as to ideas for extending their work to groups other than $GL(n)$.

The ideas described in Kutzko's article reduce many problems in the representation theory of p-adic groups to the case of unramified representations. Unramified representations are the subject of Barbasch's article on his work with Moy, especially the unitary spherical representations of classical groups. Barbasch describes large families of particularly interesting unitary representations, and introduces the technology (particularly Lusztig's theory of graded Hecke algebras) needed to study them.

The article of Johnson provides another kind of introduction to the representation theory of $GL(n)$ over a p-adic field F. Johnson considers supercuspidal representations of a special kind first constructed by Carayol. By an old result of Gelfand and Kazhdan, every supercuspidal representation of $GL(n)$ admits a Whittaker model; Johnson constructs these models explicitly for Carayol's representations.

One of the basic reasons for existence of representation theory for reductive groups over real and p-adic fields (that is, local fields) is the representation-theoretic theory of automorphic forms. Those ideas are represented here by the article of Rohlfs and Speh. One of Langlands' central conjectures is that the theory of automorphic forms is controlled by dual groups, and therefore that automorphic forms on different inner forms of the same group are closely related. Rohlfs and Speh investigate this idea in the case of inner forms of orthogonal groups, and the automorphic forms contributing to the analytic torsion of locally symmetric spaces. (Changing the inner form of the orthogonal group in this case amounts to modifying the signature of the underlying quadratic form.) Using ideas of Kottwitz about stabilizing the trace formula, and a careful study of certain special representations of real groups, they obtain very precise information about analytic torsion on different inner forms.

Juan Tirao
David A. Vogan, Jr.
Joseph A. Wolf
Editors

THE SPHERICAL DUAL FOR p-ADIC GROUPS

Dan Barbasch

0. Introduction

The local Langlands conjectures have played a very significant role in the study of the representation theory of reductive algebraic groups. Roughly they say that the parametrization of equivalence classes of irreducible representations should be given in terms of conjugacy classes of continuous homomorphisms of the Weil group $W_{\mathbb{F}}$ into the dual group ${}^LG$. When G is split, the local Langlands conjectures say basically that the parametrization of equivalence classes of irreducible representations should be given in terms of conjugacy classes of homomorphisms of the Weil group $W_{\mathbb{F}}$ into the dual group ${}^LG$. More precisely, let ${}^{\vee}G$ be the connected complex group with root data dual to the root data of G. Then consider ${}^{\vee}G$ conjugacy classes of continuous homomorphisms

$$\phi : W_{\mathbb{F}} \longrightarrow {}^LG, \tag{0.1}$$

such that the image consists of semisimple elements. In the real case, these conjectures were crucial for the classification of admissible irreducible $(\mathfrak{g}, K)$ modules in the work of Langlands, Shelstad, Knapp–Zuckerman and Vogan. In the p-adic case they play a significant role in the work of Kazhdan–Lusztig and Lusztig. There is a technical modification in that one considers maps of the Weil–Deligne–Langlands group,

$$\phi : W_{\mathbb{F}} \times SL(2) \longrightarrow {}^LG. \tag{0.2}$$

This parametrization is not very well suited for describing the unitary dual (except for the tempered part). However, motivated by global considerations, Arthur proposed that in (0.1) and (0.2) one ought to replace $W_{\mathbb{F}}$ ($W_{\mathbb{F}} \times SL(2)$ respectively) by its product with another $SL(2)$. Denoting the group in (0.1), (0.2) by $\mathcal{W}_{\mathbb{F}}$, the parametrizing space would be conjugacy classes of maps

$$\psi : \mathcal{W}_{\mathbb{F}} \times SL(2,\mathbb{C}) \longrightarrow {}^LG \qquad (\mathcal{W}_{\mathbb{F}} = W_{\mathbb{F}} \textit{ or } W_{\mathbb{F}} \times SL(2)) \tag{0.3}$$

such that the restriction to $\mathcal{W}_{\mathbb{F}}$ represents a tempered parameter (see Section 1.4 for details about the parametrization of tempered representations). We call (0.1) (or (0.2)) a *Langlands parameter* and (0.3) an *Arthur parameter.* A Langlands parameter ϕ_ψ is then obtained from ψ by the procedure

$$\phi_\psi(w) = \psi(w, \begin{bmatrix} |w|^{1/2} & 0 \\ 0 & |w|^{-1/2} \end{bmatrix}). \tag{0.4}$$

Arthur conjectures that, to each such map one should attach a family of representations with special properties, in particular they should occur as local factors in the residual spectrum of automorphic forms, and they should be *unitary.* For complex reductive groups (viewed as real groups), the representations attached to parameters satisfying $\psi|_{\mathcal{W}_\mathbb{F}} = triv$ are called *special unipotent.* They play a central role in the classification of the unitary dual.

The purpose of these lectures is to describe the progress I made (jointly with Allen Moy) on determining the unitary dual for p-adic groups. The spherical part of the unitary dual is significant for the theory of automorphic forms which deals with the decomposition of $L^2(G(\mathbb{Q})\backslash G(\mathbb{A}))$, where $\mathbb{A}$ are the adeles of a number field k. The composition factors are of the form $\otimes_v \pi_v$ where π_v is a spherical unitary representation of a quasisplit group for all but finitely many places of k.

The complete proofs of our results can be found in [BM1]–[BM3]. These papers deal with the case of representations admitting fixed vectors under an Iwahori subgroup. Papers [BM1] and [BM2] reduce the problem to an equivalent one for the Hecke algebra. Given an algebraic linear reductive p-adic group G, let $G(\mathcal{R})$ be its points defined over the integers. A representation is called spherical if it has nontrivial vectors fixed under $G(\mathcal{R})$. Then [BM3] determines the unitary spherical dual for split classical p-adic groups.

The starting point of the determination of the spherical unitary dual is the fact that the Iwahori–Matsumoto involution $\mathcal{IM}$ preserves unitarity ([BM2]). In particular if π is tempered, then it is unitary and so $\mathcal{IM}(\pi)$ is unitary. The set $\mathcal{IM}(\pi)$ gives a large class of spherical representations which are unitary and mostly isolated. It is then fairly easy to check that they are exactly the ones attached to Arthur parameters. Sections 1 and 2 provide a brief summary of the parts of [BM1] and [BM2] that are needed. In section 3 we do the special case of complementary series of representations induced from isolated Arthur parameters on Levi components of maximal parabolic subgroups. These seemed to be of special interest to C. Moeglin as part of her work on the residual spectrum. The reader interested in the most general situation and detailed proofs is referred to [BM3].

In view of the above results, a reasonable conjecture describing the spherical unitary dual would be

A spherical representation is unitary if and only if it is a complementary series formed from a unitarily induced representation of an Arthur parameter on a Levi component.

This is certainly the case for $GL(n)$ as follows from the work of Tadic. The spherical unitary dual in this case is formed of unitarily induced representations from unitary characters on Levi components and their complementary series (also called Stein complementary series).

For the other classical groups, the unitary dual is more complicated. There are complementary series off of complementary series for $GL(n)$ such that the nearest induced from a tempered representation is reducible. We give examples of this phenomenon in section 3.

As already mentioned, all results are joint work with Allen Moy. We deal mainly with the types B, C and D since the case of $GL(n)$ was already done in [T]. The results were also presented at a conference in Twente in December 1994. This version is mainly expository in nature. I hope that these notes form a useful complement to [BM3].

Finally, I would like to thank the organizers of the conference in Cordoba for their hospitality.

1. Preliminaries

1.1. Unramified representations. Let $\mathbb{F}$ be a p-adic field, and let (1.1.1)

$$\mathcal{R} = \{x \in \mathbb{F} \ : \ |x| \leq 1\},\ \mathcal{P} = \{x \in \mathbb{F} \ : \ |x| < 1\},\ \Omega = \{x \in \mathbb{F} \ : \ |x| = 1\}.$$

If G is a split reductive group over $\mathbb{F}$, denote by A a maximally split torus and fix a Borel subgroup with decomposition $B = AN$. Then $K = G(\mathcal{R})$ is a maximal compact subgroup. We choose B so that $G = KAN$. There is an exact sequence

$$1 \longrightarrow K_1 \longrightarrow K \longrightarrow G(\mathcal{R}/\mathcal{P}) \cong G(\mathbb{F}_q) \longrightarrow 1. \tag{1.1.2}$$

An unramified character of A is a character ν which is trivial on $A \cap K$.

Let $\nu \in \hat{A}$ be an unramified character. Extend it to B so that it is trivial on N. Denote by $I(\nu)$ the *unramified* principal series. Precisely, $I(\nu)$ is the space of functions $f : G \longrightarrow \mathbb{C}$ satisfying $f(gb) = \nu(b^{-1})\delta_B(b)^{-1/2}f(b)$ and with compact support modulo B. The action of G is by left translations.

It is important to study the composition series and irreducible subquotients of $I(\nu)$. This is done via the Hecke algebra attached to an Iwahori subgroup. Fix a Borel subgroup in $G(\mathbb{F}_q)$. Then its inverse image is an open compact subgroup called the Iwahori subgroup and denoted $\mathcal{I}$. The algebra $\mathcal{H}$ (under convolution) of compact supported biinvariant functions $\mathcal{H}(\mathcal{I}\backslash G/\mathcal{I})$ is called the Iwahori–Hecke algebra. It has a vector space decomposition

$$\mathcal{H}(\mathcal{I}\backslash G/\mathcal{I}) = \mathcal{H}_K \otimes_{\mathbb{C}} \mathcal{A}, \tag{1.1.3}$$

where $\mathcal{H}_K = \mathcal{H}(\mathcal{I}\backslash K/\mathcal{I})$ identifies with the finite Hecke algebra of $B(\mathbb{F}_q)$–biinvariant functions on $G(\mathbb{F}_q)$. The part denoted $\mathcal{A}$ is an abelian algebra. It is described as follows. Let $\mathcal{Y} = Hom(G_m, A)$ and $\check{\Pi} \subset \check{R}^+ \subset \check{R} \subset \mathcal{Y}$ be the simple coroots, positive coroots and coroots defined by $B = AN$. Similarly let $\mathcal{X} = \check{\mathcal{Y}} = Hom(A, G_m)$ and let $\Pi \subset R^+ \subset R \subset \mathcal{Y}$ be the simple roots, positive roots and roots. Let W be the Weyl group.

The double cosets of $\mathcal{I}$ in G are naturally parametrized by the semidirect product $W^a := W \rtimes \mathcal{Y}$. Let $s_\alpha \in W$ denote the reflection corresponding to the simple root α and $s_0 \in W^a$ be the affine reflection. A basis for $\mathcal{H}_K$ is then given by

$$T_w = \text{ characteristic function of the double coset } \mathcal{I}\tilde{w}\mathcal{I}, \tag{1.1.4}$$

where $\tilde{w}$ is a representative of $w \in W$. The algebra $\mathcal{H}_K$ is generated by T_{s_α} subject to the relations

$$(T_{s_\alpha} + 1)(T_{s_\alpha} - q) = 0. \tag{1.1.5}$$

Let

$$\mathcal{Y}^+ := \{y \in \mathcal{Y} \mid \alpha(y) \leq 0, \text{ for all } \alpha \in \Pi\}.$$

Then for $y \in \mathcal{Y}^+$ and ϖ a (fixed) prime element in $\mathbb{F}$, set

$$\theta_y := \text{ characteristic function of } \mathcal{I}y(\varpi)\mathcal{I}. \tag{1.1.6}$$

Any element $L \in \mathcal{Y}$ can then be written as $L = y_1 - y_2$ with $y_1, y_2 \in \mathcal{Y}^+$. Then

$$\theta_L := \theta_{y_1}\theta_{y_2}^{-1} \tag{1.1.7}$$

does not depend on the choice of decomposition of L. The algebra $\mathcal{A}$ is the subalgebra generated by the θ_L's.

Example 1. Suppose $G = Sp(1) \cong SL(2)$. Then we can identify

$$\mathcal{Y} = \mathbb{Z}e, \quad \mathcal{X} = \mathbb{Z}\epsilon, \tag{1.1.8}$$

where e and ϵ are dual bases. Then $\Pi = \{2\epsilon\}$. Furthermore,

$$W^a = \mathbb{Z}_2 \rtimes \mathcal{Y} \tag{1.1.9}$$

where the nontrivial element $t \in \mathbb{Z}_2$ acts by

$$t(e) = -e.$$

On the other hand, we have

$$\begin{aligned} K &= \{\begin{bmatrix} a & b \\ c & d \end{bmatrix} a, b, c, d \in \mathcal{R}, \quad ad - bc = 1\} \\ \mathcal{I} &= \{\begin{bmatrix} a & b \\ \varpi c & d \end{bmatrix} \in K \ : \ a, b, c, d \in \mathcal{R}\}. \end{aligned} \tag{1.1.10}$$

The decompositions are

(1.1.11)
$$K = \mathcal{I} \cup \mathcal{I}\begin{bmatrix} 0 & 1 \\ -1 & 0 \end{bmatrix}\mathcal{I},$$
$$G = \bigcup \mathcal{I}\begin{bmatrix} \varpi^{a_1} & 0 \\ 0 & \varpi^{a_2} \end{bmatrix}\mathcal{I} \bigcup \mathcal{I}\begin{bmatrix} \varpi^{a_1} & 0 \\ 0 & \varpi^{a_2} \end{bmatrix}\begin{bmatrix} 0 & 1 \\ -1 & 0 \end{bmatrix}\mathcal{I}.$$

Example 2. Suppose $G = SO(3) \cong PGL(2)$. Then

(1.1.12)
$$\mathcal{Y} = \mathbb{Z}e, \qquad \mathcal{X} = \mathbb{Z}\epsilon,$$

where e and ϵ are dual bases, and $\Pi = \{\epsilon\}$. Then

(1.1.13)
$$K = \{\begin{bmatrix} a & b \\ c & d \end{bmatrix} \in GL(2)\} \pmod{\mathcal{Z}[GL(2)]}$$
$$\mathcal{I} = \{\begin{bmatrix} a & b \\ \varpi c & d \end{bmatrix} \in K\}.$$

The decompositions are

(1.1.14)
$$K = \mathcal{I} \cup \mathcal{I}\begin{bmatrix} 0 & 1 \\ -1 & 0 \end{bmatrix}\mathcal{I},$$
$$G = \bigcup_{w \in W^a} \mathcal{I}w\mathcal{I},$$

where W^a is generated by the cosets

$$\begin{bmatrix} 0 & 1 \\ -1 & 0 \end{bmatrix}, \begin{bmatrix} \varpi^k & 0 \\ 0 & \varpi^{-k} \end{bmatrix}, \begin{bmatrix} \varpi & 0 \\ 0 & 1 \end{bmatrix} (\text{ in } PGL(2)).$$

Let

(1.1.15)
$\mathcal{C}(\mathcal{I}) :=$ the category of admissible finite length representations so that their subquotients are generated by $\mathcal{I}$–fixed vectors,
$\mathcal{C}(\mathcal{H}) :=$ the category of finite dimensional representations of $\mathcal{H}$.

Theorem. *([B]) The functor $V \mapsto V^{\mathcal{I}}$ is an equivalence of categories from $\mathcal{C}(\mathcal{I})$ to $\mathcal{C}(\mathcal{H})$, The inverse is given by $W \mapsto \mathcal{H}(G/\mathcal{I}) \otimes W$.*

An irreducible representation π is in $\mathcal{I}$ iff it is a subquotient of an $I(\nu)$.

A sketch of the proof can be found in [Car].

$\mathcal{H}$ also has a star operation given by

(1.1.16)
$$f^*(x) = \overline{f(x^{-1})},$$

so we can talk about hermitian and unitary modules. It is more or less clear that an admissible representation $V \in \mathcal{C}(\mathcal{I})$ of G is hermitian iff $V^{\mathcal{I}}$ is hermitian. It is also clear that if V is unitary, then so is $V^{\mathcal{I}}$. The next theorem is the main result of [BM1], [BM2], and is crucial for our approach in determining the unitary dual.

Theorem. *An irreducible representation in $C(\mathcal{I})$ is unitary if and only if $V^{\mathcal{I}}$ is unitary.*

1.2. The Generic and Graded Hecke Algebras. The Hecke algebra $\mathcal{H}$ can be described as the specialization of a generic Hecke algebra. We describe this generic algebra in terms of the root datum of the *complex* dual group ${}^\vee G$, with maximal torus ${}^\vee A$ and Borel subgroup ${}^\vee B$ containing ${}^\vee A$. Let z be an indeterminate (which we will specialize to $q^{1/2}$ to recover the Hecke algebra $\mathcal{H}$ of the p-adic group). Let $\Pi \subset R^+ \subset R$ be the simple roots, positive roots and roots corresponding to $({}^\vee B, {}^\vee A)$ and S be the simple root reflections. Let $\mathcal{Y} = Hom(G_m, {}^\vee A)$ and $\mathcal{X} = Hom({}^\vee A, G_m)$. Then the generic Hecke algebra $\mathcal{H}(z)$ is an algebra over $\mathbb{C}[z, z^{-1}]$ described in terms of the root datum $\mathcal{R} = (\mathcal{X}, \mathcal{Y}, R, \check{R}, \Pi)$. (This set is the dual data to the p-adic group G in section 1.1.) The set of generators we will use is the one first introduced by Bernstein. Write

$$\mathcal{A}(z) := \text{regular functions on } \mathbb{C}^* \times {}^\vee A. \tag{1.2.1}$$

This can be viewed as the algebra generated by $\{\theta_x\}_{x\in\mathcal{X}}$ with coefficients in $\mathbb{C}[z, z^{-1}]$, Laurent polynomials in z. Then $\mathcal{H}(z)$ is generated (over $\mathbb{C}[z, z^{-1}]$) by $\{T_w\}_{w\in W}$ and $\mathcal{A}(z)$ subject to the relations

$$\begin{aligned} T_w T_{w'} &= T_{ww'} \quad (l(w) + l(w') = l(ww')),\\ T_s^2 &= (z^2 - 1)T_s + z^2,\\ \theta_x T_s &= T_s\theta_{sx} + (z^2 - 1)\frac{\theta_x - \theta_{sx}}{1 - \theta_\alpha}. \end{aligned} \tag{1.2.2}$$

Specialized at $z = q^{1/2}$, $\mathcal{A}(z)$ gives an algebra isomorphic to $\mathcal{A}$ in (1.1.3), and $\mathcal{H}(z)$ gives $\mathcal{H}$. This realization is very convenient for detemining the center of $\mathcal{H}$ and thus parametrizing infinitesimal characters of representations. Note that W acts on $\mathcal{A}$ via the formula $w \cdot \theta_x = \theta_{wx}$.

Theorem 1. *(Bernstein–Lusztig) The center of $\mathcal{H}(z)$ is given by the Weyl group invariants in $\mathcal{A}(z)$.*

In particular, infinitesimal characters are parametrized by W–orbits $\chi = (q, t) \in \mathbb{C}^* \times {}^\vee A$. Such an infinitesimal character is called *real* if t has no elliptic part.

The study is simplified by using the graded Hecke algebra introduced by Lusztig in [L2]. Given an arbitrary infinitesimal character (q, t), decompose t into its elliptic and hyperbolic part $t = t_e t_h$. In [L2] the graded algebra $\mathbb{H}_{t_e}$ is introduced. This is done by considering the ideal $\mathcal{H}^1$ generated by $z - 1$ and $\theta_x - \theta_x(t_e)$. Then $\mathcal{H}^i$ is a filtration of ideals of $\mathcal{H}$ and one can consider the graded algebra $\mathbb{H}_{t_e}$. This algebra is generated by $t_w = T_w$ in $\mathcal{H}/\mathcal{H}^1$ and $\omega_x = \theta_x - \theta_x(t_e)$ in $\mathcal{H}/\mathcal{H}^2$ with coefficients in $\mathbb{C}[\mathbf{r}]$ where $\mathbf{r} = z - 1$ in $\mathcal{H}/\mathcal{H}^1$.

Example. Suppose we take the case of Example 1 in 1.1. Then $\mathcal{A}$ is generated by the character θ corresponding to the root. The last relation in (1.2.2) is

$$\theta T = T\theta^{-1} + (z^2-1)\frac{\theta-\theta^{-1}}{1-\theta} = T\theta^{-1} - (z^2-1)(\theta^{-1}+1). \tag{1.2.3}$$

Grading at $t_e = 1$, gives $\omega t_\alpha = -t_\alpha \omega + 4\mathbf{r}$.

In Example 2, $\mathcal{A}$ is generated by a θ such that θ^2 is the character corresponding to the root. The third relation in (1.2.2) is

$$\theta T = T\theta^{-1} + (z^2-1)\frac{\theta-\theta^{-1}}{1-\theta^2} = T\theta^{-1} - (z^2-1)\theta^{-1}. \tag{1.2.4}$$

Grading at $t_e = 1$ gives $\omega t_\alpha = -t_\alpha \omega + 2\mathbf{r}$.

On the other hand, grading (1.2.4) at $t_e = \begin{bmatrix} i & 0 \\ 0 & -i \end{bmatrix}$ means that $\theta+1 \in \mathcal{J}$, which gives the relation $\omega t = -t\omega$. This gives a very different algebra from before; namely the group algebra of the affine Weyl group. It accounts for the reducibility of a certain tempered induced module for $SL(2)$.

We summarize some properties when the grading is done at $(1,1)$. Let

$$\mathcal{J} = \{f \in \mathcal{A} \;:\; f(1,1) = 0\}. \tag{1.2.5}$$

This ideal satisfies $\mathcal{HJ} = \mathcal{JH}$, so we can introduce a filtration

$$\mathcal{H} = \mathcal{H}^0 \supset \cdots \supset \mathcal{H}^i \supset \mathcal{H}^{i+1} \supset \ldots, \tag{1.2.6}$$

and form the graded object $\mathbb{H}$. As a vector space it can be written as

$$\mathbb{H} = \mathbb{C}[\mathbf{r}]W \otimes_{\mathbb{C}} \mathbb{A}, \tag{1.2.7}$$

where $\mathbf{r} \equiv z - 1 (mod\ \mathcal{J})$, and $\mathbb{A}$ is the symmetric algebra over ${}^{\vee}\mathfrak{a} = \mathcal{X} \otimes_{\mathbb{Z}} \mathbb{C}$. The generators satisfy the relations

$$\begin{aligned} t_w t_{w'} &= t_{ww'}, \\ t_\alpha^2 &= 1, \\ t_\alpha \omega &= s_\alpha(\omega) t_s + 2\mathbf{r} < \omega, \check{\alpha} >, \quad (t_\alpha = t_{s_\alpha},\ \omega \in {}^{\vee}\mathfrak{a}). \end{aligned} \tag{1.2.8}$$

Then W acts on $\mathbb{A}$ in the usual way and the center of $\mathbb{H}$ is $\mathbb{A}^W$. In particular, infinitesimal characters are parametrized by W–orbits of elements $\overline{\chi} = (r, s) \in \mathbb{C} \times {}^{\vee}\mathfrak{a}$. Such an infinitesimal character is called *real* if s is hyperbolic.

We can specialize z to $q^{1/2}$ in the generic algebra and $\mathbf{r}$ to r in the graded Hecke algebra. We fix a choice of q which is not a root of unity (in fact, is a power of a prime in the case of a p-adic group) and an r such that $e^r = q$. The study of representations of $\mathcal{H}$ with infinitesimal character (q,t) having elliptic part t_e, is reduced to the study of the representation theory of $\mathbb{H}_{t_e}$ where $\mathbf{r}$ is specialized to r satisfying $q = e^r$. Furthermore $\mathbb{H}_{q,t_e}$ is Morita equivalent to an $\mathbb{H}$ as above, which is the graded version of some related Hecke algebra at $(1,1)$. The subject of [BM2] is to show that in fact all questions of unitarity can be reduced to the case of a Hecke algebra obtained by grading at $t_e = 1$. As a consequence, we only need to study the representation theory and unitary spectrum of $\mathbb{H}$, and at that we only need to consider real infinitesimal character (*i.e.* $t_e = 1$). The next theorem is a summary of what we need from [L2]. More details and some facts that we need from [BM2] are in 1.3.

Theorem 2. *([L2]) There is a matching $\chi \longleftrightarrow \overline{\chi}$ between real infinitesimal characters χ of $\mathcal{H}$ and real infinitesimal characters $\overline{\chi}$ of $\mathbb{H}$ so that if $\mathcal{H}_\chi$ and $\mathbb{H}_{\overline{\chi}}$ are the quotients by the corresponding ideals, then*

$$\mathcal{H}_\chi \cong \mathbb{H}_{\overline{\chi}}.$$

The classification of irreducible representations is given by the work of Kazhdan–Lusztig for $\mathcal{H}$ and Lusztig for $\mathbb{H}$:

Theorem 3. *([KL], [L4], [L5]) The irreducible representations of $\mathbb{H}$ are parametrized by ${}^\vee G$ conjugacy classes (s,e,ψ), where $s \in {}^\vee\mathfrak{g}$ is semisimple, $e \in {}^\vee\mathfrak{g}$ is nilpotent such that $[s,e] = re$ and $\psi \in \widehat{A(s,e)}$ is an irreducible representation of the component group of the centralizer of s and e. The characters ψ that appear are the same ones that occur in the Springer correspondence.*

1.3. Hermitian Modules. The * operation also transfers to the graded version. Here is a summary of what we need. We refer to §5 of [BM2] for the details.

Let $w_0 \in W$ be the longest element, t_0 be the corresponding element in $\mathbb{C}W$. Since ${}^\vee\mathfrak{a} = \mathcal{X} \otimes_{\mathbb{Z}} \mathbb{C}$, it has a conjugation coming from the complex conjugation on $\mathbb{C}$. We denote it by $^-$. Let $\iota(\omega) = (-1)^{deg\,\omega}\overline{\omega}$ and $\tilde{\omega} = w_0\iota(\omega)$.

Theorem . *(§5 in [BM2]) Let $\omega \in \mathbb{A}$. Then*

$$t_w^* = t_{w^{-1}},$$
$$\omega^* = t_0 \cdot \tilde{\omega} \cdot t_0.$$

In particular, if $\omega \in {}^\vee\mathfrak{a}$, then

$$\omega^* = -\overline{\omega} + 2r \sum_{\beta \in R^+} < \overline{\omega}, \check{\beta} > t_\beta, \tag{1.3.1}$$

where $t_\beta \in \mathbb{C}[W]$ is the reflection about β.

The main result of [BM2] can be summarized as follows. Let (e,H,f) be a Lie triple corresponding to e. Write $t = t_0 t_H$, where t_0 is an element centralizing the triple and $t_H = exp(1/2 \log qH)$. Let $t_0 = t_e t_h$ be the decomposition of t_0 into elliptic and hyperbolic parts. Denote by $\overline{t} := t_e t_h^{-1}$. An irreducible representation admits a hermitian form if and only if (t, e, ψ) is conjugate to $(\overline{t_0} t_H, e, \psi)$. An infinitesimal character (or parameter) will be called *real* if $t_e = 1$.

Corollary . *The classification of the unramified unitary dual of a split p-adic group having infinitesimal character with a given elliptic part t_e reduces to the classification of the unitary dual of the corresponding graded Hecke algebra $\mathbb{H}_{q,t_e}$. Furthermore, this is equivalent to the classification of the unitary dual of an algebra $\mathbb{H}$ obtained from a Hecke algebra by grading at $(e^r, 1)$. In other words, it is sufficient to consider the case of parameters with* real *infinitesimal character.*

The relation between s in the theorem in 1.2 and the t in this corollary is $e^s = t_h t_H$.

1.4. We summarize some of the basic results about modules of $\mathbb{H}$ related to Theorem 3 of Section 1.2. We may as well take $r = 1$.

Let (s, e, ψ) be a parameter as in theorem 3 of 1.2, and let $\{e, h, f\}$ be a Lie triple such that $s = s_0 + \frac{1}{2}h$ with s_0 in the centralizer of $\{e, h, f\}$. Denote by $\mathcal{O}$ the ${}^\vee G$ orbit of e. To each such parameter is associated a standard module $X(s, \mathcal{O})$ which decomposes into a direct sum of standard modules $X(s, \mathcal{O}, \psi)$ where ψ ranges over the characters of $A(s, e)$. Each $X(s, \mathcal{O}, \psi)$ has a unique irreducible quotient $L(s, \mathcal{O}, \psi)$. Every irreducible module is isomorphic to an $L(s, \mathcal{O}, \psi)$, and the factors of $X(s, \mathcal{O}, \psi)$ have parameters $(s, \mathcal{O}', \psi')$ such that $\mathcal{O} \subset \overline{\mathcal{O}'}$ and $\mathcal{O} \neq \mathcal{O}'$.

A parameter is called *tempered* if $s_0 = 0$. In this case the module $X(s, \mathcal{O}, \psi)$ is irreducible and corresponds to the Iwahori fixed vectors of an irreducible tempered representation of the group. The parameter is called a *discrete series* if in addition the orbit $\mathcal{O}$ of e does not meet any proper Levi component of ${}^\vee\mathfrak{g}$. Such modules correspond to the $\mathcal{I}$–fixed vectors of a *discrete series* of the p-adic group G.

Now suppose that the pair (s, e) is contained in a Levi component ${}^\vee\mathfrak{m}$. Then we can form $X_M(s, \mathcal{O})$ and $X_G(s, \mathcal{O})$. The relation between them is

$$X_G = Ind_{\mathbb{H}_M}^{\mathbb{H}}[X_M] = \mathbb{H} \otimes_{\mathbb{H}_M} X_M. \tag{1.4.1}$$

More generally, write $A(s, e, M)$ and $A(s, e, G)$ for the corresponding component groups. Then

$$Ind_{\mathbb{H}_M}^{\mathbb{H}}[X_M(s, e, \phi)] = \sum [\psi|_{A(s,e,M)} : \phi] X_G(s, e, \psi). \tag{1.4.2}$$

In other words, $A(s, e)$ plays the role of an R–group.

We can use s_0 to construct the Levi component ${}^\vee\mathfrak{m}$. We then find a tempered representation $\mathcal{W}$ equal to $X_M(h, \mathcal{O}, \psi)$ tensored with a character ν corresponding to s_0. Then

$$X_G(s, \mathcal{O}, \psi) = Ind_{\mathbb{H}_M}^{\mathbb{H}}[\mathcal{W} \otimes \nu]. \tag{1.4.3}$$

Thus we recover the usual Langlands classification. We remark that $\mathbb{H}$ is defined in terms of a fixed system of positive roots. The data s, e can be conjugated so that $s_0 = \nu$ is dominant with respect to this system; we assume that this is the case. When we want to emphasize that we are using the usual Langlands classification, the standard module will be denoted by $X(M, \mathcal{W}, \nu)$.

The W–structure of the standard modules is also known. Let $\mathcal{B}_e$ be the variety of Borel subgroups that contain e. Then $H^*(\mathcal{B}_e)$ carries an action of W called the Springer action. It is usually normalized so that for the principal nilpotent, $H^*(\mathcal{B}_e)$ is the trivial module. It commutes with the action of the component group $A(e)$. Let $d(e) = dim\mathcal{B}_e$. Then $H^{d(e)}(\mathcal{B}_e)$ decomposes according to characters of $A(e)$. Each isotypic component is irreducible as a $W \times A(e)$–module and the ensuing $\phi \leftrightarrow \sigma_{\mathcal{O},\phi}$ is called the Springer correspondence.

Results of Borho–MacPherson imply that if $[\sigma : H^*(\mathcal{B}_e)^\phi] \neq 0$, then σ is of the form $\sigma_{\mathcal{O}',\phi'}$ where $\mathcal{O}'$ contains $\mathcal{O}$ in its closure. $\mathcal{O} \neq \mathcal{O}'$ unless $\phi = \phi'$ as well and this representation occurs with multiplicity 1. We call $\sigma_{\phi,\mathcal{O}}$ a *lowest K-type* of $H^*(\mathcal{B}_e)^\phi$.

Proposition . *(Kazhdan–Lusztig) There is an isomorphism of W–modules*

$$X(s, e) \cong H^*(\mathcal{B}_e) \otimes sgn.$$

Then $A(e)$ acts on the right hand side; the action of $A(s, e)$ on the left hand side is via the natural map $A(s, e) \to A(e)$. In particular we can talk about *lowest K-types* for $X(s, e, \psi)$. They occur with multiplicity 1 and, given our discussion above, $L(s, \mathcal{O}, \psi)$ is the unique subquotient which contains the *lowest K-types* $\sigma_{\phi,\mathcal{O}} \otimes sgn$ for which the $\phi \in \widehat{A(e)}$ contains $\psi \in \widehat{A(s, e)}$ in its restriction. If $s = h/2$, then $A(s, e) = A(e)$, and $X(s, e, \psi)$ has a unique *lowest K-type* namely $\sigma_{\psi,\mathcal{O}} \otimes sgn$.

1.5. Example. Suppose ${}^\vee G = Sp(2n, \mathbb{C})$. Then nilpotent orbits in ${}^\vee\mathfrak{g}$ are parametrized by partitions of $(x_0, \dots, x_k)$ of $2n$ (with $x_i \leq x_{i+1}$) such that every odd part occurs an even number of times. The centralizer of the corresponding Lie triple is of the form

$$\prod O(r_{2a}) \times \prod Sp(s_{2b+1})$$

where r_{2a} is the number of x_i equal to $2a$ and s_{2b+1} is the number of x_i equal to $2b+1$. The component group is then $\prod \mathbb{Z}_2$. The number of $\mathbb{Z}_2$'s equals the number of (distinct) even parts occuring in the partition.

A nilpotent orbit meets a proper Levi component if and only if $x_i = x_{i+1}$ for some i. More precisely, if say $x_i = x_{i+1} = a$, then the nilpotent orbit meets the maximal Levi component $GL(a) \times Sp(2n - 2a)$. In general we will write ${}^\vee M = GL(a) \times G(n-a)$ for such Levi components. The intersection contains the nilpotent corresponding to the partition $(a) \times (x_0, \dots, \widehat{x_i}, \widehat{x_{i+1}}, \dots, x_k)$.

Here (a) denotes the principal nilpotent in $GL(a)$. Thus we can write the nilpotent orbits corresponding to discrete series as

$$(2x_0, \ldots, 2x_k) \qquad \text{with} \qquad x_i < x_{i+1}.$$

The corresponding standard modules behave as described in 1.4. Not all characters of the component group give rise to standard modules, (or equivalently occur in the Springer correspondence). The ones that do, and the corresponding lowest K-types, are described in [L3]. We won't need the explicit description of the correspondence $\psi \leftrightarrow \sigma_\psi$. But we will need the $\sigma_{\psi,\mathcal{O}}$ that occur. The information we need is described in Section 3.

2. A Unitarity Criterion

2.1 Character theory. We review some material from [BM1]. Recall the existence of the Iwahori–Matsumoto involution $\mathcal{IM}$ on the graded Hecke algebra

$$\text{(2.1.1)} \qquad \mathcal{IM}(t_w) = (-1)^{l(w)} t_w, \qquad \mathcal{IM}(\omega) = -\omega, \qquad \omega \in {}^\vee\mathfrak{a}.$$

This has the effect that it takes a spherical module into a module containing sgn. Given the infinitesimal character χ, there is a unique irreducible $L(\chi, sgn)$ (containing sgn with multiplicity 1). The classification theorem of representations of $\mathbb{H}$ then gives the following result.

Theorem . *([BM1]) Let χ be an infinitesimal character corresponding to the semisimple element $s \in {}^\vee\mathfrak{g}$ (i.e. $\chi = exp(1/2 \log qs)$). Then there is a unique orbit of maximal dimension (with representative e_{max}) such that $[s, e_{max}] = e_{max}$. Then $L(\chi, sgn)$ equals the standard module representation $X(\chi, e_{max}, \phi = triv)$. In particular if the orbit of e meets a Levi component of a proper parabolic subalgebra, then $L(\chi, sgn)$ is induced (irreducibly) from the corresponding $L_M(\chi, sgn)$.*

Modulo the characters of tempered representations (for which there is a formula in [KL]) this theorem computes the character of $L(\chi, sgn)$. Applying $\mathcal{IM}$, we get a corresponding result for spherical representations.

2.2. Unitary representations. One of the main uses of Theorem 2.1 is that it implies the unitarity of certain Arthur parameters.

Recall that the data (s, e) can be conjugated so that e can be completed to a Lie triple (e, h, f) and $s = s_0 + h/2$ where s_0 centralizes this Lie triple. A special case is the case when $s = h/2$. Then $X(\chi, sgn)$ consists of the $\mathcal{I}$–fixed vectors of a tempered representation, therefore is *unitary.* Since by [BM1] $\mathcal{IM}$ preserves unitarity, $\mathcal{IM}(X(\chi, sgn))$ is an irreducible spherical *unitary* representation. These are special cases of *Arthur parameters.* Since the representation has Iwahori fixed vectors, the map ϕ in (0.2) factors through the inertia group $\mathbb{I}_\mathbb{F}$. Since $W_\mathbb{F}/\mathbb{I}_\mathbb{F} = \mathbb{Z}$, we get a map

$$\text{(2.2.1)} \qquad \phi : \mathbb{Z} \times SL(2) \longrightarrow {}^\vee G.$$

The restriction to $\mathbb{Z}$ can be taken to be trivial in view of the fact that we are dealing with a tempered representation with real infinitesimal character. The restriction of ϕ to $SL(2)$ is the map corresponding to (e,h,f) by the Jacobson–Morozov theorem. As an Arthur parameter we take (suppressing the factor $\mathbb{Z}$)

$$\psi : SL(2) \times SL(2) \longrightarrow {}^{\vee}G \tag{2.2.2}$$

where ψ is as before on the first factor, trivial on the second factor. Since $\mathcal{IM}(X(\chi, sgn))$ is spherical, its Langlands parameter is given by a map which is trivial on the $SL(2)$ and such that $1 \mapsto h/2$. This comes from the Arthur parameter as in (2.2.2) where the two $SL(2)$'s are interchanged, *i.e.* the map corresponds to (e,h,f) on the first $SL(2)$ and is trivial on the second one. For these representations to be isolated in the unitary spectrum, it is necessary that the orbit of e must not meet any proper Levi component of ${}^{\vee}G$, but this condition is not sufficient, as the endpoints of complementary series are not isolated.

2.3. A nonunitarity criterion. This is a technique first used by Casselman [C]. In [BM3] we have recast it in terms of graded Hecke algebras and sharpened it.

Recall that the signature of a hermitian module L for $\mathbb{H}$ is the formal sum

$$Sign[L] := \sum_{\mu \in \widehat{W}} ([L:\mu]_+ - [L:\mu]_-)\mu. \tag{2.3.1}$$

Suppose $X(M, \mathcal{W}, \nu)$ is hermitian for all real ν. This module is irreducible for large ν and so the signature is independent of ν; we call it the signature at infinity. Given a representation $\sigma \in \widehat{W}$, denote by $d(\sigma)$ the lowest degree such that σ occurs in the space of harmonic polynomials on ${}^{\vee}\mathfrak{a}$.

Theorem . *Suppose μ and μ' are K-types of $X(M,\mathcal{W},\nu)$ and μ_{K_M} is a K_M-type of $\mathcal{W}$ such that*

$$[\mu : X] = [\mu' : X] = 1, \qquad [\mu_{K_M} : \mathcal{W}] = 1,$$
$$[\mu|_{W(M)} : \mu_{K_M}] = [\mu'|_{W(M)} : \mu_{K_M}] = 1.$$

Then

$$[\mu : X]_+ - [\mu : X]_- = (-1)^{d(\mu)+d(\mu')}([\mu' : X]_+ - [\mu' : X]_-).$$

In types B_n, C_n we parametrize the Weyl group representations by pairs of partitions $(\alpha;\beta)$ such that $|\alpha| + |\beta| = n$. We sometimes write the partition as $\alpha \times \beta$. In this parametrization the trivial representation corresponds to $(n) \times \emptyset$ and sgn corresponds to $\emptyset \times (1,\dots,1)$. The same parametrization holds for type D, except $(\alpha;\beta) \cong (\beta;\alpha)$; in case $\alpha = \beta$, there are two inequivalent representations $(\alpha,\alpha)_I$ and $(\alpha,\alpha)_{II}$. In types B,C the number $(-1)^{d(\sigma)}$ is equal to $(-1)^{\sum \beta_i}$. This formula is valid for type D as well, but with a particular choice between (α,β) and (β,α).

The previous theorem is our main nonunitarity criterion. In section 3, we will pick a parameter coming from a maximal parabolic subgroup; then we can think of $\nu \in \mathbb{R}^+$. Given such a parameter $(\mathcal{W},\nu)$, we will specify K-types μ, μ' and a K_M–type μ_{K_M} satisfying the conditions of the theorem and such that $(-1)^{d(\mu)+d(\mu')} = -1$. The K-type μ always occurs in $L(\mathcal{W},\nu)$. We will analyze the irreducible subquotient determined by μ' and its relation to $L(\mathcal{W},\nu)$. For a K-type occuring with multiplicity 1, we can think of the signature as a scalar ± 1. If $X(\mathcal{W},\nu)$ is irreducible, then the signature on any K-type is constant for a whole interval $[\nu,\nu+\epsilon)$. As we increase $\epsilon > 0$, the signature can only change when $\nu+\epsilon$ passes a value for which the K-type is in the kernel of the intertwining operator. We will analyze this for the interval $[0,\infty)$.

Case 1. The two K-types μ, μ' are both *lowest K-types* ; they occur in different factors at $\nu = 0$ and in the same factor for $\nu > 0$. Since μ, μ' are not in the kernel of the intertwining operator for any value $\nu > 0$, the signatures are the same over the entire half–line $\nu > 0$. The factor $L(\mathcal{W},\nu)$ has two K-types with opposite sign, therefore is *not* unitary.

Case 2. $X(\mathcal{W},\nu)$ has a single *lowest K-type* , namely μ. The representation $X(\mathcal{W},\nu)$ is irreducible for an interval $[0,\nu_0)$. At $\nu = \nu_0$ the representation becomes irreducible, and the two K-types μ, μ' are *not* in the same factor. For $\nu < \nu_0$ the signatures of μ and μ' are the same, since this is a complementary series. For $\nu > \nu_0$, there is a chance that the signatures have changed, since μ' is in the kernel of the intertwining operator at $\nu = \nu_0$. For very large ν (at infinity) the signatures are opposite by the choice of the K-types. We check that for $\nu > \nu_0$, the two K-types μ, μ' are in $L(\mathcal{W},\nu)$. Thus they have the same signature as at infinity and $L(\mathcal{W},\nu)$ is *not* unitary for $\nu > \nu_0$.

3. THE MAIN RESULT

3.1. We consider the parameters $(M,\mathcal{W}_{temp},\nu)$ where $M = GL(a)\times G(\ell) \subset G(n)$ and $\mathcal{W}_{temp} = St \otimes \mathcal{W}$ where $\mathcal{W}$ is a *discrete series* which contains *sgn*.

Theorem.

- *Suppose $X(\mathcal{W},\nu)$ is irreducible at $\nu = 0$. Then for ${}^\vee G$ of type C there is a complementary series for $-1/2 < \nu < 1/2$. No $L(\mathcal{W},\nu)$ with $|\nu| > 1/2$ is unitary. For ${}^\vee G$ of type B or D, the complementary series extends to $-1 < \nu < 1$. No $L(\mathcal{W},\nu)$ with $|\nu| > 1$ is unitary.*
- *Suppose $X(\mathcal{W},\nu)$ is reducible at $\nu = 0$. Then no $L(\mathcal{W},\nu)$ with $\nu \neq 0$ is unitary.*

The reducibility of $X(\mathcal{W},\nu)$ (at $\nu=0$) was discussed earlier. Applying $\mathcal{IM}$, we get a result about the spherical dual.

Corollary. *Suppose $M=GL(a)\times G(\ell)$ is a Levi component, and $\mathcal{L}=Triv\otimes\mathcal{L}_M$ is spherical such that $\mathcal{L}$ corresponds to an Arthur parameter which is $\mathcal{IM}$ of a discrete series. Then consider $I(\mathcal{L},\nu)$.*

- *Suppose $I(\mathcal{L},\nu)$ is irreducible at $\nu=0$. Then in type C there is a complementary series for $-1/2<\nu<1/2$. For $|\nu|>1/2$, the corresponding spherical irreducible subquotient of $I(\mathcal{L},\nu)$ is unitary if and only if it comes from an Arthur parameter. In the cases when G is of type B, D, the complementary series are in the interval $-1<\nu<1$.*
- *Suppose $I(\mathcal{L},\nu)$ is reducible at $\nu=0$. Then there is no complementary series. For $\nu\neq 0$, the spherical irreducible subquotient of $I(\mathcal{L},\nu)$ is unitary if and only if it is coming from an Arthur parameter.*

We illustrate the proof of the theorem in the case when $G=SO(2n+1)$ *i.e.* the parameter data $\mathcal{O}$ *etc.* are coming from ${}^\vee G=Sp(2n)$. We use the standard realizations of the root systems. Consider the *nilpotent* orbit with partition

$$\mathcal{O}\leftrightarrow(2x_0,\ 2x_1,\dots,2x_{2m}),\qquad 0\le x_0\le\dots\le x_i\le x_{i+1}\le\dots. \tag{3.1.1}$$

It gives rise to a (unique) irreducible *Tempered Representation* with *lowest K-type*

$$\mu_{\mathcal{O}}=(x_0,x_2,\dots,x_{2m})\times(x_1,x_3,\dots,x_{2m-1})\otimes sgn \tag{3.1.2}$$

corresponding to the trivial character of the component group of the orbit $\mathcal{O}$. According to [L3], the other Weyl group representations (*lowest K-types*) corresponding to $\mathcal{O}$ are obtained from a symbol which we call the *S-symbol* . In this case it is obtained by adding $(0,2,4\dots)$ to $(x_0,x_2,\dots)$ and putting them in the first row; then adding $(1,3,\dots)$ to $(x_1,x_3,\dots)$ and putting them in the second row:

(3.1.3)
$$\begin{pmatrix} x_0 & & x_2+2 & & \dots & & & x_{2m}+2m \\ & x_1+1 & & x_3+3 & & \dots & x_{2m-1}+2m-1 & \end{pmatrix}.$$

The symbols of the other Weyl group representations corresponding to $\mathcal{O}$ are obtained by forming all possible such symbols where

- the first row has one more entry than the second,
- the entries in each row increase by at least 2,
- all entries are permutations of the ones appearing in (3.1.3).

The infinitesimal character of the parameter is

$$(3.1.4)\qquad (1/2,\dots,x_0-1/2,\ 1/2,\dots,x_1-1/2,\ \dots\ ,1/2,\dots,x_{2m}-1/2),$$

where the first string does not appear if $x_0 = 0$. A nilpotent orbit with partition given by (3.1.1) parametrizes a *discrete series* precisely when none of the x_i are equal. If this is the case, $A(\mathcal{O}) \cong \mathbb{Z}_2^{2m-1}$ if $x_0 = 0$ and Z_2^{2m} if $x_0 \neq 0$. The number of *discrete series* representations is $\binom{2m}{m}$ if $x_0 = 0$ and $\binom{2m+1}{m}$ if $x_0 \neq 0$.

Recall $\mathcal{W}_{temp} = St \otimes \mathcal{W}$, a *discrete series* $\mathcal{W}$ on $G(\ell)$ tensored with St on $GL(a)$. We use the notation of 1.4 and 1.5. The nilpotent orbit for the parameter of $\mathcal{W}_{temp}$ has partition $(a)\times(2x_0<\dots<2x_{2m})$. The nilpotent orbit $\mathcal{O}$ corresponds to the partition $(2x_0,\dots,a,a,\dots 2x_{2m})$. The infinitesimal character is the one in (3.1.4) where we adjoin $\frac{a-1}{2},\dots,-\frac{a-1}{2}$. If $a = 2E$ is even, $A(s,\mathcal{O})$ is not the same for ${}^{\vee}M$ and ${}^{\vee}G$; the one for ${}^{\vee}G$ has an extra $\mathbb{Z}_2$. By 1.4, each $X_G = \mathbb{H}\otimes_{\mathbb{H}_M} X_M(s,e,\psi)$ decomposes into two factors. We don't need to know the correspondence $\psi\leftrightarrow\sigma_\psi$, we can just check which σ_ψ occur in $Ind_{W(M)}^{W}\sigma_\phi$.

More generally we can form a continuous family of parameters where we adjoin $\frac{a-1}{2}+\nu,\dots,-\frac{a-1}{2}+\nu$ to (3.1.4). We abbreviate this parameter as $s+\underline{\nu}$. For $\nu = 0$ we have analyzed the standard modules in the previous paragraph. For $\nu>0$, $A(s+\underline{\nu},\mathcal{O})$ is the same for ${}^{\vee}M$ and ${}^{\vee}G$. The results in (1.4) imply that the irreducible quotient of X_G must have two *lowest K-types* . We are interested in the case when $\mathcal{W}$ contains sgn. Write the nilpotent orbit as

(3.1.5a)
$$(2x_0,\dots 2x_{2k-1}<2x_{2k}=2E,2x_{2k+1}=2E,\dots),$$

or

(3.1.5b)
$$(2x_0,\dots 2x_{2k}<2x_{2k+1}=2E,2x_{2k+2}=2E,\dots).$$

These two cases really behave in exactly the same way, but the algorithm in [L3] gives different looking answers. The induced module $I(\mathcal{W},M,\nu)$ has two *lowest K-types* :

$$(3.1.6a)\qquad \begin{aligned}\mu =&(x_0,\dots,x_{2k}=E,\dots,x_{2m})\times\\ &(x_1,\dots,x_{2k+1}=E,\dots,x_{2m-1})\otimes sgn\\ \mu' =&(x_0,\dots,x'_{2k}=E+1,\dots,x_{2m})\times\\ &(x_1,\dots,x'_{2k+1}=E-1,\dots,x_{2m-1})\otimes sgn,\end{aligned}$$

or

$$\text{(3.1.6b)}\qquad \begin{aligned}\mu =&(x_0,\dots,x_{2k+2}=E,\dots)\times\\ &(x_1,\dots,x_{2k+1}=E,\dots)\otimes sgn\\ \mu' =&(x_0,\dots,x'_{2k+2}=E-1,\dots)\times\\ &(x_1,\dots,x'_{2k+1}=E+1,\dots)\otimes sgn.\end{aligned}$$

At ∞, the K-types μ and μ' have opposite signatures (Theorem 2.3). These K-types are in the same Langlands quotient for $\nu > 0$. Thus this Langlands quotient is *not unitary* for $\nu > 0$.

If on the other hand $a = 2E+1$ then $A(s, \mathcal{O})$ is the same for ${}^\vee M$ and ${}^\vee G$. The standard module $X(\mathcal{W}_{temp}, M, \nu)$ is irreducible for $\nu = 0$, and there is a complementary series up to at least $\nu = 1/2$. At this point there is at least one other factor. The tempered parameter corresponding to $\mathcal{O}'$ with partition

(3.1.7a)

$$(2x_0,\dots 2x_{2k-1} < 2x_{2k} = 2E, 2x_{2k+1} = 2E+2,\dots),$$

or

(3.1.7b)

$$(2x_0,\dots 2x_{2k}, 2x_{2k+1} = 2E, 2x_{2k+2} = 2E,\dots)$$

has the same infinitesimal character. Since sgn occurs in the induced module with multiplicity 1 and it must occur in $L(\mathcal{O}', s+\underline{1/2})$, so this factor must occur in $X(\mathcal{W}, M, 1/2)$. The *lowest K-types* of the two factors are
(3.1.8a)

$$\mu = (x_0,\dots,x_{2k}=E+1,\dots,x_{2m})\times(x_1,\dots,x_{2k+1}=E,\dots,x_{2m-1})\otimes sgn$$
$$\mu' = (x_0,\dots,x'_{2k}=E,\dots,x_{2m})\times(x_1,\dots,x'_{2k+1}=E+1,\dots,x_{2m-1})\otimes sgn,$$

or

(3.1.8b)

$$\mu = (x_0,\dots,x_{2k+2}=E,\dots,x_{2m})\times(x_1,\dots,x_{2k+1}=E+1,\dots,x_{2m-1})\otimes sgn$$
$$\mu' = (x_0,\dots,x'_{2k+2}=E+1,\dots,x_{2m})\times(x_1,\dots,x'_{2k+1}=E,\dots,x_{2m-1})\otimes sgn.$$

At infinity μ and μ' have opposite signatures. We need to show that these two K-types stay in the same factor for $\nu > 1/2$. We write out case (a) only, because (b) follows by the same argument, only the indexing is different. First observe that $\mathcal{O} \subset \overline{\mathcal{O}'}$, and there is no other nilpotent in between. Thus μ' can only be in $L(\mathcal{W}, M, \nu)$ or in a factor with parameter $\mathcal{O}'$; in fact it will be a *lowest K-type* for it. Suppose such a factor occurs. There are four cases depending on whether $x_{2k-1} \neq E, x_{2k+2} \neq E+1$, $x_{2k-1} = E, x_{2k+2} \neq E+1$, $x_{2k-1} \neq E, x_{2k+2} = E+1$, $x_{2k-1} = E, x_{2k+2} = E+1$. The first case is the easiest, in this case the only infinitesimal character for which $\mathcal{O}'$ can appear as a parameter is $s + \underline{1/2}$, and we are done. The hardest case is when

$x_{2k-1} = E, x_{2k+2} = E+1$. Parameters corresponding to $\mathcal{O}'$ can be deformed in two directions; the most general infinitesimal character is

$$(1/2,\dots,x_0-1/2,\dots,1/2,\dots,x_{2k-2}-1/2,\\ -E+1/2+\nu_1,\dots,E-1/2+\nu_1,\\ -E-1/2+\nu_2,\dots,E+1/2+\nu_2,1/2,\dots,x_{2k+3}-1/2,\dots). \tag{3.1.9}$$

This can only coincide with an infinitesimal character of the form $s+\underline{\nu}$ at $\nu=1/2, \nu_1=\nu_2=0$ and at $\nu=3/2,\ \nu_1=\nu_2=1$. The corresponding $L(\mathcal{O}')$ would have 2 *lowest K-types* :

$$\begin{aligned}\mu' &=(\dots,E,E+1,\dots)\times(\dots,E,E+1,\dots)\otimes sgn\\ \mu'' &=(\dots,E,E,\dots)\times(\dots,E+1,E+1,\dots)\otimes sgn.\end{aligned} \tag{3.1.10}$$

But μ'' does not occur in $X(\mathcal{W},M,\nu)$. Thus μ and μ' occur in the same factor for $\nu>1/2$.

Remark In the example in this section, the representation $L(\mathcal{W},\nu)$ is not unitary for $\nu = 3/2,\ 5/2,\dots$. The spherical representation at these parameters *is* unitary. It is simply *not* $\mathcal{IM}(L(\mathcal{W},\nu))$, but $\mathcal{IM}$ of a tempered representation at this infinitesimal character. For values which are not half integers, $L(\mathcal{W},\nu)=X(\mathcal{W},\nu)$ and so $\mathcal{IM}(L(\mathcal{W},\nu))$ is therefore spherical and not unitary.

3.2. We consider the example when $\ell=0$ and $G=Sp(2n)$ and ${}^{\vee}G = SO(2n+1)$. Then ${}^{\vee}M=GL(n)$. The orbit $\mathcal{O}$ attached to such a parameter corresponds to the partition $(1,n,n)$. We write the infinitesimal character as

$$s=(-\frac{n-1}{2}+\nu,\dots,\frac{n-1}{2}+\nu). \tag{3.2.1}$$

The K-structure of the standard module is

$$Ind_{W(A_{n-1})}^{W(B_n)}[sgn]=\sum_{0\le k\le n}(1^k;1^{n-k}) \tag{3.2.2}$$

where $(1^k;1^{n-k})=(\underbrace{1;\dots,1}_{k},\underbrace{1,\dots,1}_{n-k})$. The signature at ∞ is $(-1)^k$. We need to investigate the reducibility.

n=2a. There is a unique *lowest K-type* , namely

$$(1^{a+1};1^{a-1}).$$

The induced module can only become reducible if there is a larger nilpotent orbit $\mathcal{O}'={}^{\vee}G(e')$ such that $[s,e']=e'$. The only possible such ν are

$$\nu=l,\qquad 1\le l\le a. \tag{3.2.3}$$

For each such l there is only one possible choice for $\mathcal{O}'$, namely $(1, 2a-2l-1, 2a+2l+1)$. The *lowest K-types* for this orbit are

$$(1^{a-l}; 1^{a+l}), \qquad (0, 1^{2l}; 2^{a-l}), \qquad (1^{a+l+2}; 1^{a-l-2}). \tag{3.2.4}$$

Of course the third case only occurs for $a-l-2 \geq 0$.

Then s is conjugate to the semisimple element in the Lie triple corresponding to e'; so *sgn* has to occur in $L(\mathcal{O}', s, triv)$. It follows that

$$I(M, \mathcal{W}, \nu) = L(\mathcal{O}, s, triv) + L(\mathcal{O}', s, triv) + \text{ possibly } L(\mathcal{O}', s, \chi). \tag{3.2.5}$$

Thus the first reducibility point is at $\nu = 1$. The K-types $(1^{a+1}; 1^{a-1})$ and $(1^a; 1^a)$ are *not* in the same factor. At all other $\nu > 1$, they are in the same factor. Thus there is a complementary series $0 \leq \nu < 1$.

n=2a+1. There are two *lowest K-types* ,

$$(1^{a+1}; 1^{a}) \qquad (1^{a+2}; 1^{a-1}). \tag{3.2.6}$$

They have opposite signatures at ∞. If $\nu = 0$, then

$$A(\mathcal{O}) = A(\underline{\nu}, \mathcal{O}) \cong \mathbb{Z}_2,$$

but if $\nu \neq 0$, then

$$A(\mathcal{O}) = \mathbb{Z}_2, \qquad A(\underline{\nu}, \mathcal{O}) = \{1\}.$$

Thus the two *lowest K-types* are not in the same factor at $\nu = 0$ and are always in the same factor for $\nu \neq 0$. Since they have opposite signatures at ∞, there is *no* complementary series.

3.3. This is an example of complementary series off of $GL(n)$–complementary series which are *not* complementary series from Arthur parameters. Consider the case when $G = Sp(8a+4)$, ${}^{\vee}G = SO(8a+5)$. Let the Levi component be $M = GL(2a+1) \times GL(2a+1)$, and $\mathcal{W} = St \otimes St$. Then we can write the infinitesimal character as

$$s(\nu_1, \nu_2) := (-a+\nu_1, \ldots, a+\nu_1 \; ; \; -a+\nu_2, \ldots, a+\nu_2) \tag{3.3.1}$$

The *lowest K-types* are

$$(2^{a+1}; 2^{a}), \qquad (2^{a+2}; 2^{a-1}). \tag{3.3.2}$$

They have the same sign at ∞. Furthermore

$$\begin{aligned} &A(\mathcal{O}) = \mathbb{Z}_2, \quad A(s(\nu_1 = 0, \nu_2), \mathcal{O}) \cong A(s(\nu_1, \nu_2 = 0), \mathcal{O}) = \mathbb{Z}_2, \\ &A(s(\nu_1 \neq 0, \nu_2 \neq 0), \mathcal{O}) = \{1\}. \end{aligned} \tag{3.3.3}$$

Thus $I(M, \mathcal{W}, \nu_1 = \nu_2 = 0)$ is reducible. However, if we consider $\nu_1 = -\nu_2 = \nu$, we see that the intermediate induced module

$$Ind_M^{GL(4a+2)}[(St \cdot 1\!\!1_\nu) \otimes (St \cdot 1\!\!1_{-\nu})]$$

is unitary irreducible for $0 \leq \nu < 1$, by the complementary series results for $GL(n)$. An argument as in 3.2 implies that there is a region of unitarity

$$\{-1 < \nu_1,\ \nu_2 < 1\}, \tag{3.3.4}$$

but the induced module at $\nu_1 = 0,\ \nu_2 = 0$ is reducible.

To see that $L(\mathcal{O}, s, sgn)$ is not unitary outside this range, we have to argue as in 3.2 using the K-type $(2^{a+1}1; 2^{a-1}1)$ which is the *lowest K-type* for the nilpotent orbit $\mathcal{O}' = (1, 2a, 2a, 2a+2, 2a+2)$. It is crucial that there are no orbits in between $\mathcal{O}$ and $\mathcal{O}'$.

References

[B] A. Borel, *Admissible representations of a semisimple group over a local field with fixed vectors under an Iwahori subgroup*, Invent. Math. **35** (1976), 233–259.

[BM1] D. Barbasch and A. Moy, *A unitarity criterion for p-adic groups*, Invent. Math. **98** (1989), 19–38.

[BM2] ———, *Reduction to real infinitesimal character in affine Hecke algebras*, Journal of the AMS **6,** 3 (1993), 611–635.

[BM3] ———, *Unitary spherical spectrum for p-adic classical groups*, Acta Appl. Math. **44** (1996), 1-37.

[C] W. Casselman, *A new nonunitarity argument for p-adic representations*, J. Fac. Sci. Tokyo Univ. sect IA Math. **28,** 3 (1981), 907–928.

[Ca] R. Carter, *Finite groups of Lie type*, Wiley–Interscience, New York, 1985.

[Car] P. Cartier, *Representations of p-adic groups: A survey*, Automorphic Forms, Representations, and L–functions, vol. 31 part 1, American Mathematical Society, Providence, Rhode Island, 1979, pp. 111–157.

[KL] D. Kazhdan and G. Lusztig, *Proof of the Deligne–Langlands conjecture for Hecke algebras*, Inv Math. **87** (1987), 153–215.

[L1] G. Lusztig, *Characters of reductive groups over a finite field, Annals of Math. Studies*, vol. 107, Princeton University Press.

[L2] ———, *Affine Hecke algebras and their graded version*, Jour. AMS **2** (1989), 599–635.

[L3] ———, *Intersection cohomology complexes on a reductive group*, Invent. Math. **75** (1984), 205–272.

[L4] ———, *Cuspidal local systems and graded algebras I*, Publ. Math de l'IHES **67** (1988), 145–202.

[L5] ———, *Cuspidal local systems and graded algebras II* (to appear).

[T] M. Tadic, *Classification of unitary representations in irreducible representations of general linear group (nonarchimedean case)*, Ann. Scient. Ec. Norm. Sup. **19** (1986), 335–382.

Department of Mathematics, Cornell University, Ithaca NY 14853

FINITE RANK HOMOGENEOUS HOLOMORPHIC BUNDLES IN FLAG SPACES

TIM BRATTEN

INTRODUCTION

For more than forty years the study of homogeneous holomorphic vector bundles has resulted in an important source of irreducible unitary representations for a real reductive Lie group. In the mid 1950s, Harish-Chandra realized a family of irreducible unitary representations for some semisimple groups, using the global sections of homogeneous bundles defined over Hermitian symmetric spaces [6]. At about the same time Borel and Weil constructed the irreducible representations for a connected compact Lie group as global sections of line bundles defined over complex projective homogeneous spaces [3]. More than ten years later, W. Schmid in his thesis solved a conjecture by Langlands and generalized the Borel-Weil-Bott theorem to realize discrete series representations for noncompact semisimple groups [16]. This extension is nontrivial for one thing because it requires an understanding of the representations obtained on some infinite-dimensional sheaf cohomology groups.

Later, joint work of W. Schmid and J. A. Wolf [17] further generalized these results by showing that the sheaf cohomology groups of homogeneous holomorphic bundles defined over an open orbit on a full flag manifold can provide globalizations for the standard Zuckerman modules. Using different methods, H. Hecht and J. Taylor [8] established a dual result that can be related to the work of Schmid and Wolf via the combination of Serre duality [18] coupled with a corresponding duality theorem for Harish-Chandra modules [10]. Even more recently, H. Wong [22] has extended these results to open orbits of θ-stable parabolics in a generalized flag manifold (see the example at the end of Section 1 for specifics). A result of D. Vogan [19] gives some general conditions for when such representations realize irreducible unitarizable representations.

In this paper, we propose to show how the methods of localization can be used to study a finite rank homogeneous holomorphic vector bundle defined over an arbitrary open orbit in a generalized flag manifold. In particular, we will employ methods that have evolved from both the algebraic localization theory of Beilinson and Bernstein [1], as well as the the analytic localization theory of Hecht and Taylor [8]. A precise statement of our main result is labeled Theorem 1 and can be found just before the example at the end of Section 1.

As is typical for studies made in this area, a part of the development is based on certain geometric considerations, and the rest follows by tensoring arguments. In this presentation we only focus on the geometric aspects, because the tensoring argument is already spelled out in [4, Section 9].

The paper is organized as follows. In Section 1 we introduce the groups to be considered and the basic geometric constructions. At the end of Section 1 we state the main theorem to be proved (Theorem 1). In Section 2 we review a few basics of the localization theory. Then in Section 3 we sketch a proof of Theorem 1.

The author would like to thank D. Vogan and J. A. Wolf for some advice which greatly helped to clarify the exposition. The author would also like to thank Jorge Vargas for several helpful discussions. The work for this paper was completed while the author was supported in a post-doctoral position in FaMAF, Universidad Nacional de Cordoba.

SECTION 1

In this section we introduce the basic geometric setting and then finish with a statement of the main result to be proved.

Let G_0 be a real reductive group of Harish-Chandra class [7], with complexified Lie algebra $\mathfrak{g}$. Actually, to further simplify the exposition, we assume in addition that $G_0 \subset G$ is a closed subgroup of a connected complex reductive algebraic group G with Lie algebra $\mathfrak{g}$. By a *complex flag space* Y for G_0 we mean any complete homogeneous space for the group G. We recall that a maximal connected solvable subgroup $B \subset G$ is called a Borel subgroup and that a parabolic subgroup of G is any subgroup which contains a Borel subgroup. The theory of affine algebraic groups reveals that the parabolic subgroups of G are connected and Zariski closed, and that the Borel subgroups of G are all conjugate. In addition, the complex flag spaces for G_0 are precisely the spaces of the form $Y = G/P$ where $P \subset G$ is a parabolic subgroup. When the stabilizer of a point in the complex flag space is a Borel subgroup $B \subset G$, we use the notation $X = G/B$ and refer to this space as the *full flag space* for G_0.

We fix a complex flag space Y for G_0. For $y \in Y$ we let P_y denote the stabilizer of y in G and we let $\mathfrak{p}_y \subset \mathfrak{g}$ denote the Lie algebra of P_y. Since parabolic subgroups equal their own normalizers, it follows that Y can be naturally identified with the G conjugates of $\mathfrak{p}_y$. In particular, X can be identified as the variety of maximal solvable subalgebras of $\mathfrak{g}$. Given $y \in Y$, let $\mathfrak{u}_y \subset \mathfrak{p}_y$ denote the nilradical of $\mathfrak{p}_y$ and let $\mathfrak{l}_y = \mathfrak{p}_y/\mathfrak{u}_y$ denote the corresponding quotient. By a *finite-dimensional* $(\mathfrak{p}_y, G_0 \cap P_y)$*-module* V we mean any finite-dimensional continuous $G_0 \cap P_y$-module equipped with a compatible $\mathfrak{p}_y$-action. A finite-dimensional $(\mathfrak{l}_y, G_0 \cap P_y)$-module simply refers to any finite-dimensional $(\mathfrak{p}_y, G_0 \cap P_y)$-module with trivial $\mathfrak{u}_y$-action.

Fix a maximal compact subgroup $K_0 \subset G_0$ and let $K \subset G$ be the complexification of K_0. For $y \in Y$, a finite-dimensional $(\mathfrak{p}_y, K \cap P_y)$-module will be called *algebraic* if the corresponding representation for $K \cap P_y$ is a representation of algebraic groups. For our purposes, it will be important to have a certain correspondence between the finite-dimensional algebraic $(\mathfrak{l}_y, K \cap P_y)$-modules and the finite-dimensional $(\mathfrak{l}_y, G_0 \cap P_y)$-modules. This requires an additional assumption about the point y. In particular, let $\theta : G_0 \to G_0$ be a Cartan involution consistent with the choice of K_0. A point $y \in Y$ is called *special* provided that the stabilizer $G_0 \cap P_y$ of y in G_0 contains a Cartan subgroup for G_0 which is invariant under θ.

Proposition 1 *(a) If V is an irreducible finite-dimensional $(\mathfrak{p}_y, G_0 \cap P_y)$-module, then $\mathfrak{u}_y$ acts trivially on V. In particular, V is an irreducible highest weight module for $\mathfrak{l}_y$.*
(b) Suppose that $y \in Y$ is a special point. Then there is a natural equivalence of categories between the category of finite-dimensional $(\mathfrak{l}_y, G_0 \cap P_y)$-modules and the category of finite-dimensional algebraic $(\mathfrak{l}_y, K \cap P_y)$-modules.

Proof: Our proof of (b) is based on a description for the stabilizers [Lemma 1 to follow] which is adapted from the paper [14, Section 2]. The proof of (a) is elementary.

To establish (b), let $\tau : \mathfrak{g} \to \mathfrak{g}$ be the conjugation determined by the real form $\mathfrak{g}_0$. Since y is special, there exists a Cartan subalgebra $\mathfrak{c} \subset \mathfrak{p}_y$ such that $\tau(\mathfrak{c}) = \mathfrak{c}$ and $\theta(\mathfrak{c}) = \mathfrak{c}$. We recall that a *Levi factor* of $\mathfrak{p}_y$ is any subalgebra isomorphic with $\mathfrak{l}_y$ under the projection $\mathfrak{p}_y \to \mathfrak{l}_y$. Let $\mathfrak{l} \subset \mathfrak{p}_y$ be the Lie subalgebra of $\mathfrak{g}$ defined by

$$\mathfrak{l} = \mathfrak{p}_y \cap \theta\tau\mathfrak{p}_y.$$

Then, a standard exercise in the root space structure for the $\mathfrak{c}$ action on $\mathfrak{p}_y$ reveals that $\mathfrak{l}$ is a Levi factor of $\mathfrak{p}_y$. We let $L \subset P_y$ be the corresponding connected subgroup with Lie algebra $\mathfrak{l}$.

Define the following Lie subalgebras of $\mathfrak{l}_y$:

$\mathfrak{n} =$ the image of $\{\xi \in \mathfrak{l} | \theta\xi \in \mathfrak{u}_y\}$ under the projection $\mathfrak{p}_y \to \mathfrak{l}_y$;
$\bar{\mathfrak{n}} =$ the image of $\{\xi \in \mathfrak{l} | \tau\xi \in \mathfrak{u}_y\}$ under the projection $\mathfrak{p}_y \to \mathfrak{l}_y$;
$\mathfrak{r} =$ the complex subalgebra of $\mathfrak{l}_y$ generated by the image of $\mathfrak{g}_0 \cap \mathfrak{l}$ in $\mathfrak{l}_y$.

Let $U_y \subset P_y$ be the unipotent radical and consider the quotient $L_y = P_y/U_y$. To simplify notations we identify $G_0 \cap L$ and $K \cap L$ with subgroups of L_y. Let $N \subset L_y$ and $\overline{N} \subset L_y$ be the connected subgroups corresponding respectively to the Lie algebras $\mathfrak{n}$ and $\bar{\mathfrak{n}}$.

At the infinitesimal level, the following result is shown in [14, Section 2]. The extension from Lie algebras to groups is a standard gymnastic in structure theory.

Lemma 1 [Matsuki]
(a) Put $\mathfrak{q} = \mathfrak{r} + \mathfrak{n}$. *Then* $\mathfrak{q}$ *is a parabolic subalgebra of* $\mathfrak{l}_y$ *with nilradical* $\mathfrak{n}$ *and Levi factor* $\mathfrak{r}$.
(b) Put $\bar{\mathfrak{q}} = \mathfrak{r} + \bar{\mathfrak{n}}$. *Then* $\bar{\mathfrak{q}}$ *is a parabolic subalgebra of* $\mathfrak{l}_y$ *opposite to* $\mathfrak{q}$, *with nilradical* $\bar{\mathfrak{n}}$ *and Levi factor* $\mathfrak{r}$.
(c) $G_0 \cap L$ *is a real reductive group of Harish-Chandra class with maximal compact subgroup* $K_0 \cap L$.
(d) $K \cap L$ *is a complex reductive algebraic group with compact real form* $K_0 \cap L$.
(e) The image of $G_0 \cap P_y$ *in* L_y *is naturally isomorphic to a semi-direct product of* $G_0 \cap L$ *with* $\overline{N}$.
(f) The image of $K \cap P_y$ *in* L_y *is naturally isomorphic to a semi-direct product of* $K \cap L$ *with* N.

Since the Lie algebra $\mathfrak{n}$ acts nilpotently in any finite-dimensional $\mathfrak{l}_y$-module, part (b) of Proposition 1 now follows from standard facts about finite-dimensional representations of nilpotent and reductive Lie groups. □

We fix a point $y \in Y$ and assume from here on that the G_0-orbit $S = G_0 \cdot y \subset Y$ is open. Such points exist in any flag space, because G_0 has finitely many orbits on Y [21]. A fundamental result by Matsuki [13, Theorem 1 and Example 1] implies that each G_0-orbit in Y contains special points, so there is no loss of generality in assuming that y is special.

Since S is open, it follows that each finite-dimensional $(\mathfrak{p}_y, G_0 \cap P_y)$-module V determines a G_0-homogeneous holomorphic vector bundle defined over S [20, Proposition 1.19]. We let $\mathcal{O}(V)$ denote the resulting sheaf of holomorphic sections. Our main interest will be to describe the representations obtained on the compactly supported sheaf cohomology groups $H_c^p(S, \mathcal{O}(V))$.

It turns out that our methods will deliver their most effective results in case the finite-dimensional $(\mathfrak{p}_y, G_0 \cap P_y)$-module is irreducible and satisfies a certain negativity condition. In order to specify this negativity condition, we introduce the following notations. Let $U(\mathfrak{l}_y)$ denote the universal enveloping algebra of $\mathfrak{l}_y$ and suppose $Z(\mathfrak{l}_y) \subset U(\mathfrak{l}_y)$ is the center of $U(\mathfrak{l}_y)$. Then an $\mathfrak{l}_y$ *infinitesimal character* refers to a homomorphism $Z(\mathfrak{l}_y) \to \mathbb{C}$.

We utilize Harish-Chandra's method of parametrizing $\mathfrak{l}_y$ infinitesimal characters as follows. Let $\mathfrak{h}$ denote the *universal Cartan algebra* for $\mathfrak{g}$ [8, Section 1] [4, Section 2]. In particular, if $\mathfrak{b}_x$ is the Borel subalgebra corresponding to a point x in the full flag space then $\mathfrak{h}$ is canonically identified with the quotient $\mathfrak{b}_x/\mathfrak{n}_x$, where $\mathfrak{n}_x$ is the nilradical of $\mathfrak{b}_x$. In case $\mathfrak{b}_x \subset \mathfrak{p}_y$ and $\mathfrak{c} \subset \mathfrak{b}_x$ is a Cartan subalgebra, then the projection $\mathfrak{b}_x \to \mathfrak{l}_y$ identifies $\mathfrak{c}$ with a Cartan subalgebra of $\mathfrak{l}_y$. Then, modulo the right ideal in $U(\mathfrak{l}_y)$ generated by $\mathfrak{n}_x/\mathfrak{u}_y$, $Z(\mathfrak{l}_y)$ is identified with a subalgebra of the enveloping algebra $U(\mathfrak{c})$. Thus, letting $U(\mathfrak{h})$ denote the enveloping algebra of $\mathfrak{h}$ and identifying $\mathfrak{c}$ with $\mathfrak{b}_x/\mathfrak{n}_x$, we obtain the *unnormalized Harish-Chandra morphism* $Z(\mathfrak{l}_y) \to U(\mathfrak{h})$. On the other hand, using the above identification of $\mathfrak{c}$ with $\mathfrak{h}$, we can define a set of abstract positive roots $\Sigma^+ \subset \mathfrak{h}^*$ via the roots of $\mathfrak{c}$ in $\mathfrak{n}_x$. Let ρ denote one-half the sum of the elements in Σ^+. Then we say that the $\mathfrak{l}_y$-module V has infinitesimal character $\lambda \in \mathfrak{h}^*$ provided $Z(\mathfrak{l}_y)$ acts on V via the homomorphism:

$$Z(\mathfrak{l}_y) \to U(\mathfrak{h}) \xrightarrow{\lambda+\rho} \mathbb{C}.$$

An element $\lambda \in \mathfrak{h}^*$ is called *regular* provided $\check{\alpha}(\lambda)$ is not zero for each $\alpha \in \Sigma^+$. On the other hand, we say $\lambda \in \mathfrak{h}^*$ is *antidominant* provided $\check{\alpha}(\lambda)$ is not a positive integer for each $\alpha \in \Sigma^+$.

Proposition 1 implies that any finite-dimensional $(\mathfrak{l}_y, G_0 \cap P_y)$-module V determines a corresponding K-homogeneous algebraic vector bundle over the K-orbit $Q = K \cdot y$. On the other hand, suppose that a finite-dimensional algebraic representation V of $K \cap P_y$ carries a compatible $\mathfrak{l}_y$-action. Then, whenever the resulting $\mathfrak{l}_y$-module has an infinitesimal character, we can apply a certain direct image construction [5, Section 4], analogous to the direct image for $\mathcal{D}$-modules [2, Chapter VI, Section 5], to the sections of the algebraic bundle. The resulting object, $\mathcal{I}(V)$ is a K-equivariant sheaf of $\mathfrak{g}$-modules defined on all of Y and is referred to as the *standard Harish-Chandra sheaf* determined by the $(\mathfrak{l}_y, K \cap P_y)$-module V.

In Section 3, we will sketch how the following result can be established, using some standard facts in localization theory coupled with some techniques developed in [4].

Theorem 1 *Suppose that V is an irreducible finite-dimensional representation for $(\mathfrak{p}_y, G_0 \cap P_y)$. Assume that the infinitesimal character for V has a parameter $\lambda \in \mathfrak{h}^*$ that is regular and antidominant. Let $\mathcal{O}(V)$ be the sheaf of holomorphic sections for the homogeneous holomorphic bundle on S determined by V and let q denote the codimension of Q in Y. Then the compactly supported sheaf cohomology groups $H^p_c(S, \mathcal{O}(V))$ vanish except when $p = q$. The nonvanishing cohomology group $H^q_c(S, \mathcal{O}(V))$ carries a naturally defined dual nuclear Fréchet topology and a continuous G_0-action. The resulting G_0-module is topologically irreducible and in fact is a minimal globalization of the Harish-Chandra module $\Gamma(Y, \mathcal{I}(V))$, where $\mathcal{I}(V)$ is the standard Harish-Chandra sheaf determined by the corresponding $(\mathfrak{l}_y, K \cap P_y)$-module V.*

We will give a brief description of the minimal globalization [15] [11] in Section 3.

Example

The homogeneous holomorphic bundles studied in [22] and [4] are defined on open orbits that are *Levi type*. In general, the G_0-orbit $G_0 \cdot z \subset Y$ of a point $z \in Y$ is said to be *Levi type* provided that the corresponding parabolic subgroup $P_z \subset G$ has a Levi factor L such that $G_0 \cap L$ is a real form of L. When y is a special point and $S = G_0 \cdot y$ is open, then S is Levi type if and only if $\mathfrak{p}_y$ is invariant under the extension of θ to a complex involution of $\mathfrak{g}$.

We give an example to show that open orbits, that are not Levi type, do occur in nature. Let $G = SL(n,\mathbb{C})$ and let $G_0 = SL(n,\mathbb{R})$. The natural G-action on $\mathbb{C}^n - \{0\}$ defines a corresponding transitive action on complex projective space $Y = \mathbb{P}^{n-1}\mathbb{C}$. Assume n is bigger than 2. By considering the G_0-action on $\mathbb{C}^n - \{0\}$, it is easy to see that there are exactly two G_0-orbits on Y: one open and one closed. In particular, if $e_1, ..., e_n$ denotes the standard basis for $\mathbb{C}^n$, if $i \in \mathbb{C}$ denotes the principal square root of -1 and if $[v] \in \mathbb{P}^{n-1}\mathbb{C}$ denotes the point in Y corresponding to $v \in \mathbb{C}^n - \{0\}$, then the G_0-orbit of $[e_1 + ie_2]$ in Y is open and the G_0-orbit of $[e_1]$ is closed. Since $G_0 \cap P_{[e_1]}$ is a real form for the stabilizer $P_{[e_1]}$ of $[e_1]$ in G it follows that the closed orbit is Levi type and that every point inside is special, regardless of the choice of maximal compact subgroup in G_0.

We fix a maximal compact subgroup $K_0 = SO(n,\mathbb{R})$ of G_0. The corresponding involution of G is given by the inverse-transpose of a matrix. A computation produces a θ-stable Cartan subgroup of G_0 that stabilizes $[e_1 + ie_2]$ [12, Chapter 5, Section 4]. In particular, $[e_1 + ie_2]$ is special. On the other hand, it is easy to write down an element of $P_{[e_1+ie_2]}$ whose inverse-transpose does not stabilize $[e_1 + ie_2]$. It follows from the above discussion that this open orbit is not Levi type. □

Section 2

In this section we review a few facts about the localization of $\mathfrak{g}$-modules to flag spaces.

A choice of $\lambda \in \mathfrak{h}^*$ defines a $\mathfrak{g}$ infinitesimal character $\chi_\lambda : Z(\mathfrak{g}) \to \mathbb{C}$. Put $J_\lambda = \ker\chi_\lambda$ and let U_λ denote the quotient of $U(\mathfrak{g})$ by the ideal generated from J_λ. A fundamental result of Harish-Chandra implies $J_{w\cdot\lambda} = J_\lambda$ for w in the Weyl group, but for notational purposes we keep the dependence on λ specific. On the other hand, for each $\lambda \in \mathfrak{h}^*$, Beilinson and Bernstein have shown [1] how to define a certain sheaf of twisted differential operators $\mathcal{D}_\lambda^{alg}$ with regular coefficients on the full flag space X. They prove that:

$$\Gamma(X, \mathcal{D}_\lambda^{alg}) = U_\lambda \quad \text{and} \quad H^p(X, \mathcal{D}_\lambda^{alg}) = 0 \text{ for } p > 0.$$

Now suppose Y is a flag space. Given a Borel subgroup $B \subset G$ there exits a unique $z \in Y$ such that $B \subset P_z$. Thus we obtain a canonical projection $\pi : X \to Y$ from the full flag space X onto Y. Let π_* denote the direct image in the category of sheaves. If M is a U_λ module we define *the algebraic localization* of M to Y by:

$$\Delta_Y^{alg}(M) = \pi_*\mathcal{D}_\lambda^{alg} \otimes_{U_\lambda} M.$$

Let Γ_Y denote the functor of global sections on Y. Then, using the methods developed by Beilinson and Bernstein [1], one can show [5, Theorem 4.15] the following result.

Theorem 2 [Beilinson and Bernstein, Chang]
If λ is regular and antidominant then Δ_Y^{alg} and Γ_Y together determine an equivalence of categories between the category of U_λ-modules and the category of quasi-coherent $\pi_*\mathcal{D}_\lambda^{alg}$-modules.

The localization functor defined above turns out to have some nice applications to the study of Harish-Chandra modules for $(\mathfrak{g}, K)$. The idea behind the analytic localization of Hecht and Taylor, is to define a localization functor which applies more directly to the study of topological representations of the group G_0. It turns out such a program is possible, if one passes to derived categories.

In particular, a *dnF U_λ-module* is a U_λ-module M that has a dual nuclear Fréchet (= *dnF*) topology, such that the linear operators on M induced by the $\mathfrak{g}$-action are continuous. A morphism of two *dnF* U_λ-modules is a continuous, equivariant linear map. The corresponding category $\mathcal{M}_{dnF}(U_\lambda)$, although not abelian, determines (using a construction modeled after the construction for an abelian category) a derived category $D(\mathcal{M}_{dnF}(U_\lambda))$ [8, Proposition 5.1]. We observe that any finitely generated $\mathfrak{g}$-module with infinitesimal character λ has a unique topology making it a *dnF* U_λ-module. Another simple, although important, observation is that whenever M is a *dnF* space, then the algebraic tensor product $U_\lambda \otimes M$ is a *dnF* U_λ-module.

Let $\mathcal{D}_\lambda$ denote the twisted sheaf of differential operators on X with holomorphic coefficients and let $\mathcal{O}_Y$ be the sheaf of holomorphic functions on Y. Given a point $y \in Y$, we let $\mathcal{O}_{Y,y}$ denote the corresponding stalk of $\mathcal{O}_Y$ at y. For a sheaf $\mathcal{F}$ of modules over $\mathcal{O}_Y$, the *geometric fiber* of $\mathcal{F}$ at y is the complex vector space defined by

$$T_y(\mathcal{F}) = \mathbb{C} \otimes_{\mathcal{O}_{Y,y}} (\mathcal{F}_y).$$

Because $\pi_*\mathcal{D}_\lambda$ is locally free as an $\mathcal{O}_Y$-module, and because its geometric fibers have countable dimension, it turns out that the sections of $\pi_*\mathcal{D}_\lambda$ over compact sets are topological algebras with a *dnF* topology. A *sheaf of dnF $\pi_*\mathcal{D}_\lambda$-modules $\mathcal{F}$* is a sheaf of $\pi_*\mathcal{D}_\lambda$-modules that carries a compatible *dnF* topology over the compact subsets of Y. As above we obtain a category $\mathcal{M}_{dnF}(\pi_*\mathcal{D}_\lambda)$ of *dnF* $\pi_*\mathcal{D}_\lambda$-modules and a corresponding derived category $D(\mathcal{M}_{dnF}(\pi_*\mathcal{D}_\lambda))$ [8, Proposition 5.1] [4, Section 3]. One simple, although crucial, construction is that given a *dnF* space M we can form a completed tensor product

$$\pi_*\mathcal{D}_\lambda \hat{\otimes} M$$

, which defines an object in the category $\mathcal{M}_{dnF}(\pi_*\mathcal{D}_\lambda)$ (if M has a countable basis then the completed tensor product of M and $\pi_*\mathcal{D}_\lambda$ coincides with the algebraic tensor product) [8, Sections 2 and 3]. A more subtle, but equally crucial point is that the category $\mathcal{M}_{dnF}(\pi_*\mathcal{D}_\lambda)$ has enough acyclics for the functor of global sections [8, Section 3]. In particular, we obtain a derived functor

$$R\Gamma_Y : D(\mathcal{M}_{dnF}(\pi_*\mathcal{D}_\lambda)) \to D(\mathcal{M}_{dnF}(U_\lambda)).$$

To define the analytic localization $L\Delta_Y(M)$ of a dnF U_λ-module M we proceed as follows. We begin with a certain canonical free resolution (called the Hochschild resolution) $F.(M)$, where

$$F_p(M) = \otimes^{p+1}U_\lambda \otimes M.$$

Next we apply the functor $\pi_*\mathcal{D}_\lambda \otimes_{U_\lambda} (\ )$ to the complex $F.(M)$ and then complete the tensor products in the resulting complex. This last complex, when viewed as an object in $D(\mathcal{M}_{dnF}(\pi_*\mathcal{D}_\lambda))$ is called the *analytic localization* of M. The construction just described, in fact leads to the definition of a derived functor

$$L\Delta_Y : D(\mathcal{M}_{dnF}(U_\lambda)) \to D(\mathcal{M}_{dnF}(\pi_*\mathcal{D}_\lambda)).$$

The analog of Theorem 2 for the analytic localization functor is the following [8, Theorem 5.4]:

Theorem 3 [Hecht and Taylor]
Suppose that λ is regular and X is the full flag space. Then $L\Delta_X$ and $R\Gamma_X$ together define an equivalence of categories between $D(\mathcal{M}_{dnF}(U_\lambda))$ and $D(\mathcal{M}_{dnF}(\mathcal{D}_\lambda))$.

It turns out that the analytic localization carries relevant information about analytic representations of the group G_0. In particular, an *analytic G_0-module* is a dnF space M equipped with continuous G_0-action such that each vector in M is an analytic vector. If the operators defined by the corresponding $\mathfrak{g}$-action are continuous on M, then we call M an *analytic $(\mathfrak{g}, G_0)$-module.*

Suppose M is an analytic $(\mathfrak{g}, G_0)$-module with infinitesimal character λ. Then, one of the key ideas in [8] was the realization that in order to understand the analytic localization of M to Y, it is enough to understand the corresponding geometric fibers [8, See Proposition 8.3 together with Proposition 8.7, as well as Lemma 4.3 (which is central to the proof of Theorem 3 above). Also consider the proof of Proposition 1.3]. In particular, let $\mathcal{M}_{dnF}(U_\lambda(\mathfrak{l}_y))$ denote the category of dnF $\mathfrak{l}_y$-modules with infinitesimal character $\lambda \in \mathfrak{h}^*$. Using the fact that $U_\lambda(\mathfrak{l}_y)$ acts on the geometric fiber of a sheaf of $\pi_*\mathcal{D}_\lambda$-modules, one can show that the geometric fiber determines a derived functor [4, Propostion 3.3]

$$LT_y : D(\mathcal{M}_{dnF}(\pi_*\mathcal{D}_\lambda)) \to D(\mathcal{M}_{dnF}(U_\lambda(\mathfrak{l}_y)).$$

Since a complex of dnF spaces with finite-dimensional homology has Hausdorff homology, the following result follows from [8, Propositions 8.3 and 8.7] [4, See the discussion in the neighborhood of Lemma 6.3].

Proposition 2 *Let M be an analytic $(\mathfrak{g}, G_0)$-module with infinitesimal character λ and suppose that $U \subset Y$ is a G_0-invariant open set. Assume that for each G_0-orbit $S \subset U$ there exits a point $z_S \in S$ such that $LT_{z_S} \circ L\Delta_Y(M)$ has finite-dimensional homology groups in all degrees. For each integer p let $h_p(LT_{z_S} \circ L\Delta_Y(M))$ denote the p-th homology of $LT_{z_S} \circ L\Delta_Y(M)$. Then:*

(a) If $h_p(LT_{z_S} \circ L\Delta_Y(M)) \simeq 0$ for some p and for one of the points z_S then $L_p\Delta_Y(M)$ is zero on the corresponding G_0-orbit S.

(b) More generally, suppose there exits a finite-rank G_0-equivariant holomorphic vector bundle $\mathbb{V}$ defined over U. For each of the points $z_S \in U$ let $\mathbb{V}_{z_S}$ denote the fiber of the bundle $\mathbb{V}$ as $(\mathfrak{p}_{z_S}, G_0 \cap P_{z_S})$-module. Suppose that $\mathfrak{u}_{z_S}$ acts trivially on $\mathbb{V}_{z_S}$. Suppose in addition, that there exists a p such that $h_p(LT_{z_S} \circ L\Delta_Y(M) \simeq \mathbb{V}_{z_S}$ for each of the points z_S. Then the restriction of $L_p\Delta_Y(M)$ to U is isomorphic with the sheaf of holomorphic sections of $\mathbb{V}$.

We conclude this section by stating a simple result that relates the analytic and algebraic localizations of a finitely generated U_λ-module. The proof follows easily from the definitions. For simplicity, we assume λ is regular. Let $\mathcal{O}_Y^{alg}$ denote the regular functions on Y. If $\mathcal{F}$ is a sheaf of modules for $\mathcal{O}_Y^{alg}$, we let $T_y^{alg}(\mathcal{F}) = \mathbb{C} \otimes_{\mathcal{O}^{alg}_{Y,y}} (\mathcal{F}_y)$ denote the corresponding geometric fiber. Then T_y^{alg} determines a derived functor LT_y^{alg} from the derived category of quasi-coherent $\pi_*\mathcal{D}_\lambda^{alg}$-modules to the derived category of $U_\lambda(\mathfrak{l}_y)$-modules. On the other hand, there is a derived functor $L\Delta_Y^{alg}$ defined on the derived category $D(\mathcal{M}(U_\lambda))$ of U_λ-modules. Let $\mathcal{M}_{fg}(U_\lambda)$ be the category of finitely generated U_λ-modules. There are fully faithful functors I_1 and I_2 embedding $\mathcal{M}_{fg}(U_\lambda)$ into the respective derived categories $D(\mathcal{M}(U_\lambda)$ and $D(\mathcal{M}_{dnF}(U_\lambda))$. Thus, using the composition of functors, we can view $L\Delta_Y^{alg}$ and $L\Delta_Y$ as functors defined on the category $\mathcal{M}_{fg}(U_\lambda)$.

Proposition 3 *$LT_y^{alg} \circ L\Delta_Y^{alg} \circ I_1$ defines a functor from $\mathcal{M}_{fg}(U_\lambda)$ into $D(\mathcal{M}_{dnF}(U_\lambda(\mathfrak{l}_y)))$ naturally isomorphic with $LT_y \circ L\Delta_Y \circ I_2$.*

We remark that is is not much more difficult to see that the above result continues to hold when we view $L\Delta_Y^{alg}$ and $L\Delta_Y$ as functors defined on the larger category $\mathcal{M}_{dnF}(U_\lambda)$.

Section 3

We now proceed with the notations introduced in Section 1 and sketch a proof of Theorem 1. Let V be an irreducible finite-dimensional $(\mathfrak{p}_y, G_0 \cap P_y)$-module and assume that V has a regular antidominant infinitesimal character $\lambda \in \mathfrak{h}^*$. Since we are assuming that the G_0-orbit $S = G_0 \cdot y$ is open and that y is special, it follows by a result of J. A. Wolf [21] [13, See Section 3] that the corresponding K-orbit $Q = K \cdot y$ is closed.

Let $\mathcal{I}(V)$ denote the standard Harish-Chandra sheaf on Y determined by the algebraic $(\mathfrak{l}_y, K \cap P_y)$-module V. Then $\mathcal{I}(V)$ is a K-equivariant sheaf of $\pi_*\mathcal{D}_\lambda^{alg}$-modules on Y. Hence $M = \Gamma(Y, \mathcal{I}(V))$ is a $\mathfrak{g}$-module with infinitesimal character λ. On the other hand, since Q is closed, it follows that Kashiwara's equivalence of categories [2, Theorem 7.11] [5, Theorem 4.12] holds for the direct image functor used to define $\mathcal{I}(V)$. In particular, the direct image functor gives an equivalence of categories between a certain category of K-equivariant sheaves defined on Q and the category of K-equivariant quasi-coherent $\pi_*\mathcal{D}_\lambda^{alg}$-modules supported on Q. Therefore, since V is an irreducible algebraic $(\mathfrak{l}_y, K \cap L)$-module, $\mathcal{I}(V)$ is an irreducible K-equivariant sheaf of $\pi_*\mathcal{D}_\lambda^{alg}$-modules supported on Q. Now Theorem 2 implies M is an irreducible $(\mathfrak{g}, K)$-module.

We now introduce the *minimal globalization* M_{min} of M [15] [11]. We begin by sketching a few relevant details. Suppose that N is a Harish-Chandra module for $(\mathfrak{g}, K)$. Then a *globalization* N_{glob} *of* N is a complete locally convex space equipped with a continuous G_0-action, that contains N as the underlying set of K_0-finite vectors. The *minimal globalization* N_{min} *of* N is a functorially defined globalization of N on a *dnF* space that is uniquely characterized by the following condition. Namely, given any globalization N_{glob} of N and any $(\mathfrak{g}, K_0)$-equivariant inclusion $N \xrightarrow{\phi} N_{glob}$ onto the K_0-finite vectors in N_{glob}, then ϕ lifts to a continuous G_0-equivariant inclusion $N_{min} \to N_{glob}$. As pointed out by W. Schmid [15], it is not difficult to see that the existence of N_{min} is implied by the existence of some globalization. A much deeper result [15] [11] is that the functor of minimal globalization is exact and can be identified with the analytic vectors in a Banach space globalization. In particular, M_{min} is a topologically irreducible analytic $(\mathfrak{g}, G_0)$-module with infinitesimal character λ. Hence, we can consider the analytic localization of M_{min} to Y.

An important point for our purposes is that the hypercohomology of the complex representing $L\Delta_Y(M_{min})$ is equal to M_{min} [8, Proposition 5.2] [4, Proposition 3.2]. Formally:

$$R\Gamma_Y \circ L\Delta_Y \simeq Id \tag{1}$$

where Id denotes the identity on $D(\mathcal{M}_{dnF}(U_\lambda))$.

Let $\mathcal{O}(V)_{S,Y}$ denote the extension by zero of $\mathcal{O}(V)$ to Y and suppose that $\mathcal{O}(V)_{S,Y}[q]$ denotes the sheaf $\mathcal{O}(V)_{S,Y}$, when viewed as complex with homology concentrated in degree q. The point of our argument is to show that $L\Delta_Y(M_{min})$ is isomorphic with (in the sense of derived categories) $\mathcal{O}(V)_{S,Y}[q]$. When this is done, then equation (1) will imply

$$R\Gamma_Y(\mathcal{O}(V)_{S,Y}[q]) \simeq M_{min}. \tag{2}$$

Theorem 1 follows from this last equation, because the left hand side computes the compactly supported cohomology (on S) of $\mathcal{O}(V)$.

Thus, according to Proposition 2, in order to establish Theorem 1, it is enough to demonstrate the following two points:

(A) If z is any point in Y that is not in S then $LT_z \circ L\Delta_Y(M_{min}) \simeq 0$.
(B) $LT_y \circ L\Delta_Y(M_{min}) \simeq V[q]$.

Here $V[q]$ is the module V viewed as a complex with homology concentrated in degree q.

In order to examine the geometric fibers of the analytic localization, we will make a certain reduction, utilizing the fibration $\pi : X \to Y$. For $z \in Y$ let $X_z = \pi^{-1}(\{z\}) \subset X$ be the fiber over z and let $i : X_z \to X$ denote the inclusion. The sheaf $\mathcal{D}_\lambda$ induces a twisted sheaf of differential operators $\mathcal{D}^i_\lambda$ defined on the flag manifold X_y such that $\Gamma(X_y, \mathcal{D}^i_\lambda) \simeq U_\lambda(\mathfrak{l}_y)$. In addition, the pull back i^*, in the category of $\mathcal{O}$-modules determines a corresponding derived functor [4, Proposition 3.3]

$$Li^* : D(\mathcal{M}_{dnF}(\mathcal{D}_\lambda)) \to D(\mathcal{M}_{dnF}(\mathcal{D}^i_\lambda)).$$

Let $R\Gamma_{X_z}$ be the derived functor for the global sections on X_z. Then we have the following base change formula [4, Proposition 3.3]:

$$LT_z \circ L\Delta_Y \simeq R\Gamma_{X_z} \circ Li^* \circ L\Delta_X. \tag{3}$$

This formula suggests we should try to understand the analytic localization of M_{min} to X. To approach this problem, we first utilize equation (3) to understand the localization of M to X. Next, the comparison theorem of Hecht and Taylor [9] and a second application of Proposition 2, will allow us to relate this information to $L\Delta_X(M_{min})$.

We apply the following line of reasoning to understand the left hand side of equation (3) for M. Since λ is antidominant and regular, Theorem 2 implies

$$LT_z^{alg} \circ L\Delta_Y^{alg}(M) \simeq LT_z^{alg}(\mathcal{I}(V)).$$

Thus, according to Proposition 3:

$$LT_z \circ L\Delta_Y(M) \simeq LT_z^{alg}(\mathcal{I}(V)).$$

Now we apply the *algebraic base change formula* [2, Theorem 8.4] [4, Proposition 4.4] to the right hand side of the above equation. Since Q is closed, we obtain:

$$LT_z \circ L\Delta_Y(M) \simeq 0 \text{ if } z \text{ is not in } Q. \tag{4}$$

$$LT_y \circ L\Delta_Y(M) \simeq V[q]. \tag{5}$$

We use this information to compute the geometric fibers of the localization of M to X. Let $x \in X$ be a special point and suppose that $z = \pi(x)$ does not belong to S. It follows from Matsuki duality for Y [14] that z does not belong to Q (to see how this follows from [14] consider [4, Section 5]). In fact, in our special case, since Q is closed, it follows that $Q \subset S$ (because K_0 acts transitively on any complete homogeneous space for K). Hence equation (4) and equation (3) imply that $R\Gamma_{X_z} \circ Li^* \circ L\Delta_X(M) \simeq 0$. Since λ is regular, Theorem 3 implies that $Li^* \circ L\Delta_X(M) \simeq 0$. And this last equation in turn implies that $LT_x \circ L\Delta_X(M) \simeq 0$.

On the other hand, suppose that $x \in X$ is a special point and that $y = \pi(x)$. Then equation (5) and equation (3) imply that $R\Gamma_{X_y} \circ Li^* \circ L\Delta_X(M) \simeq V[q]$. According to the Borel-Weil theorem [3] [20, Theorem 1.24], there is a holomorphic line bundle on X_y with sheaf of sections $\mathcal{O}_{X_y}(\lambda)$ such that $R\Gamma_{X_y}(\mathcal{O}_{X_y}(\lambda)) \simeq V$. Once again using the fact that $R\Gamma_{X_y}$ defines an equivalence of categories, we conclude that $Li^* \circ L\Delta_X(M) \simeq \mathcal{O}_{X_y}(\lambda)[q]$. ¿From this it follows that $LT_x \circ L\Delta_X(M) \simeq \mathbb{C}_\lambda[q]$, where $\mathbb{C}_\lambda$ is the one-dimensional $(\mathfrak{b}_x, K \cap B_x)$-module naturally determined by the geometric fiber of $\mathcal{O}_{X_y}(\lambda)$.

Summing up the previous analysis:

Lemma 2 *(a) If $x \in X$ is special and does not belong to $\pi^{-1}(S)$ then $LT_x \circ L\Delta_X(M) \simeq 0$.*
(b) If $x \in X_y$ is special then $LT_x \circ L\Delta_X(M) \simeq \mathbb{C}_\lambda[q]$.

The geometric version of the comparison theorem of H. Hecht and J. Taylor [9] says the following:

Theorem 4 [Hecht and Taylor]
Suppose N is a Harish-Chandra module for $(\mathfrak{g}, K)$ with regular infinitesimal character and $x \in X$ is a special point. Then there is a natural isomorphism:

$$LT_x \circ L\Delta_X(N) \simeq LT_x \circ L\Delta_X(N_{min})$$

Suppose $j : X_y \to \pi^{-1}(S)$ is the inclusion and let j^* denote the corresponding pull back in the category of $\mathcal{O}$-modules. Since the group $G_0 \cap P_y$ acts on the holomorphic bundle with sections $\mathcal{O}_{X_y}(\lambda)$, there exists a G_0-equivariant holomorphic line bundle, whose sheaf of holomorphic sections we denote by $\mathcal{O}(\lambda)$, defined on the open set $\pi^{-1}(S)$, such that $j^*\mathcal{O}(\lambda) \simeq \mathcal{O}_{X_y}(\lambda)$. Using the notations adopted before, we let $\mathcal{O}(\lambda)_{\pi^{-1}(S),X}$ denote the extension to X by zero. Applying Lemma 2, Theorem 4 and Proposition 2, we immediately obtain the following result:

Lemma 3 $L\Delta_X(M_{min}) \simeq \mathcal{O}(\lambda)_{\pi^{-1}(S),X}[q]$

Now (A) and (B) follow directly from Lemma 3 and equation (3).

References

[1] A. Beilinson and J. Bernstein, *Localization de $\mathfrak{g}$ modules*, C.R. Acad. Sci. Paris, 292 (1981), pp. 15–18.

[2] A. Borel, et al., *Algebraic D-Modules*, no. 2 in Perspectives in Mathematics, Academic Press, Inc., 1987.

[3] R. Bott, *Homogeneous vector bundles*, Ann. of Math., 66 (1957), pp. 203–248.

[4] T. Bratten, *Realizing representations on generalized flag manifolds*, Preprint, (1995). to appear in Compositio Math.

[5] J. Chang, *Special K-types, tempered characters and the Beilinson-Bernstein realization*, Duke Math. J., 56 (1988), pp. 345–383.

[6] Harish-Chandra, *Representations of semisimple Lie groups VI*, Amer. J Math., 78 (1956), pp. 564–628.

[7] ——, *Harmonic analysis on real reductive groups I*, J. Func. Anal., 19 (1975), pp. 104–204.

[8] H. Hecht and J. Taylor, *Analytic localization of group representations*, Advances in Math., 79 (1990), pp. 139–212.

[9] ——, *A comparison theorem for $\mathfrak{n}$ homology*, Composito Math., 86 (1993), pp. 189–207.

[10] H. Hecht, D. Miličić, W. Schmid and J. A. Wolf, *Localization and standard modules for semisimle Lie groups I: the duality theorem*, Invent. Math., 90 (1987), pp. 297–332.

[11] M. Kashiwara and W. Schmid, *Quasi-equivariant $\mathcal{D}$-modules, equivariant derived category and representations of reductive Lie groups*, Research anouncement, Research Institute for Mathematical Sciences, Kyoto University (1994).

[12] A. Knapp, *Representation Theory of Semisimple Groups*, Princeton University Press, 1986.

[13] T. Matsuki, *The orbits of affine symmetric spaces under the action of minimal parabolic subgroups*, J. Math. Soc. Japan, 31 (1979), pp. 331–357.

[14] ——, *Orbits on affine symmetric spaces under the action of parabolic subgroups*, Hiroshima Math. J., 12 (1982), pp. 307–320.

[15] W. Schmid, *Boundary value problems for group invariant differential equations*, Proc. Cartan Symposium, Astérique, (1985).

[16] ——, *Homogeneous complex manifolds and representations of semisimple Lie groups, in Representation Theory and Harmonic Analysis on Semisimple Lie Groups*, no. 31 in Math. Surveys and Monographs, Amer. Math. Soc., 1989.

[17] W. Schmid and J. A. Wolf, *Geometric quantization and derived functor modules for semisimple Lie groups*, J. Func. Anal., 90 (1990), pp. 48–112.

[18] J. Serre, *Un théoréme de dualité*, Comment. Math. Helv., 29 (1955), pp. 9–26.

[19] D. Vogan, *Unitarizability of certain series of representations*, Ann. of Math., 120 (1984), pp. 141–187.

[20] ——, *Unitary Representations of Reductive Lie Groups*, no. 118 in Annals of Math. Studies, Princeton Univ. Press, 1987.

[21] J. A. Wolf, *The action of a real semi-simple group on a complex flag manifold,I: orbit structure and holomorphic arc components*, Bull. Amer. Math. Soc., 75 (1969), pp. 1121–1237.

[22] H. Wong, *Dolbeault cohomologies associated with finite rank representations*, Ph.D thesis, Harvard University, (1991).

Facultad de Ciencias Exactas
Universidad del Centro de la Provincia de Buenos Aires
Campus Universitario
Paraje Arroyo Seco, 7000 Tandil, Argentina

ETALE AFFINE REPRESENTATIONS OF LIE GROUPS

DIETRICH BURDE

1. INTRODUCTION

Let G be a finite-dimensional connected Lie group with Lie algebra $\mathfrak{g}$. Denote by E a real vector space and by $\mathbf{Aff}(E)$ the group of affine automorphisms

$$\mathbf{Aff}(E) = \left\{ \begin{pmatrix} A & b \\ 0 & 1 \end{pmatrix} \mid A \in \mathbf{GL}(E),\ b \in E \right\}.$$

Let $\mathfrak{aff}(E)$ be the Lie algebra of $\mathbf{Aff}(E)$. An affine representation $\alpha : G \to \mathbf{Aff}(E)$ of G is called *étale* if there exists a $v \in E$ whose stabilizer G_v is discrete in G, and whose G orbit $G \cdot v$ is open in E. Its differential $\varrho : \mathfrak{g} \to \mathfrak{aff}(E)$ is a Lie algebra homomorphism such that the *evaluation map* $\mathrm{ev}_p : \mathfrak{g} \to E$, $x \mapsto \varrho(x)p = \theta(x)p + u(x)$ is an isomorphism for some $p \in E$, where $\theta : \mathfrak{g} \to \mathfrak{gl}(E)$ is a linear representation and u is the translational part of ϱ. Such a Lie algebra representation is called *étale* again. In this case it follows that $\dim E = \dim G$. We are interested in the following question:

(1) *Which Lie groups admit étale affine representations?*

Etale affine representations of a Lie group arise in the theory of affine manifolds and affine crystallographic groups; see [MIL]. Here the most difficult case is when G is nilpotent. If G is *reductive*, étale affine representations can be studied by methods of invariant theory of affine algebraic varieties; see [BAU], [BU2]. The following has been proved: A semisimple Lie group G does not admit *any* étale affine representation. If G is reductive such that its Lie algebra $\mathfrak{g} = \mathfrak{s} \oplus \mathfrak{z}$ has a 1- dimensional center $\mathfrak{z}$ and $\mathfrak{s}$ is simple, then G admits étale affine representations iff $\mathfrak{s}$ is of type A_ℓ, i.e., if G is $\mathbf{GL}(n)$. For $\mathbf{GL}(n)$ all such representations can be classified; see [BU2].

There is a canonical one-to-one correspondence between étale affine representations of G (up to conjugacy in $\mathbf{Aff}(G)$) and left-invariant affine structures on G (up to affine equivalence); see Definition 1. Given such a structure on G, we can construct many examples of affine manifolds. If G has a left-invariant affine structure and Γ is a discrete subgroup of G, then the homogeneous space $\Gamma\backslash G$ of right cosets inherits an affine structure. If G is nilpotent, then $\Gamma\backslash G$ is called an *affine nilmanifold.* Any compact complete affine manifold with a nilpotent fundamental group is already an affine nilmanifold ([FGH]).

Left-invariant affine structures also play an important role in the study of affine crystallographic groups (ACGs), and of fundamental groups of affine manifolds; see [MIL]. A group $\Gamma \leq \mathbf{Aff}(E)$ is called ACG if it acts properly discontinuously on E with compact quotient. There is the following well-known conjecture by Auslander: *An* ACG *is virtually polycyclic.* This may be restated as follows: The fundamental group of a compact complete affine manifold is virtually polycyclic. The conjecture is still open, though Abels, Margulis and Soifer recently made some progress proving the conjecture up to dimension 6 (see [AMS]).

Milnor proved that a finitely generated torsion-free virtually polycyclic group Γ can be realized as a subgroup of $\mathbf{Aff}(E)$ acting properly discontinuously. Hence it is the fundamental group of a complete affine manifold. Auslander's conjecture is equivalent to the following: A compact complete affine manifold is finitely covered by quotients of solvable Lie groups with complete left-invariant affine structures.

Milnor asked in this context ([MIL]):

(2) *Which Lie groups admit left-invariant affine structures?*

Of course, this is equivalent to our question (1). As said before, this question is particularly difficult for nilpotent Lie groups. There was much evidence that *every* nilpotent Lie group admits left-invariant affine structures. Milnor conjectured this to be true even for solvable Lie groups ([MIL]). Recently, however, counterexamples were discovered ([BGR] and [BEN]). There are nilmanifolds which are not affine. The key step here is to find n-dimensional nilpotent Lie algebras having no faithful representations in dimension $n+1$, hence no affine representation which could arise from a left-invariant affine structure on the Lie group G. We will present some new examples here. There are, however, no counterexamples for the Auslander conjecture.

Left-invariant affine structures on G also correspond to *left-symmetric* algebra structures on $\mathfrak{g}$ (in short, LSA-structures; see Definition 2). Given a Lie algebra $\mathfrak{g}$ over a field of *arbitrary* characteristic, the question of existence of LSA-structures on $\mathfrak{g}$ makes sense and leads to interesting structures. In the case that $\mathfrak{g}$ is a classical simple Lie algebra over a field k of prime characteristic, LSA-structures on $\mathfrak{g}$ are closely related to the first cohomology groups $H^1(G_1, L(\lambda))$, where G_1 is the first Frobenius kernel of a simple algebraic group G with $\mathrm{Lie}(G) = \mathfrak{g}$ and $L(\lambda)$ is a highest weight module of dimension less than or equal to $\dim G$. We have the following result (see [JAN], [BU1]):

Let G be a connected semisimple algebraic group of type $A_l\,(l \geq 1), B_l\,(l \geq 3), C_l(l \geq 2), D_l\,(l \geq 4), G_2, F_4, E_6, E_7, E_8$ over an algebraically closed field k of characteristic $p > 2$. Let $X_1(T)$ denote the set of restricted dominant weights and let $\mathfrak{g} = \mathrm{Lie}(G)$. Assume that

(1) $p > 3$, if G is of type G_2, F_4, E_6, A_1
(2) $p \not| \ l+1$, if G is of type A_l
(3) $p \not| \ l$, if G is of type C_l

Then $H^1(G_1, L(\lambda)) = 0$ for all $\lambda \in X_1(T)$ with $\dim\, L(\lambda) \leq \dim G$. *Furthermore, if $\mathfrak{g}$ admits an LSA-structure, then $p \mid \dim \mathfrak{g}$.*

It is not known in general whether $p \mid \dim \mathfrak{g}$ implies the existence of LSA-structures on such Lie algebras. However, it is true for $\mathfrak{sl}(2,k)$ and $\mathfrak{sl}(3,k)$. In the case of $\mathfrak{sl}(2,k)$, all LSA-structures have been classified ([BU1]). Note that it follows from the proof of the above result that semisimple Lie algebras over characteristic zero do not admit LSA-structures. Hence semisimple Lie groups do not admit étale affine representations.

2. Preliminaries

We consider *affine structures* on a connected Lie group G. Therefore we recall the following definition (see [MIL]):

Definition 1.

Let M denote an n-dimensional manifold. An *affine atlas* on M is a covering of M by coordinate charts such that each coordinate change between overlapping charts is *locally affine*, i.e., extends to an affine automorphism $x \mapsto Ax + b$, $A \in \mathbf{GL}_n(\mathbb{R})$, of some n-dimensional real vector space E. A maximal affine atlas is an *affine structure* on M, and M together with an affine structure is called an *affine manifold.*

Affine manifolds are *flat*; there is a natural correspondence between affine structures on M and *flat torsion-free affine connections* ∇ on M. Such an affine connection is a connection in the tangent bundle with zero torsion and zero curvature.

Subclasses of affine manifolds are *Riemannian-flat* and *Lorentz-flat* manifolds. Note that a manifold does not always admit an affine structure: A closed surface admits affine structures if and only if its Euler characteristic vanishes, i.e., if it is a torus. For higher dimensions ($n \geq 3$) it is in general difficult to decide whether the manifold admits affine structures or not (see [SMI] for more information).

Many examples of affine manifolds come from *left-invariant affine structures on Lie groups*: For a Lie group G, an affine structure on G is *left-invariant*, if for each $g \in G$ the left-multiplication by g, $L_g : G \to G$, is an automorphism of the affine structure. For G simply connected there is a canonical affine immersion $D : G \to E$, called the *developing map* (see [FGH]). Then there exists for each $g \in G$ a unique affine automorphism $\alpha(g)$ of E, such that $\alpha(g) \circ D = D \circ L_g$. In that case $\alpha : G \to \mathbf{Aff}(E)$ is an affine representation.

It is not difficult to see ([FGH]) that G admits a complete left-invariant structure if and only if G acts *simply transitively* on E as affine transformations. By a result of Auslander, G then must be solvable ([AUS]).

Definition 2.

A *left-symmetric algebra structure* (or *LSA-structure*, in short) *on* $\mathfrak{g}$ *over a field* k is a k-bilinear product $\mathfrak{g} \times \mathfrak{g} \to \mathfrak{g}$, $(x,y) \mapsto x \cdot y$ satisfying the conditions $x \cdot y - y \cdot x = [x,y]$ and $(x,y,z) = (y,x,z)$ for all x,y,z, where $(x,y,z) = x \cdot (y \cdot z) - (x \cdot y) \cdot z$ denotes the associator of $x,y,z \in \mathfrak{g}$.

The main definitions given so far are quite related:

Lemma 1. *There is a canonical one-to-one correspondence between the following classes of objects (up to suitable equivalence):*

(a) {*Etale affine representations of* G}
(b) {*Left-invariant affine structures on* G}
(c) {*Flat torsionfree left-invariant affine connections* ∇ *on* G}
(d) {*LSA-structures on* $\mathfrak{g}$}

Proof. This is well known; see [BU3],[SEG],[KIM]. We will give some arguments in order to establish notations. If we have any LSA-structure on $\mathfrak{g}$ with product $(x,y) \mapsto x \cdot y$, then denote by $\lambda : x \mapsto \lambda(x)$ the left-regular representation on the LSA $(\mathfrak{g}, \cdot)$: $\lambda(x)y = x \cdot y$. It is a Lie algebra representation: $\lambda : \mathfrak{g} \to \operatorname{End}(\mathfrak{g})$, $[\lambda(x), \lambda(y)] = \lambda([x,y])$. Denote the corresponding $\mathfrak{g}$-module by $\mathfrak{g}_\lambda$. Furthermore, the identity map $\mathbf{1} : \mathfrak{g} \to \mathfrak{g}_\lambda$ is a 1-cocycle in $Z^1(\mathfrak{g}, \mathfrak{g}_\lambda)$: $\mathbf{1}([x,y]) = \mathbf{1}(x) \cdot y - \mathbf{1}(y) \cdot x$. Let $\mathfrak{aff}(\mathfrak{g})$ be the Lie algebra of $\mathbf{Aff}(G)$, i.e.,

$$\mathfrak{aff}(\mathfrak{g}) = \left\{ \begin{pmatrix} A & b \\ 0 & 0 \end{pmatrix} \mid A \in \mathfrak{gl}(\mathfrak{g}),\ b \in \mathfrak{g} \right\}$$

which we identify with $\mathfrak{gl}(\mathfrak{g}) \oplus \mathfrak{g}$. Denote the linear part by $\ell(A,b) = A$ and the translational part by $t(A,b) = b$. Now we associate to the LSA $(\mathfrak{g}, \cdot)$ the map $\alpha = \lambda \oplus \mathbf{1} : \mathfrak{g} \to \mathfrak{aff}(\mathfrak{g})$. This is an affine representation of $\mathfrak{g}$. We have $\lambda = \ell \circ \alpha$ and $t \circ \alpha = \mathbf{1}$. The corresponding affine representation of G is étale; see [SEG].

3. Affine representations of reductive Lie groups

Let k be an algebraically closed field of characteristic zero. A Lie algebra $\mathfrak{g}$ is said to be *reductive* if its solvable radical $\mathfrak{r}(\mathfrak{g})$ coincides with the center $\mathfrak{z} = \mathfrak{z}(\mathfrak{g})$. Then the Lie algebra $\mathfrak{s} = [\mathfrak{g}, \mathfrak{g}]$ is semisimple and we have $\mathfrak{g} = \mathfrak{s} \oplus \mathfrak{z}$. A Lie group G is said to be reductive if its Lie algebra is reductive. Assume that $(\mathfrak{g}, \cdot)$ is an LSA-structure on $\mathfrak{g}$. The first cohomology groups of a reductive Lie algebra do not vanish in general. However, if the center is one-dimensional and the $\mathfrak{g}$- module is $\mathfrak{g}_\lambda$ arising from an étale affine representation of G, then we are able to prove (see [BU2]):

Proposition 1. *Let* $(\mathfrak{g}, \cdot)$ *be an LSA-structure on* $\mathfrak{g}$. *If* $\dim \mathfrak{z} = 1$ *then* $H^0(\mathfrak{g}, \mathfrak{g}_\lambda) = 0$ *and* $H^1(\mathfrak{g}, \mathfrak{g}_\lambda) = 0$.

Proposition 2. *Let* $\mathfrak{g} = \mathfrak{s} \oplus \mathfrak{z}$ *be a reductive Lie algebra such that* $\dim \mathfrak{z} = 1$ *and* $\mathfrak{s}$ *is of type* A_ℓ, B_ℓ, C_ℓ, D_ℓ, G_2, F_4, E_6, E_7, E_8. *Then* $\mathfrak{g}$ *admits an LSA-structure if and only if* $\mathfrak{s}$ *is of type* A_ℓ.

Here is a brief outline of the proof of Proposition 2. Let $\dim \mathfrak{s} = n$. The $\mathfrak{g}$-module $\mathfrak{g}_\lambda$ is completely reducible as an $\mathfrak{s}$-module and has *no invariants* by Proposition 1, i.e., the trivial module k is not a summand in the decomposition of $\mathfrak{g}_\lambda$. Hence we know that $\mathfrak{g}_\lambda = \oplus_i V_i$ and $\sum_i \dim V_i = n+1$, where V_i are irreducible $\mathfrak{s}$-modules with $2 \le \dim V_i \le \dim \mathfrak{g} = n+1$. On the other hand, there are not many irreducible $\mathfrak{s}$-modules of dimension smaller than or equal to $n+1$. It is possible to classify them. For a given type of $\mathfrak{s}$ the dimensions of these modules have to add up to $\dim \mathfrak{g}_\lambda$. However, in most cases this is possible only if $\mathfrak{s}$ is of type A_l. This argument only fails in the case of type B_3, D_5, D_7, where the modules are

$$\mathfrak{g}_\lambda = L(\omega_1) \oplus L(\omega_1) \oplus L(\omega_3) \text{ for } B_3,$$
$$\mathfrak{g}_\lambda = L(\omega_1) \oplus L(\omega_1) \oplus L(\omega_1) \oplus L(\omega_5) \text{ for } D_5,$$
$$\mathfrak{g}_\lambda = L(\omega_1) \oplus L(\omega_1) \oplus L(\omega_7) \text{ for } D_7.$$

Here $\omega_1, \dots, \omega_\ell$ denote the fundamental weights and $L(\omega_i)$ the highest weight module to ω_i. The dimensions satisfy $22 = 7+7+8, 46 = 10+10+10+16$ and $92 = 14+14+64$ respectively.

To prove the result in these cases, we use invariant theory: Let $\varrho : \mathfrak{g} \to \mathfrak{aff}(\mathfrak{g})$ be an étale affine representation arising from an LSA-structure. Let S be the simply connected semisimple algebraic group with Lie algebra $\mathfrak{s}$. The linear part of ϱ is the differential of a rational representation $\rho : S \to \mathbf{Aff}(V)$. Thus we may regard V as an algebraic S-variety. If the center of $\mathfrak{g}$ is one-dimensional, we know that V is isomorphic to a *linear* S-variety. Since ϱ is étale, we have $\dim V = \dim S + 1$ and V has an S orbit of codimension 1. However, it is easy to see that the above modules (where S is an orthogonal group) do *not* have an S orbit of codimension 1. If the center of $\mathfrak{g}$ is higher-dimensional, then the situation becomes more complicated (see [HEL], [BU2]).

As mentioned before, in the case of $\mathbf{GL}(n)$ we can classify all étale affine representations, i.e., all LSA-structures on $\mathfrak{gl}(n)$.

Let $\mathcal{A} = (\mathfrak{g}, \cdot)$ be an LSA-structure on $\mathfrak{g}$. Denote by $\mathrm{End}_*(\mathfrak{g})$ the set $\{\tau \in \mathrm{End}(\mathfrak{g}) \mid (\mathbf{1}-\tau)^{-1} \text{ exists and } \tau(\mathcal{A}) \subset k(\mathcal{A})\}$ where

$$k(\mathcal{A}) := \{a \in \mathcal{A} \mid [\lambda(b), \varrho(a)] = 0 \quad \forall\, b \in \mathcal{A}\}.$$

Here λ and ϱ denote left and right multiplication in $\mathcal{A}$. Let $\tau \in \mathrm{End}_*(\mathfrak{g})$ with $\phi = (\mathbf{1}-\tau)^{-1}$. Then $\lambda_\tau(a) := \phi \circ (\lambda(a) - \varrho(\tau(a))) \circ \phi^{-1}$ defines an LSA-structure on $\mathfrak{g}$. We call $\mathcal{A}_\tau$ the *τ-deformation* of $\mathcal{A}$. The result is ([BAU],[BU2]):

Proposition 3. *The τ deformations of the full matrix algebra exhaust all possible LSA-structures on $\mathfrak{gl}_n(k)$ for $n > 2$. Their isomorphism classes are parametrized by the conjugacy classes of elements $X \in \mathfrak{gl}_n(k)$ with* $\mathrm{tr}\,(X) = n$. *In the case of $\mathfrak{gl}(2,k)$ we have one more isomorphism class.*

4. Affine representations of nilpotent Lie groups

Milnor conjectured in [MIL] that every nilpotent Lie group G admits étale affine representations, i.e., its Lie algebra $\mathfrak{g}$ admits LSA-structures. Indeed, many classes of nilpotent Lie algebras admit LSA-structures (see [BU3]):

Proposition 4. *Let $\mathfrak{g}$ be a nilpotent Lie algebra of characteristic zero satisfying one of the following conditions:*

(1) $\dim \mathfrak{g} < 8$.
(2) *$\mathfrak{g}$ is p-step nilpotent with $p < 4$.*
(3) *$\mathfrak{g}$ is $\mathbb{Z}$-graded.*
(4) *$\mathfrak{g}$ possesses a nonsingular derivation.*
(5) *$\mathfrak{g}$ is filiform nilpotent and a quotient of a higher dimensional filiform nilpotent Lie algebra.*
(6) *$\mathfrak{g}$ possesses a nonsingular 1- cocycle in $Z^1(\mathfrak{g}, \mathfrak{g}_\theta)$, where $\theta : \mathfrak{g} \to \mathfrak{gl}(\mathfrak{g})$ is a representation.*

Then $\mathfrak{g}$ admits an LSA-structure.

However, there are nilpotent Lie algebras without any LSA-structure. To construct such examples we use

Lemma 2. *If $\mathfrak{g}$ admits an LSA-structure then $\mathfrak{g}$ has a faithful representation of dimension* $\dim \mathfrak{g} + 1$.

Proof. The LSA-structure on $\mathfrak{g}$ induces (see Lemma 1) a faithful affine representation $\alpha : \mathfrak{g} \to \mathfrak{aff}(\mathfrak{g})$ called the *affine holonomy representation.* If $\dim \mathfrak{g} = n$ then $\mathfrak{aff}(\mathfrak{g}) \subset \mathfrak{gl}(n+1)$ and we obtain a faithful linear representation of dimension $n + 1$.

Definition 3.

Let $\mathfrak{g}$ be a finite-dimensional Lie algebra over a field k. Define

$$\mu(\mathfrak{g}, k) := \min \{\dim_k M \mid M \text{ is a faithful } \mathfrak{g}\text{-module}\}$$

By Ado's theorem (and Iwasawa's in prime characteristic) we know that μ is integer valued. It seems that there is not much known about μ in the literature. We list a few properties proved in [BU4]. Let $\lceil x \rceil$ denote the ceiling of x :

Proposition 5. *Let $\mathfrak{g}$ be a Lie algebra of dimension n over $\mathbb{C}$.*

(1) *If $\mathfrak{g}$ is abelian then $\mu(\mathfrak{g}) = \lceil 2\sqrt{n-1} \rceil$.*
(2) *If $\mathfrak{g}$ has trivial center then $\mu(\mathfrak{g}) \leq n$.*
(3) *If $\mathfrak{g}$ is a Heisenberg Lie algebra $\mathfrak{h}_{2m+1}$ of dimension $2m + 1$, then $\mu(\mathfrak{g}) = m + 2$.*

(4) *If* $\mathfrak{g}$ *is solvable then* $\mu(\mathfrak{g}) < 2^n$.
(5) *If* $\mathfrak{g}$ *is filiform nilpotent with abelian commutator algebra then* $\mu(\mathfrak{g}) = n$.
(6) *If* $\mathfrak{g}$ *is filiform nilpotent then* $n \le \mu(\mathfrak{g}) < \sqrt{1/n}\exp(\pi\sqrt{2(n-1)/3})$.
(7) *If* $\mathfrak{g}$ *admits an LSA-structure then* $\mu(\mathfrak{g}) \le n+1$.
(8) *If* $\mathfrak{g}$ *is a quotient of a filiform nilpotent Lie algebra* $\mathfrak{g}'$ *with* $\dim \mathfrak{g}' > \dim \mathfrak{g} = n$ *then* $\mu(\mathfrak{g}) = n$.
(9) *If* $\mathfrak{g}$ *is filiform nilpotent of dimension* $n < 10$ *then* $\mu(\mathfrak{g}) = n$.

Note that even the abelian case (1) is nontrivial. Here $\mu(\mathfrak{g}) = n$ is *not* true if $n > 4$: Let ϕ be a faithful representation of $\mathfrak{g}$ into $\mathfrak{gl}(V)$ where V is a d-dimensional vector space. Then $\phi(\mathfrak{g})$ is an n-dimensional commutative subalgebra of the matrix algebra $M_d(\mathbb{C})$. Hence its dimension is bounded by $[\frac{d^2+4}{4}]$ by a well-known result of Schur, i.e., $n \le [\frac{d^2+4}{4}]$ and $d \ge \lceil 2\sqrt{n-1}\rceil$.

The key step for the construction of the counterexamples to the Milnor conjecture is to determine Lie algebras with $\mu(\mathfrak{g}) > \dim \mathfrak{g} + 1$. In the following we will construct filiform Lie algebras in dimensions 10 and 11 with that property. These algebras have no extension by any filiform Lie algebra of higher dimension.

Let $\mathfrak{g}$ be a p-step nilpotent Lie algebra and let $\mathfrak{g}^0 = \mathfrak{g}, \quad \mathfrak{g}^k = [\mathfrak{g}^{k-1}, \mathfrak{g}]$. The series $\mathfrak{g} = \mathfrak{g}^0 \supset \mathfrak{g}^1 \supset \ldots \supset \mathfrak{g}^{p-1} \supset \mathfrak{g}^p = 0$ is called *lower central series.* Recall that a p-step nilpotent Lie algebra of dimension n is called *filiform nilpotent* if $p = n-1$.

Definition 4.

Let $L = L(n)$ be the Lie algebra generated by $e_0, \ldots, e_n$ with Lie brackets $[e_0, e_i] = e_{i+1}$ for $i = 1, 2, \ldots, n-1$ and the other brackets zero. L is called the *standard graded filiform of dimension* $n+1$.

Consider the affine algebraic variety of all Lie algebra structures in dimension n over $\mathbb{C}$. In particular, we have the subvariety of nilpotent filiform Lie algebra structures. The following result is due to Vergne ([VER]):

Proposition 6. *Every filiform nilpotent Lie algebra of dimension* $n+1 \ge 4$ *is isomorphic to an infinitesimal deformation of the standard graded* $(n+1)$-*dimensional filiform* L. *More precisely, it is isomorphic to an algebra* $(L)_\psi$ *where* ψ *is an integrable* 2-*cocycle whose cohomology class lies in*

$$\begin{array}{lll} F_1H^2(L,L) & & if \quad n \equiv 0(2) \\ F_1H^2(L,L) & + \; < \psi_{\frac{n-1}{2},n} > & if \quad n \equiv 1(2) \end{array}$$

Here the algebra $\mathfrak{g}_\psi = (L)_\psi$ is defined by the bracket $[a,b]_\psi = [a,b]_L + \psi(a,b)$. The fact that ψ is integrable means that this bracket satisfies the Jacobi identity, i.e., $\psi(a,\psi(b,c)) + \psi(b,\psi(c,a)) + \psi(c,\psi(a,b)) = 0$. For the definition of $F_1H^2(L,L)$ see [HAK]. Here we determine a canonical basis for this space (see [BU3]):

Proposition 7. *Define canonical* 2- *cocycles* $\psi_{k,s}$ *by* $\psi_{k,s}(e_i, e_{i+1}) = \delta_{ik} e_s$ *for pairs* (k,s) *with* $1 \le k \le n-1$ *and* $2k \le s \le n$. *The cohomology classes*

of the cocycles $\psi_{k,s}$ with $1 \leq k \leq [n/2] - 1,\ 2k+2 \leq s \leq n$ form a basis of $F_1H^2(L,L)$. This space has dimension $\frac{(n-2)^2}{4}$ if n is even and dimension $\frac{(n-3)(n-1)}{4}$ if n is odd. The following formula holds:
$\psi_{k,s}(e_i,e_j) = (-1)^k\binom{j-k-1}{k-i}(\mathrm{ad}e_0)^{i+j-2k-1}e_s$ for $1 \leq i < k < j-1 \leq n-1$. In the case $i > k, \psi_{k,s}(e_i,e_j) = 0$ and $\psi_{k,s}(e_k,e_j) = e_{s+j-k-1}$ for $k < j$.

4.1 Filiform Lie algebras of dimension 10

Let $L = L(9) =< e_0, e_1, \dots, e_9 >$ be the standard graded filiform Lie algebra of dimension 10. According to Proposition 6 every filiform nilpotent Lie algebra of dimension 10 is isomorphic to $\mathfrak{g}_\psi = (L)_\psi$ for some $\psi \in F_1H^2(L,L) + < \psi_{4,9} >$. In terms of the basis of this cohomology space we may write

$$\begin{aligned}\psi = \ & \alpha_1\psi_{1,4} + \alpha_2\psi_{1,5} + \dots + \alpha_6\psi_{1,9} \\ & + \alpha_7\psi_{2,6} + \dots + \alpha_{10}\psi_{2,9} \\ & + \alpha_{11}\psi_{3,8} + \alpha_{12}\psi_{3,9} \\ & + \alpha_{13}\psi_{4,9}\end{aligned}$$

The cocycle ψ is integrable if and only if $[a,b]_\psi = [a,b]_L + \psi(a,b)$ satisfies the Jacobi identity. This is equivalent to the following equations:

$$(1) \quad \alpha_{13}(2\alpha_3 + \alpha_9) - \alpha_{12}(2\alpha_1 + \alpha_7) - 3\alpha_{11}(\alpha_2 + \alpha_8) + 7\alpha_7\alpha_8 = 0$$
$$(2) \quad \alpha_{11}(2\alpha_1 + \alpha_7) - 3\alpha_7^2 = 0$$
$$(3) \quad \alpha_{13}(2\alpha_1 - \alpha_7 - \alpha_{11}) = 0.$$

Using these simple conditions we obtain the following classes of filiform Lie algebras $\mathfrak{g}_\psi$ with bracket $[a,b]_\psi$:

Case A: $2\alpha_1 + \alpha_7 \neq 0$:

Class $(A1)$: $\alpha_1 \neq 0,\ \alpha_7 = -\alpha_1,\ \alpha_{11} = 3\alpha_1$.
Class $(A2)$: $\alpha_1 \neq 0,\ \alpha_{11} = \alpha_7 = \alpha_1$.
Class $(A3)$: $\alpha_1 \neq 0,\ \alpha_7^2 \neq \alpha_1^2,\ \alpha_{11} = 3\alpha_7^2/(2\alpha_1 + \alpha_7), \alpha_{13} = 0$.

Case B: $2\alpha_1 + \alpha_7 = 0$:

Class $(B1)$: $\alpha_{13} = \alpha_7 = \alpha_1 = 0,\ \alpha_{11}(\alpha_2 + \alpha_8) = 0$.
Class $(B2)$: $\alpha_{13} \neq 0,\ \alpha_{11} = \alpha_7 = \alpha_1 = 0,\ \alpha_9 = -2\alpha_3$.

In case A, α_{12} is uniquely determined by equation (1). We want to know the minimal dimension of faithful modules for these classes of Lie algebras. The result is

Proposition 7. *If* $\mathfrak{g}_\psi$ *is a filiform Lie algebra of class* $A3$, $B1$, $B2$ *then* $\mu(\mathfrak{g}_\psi) = 10$; *if* $\mathfrak{g}_\psi$ *is of class* $A1$ *satisfying the additional condition* $3\alpha_2 + \alpha_8 = 0$, *or is of class* $A2$, *then* $\mu(\mathfrak{g}_\psi) = 10$ *or* 11.

The class excluded above indeed provides counterexamples to Milnor's conjecture:

Proposition 8. *Let* $\mathfrak{g}_\psi = \mathfrak{g}(\alpha_1, \dots, \alpha_{13})$ *be a Lie algebra of class* $A1$, *satisfying* $3\alpha_2 + \alpha_8 \neq 0$. *Then* $12 \leq \mu(\mathfrak{g}_\psi) \leq 22$.

The proof is given in [BU3]. The rough idea is as follows: Let $\mathfrak{g}_\psi$ be a filiform nilpotent Lie algebra of dimension 10. Suppose there is *any* faithful module M of dimension $m < 12$. By Lemma 3.2. in [BEN] we may assume that M is nilpotent and is of dimension 11. For such modules we construct a *combinatorical type*, thereby classifying such modules. Note that the faithfulness is a strong condition which excludes many types of modules. For each type we check the conditions for M to be a faithful nilpotent module of dimension $m < 12$. This means certain equations in the α_i. The crucial equation is $3\alpha_2 + \alpha_8 = 0$. On the other hand, we construct a faithful module of dimension 22 for all filiform Lie algebras of dimension 10.

Remark 1. Let G be the connected simply connected Lie group with filiform nilpotent Lie algebra as in Proposition 8. Then G does not admit an étale affine representation. There is the question whether the Lie groups corresponding to the other classes (see Proposition 7) *do* admit such representations. This seems not to be true in general. However, for the classes $A3, B1, B2$ the answer is positive.

4.2 Filiform Lie algebras of dimension 11

Let $L = L(10) = < e_0, e_1, \dots, e_{10} >$ be the standard graded filiform Lie algebra of dimension 11. Then every filiform nilpotent Lie algebra of dimension 11 is isomorphic to $\mathfrak{g}_\psi = (L)_\psi$ for some $\psi \in F_1 H^2(L, L)$. In terms of the basis of this cohomology space we may write

$$\begin{aligned}\psi \; = \; & \alpha_1 \psi_{1,4} + \alpha_2 \psi_{1,5} + \dots + \alpha_7 \psi_{1,10} \\ & + \alpha_8 \psi_{2,6} + \alpha_9 \psi_{2,7} \dots + \alpha_{12} \psi_{2,10} \\ & + \alpha_{13} \psi_{3,8} + \dots + \alpha_{15} \psi_{3,10} \\ & + \alpha_{16} \psi_{4,10}\end{aligned}$$

The integrability of ψ is determined by four equations. We are interested here in the case $\alpha_1 \neq 0$. We have the following result, using the same methods as above (see also [BGR]):

Proposition 9. *Let* $\mathfrak{g}_\psi$ *be a filiform nilpotent Lie algebra of dimension* 11 *satisfying* $\alpha_1 \neq 0$. *Then* $\mu(\mathfrak{g}_\psi) \leq 12$ *if and only if* $\alpha_8 = 0$ *or* $10\alpha_8 = \alpha_1$ *or* $5\alpha_8^2 = 2\alpha_1^2$ *or* $4\alpha_1^2 - 4\alpha_1\alpha_8 + 3\alpha_8^2 = 0$.

References

[AMS] H. ABELS, G.A. MARGULIS AND G.A. SOIFER, On the Zariski closure of the linear part of a properly discontinuous group of affine transformations, *Preprint* (1995).

[AUS] L. AUSLANDER, Simply transitive groups of affine motions, *Amer. J. Math.* **99** (1977), 809-826.

[BAU] O. BAUES, *Flache Strukturen auf* **GL**(n) *und zugehörige linkssymmetrische Algebren*, Dissertation, Düsseldorf 1995.

[BEN] Y. BENOIST, Une nilvariété non affine, *J. Diff. Geometry* **41** (1995), 21-52.

[BGR] D. BURDE AND F. GRUNEWALD, Modules for certain Lie algebras of maximal class, *J. of pure and appl. Algebra*, **99** (1995), 239-254.

[BU1] D. BURDE, Left-symmetric structures on simple modular Lie algebras, *J. Algebra* **169** (1994), 112-138.

[BU2] D. BURDE, Left-invariant affine structures on reductive Lie groups, *J. of Algebra* **181** (1996), 884-902.

[BU3] D. BURDE, Affine structures on nilmanifolds, *International Journal of Math.* **7**: 5 (1996), 599-616.

[BU4] D. BURDE, On a Refinement of Ado's Theorem, to appear in *Archiv der Mathematik.*

[FGH] D. FRIED, W. GOLDMAN AND M.W. HIRSCH, Affine manifolds with nilpotent holonomy, *Comment. Math. Helv.* **56** (1981), 487-523.

[HAK] Y. B. HAKIMJANOV, Variété des lois d'àlgèbres de Lie nilpotentes *Geometriae Dedicata* **40** (1991), 269-295.

[HEL] J. HELMSTETTER, Algèbres symétriques à gauche, *C.R. Acad. Sc. Paris* **272** (1971), 1088-1091.

[JAN] J. JANTZEN, First cohomology groups for classical Lie algebras, *Prog. Math.* **95** (1991), 289-315.

[KIM] H. KIM, Complete left-invariant affine structures on nilpotent Lie groups, *J. Differential Geometry* **24** (1986), 373-394.

[MIL] J. MILNOR, On fundamental groups of complete affinely flat manifolds, *Advances in Math.* **25** (1977), 178-187.

[SEG] D. SEGAL, The structure of complete left-symmetric algebras, *Math. Ann.* **293** (1992), 569-578.

[SMI] J. SMILLIE, An obstruction to the existence of affine structures, *Inventiones Math.* **64** (1981), 411-415.

[VER] M. VERGNE, Cohomologie des algèbres de Lie nilpotentes. Application à l'étude de la variété des algèbres de Lie nilpotentes, *Bull. Math. Soc. France* **78** (1970), 81-116.

Institut für Mathematik
Heinrich-Heine Universitat Dusseldorf
D-40225 Dusseldorf
Germany

COMPATIBILITY BETWEEN A GEOMETRIC CHARACTER FORMULA AND THE INDUCED CHARACTER FORMULA

ESTHER GALINA[1]

ABSTRACT. here is a geometric formula for characters of representations of a semisimple Lie group in terms of cohomologies of complexes of sheaves associated to the representation. We explicitly prove the compatibility of this formula with the known induced character formula. The tools we use in the proof are some geometric properties of generalized flag manifolds and their orbits.

1. INTRODUCTION

Around 1987 Kashiwara conjectured a geometric formula for characters of representations of a semisimple Lie group. This formula, which originates from Kashiwara's ideas on characteristic cycles, was proved by Schmid and Vilonen [10]. An alternative proof, based on a different circle of ideas, was given by Hecht and Taylor [4].

In [10] the authors observed that for an induced representation the geometric formula is compatible with the Harish-Chandra–Hirai–Wolf induced character formula ([2], [6], [13]). The purpose of this work is to make this compatibility more explicit taking the point of view of the geometry of orbits. Moreover, we reobtain the character of an induced representation of a connected real semisimple Lie group G_o from a representation of any parabolic subgroup P_o of G_o. Note that in [13] the formula of the character is given only for induced representations where the inducing group P_o is a cuspidal parabolic subgroup of G_o. The induced character formula shows that the character of the induced representation restricted to the regular elements of a Cartan subgroup C_o of G_o vanishes unless C_o is conjugate under G_o to a Cartan of the Levi factor of P_o. We obtain this by analyzing the geometry of orbits on generalized flag varieties. The rest of the techniques we use come from homological algebra and from analytic $\mathcal{D}_\lambda$-modules [3].

In the next section we introduce the necesary notation. In the third section we prove some geometric properties of the closed G_o-orbit in a generalized flag manifolds. Propositions 3.2 and 3.3 are interesting by themselves. They are the key points of the proof of the main result included in section four.

This article is an expanded version of a talk given at the Fifth Workshop on Representation Theory of Lie Groups and Its Applications. I want to thank

[1]Supported by CONICET, CONICOR and Univ. Nacional de Córdoba, Argentina.

the Scientific Organizing Commitee of this workshop for the invitation to give this talk. I also want to express my gratitude to Henryk Hecht for all the mathematics I learned from him during my stay at the University of Utah.

2. Notation

Let G_o be a connected real semisimple Lie group with finite center. Denote by $\mathfrak{g}_o$ its Lie algebra and by $\mathfrak{g}$ its complexification. In general we will use a zero subscript for real groups or real algebras and the same letter without it for its complexification.

Let Y be a generalized flag variety, that is, a conjugacy class of parabolic subalgebras of $\mathfrak{g}$. For each point y of Y, $\mathfrak{p}_y$ denotes the corresponding parabolic subalgebra, $\mathfrak{u}_y$ its nilpotent radical, and $\mathfrak{l}_y = \mathfrak{p}_y/\mathfrak{u}_y$ its Levi "factor subalgebra". Given $y \in Y$, we can express Y as the quotient G/P_y of any connected complex Lie group G that acts transitively on Y and the corresponding parabolic subgroup P_y that stabilizes y. We fix a group G such that it is a complexification of G_o; that is, there is a homomorphism from G_o into G with derivative the inclusion of $\mathfrak{g}_o$ into $\mathfrak{g}$. We regard Y as a complex manifold and denote by $\mathcal{O}_Y$ its structure sheaf.

Let $\mathcal{G}_Y = \mathcal{O}_Y \otimes_{\mathbb{C}} \mathfrak{g}$ be the sheaf of local sections of the trivial bundle with fiber $\mathfrak{g}$. The collections $\{\mathfrak{p}_y\}$, $\{\mathfrak{u}_y\}$ and $\{\mathfrak{l}_y\}$ form holomorphic G-homogeneous vector bundles on Y. The corresponding sheaves of local sections are subsheaves of $\mathcal{G}_Y$. Call them $\mathcal{P}$, $\mathcal{U}$ and $\mathcal{L}$, respectively.

When $\{\mathfrak{p}_y\}$ is the set of Borel subalgebras of $\mathfrak{g}$, is the ordinary flag variety, that we denote by X, $\mathfrak{b}_x$ denotes the Borel subalgebra associated to $x \in X$, and $\mathfrak{n}_x$ its nilpotent radical. The vector bundle over X associated to the collection $\{\mathfrak{b}_x/\mathfrak{n}_x\}$ is trivial because B_x acts on its fiber at x as the identity transformation. Denote by $\mathcal{H}$ the sheaf of local sections of such a vector bundle. The space $\mathfrak{h}$ of constant sections of $\mathcal{H}$ is called the *abstract* Cartan subalgebra of $\mathfrak{g}$. For each x, $\mathfrak{h}$ is isomorphic to $\mathfrak{b}_x/\mathfrak{n}_x$. Consequently, if $\mathfrak{c} \subset \mathfrak{b}_x$ is a Cartan subalgebra there is a natural isomorphism $\mathfrak{h} \to \mathfrak{c}$, which is called the *specialization* of $\mathfrak{h}$ to $\mathfrak{c}$ at x. It is also possible to specialize $\mathfrak{h}^*$ at x by duality, that is, for $\lambda \in \mathfrak{h}^*$, λ_x is a linear form on $\mathfrak{c}$. In this context we can define a root system Δ, and a positive root system Δ^+ of $\mathfrak{h}$ that correspond to the pullback of the data $(\mathfrak{c}, \mathfrak{b}_x, \mathfrak{g})$.

Let $\pi\colon X \to Y$ the projection which assigns to each Borel subalgebra $\mathfrak{b}$ the unique parabolic subalgebra $\mathfrak{p}$ in Y that contains $\mathfrak{b}$. This projection is G-invariant. If B and P denote the analytic subgroups of G with Lie algebra $\mathfrak{b}$ and $\mathfrak{p}$ respectively, π becomes the natural projection $\pi\colon G/B \to G/P$. Its fiber over $\mathfrak{p}$ is equal to P/B, and it is the flag variety of the Levi factor $\mathfrak{l} = \mathfrak{p}/\mathfrak{b}$. We remark that $\mathfrak{l}$ is a reductive Lie algebra.

Let Π be an admissible, continuous Hilbert space representation of G_o with infinitesimal character. Let M be the space of analytic vectors of Π. It is a dense G_o-invariant subset of the representation space and it coincides with the minimal globalization of the Harish-Chandra module of Π in the sense of

Schmid in [9]. According with the notation of [3] and [4], we will call it the *admissible analytic module of* Π.

Let W be the Weyl group of Δ. Then a character θ of the center $\mathcal{Z}(\mathfrak{g})$ of the universal enveloping algebra $\mathfrak{A}(\mathfrak{g})$ of $\mathfrak{g}$ can be identified with a Weyl group orbit in $\mathfrak{h}^*$. Let $\mathfrak{A}_\theta = \mathfrak{A}(\mathfrak{g})/\mathfrak{A}(\mathfrak{g})\ker(\theta)$ where $\ker(\theta)$ is the kernel of the character θ. For each $\lambda \in \mathfrak{h}^*$ regular and antidominant, Beilinson–Bernstein localization theory asserts that the global section functor provides an equivalence of the category of $\mathfrak{A}_\theta$-modules and the category of quasicoherent modules over the $\mathcal{D}_\lambda^{alg}$-sheaf of algebraic twisted differential operators. If λ is only regular, there is an equivalence of categories after passing to the derived categories and derived functors, always from an algebraic point of view.

Another localization theory, called analytic localization, was developed by Hecht and Taylor in [3]. There they replace $\mathcal{D}_\lambda^{alg}$ by the sheaf $\mathcal{D}_\lambda$ of twisted differential operators with holomorphic coefficients on X. For λ regular they proved an equivalence between the derived category of topological $\mathfrak{A}_\theta$-modules with a structure of dual of a nuclear Fréchet (DNF) space and the derived category $D\mathfrak{M}(\mathcal{D}_\lambda)$ of DNF $\mathcal{D}_\lambda$-modules. Moreover, if $D\mathfrak{M}(\mathcal{D}_\lambda, G_o)$ denotes the full triangulated subcategory generated by G_o-equivariant $\mathcal{D}_\lambda$-modules whose restrictions to any G_o-orbit are locally free of finite rank over $\mathcal{O}_X$; then, they proved that any admissible analytic $\mathfrak{A}_\theta$-module M is isomorphic to $H^0(X, \mathcal{M})$ where $\mathcal{M}$ is a certain element of $D\mathfrak{M}(\mathcal{D}_\lambda, G_o)$ with hypercohomology concentrated in degree zero. The complex of sheaves $\mathcal{M}$ is called the *analytic localization* of M. It has the property that along each G_o-orbit the cohomology sheaves are given by local sections of holomorphic G_o-equivariant vector bundles.

We will consider some special sheaves on Y called *induced analytic sheaves.* See [3] for a general definition of them. A special description for the case where G is a complexification of G_o is the following. Let S be a G_o-orbit of the action of G_o on Y, and let $P_{y,o}$ be the isotropy group of a point y of S in G_o. Consider (ω, V) an analytic DNF $(\mathfrak{p}_y, P_{y,o})$-module where the unipotent radical of $P_{y,o}$ and the nilpotent radical $\mathfrak{u}_y$ of $\mathfrak{p}_y$ act trivially. By inducing V we obtain a G_o-homogeneous vector bundle on S. Since V is an analytic module, the resulting vector bundle is a real analytic G_o-homogeneous vector bundle over S. Denote by $\mathcal{S}(\mathbb{V})$ the sheaf of real analytic sections of this bundle. Let $\phi\colon G_o \to S$ be the projection $g \to gy$. A section s of $\mathcal{S}(\mathbb{V})$ over an open set U of S can be identified with a function $s\colon \phi^{-1}(U) \to V$ such that $s(gb) = \omega(b^{-1})s(g)$ for all $b \in P_{y,o}$.

There are two actions of the sheaf $\mathcal{P}$ on $\mathcal{S}(\mathbb{V})$. Let $\sigma \in \mathcal{P}$ be defined over an open neighbordhood U in Y of a point $z = gy$ such that g is in G_o, and let $s \in \mathcal{S}(\mathbb{V})$ be defined over a neighborhood U on S of z. One action is the left action $\sigma_l s(g) = (\sigma(gy)_l s)(g)$. The left action of $Z = Z_1 + iZ_2 \in \mathfrak{g}$ on s at $g \in G_o$ is defined by

$$Z_l s(g) = \left.\frac{d}{dt}\right|_{t=0} s(exp(-t\ Z_1)g) + i\left.\frac{d}{dt}\right|_{t=0} s(exp(-t\ Z_2)g).$$

The other action is $(\sigma * s)(g) = \omega(\mathrm{Ad}(g^{-1}\sigma(gy)))s(g)$. Now, extend by zero to all of Y the subsheaf of sections s of $\mathcal{S}(\mathbb{V})$ that satisfy the equation

$$\sigma_l s - \sigma * s = 0. \tag{2.1}$$

This resulting sheaf is by definition an induced analytic sheaf. We denote it by $\mathcal{I}(y, V)$. It turns out to be a $(\mathcal{G}_Y, G_o)$-module. Moreover, if the inducing $(\mathfrak{p}_y, P_{y,o})$-module V has infinitesimal character λ, then $\mathcal{I}$ is a $(\pi_*(\mathcal{D}_\lambda), G_o)$-module [3].

If we add the condition that the G_o-orbit S is closed, the description above of $\mathcal{I}(y, V)$ is simpler because every section of $\mathcal{S}(\mathbb{V})$ satisfies the equation (2.1). Then, the induced analytic sheaf $\mathcal{I}(y, V)$ is equal to the extention of the sheaf $\mathcal{S}(\mathbb{V})$ by zero to all of Y.

Let X be the flag variety and Y a generalized flag variety. Let C_o be a Cartan subgroup of G_o with complexified Lie algebra $\mathfrak{c}$. Fix a point $x \in X$ stabilized by C_o. Let N_x be the analytic group corresponding to the nilpotent subalgebra $\mathfrak{n}_x$ of the Borel subalgebra $\mathfrak{b}_x$ associated to x.

If $u = \pi(x)$ for some $x \in X$, define

$$\Delta(\mathfrak{p}_u, \mathfrak{c}) = \{\alpha \in \Delta_x \,/ \mathfrak{g}_\alpha \subset \mathfrak{p}_u\}$$
$$\Delta(\mathfrak{u}_u, \mathfrak{c}) = \{\alpha \in \Delta_x \,/ \mathfrak{g}_\alpha \subset \mathfrak{u}_u\}.$$

The set of N_x-orbits of Y are parametrized by the set W/W_L, where W_L is described as follows. Take the Weyl group generated by the root subsystem $\Delta(\mathfrak{l}_u, \mathfrak{c}) = \Delta(\mathfrak{p}_u, \mathfrak{c}) \setminus \Delta(\mathfrak{u}_u, \mathfrak{c})$ for $u = \pi(x)$. It is a subgroup of the specialization of W at x. Then W_L is its pullback to W. Note that W_L is well defined and it doesn't depend on the choice of $x \in \pi^{-1}(u)$ fixed by C_o. The N_x-orbits in X are parametrized by W as follows: the N_x-orbit Q_v is the set of points z given by the data $(\mathfrak{t}, v_x^{-1}\Delta_x^+)$ where $\mathfrak{t}$ is any Cartan subalgebra of $\mathfrak{b}_x$, Δ_x^+ is the positive root system associated to $(\mathfrak{t}, \mathfrak{b}_x, \mathfrak{g})$, and v_x^{-1} is the specialization at x of the element v of the abstract Weyl group W. Hence, the parabolic subalgebra associated to $y = \pi(z)$ for $z \in Q_v$ is

$$\mathfrak{p}_y = \mathfrak{t} + \sum_{\alpha_z \in \Delta(\mathfrak{p}_y, \mathfrak{t})} \mathfrak{g}_{\alpha_z} = \mathfrak{t} + \sum_{\alpha_x \in \Delta(\mathfrak{p}_{\pi(x)}, \mathfrak{t})} \mathfrak{g}_{v\alpha_x}.$$

But, if we change v by vw with $w \in W_L$, the last equality does not change. This means that $y = \pi(z)$ for all $z \in X$ given by $\mathfrak{b}_z = (\mathfrak{t}, (vw)_x^{-1}\Delta_x^+)$ where w runs through W_L. As π is G-invariant and onto, every N_x-orbit $\tilde{Q}$ in Y is an image of an N_x-orbit in X. If $\pi(Q_v) = \tilde{Q}$, then $\pi^{-1}(\tilde{Q})$ is the union of the N_x-orbits Q_{vw} in X with $w \in W_L$. It is known that each Q_v in X has a unique C_o-fixed point given by $(\mathfrak{c}, v_x^{-1}\Delta_x^+)$. So, the points fixed by C_o in $\pi^{-1}(\tilde{Q})$ have the same image in Y. Therefore, $\tilde{Q}$ is associated to a class of W/W_L and contains only one C_o-fixed point.

Another relevant fact is the following. If $y \in Y$ is fixed by C_o, there exists a C_o-fixed point z_o in $\pi^{-1}(y)$ such that

$$\mathfrak{n}_x \cap \mathfrak{p}_y = \mathfrak{n}_x \cap \mathfrak{n}_{z_o}. \tag{2.2}$$

This means that there exists an N_x-orbit in X, the one associated to z_o, with complex dimension equal to the complex dimension of $\tilde{Q}$. This dimension is exactly the length of the element of W associated to z_o.

Whenever $f\colon Z \to Z'$ is a continuous function between two varieties, f^{-1} denotes the inverse-image functor in the category of sheaves and $f_!$ the direct-image functor of compact supports in the same category. If Z is a complex submanifold of Z', $\mathcal{O}_Z$ denotes the structure sheaf of Z. Now, f^* and f_* are the functors of inverse and direct image in the category of $\mathcal{O}$-modules. When the function f is proper the functors f_* and $f_!$ are equal. If we consider derived categories, Df^* and Df_* denote the derived functors of f^* and f_*, respectively, between the categories of complexes of sheaves.

Let $\mathcal{T}$ be the tangent sheaf to an N_x-orbit Q in X with geometric fiber $T_z(Q)$ at z isomorphic to $\mathfrak{n}_x/\mathfrak{n}_x \cap \mathfrak{n}_z$. Denote by i_Q the embedding of Q in X. If $\mathcal{M}$ is a $\mathcal{D}_\lambda$-module, $i_Q^*\mathcal{M}$ turns to be a $\mathcal{T}$-module because the elements of a geometric fiber of $\mathcal{T}$ act as differential operators. Hence, we can define the de Rham complex of $\mathcal{M}$ along Q by $\mathrm{DR}_Q\mathcal{M} = \mathrm{Hom}_{\mathcal{O}_Q}(\bigwedge \mathcal{T}, i_Q^*\mathcal{M})$. It is a complex of sheaves on Q. It may be regarded as a complex on X by defining it to be equal to zero off Q. If $i_Q^*\mathcal{M}$ is a connection on Q, that is a $\mathcal{D}_Q$-module which is coherent as an $\mathcal{O}_Q$-module, then its associated de Rham complex is exact except in degree zero, and its zero-th cohomology is equal to the sheaf $(i_Q^*\mathcal{M})^{\mathcal{T}}$ of $\mathcal{T}$-invariants, or equivalently, of $\mathfrak{n}_x$-invariants. Thus, $\mathrm{DR}_Q(i_Q^*\mathcal{M})$ is quasi-isomorphic to the complex which is $(i_Q^*\mathcal{M})^{\mathcal{T}}$ in degree zero and zero in other degrees. As $(i_Q^*\mathcal{M})^{\mathcal{T}}$ is a locally constant C_o-equivariant sheaf of finite rank along its support and is zero elsewhere, the cohomology groups $H^*(X, \mathrm{DR}_Q\mathcal{M})$ are finite dimensional C_o-modules. Now, if $\mathcal{M}$ is a complex of sheaves of $\mathcal{O}_Q$-modules the de Rham complex of $\mathcal{M}$ is also well defined by

$$\mathrm{DR}_Q\mathcal{M} = \mathrm{Hom}_{\mathcal{O}_Q}(\bigwedge \mathcal{T}, Di_Q^*\mathcal{M}). \tag{2.3}$$

3. Some properties of the closed G_o-orbit of Y

By Wolf's result in [12] there is only one closed G_o-orbit in a generalized flag manifold Y. In this section we prove some properties of this orbit S for some special Y, and some relations between the fixed points on S by a Cartan subgroup C_o of G_o.

Let σ be the involution of $\mathfrak{g}$ that fixes the real part $\mathfrak{g}_o$. This involution induces an involution in G that we also denote by σ. We say that a point $y \in Y$ is σ-stable if $\mathfrak{p}_y$, or equivalently, P_y is σ-stable.

Proposition 3.1. *Let Y be a generalized flag manifold that contains at least one σ-stable point. Then, the closed G_o-orbit is exactly the set of σ-stable points of Y.*

Proof. Let $y \in Y$ be a σ-stable point. Hence, the stabilizer $P_{y,o}$ of y in G_o is equal to the real part of P_y that is a parabolic subgroup of G_o. Then, the G_o-orbit S of y is isomorphic to $G_o/P_{y,o}$ which is a real closed submanifold of Y. In fact, there is an Iwasawa decomposition of $G_o = K_oA_oN_o$ such that the compact group K_o acts transitively on S; that is S is a compact subset of Y. Then, S is closed in Y. By [12] there is only one closed G_o-orbit, so S is the closed G_o-orbit of Y.

On the other hand, as S contains a σ-stable point $y \in Y$, any other point of S is σ-stable. □

Now, we will characterize the intersection of N_x-orbits in Y and S, for Y with a σ-stable point and some special $x \in X$.

Let g be an element of the subset G_o^{reg} of regular points of G_o, it is contained in a unique Cartan subgroup C_o of G_o. There exists a fixed point x by g in X, called an *expansion point for* g such that all the eigenvalues of $\mathrm{Ad}_g\colon \mathfrak{n}_x \to \mathfrak{n}_x$ have absolute value ≤ 1, or equivalently that $g\colon \mathbf{T}^*_x(X) \to \mathbf{T}^*_x(X)$ has all its eigenvalues of absolute value ≥ 1. For each $y \in Y$ fixed by C_o we can choose a special expansion point x for g as follows. Take $z_o \in \pi(y)^{-1}$ fixed by C_o such that $|e^\alpha(g)| \leq 1$ for all $\alpha \in \Delta(\mathfrak{n}_{z_o}/\mathfrak{u}_y, \mathfrak{c})$. Now, define x by the expansion point for g in X nearest to z_o, that is

$$\Delta_x^+ = \{\alpha \in \Delta_{z_o}^+ \colon |e^\alpha(g)| \leq 1\} \cup \{-\alpha \colon \alpha \in \Delta_{z_o}^+, |e^\alpha(g)| > 1\}. \tag{3.1}$$

The element z_o is not unique, hence x depends on the choice of z_o. But we will only need to know that there exists an expansion point x for g such as this one in (3.1).

Let ϕ be the subset of roots given by

$$\phi = \{\alpha \in \Delta_x^+ \colon |e^\alpha(g)| < 1\}. \tag{3.2}$$

The involution σ also induces an involution on the set of roots Δ_x: $\sigma(\alpha)(H) = \overline{\alpha(\sigma(H))}$ for all $H \in \mathfrak{c}$, where the bar denotes the conjugation in $\mathbb{C}$. So every root can be decomposed as a sum $\alpha = \alpha_{\mathbb{R}} + \alpha_I$ where the first summand is in the eigenspace of σ associated to the eigenvalue 1, and the second one to -1. Using this notation we have that $\sigma(\alpha) = \alpha_{\mathbb{R}} - \alpha_I$.

Note that the set ϕ is σ-stable. In fact, let $\alpha \in \phi$, as g is in C_o, $|e^{\sigma(\alpha)}(g)| = |e^{\alpha_{\mathbb{R}}}(g)| = |e^\alpha(g)|$; then, $\sigma(\alpha)$ is in ϕ.

If ψ is some set of roots, denote by $\mathfrak{n}_\psi$ the subalgebra of $\mathfrak{g}$ generated by the root spaces associated to the roots in ψ, and N_ψ the analytic subgroup associated to $\mathfrak{n}_\psi$.

Observe that ϕ is an ideal of Δ_x^+, i.e. any root that is a sum of a root of ϕ plus a root of Δ_x^+ is in ϕ. Therefore, there is a decomposition $N_x = N_\phi N_{(\Delta_x \setminus \phi)}$. As the group $N_{(\Delta_x \setminus \phi)}$ is contained in the stabilizer of y,

$$N_x y = N_\phi y. \tag{3.3}$$

Proposition 3.2. *Let $y \in Y$ be fixed by C_o, x as in (3.1) and ϕ as in (3.2). Suppose that y is not in the closed G_o-orbit S of Y. Then*

$$N_x y \cap S = N_\phi y \cap S = \emptyset.$$

Proof. Suppose there exists $n \in N_\phi$ such that the point ny is in S. By Proposition 3.1 this means that P_{ny} is σ-stable, and $\sigma(n)\sigma(P_y)\sigma(n^{-1}) = \sigma(P_{ny}) = P_{ny} = nP_yn^{-1}$. Thus, $\sigma(P_y) = \sigma(n^{-1})nP_y(\sigma(n^{-1})n)^{-1}$. Note that the Lie algebra $\mathfrak{n}_\phi$ is σ-stable. In fact, the set ϕ is itself σ-stable, and the involution σ is a Lie algebra isomorphism that sends the root space $\mathfrak{g}_\alpha$ associated to α into the root space $\mathfrak{g}_{\sigma(\alpha)}$. This implies that the element $m = \sigma(n^{-1})n$ is in N_ϕ. But this means that $\sigma(P_y)$ is the stabilizer of the point $my \in N_\phi y$. So, as $C_o \subset \sigma(P_y)$, the point my is fixed by C_o. Therefore, $my = y$ because there is only one point in $N_x y$ fixed by C_o (see §2). Hence, $P_{my} = \sigma(P_y) = P_y$, contradicting the assumption that y is not σ-stable. □

In the following proposition we consider the case when y is in S.

Proposition 3.3. *Let $y \in Y$ be fixed by C_o. Let $x \in X$ be the expansion point for $g \in C_o$ as in (3.1). If y is in the closed G_o-orbit S of Y, then*

$$N_x y \cap S \cong \mathbb{R}^{\dim_{\mathbb{C}} N_x y}.$$

Proof. Let $\mathfrak{n}_\phi$ the Lie algebra associated to ϕ. It is σ-stable. Take $z_o \in \pi(y)^{-1}$ fixed by C_o such that $|e^\alpha(g)| \leq 1$ for all $\alpha \in \Delta(\mathfrak{n}_{z_o}/\mathfrak{u}_y, \mathfrak{c})$. If x is the expansion point nearest to z_o as in (3.1), and ϕ is as in (3.2), then

$$\mathfrak{n}_\phi \cap \mathfrak{p}_y = \sum_{\substack{\alpha\in\Delta^+_{z_o}\\ |e^\alpha(g)|<1}} \mathfrak{g}_\alpha.$$

It is σ-stable and has a σ-stable complement $\mathfrak{n}^c_\phi$ in $\mathfrak{n}_\phi$ given by

$$\mathfrak{n}^c_\phi = \sum_{\substack{\alpha\in\Delta^+_{z_o}\\ |e^\alpha(g)|>1}} \mathfrak{g}_{-\alpha}.$$

Let N^c_ϕ be the analytic subgroup of N_ϕ with Lie algebra $\mathfrak{n}^c_\phi$. As the intersection $N^c_\phi \cap \exp(\mathfrak{n}_\phi \cap \mathfrak{p}_y) = 1$, there is a diffeomorphism

$$\begin{aligned} \mathfrak{n}^c_\phi \oplus (\mathfrak{n}_\phi \cap \mathfrak{p}_y) &\to N^c_\phi \exp(\mathfrak{n}_\phi \cap \mathfrak{p}_y) \\ (X, Y) &\to \exp X \exp Y \end{aligned}$$

Thus, $N_\phi y = N^c_\phi y$. Denote by $\mathfrak{n}^c_{\phi,o}$ and $\mathfrak{n}^c_{\phi,i}$ the eigenspaces of σ in $\mathfrak{n}^c_\phi$ associated to 1 and -1 respectively. Hence, as before, there is a diffeomorphism between the direct sum of these subspaces and the submanifold of N^c_ϕ given by the product $N^c_{\phi,o}N^c_{\phi,i}$ of the images under the exponential map of $\mathfrak{n}^c_{\phi,o}$ and $\mathfrak{n}^c_{\phi,i}$ respectively. This means that $N^c_\phi y = N^c_{\phi,o}N^c_{\phi,i}y$.

Let $ny \in S$ with $n = n_o n_i \in N_\phi$, according to the decomposition above. Therefore, $n_i y$ is also in S. Then, by Proposition 3.1,

$$P_{n_i y} = \sigma(P_{n_i y}) = \sigma(n_i)P_y\sigma(n_i^{-1}) = n_i^{-1}P_y n_i.$$

This implies that $n_i n_i = \exp(2Z) \in P_y$ for some $Z \in \mathfrak{n}^c_{\phi,i}$, or equivalently, that $n_i \in P_y$. Then, by definition of N^c_ϕ, $n_i = 1$. This means that $N^c_\phi y = N^c_{\phi,o} y$ which is isomorphic to $\mathbb{R}^{\dim_{\mathbb{C}} N_x y}$. □

Finally, we relate the different points $y \in S$ fixed by C_o. We know that for each pair of such points there exists an element in $N_G(C)$ that relates one with the other by conjugation. Fix one of these points and denote it by y_o. Let $P_o = P_{y_o,o}$ and fix a Levi factor L_o of $P_o = L_o U_o$ such that C_o is contained in L_o. Now choose some representatives $C_o = C_{0,o}, C_{1,o}, \ldots, C_{r,o}$ from the several L_o-conjugacy classes of Cartan subgroups in L_o, which are G_o-conjugated. Let $W(G_o, C_{i,o}) = N_{G_o}(C_{i,o})/C_{i,o}$ be the Weyl group of $C_{i,o}$ in G_o. The action of $W(G_o, C_{i,o})$ on $C_{i,o}$ makes sense up to conjugation by some elements of $C_{i,o}$. Fix also some elements of G_o, $1 = h_0, \ldots, h_r$ such that $h_i C_o h_i^{-1} = C_{i,o}$.

Proposition 3.4. *Let $y_o \in S$ be fixed by C_o. The point $y = ny_o$ with $n \in N_G(C)$ is in S if and only if $y = wh_i^{-1} y_o$ for some $w \in N_{G_o}(C_o)$ and some h_i. Or equivalently, $y = h_i^{-1} w y_o$ for some $w \in N_{G_o}(C_{i,o})$.*

Proof. Suppose that $y = ny_o \in S$, that is, there exists $h \in G_o$ such that $y = ny_o = hy_o$. This means that $n = hq$ for some $q \in P_{y_o}$. As $n \in N_G(C)$, $h^{-1}Ch = qCq^{-1} \subset P_{y_o}$. Hence, $h^{-1}C_o h$ is a Cartan subgroup in $P_{y_o,o}$. Then, $h^{-1}C_o h$ is conjugated by $P_{y_o,o}$ to a Cartan subgroup of $L_{y_o,o} = L_o$. That is, $k^{-1}h^{-1}C_o hk$ is a Cartan subgroup of L_o for some $k \in P_{y_o,o}$. But it is conjugated by L_o to some $C_{i,o}$. Therefore, $hk \in wh_i^{-1} L_o$ for some $w \in N_{G_o}(C_o)$, or equivalently,

$$y = ny_o = hy_o = hky_o = wh_i^{-1} y_o. \tag{3.5}$$

Conversely, assume that (3.5) happens. Then, as $wh_i^{-1} \in G_o$, $y \in S$. □

4. Compatibility of the character formulas

On the level of G_o-modules, the distributional character is an invariant of the admissible analytic module M of analytic vectors of the representation space M_Π (minimal globalization of the underlying Harish-Chandra module M^{alg} in the sense of Schmid [9]). The Hecht–Taylor character formula for an admissible analytic module M with regular infinitesimal character θ [3] can be regarded as the character of $\mathcal{M}$, the analytic localization of M in $D\mathfrak{M}(\mathcal{D}_\lambda, G_o)$, with a fixed $\lambda \in \theta$ (see §2). Denote this character formula by $\Theta^{G_o}(\mathcal{M})$.

If $g \in G_o^{reg}$, and C_o its centralizer in G_o, then, for each point $z \in X$ fixed by C_o we can choose an expansion point $x = x_z$ for g. As in (2.3), $\mathrm{DR}_Q(\mathcal{M})$ denotes the de Rham complex of $\mathcal{M}$ along the Bruhat cell Q. One has

$$\Theta^{G_o}(\mathcal{M})(g) = \sum_{\{z \in X:\ C_o z = z\}} \frac{\sum_p (-1)^p \mathrm{tr}\,(g\colon H^p(X, \mathrm{DR}_Q(\mathcal{M})))}{\det(1 - g\colon \mathbf{T}_z^*(X))}, \tag{4.1}$$

where $Q = N_{x_z} z$ and $g\colon \mathbf{T}_z^*(X) \to \mathbf{T}_z^*(X)$ denotes the action of g on the holomorphic cotangent space $\mathbf{T}_z^*(X)$ at z.

The known Hirai–Wolf induced character formula for $I = \mathrm{Ind}_{P_o}^{G_o}(V)$ — the induced representation of G_o from the representation V of the parabolic subgroup P_o of G_o — can be formulated as follows. Remember the notation of Proposition 3.4 for L_o a Levi factor of P_o and C_o a Cartan subgroup of L_o. Let $g \in C_o \cap G_o^{reg}$,

$$\Theta_I^{G_o}(g) = \sum_{i=1}^{r} \frac{1}{c_i} \sum_{v \in W(G_o, C_{i,o})} \frac{|\det(1 - \mathrm{Ad}_{\mathfrak{n}} g)|}{|\det(1 - \mathrm{Ad}_{\mathfrak{m}} g)|} \; \Theta_V^{L_o}(v(h_i g h_i^{-1})) \tag{4.2}$$

where c_i is the cardinality of the group $W(G_o, C_{i,o})$, $\mathfrak{n}$ is a maximal nilpotent subalgebra of $\mathfrak{g}$ and $\mathfrak{m} \subset \mathfrak{n}$ is a maximal nilpotent subalgebra of the Lie algebra $\mathfrak{l}$ of L_o.

Formula (4.2) was proved in [13] only for cuspidal parabolic subgroups P_o. We will reobtain it from (4.1) without restriction on P_o, expressing an induced representation in terms of a complex of sheaves. Let $\mathrm{Ind}_{P_o}^{G_o}(V)$ be as above. Consider the case when V is trivial as a U_o-module, where U_o is the unipotent radical of P_o. Hirai and Wolf formulas are for this kind of induced representations.

Now, $\mathrm{Ind}_{P_o}^{G_o}(V)$ can be seen as the 0th-cohomology group of an induced analytic sheaf. In fact, let Y be the generalized flag manifold isomorphic to G/P, where G and P denotes the complexification groups of G_o and P_o respectively. Then, there exist a point $y_o \in Y$ whose stabilizer P_{y_o} under the action of G in Y is exactly P. As V is a $(\mathfrak{p}_{y_o}, P_o)$-module, we can define the induced analytic sheaf $\mathcal{I}(y_o, V)$ as in §2. By construction, this sheaf only has global sections and the cohomology groups of nonzero grade are trivial [3]. Moreover,

$$\mathrm{Ind}_{P_o}^{G_o}(V) = H^0(Y, \mathcal{I}(y_o, V)). \tag{4.3}$$

By Proposition 3.1, the support of $\mathcal{I}(y_o, V)$ is the closed G_o-orbit S of Y.

We can say more: if Y is a generalized flag manifold with a σ-stable point, then $H^0(Y, \mathcal{I})$ is an induced representation of G_o for every induced analytic sheaf $\mathcal{I}$ as in §2, whose support is the closed orbit S. The inducing $P_{y,o}$-module is the geometric fiber of $\mathcal{I}$ at $y \in S$. It does not depend on the choice of $y \in S$.

The derived category of complex of sheaves over Y generated by the induced analytic sheaves that are $\pi_*(\mathcal{D}_\lambda)$-modules with support the closed orbit S is equivalent to a derived category of complex of sheaves over X with support $\pi^{-1}(S)$ [1]. Every element in the category over Y can be seen as the image of a complex of sheaves of $\mathcal{D}_\lambda$-modules over X with support on $\pi^{-1}(S)$.

Now we can formulate the main theorem.

Theorem 4.1. *Let $\mathcal{M}$ be a complex in the derived category $D\mathfrak{M}(\mathcal{D}_\lambda, G_o)$ over X with support on $\pi^{-1}(S)$. Let $g \in C_o \cap G_o^{reg}$; then*

$$\Theta^{G_o}(\mathcal{M})(g) = \sum_{\{y \in S:\ C_o y = y\}} |\det(1 - \mathrm{Ad}_{\mathfrak{u}_y} g)|^{-1} \Theta^{L_{y,o}}(Di_y^* \mathcal{M})(g)$$

where i_y is the natural embedding of the flag manifold $X_y = \pi^{-1}(y)$ for $L_{y,o}$ into X.

From this theorem it is easy to see that the second term of the equality is exactly the induced character formula $\Theta_I^{G_o}(g)$ for $I = \mathrm{Ind}_{P_o}^{G_o}(V)$ in g if $\mathcal{M}$ is such that $D\pi_*(\mathcal{M}) = \mathcal{I}(y_o, V)$ and $P_{y_o,o} = P_o$. In fact, by Proposition 3.4 we can replace the index set of the sum by the set $\{(i,w) \colon i = 0, \ldots, r\ ,\ w \in W_i\}$ with $W_i = W(G_o, C_{i,o})/W(L_o, C_{i,o})$. This means that the character restricted to $C_o \cap G_o^{reg}$ vanishes unless C_o is conjugate under G_o to a Cartan of a Levi factor of P_o. On the other hand, the automorphism on X given by left multiplication by $h_i^{-1}w$ with $w \in W(G_o, C_{i,o})$ induces isomorphisms between $Di_{y_o}^*\mathcal{M}$ and $Di_{h_i^{-1}wy_o}^*\mathcal{M}$, and between $L_{y,o}$ and $L_{h_i^{-1}wy,o}$. Then, in this case we can replace the second term of the equality of the Theorem 4.1 by

$$\sum_{i=0}^{r} \sum_{w \in W_i} |\det(1 - \mathrm{Ad}_{\mathfrak{u}_{y_o}}(w(h_i g h_i^{-1})))|^{-1} \Theta^{L_o}(Di_{y_o}^*\mathcal{M})(w(h_i g h_i^{-1})).$$

But $D\pi_*(\mathcal{M}) = \mathcal{I}(y_o, V)$, and as π is proper, $D\pi_* = D\pi_!$. This permits us to do a base change [8]. The commutative diagram of functions is

$$\begin{array}{ccc} X_{y_o} & \xrightarrow{i_{y_o}} & X \\ \tilde{\pi}\downarrow & & \downarrow\pi \\ \{y_o\} & \xrightarrow[j]{} & Y \end{array}.$$

Hence, we obtain

$$\begin{aligned} H^k(X_{y_o}, Di_{y_o}^*\mathcal{M}) &= H^k(\{y_o\}, D\tilde{\pi}_* Di_{y_o}^*\mathcal{M}) = H^k(\{y_o\}, j^* D\pi_*\mathcal{M}) \\ &= H^k(\{y_o\}, j^*\mathcal{I}(y_o, V)) \ = \begin{cases} 0 & k \neq 0 \\ V & k = 0 \end{cases}. \end{aligned}$$

This implies that $\Theta^{L_o}(Di_{y_o}^*\mathcal{M}) = \Theta_V^{L_o}$; or equivalently, that the formula (4.1) coincides with (4.2) when $\mathcal{M}$ is such that $D\pi_*(\mathcal{M}) = \mathcal{I}(y_o, V)$. This proves the compatibility of the formulas.

Proof of Theorem 4.1. By the observation before the theorem, it is enough to prove the result for any induced analytic sheaf (or standard analytic sheaf [3],[4]) $\mathcal{I}$ on X with support contained in $\pi^{-1}(S)$ that is a $\mathcal{D}_\lambda$-module.

First, we have

$$\Theta^{G_o}(\mathcal{I})(g) = \sum_{\substack{y \in Y \\ C_o y = y}} \sum_{\substack{z \in X_y \\ C_o z = z}} \frac{\sum_p (-1)^p \mathrm{tr}\,(g \colon H^p\,(Y, D\pi_* \mathrm{DR}_{Q_z}(\mathcal{I})))}{\det(1 - g \colon \mathbf{T}_z^*(X))}. \tag{4.4}$$

The support of $D\pi_*\mathrm{DR}_{Q_z}\mathcal{I}$ is $\pi(Q_z)\cap S$. Then, by Propositions 3.2 and 3.3, we can choose a special expansion point $x = x_y$ for g, for each y, such that $\pi(Q_z)\cap S = N_x y \cap S$ is contractible if $y \in S$ and the empty set if $y \notin S$. This expansion point x_y satisfies (3.1) and (3.3) for ϕ as in (3.2). On the other hand, as π is proper $D\pi_*$ preserves constructibility [8], that is, $D\pi_*\mathrm{DR}_{Q_z}(\mathcal{I})$ is locally constant for each $y \in S$ fixed by C_o. Therefore, by Poincaré duality

$$H^p(Y, D\pi_*\mathrm{DR}_{Q_z}(\mathcal{I})) = \begin{cases} H^{\dim N_x y}(N_x y \cap S, \mathbb{C}) \otimes (D\pi_*\mathrm{DR}_{Q_z}(\mathcal{I}))_y & p = \dim N_x y \\ 0 & p \neq \dim N_x y \end{cases}$$

where $(.)_y$ denotes the stalk of $(.)$ over y. But,

$$(D^q\pi_*\mathrm{DR}_{Q_z}(\mathcal{I}))_y = (D^q\pi_*(i^*_{Q_z}\mathcal{I})^{\mathfrak{n}_x})_y = H^q(X_y, i_y^{-1}(i^*_{Q_z}\mathcal{I})^{\mathfrak{n}_x}).$$

Note that these hypercohomologies are nonzero only for $q = 0$. Now, as there exists an x' in the flag manifold X_y with associated maximal nilpotent subalgebra $\mathfrak{m}_{x'} = \mathfrak{n}_x/\mathfrak{u}_y$, and the orbit $N_{x'}z = Q_z \cap X_y$ is a Bruhat cell of X_y,

$$i_y^{-1}(i^*_{Q_z}\mathcal{I})^{\mathfrak{n}_x} = (i^*_y i^*_{Q_z}\mathcal{I})^{\mathfrak{m}_{x'}} = (i^*_{Q_z\cap X_y} i^*_y\mathcal{I})^{\mathfrak{m}_{x'}} = \mathrm{DR}_{Q_z\cap X_y}(i^*_y\mathcal{I}).$$

Then, the second term of equality (4.4) turns out to be

$$\sum_{\substack{y\in Y\\ C_o y = y}} (-1)^{\dim N_x y} \sum_{\substack{z\in X_y\\ C_o z = z}} \frac{\sum_q (-1)^q \mathrm{tr}\,(g\colon H^q(X_y, \mathrm{DR}_{Q_z\cap X_y}(i^*_y\mathcal{I})))}{\det(1 - g\colon \mathbf{T}^*_z(X))}.$$

But, $\det(1 - g\colon \mathbf{T}^*_z(X)) = \det(1 - \mathrm{Ad}_{\mathfrak{u}_y} g)\det(1 - g\colon \mathbf{T}^*_z(X_y))$. So, applying the formula (4.1) for the groups $L_{y,o}$ and the flag varieties X_y, the above expression is equal to

$$\sum_{\{y\in S\colon C_o y = y\}} \frac{(-1)^{\dim_{\mathbb{C}} N_x y}}{\det(1 - \mathrm{Ad}_{\mathfrak{u}_y} g)} \Theta^{L_{y,o}}(Di^*_y\mathcal{M})(g).$$

Thus, it remains to prove the equality

$$|\det(1 - \mathrm{Ad}_{\mathfrak{u}_y} g)|^{-1} = \frac{(-1)^{\dim_{\mathbb{C}} N_x y}}{\det(1 - \mathrm{Ad}_{\mathfrak{u}_y} g)}.$$

Note that $\dim_{\mathbb{C}} N_x y$ is equal to the cardinality of $\{\alpha \in \Delta(\mathfrak{u}_y, \mathfrak{c})\colon |e^\alpha(g)| > 1\}$; this is true because (3.1) characterizes x. On the other hand,

$$\text{(4.5)}\quad \frac{\det(1 - \mathrm{Ad}_{\mathfrak{u}_y} g)}{|\det(1 - \mathrm{Ad}_{\mathfrak{u}_y} g)|} = \left(\prod_{\alpha\in\Delta(\mathfrak{u}_y,\mathfrak{c})} (1 - e^\alpha(g))\right)\left|\prod_{\alpha\in\Delta(\mathfrak{u}_y,\mathfrak{c})} (1 - e^\alpha(g))\right|^{-1}.$$

But this can admit only the values ± 1. In fact, as $y \in S$, the set of roots $\Delta(\mathfrak{u}_y, \mathfrak{c})$ is σ-stable and can only contain real or complex roots. Hence, if α is complex

$$(1 - e^{\alpha}(g))(1 - e^{\sigma\alpha}(g)) = \left(1 + e^{2\alpha_{\mathbb{R}}}(g) - e^{\alpha_{\mathbb{R}}}(e^{\alpha_I} + e^{-\alpha_I})(g)\right).$$

But this is a positive real number because $e^{\alpha_I}(g)$ and $e^{-\alpha_I}(g)$ are conjugated complex numbers of module one. If α is real, $1 - e^{\alpha}(g) < 0$ only if $e^{\alpha}(g) > 1$. This implies that the right term of (4.5) is equal to $(-1)^r$ where r is the number of real roots of $\Delta(\mathfrak{u}_y, \mathfrak{c})$ that satisfy $|e^{\alpha}(g)| > 1$. But the number of complex roots in $\{\alpha \in \Delta(\mathfrak{u}_y, \mathfrak{c}) \colon |e^{\alpha}(g)| > 1\}$ is even. So, $(-1)^{\dim_{\mathbb{C}} N_x y} = (-1)^r$. This concludes the proof of the theorem. □

References

1. T. Bratten, Ph. D. Thesis, University of Utah, 1993.
2. Harish-Chandra, *Discrete series for semisimple Lie groups II. Explicit determination of the characters*, Acta Math. **116** (1966), 1–111.
3. H.Hecht and J.L.Taylor, *Analytic localization of group representations*, Advances in Mathematics **79** (1990), 139–212.
4. ———, *A geometric formula for characters of semisimple groups* (1992).
5. H.Hecht and W.Schmid, *Characters, asymtotics and $\mathfrak{n}$-homology of Harish-Chandra modules*, Acta Math. **151** (1983), 49–151.
6. T.Hirai, *The characters of some induced representations of semisimple Lie groups*, J.Math. Kyoto Univ **8** (1968), 313–369.
7. M. Kashiwara, *Character, character cycle, fixed point theorem and group representation*, Adv. Stud. Pure Math. **14** (1988), 369–378.
8. M. Kashiwara and P. Shapira, *Sheaves on manifolds*, Springer-Verlag, 1990.
9. W. Schmid, *Boundary value problems for group invariant differential equations*, Elie Cartan et les mathématiques d'aujourd'hui, Astérisque (1985), 311–322.
10. W. Schmid and K. Vilonen, *Characters, fixed points and Osborne's conjecture*, Expanded version preprint (1992).
11. G. Warner, *Harmonic analysis on semisimple Lie groups I*, Springer-Verlag, 1974.
12. J.A. Wolf, *The action of real semisimple group on a complex flag manifold. I: orbit structure and holomorphic arc components*, Bull. Amer. Math. Soc. **75** (1969), 1121–1237.
13. ———, *Unitary representations on partially holomorphic cohomology spaces*, Amer. Math. Soc. Memoir **138** (1974).

FAMAF, Ciudad Universitaria, (5000) Córdoba, Argentina.
email: galina@mate.uncor.edu

AN ACTION OF THE R-GROUP ON THE LANGLANDS SUBREPRESENTATIONS

EUGENIO GARNICA VIGIL

ABSTRACT. We use the Vogan calculus for a real reductive Lie group to constuct an action of the R-group on the set of indecomposable subrepresentations of any given standard representation. This action allows us to describe a formula for the decomposition of the standard representation.

INTRODUCTION

To begin with, we describe the type of real reductive Lie groups we will be working with in this paper. From [Ga], we say that a real Lie group G is in the Harish-Chandra class if there exists a connected reductive algebraic group $G_{\mathbb{C}}$ over $\mathbb{C}$ such that G is a finite covering group for some subgroup of a real form of $G_{\mathbb{C}}$ containing the identity component. The group G must satisfy the property

$$Ad(G) \subseteq Ad(G_{\mathbb{C}})$$

for the corresponding adjoint representations. A group G in the Harish-Chandra class possesses a Cartan decomposition with Cartan involution θ, as well as an Iwasawa decomposition $G = KAN$, where K is a maximal compact subgroup of G. We identify $\mathfrak{g} = \text{Lie}(G)$, the Lie algebra of $G_{\mathbb{C}}$, with the complexification of the Lie algebra of G. Our choice of this type of group is useful to avoid cases when we constuct representations using either real parabolic induction or cohomological induction because the Levi factor of a parabolic subgroup of G and the normalizer in G of a θ-stable parabolic subalgebra of $\mathfrak{g}$ are in the Harish-Chandra class.

For a group G in the Harish-Chandra class the Langlands classification holds. That is, any irreducible representation in the category of admissible representations of G may be realized as a summand of the socle, the maximal completely reducible subrepresentation, of some standard representation. We call such a socle Langlands subrepresentation.

The Langlands classification is stated in [Green] in two different ways: using real parabolic induction on the one hand, and by Vogan–Zuckerman theory of cohomological induction on the other hand. Although the groups considered there are linear, the result and procedure can be extended for groups in the Harish-Chandra class without any change. Here we also have the disjointness property; two Langlands subrepresentations have a common summand if and only if the corresponding standard representations are isomorphic.

Thus, it is of great importance to find the composition series for any standard representation. When the standard representation has an integral infinitesimal character, the composition factors may be computed as result of the Vogan calculus in [Green, Ch. VII]; however, a general answer has not been found. Along this line, we formulate our main result. By means of the Vogan calculus we define an action of the R-group of Knapp and Stein on the set of indecomposable subrepresentations of a standard representation. This action is suitable for describing a formula for the decomposition of the standard representation. Moreover, this R-group action allows us to compute the multiplicity of any indecomposable constituent. In particular, this formula also describes both the decomposition of any Langlands subrepresentation as the direct sum of its irreducible subrepresentations and the classification of tempered representations.

Using the Harish-Chandra correspondence [Ga, 4.15], we identify an admissible representation of G with the corresponding $(\mathfrak{g}, K)$-module. We will be working with the category of admissible $(\mathfrak{g}, K)$-modules of finite length. Here, the standard representations will be built by means of the Vogan-Zuckerman cohomological induction. In fact, we will be dealing with more general kinds of representations, those cohomologically induced from limit characters.

I. LIMIT CHARACTERS AND STANDARD REPRESENTATIONS.

(1) Notation.

Let G be a group in the Harish-Chandra class as specified in the introduction. Let $\mathfrak{g}$ be the complexification of $\mathfrak{g}_o = \mathrm{Lie}(G)$, the Lie algebra of G. Let K be a maximal compact subgroup of G. Analogously, $\mathfrak{k} = \mathfrak{k}_o \otimes \mathbb{C}$ where $\mathfrak{k}_o = \mathrm{Lie}(K)$. We write $\widehat{A}$ to denote the irreducible classes of representations of A. Set $\mathfrak{F}(\mathfrak{g}, K)$ to denote the the category of $(\mathfrak{g}, K)$-modules of finite length. To denote multiplicity of $\delta \in A$ in some representation $\mu \in B$, with $A \subseteq B$, let us write $m(\delta, \mu)$.

Let $H = TA$ be a θ-stable Cartan subgroup of G, and let $MA = G^A$ be the Langlands decomposition of the centralizer of A in G. Write $\mathfrak{m} = \mathrm{Lie}(M) \otimes \mathbb{C}$, $\mathfrak{h} = \mathrm{Lie}() \otimes \mathbb{C}$, and $\mathfrak{t} = \mathrm{Lie}(T) \otimes \mathbb{C}$.

(2) A set $(H, \Gamma, \overline{\gamma}, \Delta^+_{im})$, denoted by $(H, \underline{\gamma})$ is called a *limit character* (resp. *regular character*) if it satisfies

(*a*) Γ is in $\widehat{H}$ and Δ^+_{im} is a positive system for $\Delta(\mathfrak{m}, \mathfrak{t})$;

(*b*) $\overline{\gamma} \in \mathfrak{h}^*$ is such that for any $\alpha \in \Delta^+_{im}$, $< \alpha, \overline{\gamma} > \geq 0$ (resp. $< \alpha, \overline{\gamma} > > 0$);

and when α is simple compact root the inequality is strict.

(*c*) $d\Gamma = \overline{\gamma} + \rho_{\mathfrak{m}} - 2\rho_{\mathfrak{m}\cap\mathfrak{k}}$ (where $\rho_{\mathfrak{m}} = \rho(\Delta^+_{im})$ and $\rho_{\mathfrak{m}\cap\mathfrak{k}} = \rho(\Delta^+_{im,c})$.

Here $\Delta^+_{im,c}$ denotes the set of compact roots in Δ^+_{im}. Note that the choice of Δ^+_{im} is not superfluous in the previous definition, except for regular characters.

Next, let $\mathfrak{g} = \mathfrak{k} + \mathfrak{p}$ be a Cartan decomposition of Lie$(G) \otimes \mathbb{C}$. It is a known fact that for any element λ in the dual $\mathfrak{t}^*$, there is a unique θ-stable parabolic subalgebra $\mathfrak{q}$ of $\mathfrak{g}$ associated to λ, with Levi decomposition $\mathfrak{q} = \mathfrak{l} + \mathfrak{u}$. Let us consider sets of data $(\mathfrak{q}, H, \Gamma_1, \overline{\gamma}_1, \Delta^+(\mathfrak{m} \cap \mathfrak{l}))$ where H is as above, $\overline{\gamma}_1 \in \mathfrak{h}^*$, and $\mathfrak{q}$ is the Lie algebra associated to $\overline{\gamma}_1|_{\mathfrak{t}}$. Denote $(\mathfrak{q}, H, \Gamma_1, \overline{\gamma}_1, \Delta^+(\mathfrak{m}\cap\mathfrak{l}))$ as $(\mathfrak{q}, (H, \underline{\gamma}_1))$. We call a set of data $(\mathfrak{q}, (H, \Gamma_1, \overline{\gamma}_1, \Delta^+(\mathfrak{m}\cap\mathfrak{l})))$ *θ-stable limit data*, or just *θ-stable data* for G according to whether $(H, \Gamma_1, \overline{\gamma}_1, \Delta^+(\mathfrak{m}\cap\mathfrak{l})) = (H, \underline{\gamma}_1)$ is a limit character or regular character for L, the normalizer of $\mathfrak{q}$ in G.

In analogy to ([Green], thm. 6.6.2), we state the following result:

Theorem 2.1. *In the setting of the previous remark and definition, there is a bijection between limit characters of G and sets of θ-stable limit data for G that preserves conjugacy under K. Restricted, this correspondence associates regular characters of G to sets of θ-stable data.* □

For any limit character $(H, \underline{\gamma}) = (H, \Gamma, \overline{\gamma}, \Delta^+_{im})$ of G and corresponding set of θ-stable limit data $(\mathfrak{q}, (H, \underline{\gamma_1})) = (\mathfrak{q}, H, \Gamma_1, \overline{\gamma}_1, \Delta^+(\mathfrak{m} \cap\ lx))$ under the bijection of Theorem (2.1), we can build up $(\mathfrak{g}, K)$-modules. To do this consider first the discrete series or limit of discrete series representation of M (resp. of $L\cap M$), denoted ω (resp. ω_1) associated to the parameter $\Gamma|_T$ (resp. $\Gamma_1|_T$). Next, let $P_1 = M_1A_1N_1$ be a parabolic subgroup of G containing a minimal parabolic subgroup for which the character $\nu := \Gamma|A$ in $\widehat{A}$ is negative. Thus, $P_2 := P_1 \cap L$ is a parabolic subgroup of L containing a minimal parabolic subgroup for which the character $\nu_1 := \Gamma_1|_A$ of $\widehat{A}$ is negative. Put $K_L = K \cap L$. Then we obtain two induced representations whose Harish-Chandra modules are $X_G(P_1, \omega \otimes \nu) \in \mathfrak{F}(\mathfrak{g}, K)$ and $X_L(P_2, \omega_1 \otimes \nu) \in \mathfrak{F}(\mathfrak{l}, K_L)$ respectively. This is so for $\mathfrak{F}(\mathfrak{g}, K)$ and $\mathfrak{F}(\mathfrak{l}, K_L)$, the categories of admissible representations of finite length associated to the pairs $(\mathfrak{g}, K)$ and $(\mathfrak{l}, K_L)$. Now, to apply cohomological induction, we need the Zuckerman functors

$$\mathfrak{R}^i_{\mathfrak{q}}(\ ) : \mathfrak{F}(\mathfrak{l}, K_L) \to \mathfrak{F}(\mathfrak{g}, K).$$

It turns out that $S = \dim(\mathfrak{u} \cap \mathfrak{k})$ is a number such that

the $(\mathfrak{g}, K)$-module $\mathfrak{R}^i_{\mathfrak{q}}(X_L(P_2, \omega_1 \otimes \nu))$ is zero if and only if $i \neq S$.

The nonzero-module $\mathfrak{R}^S_{\mathfrak{q}}(X_L(P_2, \omega_1 \otimes \nu))$ is isomorphic precisely to $X_G(P_1, \omega \otimes \nu)$ ([Green, thm (6.5.9)] and [A-B-V, prop. 11.9]).

(2.2) In the previous construction we will refer to $X_G(P_1, \omega \otimes \nu)$ as the representation *induced from the limit character* $(H, \underline{\gamma}) = (H, \Gamma, \overline{\gamma}, \Delta^+_{im})$ of G. To make explicit the limit character, let us write

$$X_G(P_1, \omega \otimes \nu) = X_G(H, \underline{\gamma}) = \mathfrak{R}^S_{\mathfrak{q}}(X_L(P_2, \omega_1 \otimes \nu)) = \mathfrak{R}^S_{\mathfrak{q}}(X_L(H, \underline{\gamma_1})).$$

Also $\overline{X}_G(P,\omega\otimes\nu)=\overline{X}_G(H,\underline{\gamma})$ denotes the maximal completely reducible subrepresentation of $X_G(H,\underline{\gamma})$.

As an observation, standard representations of G are given by $X_G(H,\underline{\gamma})$ for regular characters $(H,\underline{\gamma})$. In this case, $\overline{X}_G(H,\underline{\gamma})$ are *the Langlands subrepresentations.* On the other hand, a limit character $(H,\underline{\gamma})$ is called a *final character* if, whenever there is a real root $\alpha\in\Delta(\mathfrak{g},\mathfrak{h})$ orthogonal to $\overline{\gamma}$, it does not satisfy the parity condition in [S-V].

The following result on final characters is proved in ([A-B-V], prop. 11.18).

Theorem 2.2.1. *If $(H,\underline{\gamma})$ is a final character then the corresponding induced representation $X_G(H,\underline{\gamma})$ has a unique nonzero irreducible subrepresentation. In particular, if $X_G(H,\underline{\gamma})$ is a unitary representation induced from a final character, it is irreducible.* □

It is not difficult to prove that the correspondence in Theorem 2.1 sends a final character $(H,\underline{\gamma})$ for G to a θ-limit data $(\mathfrak{q},(H,\underline{\gamma_1}))$ with the property that $(H,\underline{\gamma_1})$ is a final character of L.

By means of the Schmid identities we can prove that any induced representation from a limit character can be decomposed as a direct sum of indecomposable representations arising from final characters. Here, however, we will construct directly a set of limit characters that are not necessarily final such that these give us the indecomposable summands. Of course, if we apply Schmid identities to these characters then we get final characters, but there is no more reducibility. The set of characters obtained will actually be an orbit of the R-group.

(3) We fix a regular character $(H,\underline{\gamma})$ with a corresponding set of θ-stable data $(\mathfrak{q},(H,\underline{\gamma_1}))$ under the bijection in Theorem 2.1. Therefore, $(H,\Gamma_1,\overline{\gamma}_1,\Delta^+(\mathfrak{m}\cap\mathfrak{l}))$ is a regular character $(H,\underline{\gamma_1})$ for L. The group L is quasisplit, and the Cartan subgroup $H=TA$ is maximally split; that is, T is the centralizer M of A in $K\cap L$. Write, $\delta=\Gamma_1|_T$, and $\nu=\Gamma_1|_A$. Thus, the representation δ is M-fine. Denote $\overline{\Delta}_L$ for the reduced system of roots given by the following restricted roots:

$$\{\alpha|_\mathfrak{a}\mid(\alpha|_\mathfrak{a}\neq 0\text{ with }\alpha\in\Delta(\mathfrak{l},\mathfrak{h})),\text{ and }\nexists\beta\text{ in }\Delta(\mathfrak{l},\mathfrak{h})\text{ such that }\tfrac{1}{2}\beta|_\mathfrak{a}=\alpha|_\mathfrak{a}\}.$$

Next we consider the root system $\overline{\Delta}_\delta$ formed by the good roots for δ in $\overline{\Delta}_L$. A positive system of roots for $\overline{\Delta}_L$ determines a minimal parabolic subgroup of L negative for ν if it contains the set of roots

$$\overline{\Delta}_{\nu<0}=\{\alpha\in\overline{\Delta}_L\mid\text{ either }<\alpha,Re(\nu)><0\text{ or }(<\alpha,Re(\nu)>=0\text{ and }<\alpha,\ Im(\nu)><0)\ \}.$$

We need to take care of this condition. Let us choose a positive system $\overline{\Delta}_\delta^+$ for $\overline{\Delta}_\delta$ containing the set of roots $\overline{\Delta}_{\nu<0}\cap\overline{\Delta}_\delta$. Then we write the semi-sum of positive roots $\rho_\delta=\rho(\overline{\Delta}_\delta^+)$. Set Φ for the root system defined by $\{\alpha\in\overline{\Delta}_L\mid\ <\alpha,\nu>=0\}$. It will be necessary to choose two positive systems $\overline{\Delta}_L^{+_o}$ and $\overline{\Delta}_L^{+_1}$ for $\overline{\Delta}_L$. Elect these among those satisfying

$\overline{\Delta}_L^{+_o} \supseteq \{\alpha \in \overline{\Delta}_L \mid \ <\alpha, \rho_\delta> \ > 0\}$
$\overline{\Delta}_L^{+_1} \supseteq \overline{\Delta}_{\nu<0} \cup \{\alpha \in \overline{\Delta}_L \mid \ <\alpha, \rho_\delta> \ > 0 \text{ and } <\alpha, \nu> = 0\}$.

Computed in [Green], $\overline{\Delta}_s^+ = \{\alpha \in \overline{\Delta}_L^{+_o} \mid \ <\alpha, \rho_\delta> \ = 0\}$ is a set of real roots that are strongly orthogonal. To apply induction by stages let us use $\overline{\Delta}_L^{+_1}$. Let Π be the set of simple roots in $\overline{\Delta}_L^{+_1}$. Finally, define $E = \Pi \cap \overline{\Delta}_s^+$, and $F = \Pi \cap (\overline{\Delta}_s^+ \cap \Phi)$.

(3.1) Now, fix a minimal parabolic subgroup $P = MAN$ of L corresponding to the positive root system $\overline{\Delta}^{+_1}$. Suppose C is either E or F. Associated to C, we have a parabolic subgroup $P_C = M_C A_C N_C$ of L containing P. Note that P is negative for ν, $H = MA$, and $P \cap M_C = MAN' = P'$ is a minimal parabolic subgroup of M_C. Although the principal series representation $X_L(H, \underline{\gamma_1})$ of L is in the fine setting established by Vogan, we are in a finer situation with (M_C, δ) because M_C is a Lie group with compact center having a Lie algebra of the form

$$\text{Lie}(M_C) \cong sl(2,\mathbb{R}) + sl(2,\mathbb{R}) + \cdots + sl(2,\mathbb{R}) + \mathfrak{c}_o,$$

where $\mathfrak{c}_o$ is the center of Lie(M_C), and $sl(2,\mathbb{R})$ appears $|C|$-times.

Let us give some notation for our main tools that suggest in which stage of the induction process we are. Assume C is any of the sets L, E or F, put $K_C = K \cap L$ or $= K \cap M_C$ accordingly. Then M'_C will denote the normalizer of A in K_C, and W_C will be the Weyl group M'_C/M. We define $A^C(\delta)$ to be the set of minimal K_C-types in $X_{M_C}(\delta \otimes 1)$ for C either E or F, or in $X_L(\delta \otimes 1)$ for $C = L$. The set $A^C(\delta)$ consists of K_C-fine representations. This means that an element of $A^C(\delta)$ is determined by its restriction to M'_C, giving an irreducible representation. In fact the restriction map gives a bijection. Thus, we can regard the elements of $A^C(\delta)$ as irreducible representations in $\widehat{M'}_C$ ([Ga, cor. 3.3]).

Suppose C is L, E or F. By conjugation of M'_C on M, both M'_C and W_C may act on $\widehat{M}$. We denote by $(M'_C)_\delta$ and $(W_C)_\delta$ for the stabilizers of any $\delta \in \widehat{M}$. Similar to the construction of the set $\overline{\Delta}_L$, we find systems of roots $\overline{\Delta}_E$ and $\overline{\Delta}_F$. Write $\overline{\Delta}_\delta(C)$ for the corresponding set of good roots of δ in $\overline{\Delta}_C$. It turns out that $(W_C)_o := W(\overline{\Delta}_\delta(C))$, the Weyl group of the root system $\overline{\Delta}_\delta(C)$, is a normal subgroup of $(W_C)_\delta$. Now, we introduce the R-group of Knapp and Stein. This is given as

$$R_\delta(C) = (W_C)_\delta / W(C)_o.$$

Both sets E and F have been chosen such that the set $\overline{\Delta}_\delta(C)$ is empty. Therefore, we obtain $R_\delta(C) \cong (W_C)_\delta$ for C either E or F. In any case the R-group $R_\delta(C)$ has the property of being a finite direct sum of copies of $\mathbb{Z}/2\mathbb{Z}$. This is clear for E and F, and for L may be deduced from the fact that $R_\delta(L)$ and $R_\delta(E)$ are isomorphic, [Ga, Lemma 4.3.3].

Proposition 3.2. *By the property of fine representations we identify $A^C(\delta)$ as a subset of $\widehat{M_C}$, as in (3.1). If C is either L, E or F then $\widehat{R_\delta}(C)$ acts transitively on $A^C(\delta)$ as follows:*

$$\chi \cdot \mu := Ind^{M'_C}_{(M'_C)_\delta}(\chi \otimes \mu_\delta),$$

where$\chi \in \widehat{R_\delta}(C)$, and μ_δ is the δ-primary part of $\mu \in A^C(\delta)$. □

The induction of compact groups in Proposition 3.2 from some δ-primary part gives an irreducible representation. This is can be read carefully in [Ga, Ch. II]. Proposition (3.2) may be deduced after analyzing the decomposition for the principal series $X_{M_{C\#}}(\delta_o \otimes 1)$ for C either E or F where $M_{C\#} = Ad^{-1}(Ad(M_{C_o}))$ for the identity component M_{C_o}, and δ_o is any of the irreducible representations of $M_\# = M \cap M_{C\#}$ occurring in the restriction $\delta|_{M_\#}$.

Write $A^\#(\delta_o)$ for the set of $K_\#$-fine representations, with $K_\# = K \cap M_{C\#}$. Thus if $M'_\# := M'_C \cap M_{C\#}$, we have $W_C \cong M'_\#/M_\#$. Therefore there is a natural action of $\widehat{W}_C$ on the set $A^\#(\delta_o)$, identified with a subset of $\widehat{M'}_\#$ as in (3.1), by means of the tensor product. This action turns out to be simple and transitive. Moreover, for any η_o element in $A^\#(\delta_o)$ we obtain

$$X_{M_{C\#}}(\delta \otimes 1) = \sum_{\chi \in \hat{W}_D} I_{\chi \cdot \eta_o}$$

where $I_{\chi \cdot \eta_o}$ is the discrete series representation of M_C determined by the minimal $K_\#$-type $\chi \cdot \eta_o$. The multiplicity of $I_{\chi \cdot \eta_o}$ in $X_{M_{C\#}}(\delta \otimes 1)$ is exactly one. It is possible to show $m(\delta_o, \delta) = m(\delta, \eta)$, and $m(\eta_o, \eta) = 1$ (identities about multiplicities), in comparing this special situation with the one for (M_C, δ); assuming $\eta \in A^C(\delta)$ is such that η_o occurs in $\eta|_{M'_\#}$. Hence, $\eta = Ind^{K_C}_{K_\#}(\eta_o)$ is an irreducible representation.

The group $\widehat{W}_C$ acts also on $A^C(\delta)$ by tensor product. In fact,

$$\chi \cdot \eta = Ind^{K_C}_{K_\#}(\chi \cdot \eta_o)$$

for η_o occurring in $\eta|_{M'_\#}$, $\eta \in A^C(\delta)$, and $\chi \in \widehat{W}_C$.

In the first part of [Ga], we study representations of finite groups of characters. From there ([Ga; Corollary 3.9]), we have the next isomorphism for any δ-primary part η_δ,

$$\widehat{W}_C/(\widehat{W}_C)_\eta \cong \widehat{R_\delta}(C)/(\widehat{R_\delta}(C))_{\eta_\delta}$$

which allows us to substitute the R-group for W_C. Because the action of $\widehat{R_\delta}(C)$ is on a subset of $\widehat{M'}_C$ we have $(\widehat{R_\delta}(C))_{\eta_\delta} = (\widehat{R_\delta}(C))_\eta$. This means also that $\widehat{R_\delta}(C)$ acts transitively on $A^C(\delta)$.

The formula for $C = E$ is useful to give the decomposition of $X_L(\delta \otimes 1)$. In particular the next proposition implies Proposition 3.2.

Proposition 3.3. *([Ga, Lemma 4.3.3]) Assume $\nu = 1$.*

(a) The restriction map $\mu \mapsto \mu|$ going from $A^L(\delta)$ to $A^E(\delta)$ is a bijection.

(b) There are natural isomorphisms, $\widehat{R_\delta}(L) \cong \widehat{R_\delta}(E)$, and

$$\widehat{R_\delta}(L)/(\widehat{R_\delta}(L))_\mu \cong \widehat{R_\delta}(E)/(\widehat{R_\delta}(E))_{\mu|}$$

(which implies $m(\delta, \mu|) = m(\delta, \mu)$).

(c) The value $m(\delta, \mu) = \mid Hom_M(\delta, \mu|_M) \mid$ is independent of $\mu \in A^L(\delta)$. Moreover, we get the decomposition

$$X_L(\delta \otimes 1) = m(\delta, \mu) \sum_{\overline{\chi} \in \hat{R}_\delta(L)/(\hat{R}_\delta(L))_\mu} I_{\overline{\chi} \cdot \mu}$$

where $I_{\chi \cdot \mu}$ is the unique irreducible representation of M_C determined by the minimal K_L-type $\chi \cdot \mu$. The multiplicity of $I_{\chi \cdot \mu}$ in $X_L(\delta \otimes 1)$ is exactly $m(\delta, \mu)$. □

Take η in $A^F(\delta)$. We define for C being either E or L the following set,

$$A^C(\eta) = \{\mu \in A^C(\delta) \mid \eta \text{ occurs in the restriction } \mu|_{K_F}\}.$$

The restriction map in Proposition 3.3 also gives a bijection between $A^E(\eta)$ and $A^L(\eta)$. Now observe that if C is either E or L, and η is varied, we get a partition of $A^C(\delta)$. To control such a partition we need the following sets of characters:

$$\begin{aligned} Q(\delta, C) &= \{\chi \in \widehat{R_\delta}(\delta) \mid \chi|_{R_\delta(F)} \cdot \eta = x \cdot \eta \quad \text{for some } x \in K_C\} \\ Q(\delta, C)_o &= \{\chi \in \widehat{R_\delta}(\delta) \mid \chi|_{R_\delta(F)} \cdot \eta = \eta\} \end{aligned}$$

Let us write the elements in $\widehat{R_\delta}(C)/Q(\delta, C)_o$ with one line over and for the elements in $\widehat{R_\delta}(C)/Q(\delta, C)$ with two lines over. For any $\eta \in A^F(\delta)$ and $\chi \in \widehat{R_\delta}(C)$, we have $A^C(\chi| \cdot \eta) = A^C(\eta)$ if and only if $\chi \in Q(\delta, C)$. Moreover, we obtain the disjoint union

$$A^C(\delta) = \bigsqcup_{\overline{\overline{\chi}} \in \hat{R}_\delta(C)/Q(\delta, C)} \overline{\overline{\chi}} \cdot A_C(\eta).$$

(4) Suppose $(\mathfrak{q}, (H, \underline{\gamma_1}))$ is the θ-stable data given from the fixed character $(H, \underline{\gamma})$. Consider F the set of simple roots given in (3.1). Let $H_F = T_F A_F$ be a θ-stable Cartan subgroup arising from application of Cayley transforms on H with respect to F such that $T_o \subseteq T_F$ and $A_F \subseteq A$. Fix $\eta \in A^F(\delta)$. T_F is a compact Cartan subgroup of M_F. Fix any irreducible T_F-module η_o occurring in the restriction $\eta|_{T_F}$. We define a set of θ-stable limit data $(\mathfrak{q}, (H_F, \underline{\gamma_1}_F)) = (\mathfrak{q}, (H_F, \Gamma_{1F}, \overline{\gamma_1}_F, \Delta^+_{F,im}))$ for G as follows:

(*i*) Γ_{1F} is in $\widehat{H}_F$ given by $\Gamma_{1F}|_{T_F} = \eta_o$ and $\Gamma_{1F}|_{A_F} = \Gamma_1|_{A_F} = \nu_F$.

(*ii*) $\overline{\gamma_1}_F \in \mathfrak{h}_F^*$ is given by $\overline{\gamma_1}_F|_{\mathfrak{t}} = d\delta$, $\overline{\gamma_1}_F|_{\mathfrak{t}^\perp} \equiv 0$ and $\overline{\gamma_1}_F|_{\mathfrak{a}_F} = \nu_F$; where $\mathfrak{t}_F = \mathfrak{t} + \mathfrak{t}^\perp$.

(*iii*) $\Delta^+_{F,im}$ is the unique positive system of $\Delta(\mathfrak{m}_F, \mathfrak{t}_F)$ such that $d\eta_o$ is dominant.

We form the following set of θ-stable limit data:

$\Omega_{1F} := \{ \ (\mathfrak{q}, (H_F, \underline{\gamma_1}_F)) \mid (H_F, \underline{\gamma_1}_F)$ is built for some $\eta \in A^F(\delta)\}$.

From the bijection between limit characters and sets of θ-stable limit data in Theorem 2.1, we define,

$\Omega_F := \{(H_F, \underline{\gamma_F}) \mid (H_F, \underline{\gamma_F})$ is associated to some $(\mathfrak{q}, (H_F, \underline{\gamma_{1F}}))$ in $\Omega_{1F}\}$.

In the same way part (*c*) of Proposition 3.3 is proved, we obtain the formula

$$X_{M_F}(\delta \otimes 1) = m(\delta, \eta) \sum_{\overline{\chi} \in \hat{R}_\delta(F)/(\hat{R}_\delta(F))_\eta} I_{\overline{\chi} \cdot \eta}.$$

4.1 The construction of $(H_F, \underline{\gamma_F})$, up to conjugation, depends on η and not on the choice of η_o. We note that $X_G(P_F, I_\eta \otimes \nu_F)$, the Harish-Chandra module associated to $Ind_{P_F}^G(I_\eta \otimes \nu_F)$, is the representation induced from the limit character $(H_F, \underline{\gamma_F})$. Furthermore, we have

Lemma 4.2. *Assume* $(\mathfrak{q}, (H_F, \underline{\gamma_1}_F)) \in \Omega_{1F}$ *and* $(H_F, \underline{\gamma_F}) \in \Omega_F$. *Then in the previous notation, each of the representations* $X_L(P_F, I_\eta \otimes \nu_F) = X_L(H_F, \underline{\gamma_{1F}})$, *and* $X_G(H_F, \underline{\gamma_F})$ *has a unique irreducible subrepresentation.* □

4.3 Define an action of $\widehat{R_\delta}(F)$ on Ω_{1F} as follows: for any limit character in Ω_{1F}, say $(H_F, \underline{\gamma_1}_F) = (H_F, \Gamma_{1F}, \overline{\gamma_1}_F, \Delta^+_{F,im})$ and any character $\chi \in \widehat{R_\delta}(F)$, we set

$$\chi \cdot (H_F, \underline{\gamma_1}_F) = (H_F, \Gamma'_{1F}, \overline{\gamma'_1}_F, \chi \cdot \Delta^+_{F,im})$$

where $\Gamma'_F|_{T_F} = \chi \cdot \Gamma_F|_{T_F}$ and $\Gamma'_F|_{A_F} = \nu_F$ and $\chi \cdot \Delta^+_{F,im}$ is the unique positive system such that $d(\chi \cdot \Gamma_F)$ is dominant.

Since the the induction of a module $X_L(H_F, \underline{\gamma_1}_F)$ depends only on the class of conjugation of $(H_F, \underline{\gamma_1}_F)$ by K_L, given $(H_F, \underline{\gamma_F}) \in \Omega_F$ and $\chi \in \widehat{R_\delta}(F)$, we can deduce that $X_L(\chi \cdot (H_F, \underline{\gamma_1}_F)) = X_L(H_F, \underline{\gamma_1}_F)$ if and only if $\chi \cdot \Gamma_F|_{T_F} = x \cdot \Gamma_F|_{T_F}$ for some $x \in M'_F$. This is the case if and only if $\chi \cdot \eta = \eta$, with $\eta = Ind_{T_F}^{K_F}(\Gamma_F|_{T_F})$. Since the group $\widehat{R_\delta}(F)$ acts transitively on $A^F(\delta)$, it does it so on Ω_{1F}. Therefore, using the isomorphisms coming from restrictions

$$\widehat{R_\delta}(L)/Q(\delta, L)_o \cong \widehat{R_\delta}(E)/Q(\delta, E)_o \cong \widehat{R_\delta}(F)/(\widehat{R_\delta}(F))_\eta,$$

we are allowed to continue the action to $\widehat{R_\delta}(L)/Q(\delta, L)_o$, which is simple and transitive.

Next let us lift the action on characters to representations. Let $\overline{\chi}$ be any character in $\widehat{R_\delta}(F)/(\widehat{R_\delta}(F))_\eta$. We define $\overline{\chi}\cdot X_L(H_F,\underline{\gamma_1}_F)$ to be the $(\mathfrak{l},K_L)$-representation $X_L(\overline{\chi}\cdot(H_F,\underline{\gamma_1}_F))$. By Lemma 4.2, let $\overline{X}_L(H_F,\underline{\gamma_1}_F)$ be the unique irreducible submodule of $X_L(H_F,\underline{\gamma_1}_F)$. We define, naturally, $\overline{\chi}\cdot\overline{X}_L(H_F,\underline{\gamma_1}_F)$ to be the unique irreducible $(\mathfrak{l},K_L)$-subrepresentation of $\overline{\chi}\cdot X_L(H_F,\underline{\gamma_1}_F)$. Define $\overline{\chi}\cdot X_G(H_F,\underline{\gamma}_F)$ and $\overline{\chi}\cdot\overline{X}_G(H_F,\underline{\gamma}_F)$ similarly.

The bijection between Ω_{1F},and Ω_F and lemma (4.2) imply that each indecomposable constituent in the decomposition of $X_G(H,\underline{\gamma})$ is induced from some limit character in Ω_F. Moreover, we have a transitive action of $\widehat{R_\delta}(L)/Q(\delta,L)_o$ on the set of representations induced from limit characters in Ω_F.

Finally we need to analyze the set of minimal K-types of $X_G(H,\underline{\gamma})$, denoted by $A^K(\delta)$, for the regular character $(H,\underline{\gamma})$. Denote by $A^K(\eta)$ the set of minimal K-types of $X_G(H_F,\underline{\gamma}_F)=\mathfrak{R}^S_\mathfrak{q}(X_L(H_F,\underline{\gamma_1}_F))$ with $(H_F,\underline{\gamma_1}_F)$, as in the beginning of subsection (4) for $\eta\in A^F(\delta)$. Multiply $\otimes\Lambda^{dim\mathfrak{u}\cap\mathfrak{p}}(\mathfrak{u}\cap\mathfrak{p})$ by elements μ in $A^L(\delta)$. There exists a unique K-irreducible representation whose restriction to K_L contains $\mu\otimes\Lambda^{dim\mathfrak{u}\cap\mathfrak{p}}(\mathfrak{u}\cap\mathfrak{p})$ having the same highest weight ([Ga, cor. 4.2.8]). This gives a bijection between $A^L(\delta)$ and $A^K(\delta)$ which in turn establishes a bijection between $A^L(\eta)$ and $A^K(\eta)$ ([Green, lemma 5.3.23]).

We can now consider the action of $\widehat{R_\delta}(L)/Q(\delta,L)$ on subsets $A^L(\eta)$ on the subsets $A^K(\eta)$ (see 3.3). Moreover,

$$A^K(\delta)=\bigsqcup_{\overline{\overline{\chi}}\in\hat{R}_\delta(L)/Q(\delta,L)}\overline{\overline{\chi}}\cdot A^K(\eta).$$

The identity $m(\delta,\eta)\cdot m(\eta,\mu)|Q(\delta,L)/Q(\delta,L)_o|=m(\delta,\mu)$ is valid for any element $\mu\in A^L(\delta)$ (proved in [Ga, lemma 3.11]). Here we write the consequence of the action of the R-group.

Theorem 4.4. *([Ga, Thm. 5.3.5]) Assume $(H,\underline{\gamma})$ is a regular character for G. Let $(H_F,\underline{\gamma_F})$ be the limit character constructed as mentioned in (4.1), with respect to $(H,\underline{\gamma})$, some $\eta\in A^F(\delta)$, and $\mu\in A^L(\delta)$. We obtain the following:*

(a) The decompositions

$$X_G(H,\underline{\gamma})=m(\delta,\eta)\sum_{\overline{\chi}\in\hat{R}_\delta(L)/Q(\delta,L)_o}\overline{\chi}\cdot X_G(H_F,\underline{\gamma_F}).$$

(b) The set of minimal K-types of the $(\mathfrak{g},K)$-module $\overline{\chi}\cdot X_G(H_F,\underline{\gamma_F})$ is precisely $\overline{\overline{\chi}}\cdot A^K(\eta)$.

(c) $\overline{\overline{\chi}}\cdot A^K(\eta)=A^K(\eta)$ if and only if $\overline{\overline{\chi}}=1$.

(d) $\overline{\chi}\cdot X_G(H_F,\underline{\gamma_F})\cong X_G(H_F,\underline{\gamma_F})$ if and only if $\overline{\chi}=1$.

(e) The multiplicity of $\overline{\chi}\cdot X_G(H_F,\underline{\gamma_F})$ in $X_G(H,\underline{\gamma})$ is $m(\delta,\eta)$.

(f) The multiplicity of any $\omega\in\overline{\overline{\chi}}\cdot A^K(\eta)$ in $\overline{\chi}\cdot X_G(H_F,\underline{\gamma_F})$ is $m(\eta,\mu)$.

(*g*) $m(\delta,\eta)\cdot m(\eta,\mu) = 2^m$ *for some* $m \in \mathbb{N}$. □

As we mentioned, $(H_F, \underline{\gamma_F})$ in Theorem 4.4 might not be a final character; however the module $X_G(H_F, \underline{\gamma_F})$ is indecomposable. On the other hand, the statements in Theorem 4.4 do not change if we substitute the $(\mathfrak{g}, K)$-modules there for the corresponding socles, which possess the minimal K-types if $X_G(H, \underline{\gamma})$ is unitary.

Corollary. *We have obtained both the decomposition of any Langlands subrepresentation and the classification of tempered representations.* □

To finish, we mention that there exists a group G in the Harish-Chandra class such that its identity component G_o is given by

$$G_o = \frac{(Sl(2,\mathbb{R}) \times Sl(2,\mathbb{R}))}{\left\{\pm \left(\begin{smallmatrix} I & 0 \\ 0 & I \end{smallmatrix}\right)\right\}}$$

and if $P = MAN$ is a minimal parabolic subgroup of G then M is isomorphic to the quaternion group. Write $M^o := M \cap G_o$; hence $M^o \cong \mathbb{Z}/2\mathbb{Z}$. Consequently, we have $G/G_o \cong M/M^o \cong \mathbb{Z}/2\mathbb{Z} + \mathbb{Z}/2\mathbb{Z}$. Let δ be the two-dimensional representation of M. It is known that this representation, restricted to M^o, can be written as the direct sum $\delta|_{M^o} \cong \delta_o + \delta_o$ where δ_o is the nontrivial representation of $\mathbb{Z}/2\mathbb{Z}$; that is, we have the multiplicity $m(\delta_o, \delta) = 2$. Using Remark 4.2, we conclude that $\mathrm{Ind}_P^G(\delta \otimes 1) \cong I + I$ where I is an irreducible representation of G appearing with multiplicity two; there is only one minimal K-type in the standard representation appearing with multiplicity two.

References

[A-B-V] J. Adams, D. Barbasch, and D. A. Vogan, *The Langlands classification and irreducible characters for real reductive groups*, Progress in Math., Birkhäuser, Boston, 1992.

[B-W] A. Borel and N. Wallach, *Continuous cohomology, discrete subgroups, and representations of reductive groups*, Princeton University Press, Princeton, 1980.

[Ga] E. Garnica Vigil, *On the decomposition of Langlands surepresentations for a group in the Harish-Chandra class*, Trans. Amer. Math. Soc. **347** (1995), 1609–1648.

[Green] D. A. Vogan, *Representations of real reductive Lie groups*, Progress in Math., Birkhäuser, Boston, 1981.

[H-Ch 1] Harish-Chandra, *Representations of semi-simple Lie groups* I, Trans. Amer. Math. Soc. **75** (1953), 125–243.

[H-Ch 2] Harish-Chandra, *Representations of semi-simple Lie groups* III, Trans. Amer. Math. Soc. **76** (1954), 234–253.

[H-Ch 3] Harish-Chandra, *Supertempered distributions on a real reductive group*, Collected Papers, Vol IV. Springer-Verlag, 447–461.

[K] A. W. Knapp, *Representation theory of semi-simple groups. An overview based on examples*, Princeton University Press, Princeton, 1986.

[K-Z] A. W. Knapp and G. J. Zuckerman, *Classification of irreducible tempered representations of semi-simple Lie groups*, Ann. of Math. **116** (1982), 389–501.

[K-V] A. W. Knapp and D. A. Vogan, *Cohomological induction and unitary representations*, Princeton University Press, Princeton, 1995.

[M] I. Mirković, *Classification of irreducible tempered representations of semi-simple groups*, Ph.D. thesis, Univ. Utah, 1986.

[S-V] B. Speh and D. A. Vogan, *Reducibility of generalized principal series representations*, Acta Math. **145** (1980), 227–299.

[V-I] D. A. Vogan, *The algebraic structure of the representations of semi-simple Lie groups* I, Ann. of Math. **109** (1976), 1–60.

[V-II] D. A. Vogan, *Irreducible characters of semi-simple Lie groups* I, Duke Math. J **46** (1979), 61–108.

[W-1] N. Wallach, *Representations of semi-simple Lie groups and Lie algebras*, Proc. Canad. Math. Cong. (1977), 154–245.

[W-2] N. Wallach, *Real reductive groups* I, Academic Press, Inc., 1988.

[W-3] N. Wallach, *Real reductive groups* II, Academic Press, Inc., 1992.

Facultad de Ciencias UNAM, México 04510
email: garnica@redvax1.dgsca.unam.mx

GEOMETRIC QUANTIZATION FOR NILPOTENT COADJOINT ORBITS

WILLIAM GRAHAM* AND DAVID A. VOGAN, JR. †

CONTENTS

1. INTRODUCTION

Suppose G is a Lie group, $\mathfrak{g}$ its Lie algebra, and $\mathfrak{g}^*$ the dual vector space. It is a classical idea of Kirillov and Kostant (see [10] and [11]) that irreducible unitary representations of G are related to the orbits of G on $\mathfrak{g}^*$. This idea finds its purest form in

Theorem 1.1 (Kirillov [10]). *Suppose G is a connected and simply connected nilpotent Lie group. Then there is a bijection from the set $\mathfrak{g}^*/G$ of coadjoint orbits of G to the set $\widehat{G}$ of equivalence classes of irreducible unitary representations of G.*

For other groups there are complications even with regard to what is true (never mind what one can prove). We recall very briefly a few of these. First, not every coadjoint orbit can correspond to a representation: one has to impose an appropriate "integrality" requirement on the orbit. This complication occurs already for G the circle group.

Second, one needs a little more information beyond the orbit itself: very roughly speaking, something like a local coefficient system on the orbit. This complication occurs in semidirect product groups $K \times V$ (with K compact and V a vector group on which K acts) whenever the isotropy groups of the K action on V can be disconnected. More serious problems in this direction were found by Rothschild and Wolf in [17]. They gave an example in the split real group of type G_2 in which two representations of different infinitesimal characters could be attached to the same coadjoint orbit.

* Supported in part by an NSF postdoctoral fellowship
† Supported in part by NSF grant DMS-9402994

Third, the same unitary representation may arise from each of several coadjoint orbits. Exactly when this complication occurs depends on exactly how the correspondence from orbits to representations is defined. In the approach we will follow (due mostly to Duflo), the simplest example has $G = SU(2)$; the trivial representation is attached to the orbit $\{0\}$ and also to the orbit of the half-sum of positive roots.

Fourth, some unitary representations are not attached to any coadjoint orbit. This complication appears first for $SL(2,\mathbb{R})$, where the complementary series representations are not attached to orbits.

Fifth, the unitary representations attached to some coadjoint orbits are not irreducible. The simplest way that this happens is that the representation is zero. With Duflo's version of the correspondence, this happens for $G = U(3)$, in the following way. Coadjoint orbits for $U(3)$ are parametrized by decreasing sequences of three real numbers. Consider the orbit $\mathcal{O}_b$ parametrized by $(b, -1/2, -1/2)$, with b a real number greater than $-1/2$. These orbits are all isomorphic to $\mathbb{CP}^2$, the complex projective plane. Duflo's integrality condition (admissibility, as defined in section 6) amounts to $b \in \mathbb{Z}$. The representation Duflo attaches to $\mathcal{O}_b$ is the space of holomorphic sections of the $b-1$ tensor power of a standard line bundle. For $b - 1 \geq 0$, this space is $S^{b-1}(\mathbb{C}^3)$, a symmetric power of the standard representation of $U(3)$; but for $b = 0$ there are no holomorphic sections, so the representation is zero. Of course a more interesting possibility is that the representation is actually reducible. The simplest example I know of this phenomenon occurs in the complex form of the exceptional group G_2, for a singular elliptic coadjoint orbit. The representation attached to such an orbit is always a unitary degenerate series representation (induced from a non-trivial one-dimensional character of a parabolic subgroup). One such induced representation is reducible.

Finally, when G is not of type I there are terrible complications, because irreducible unitary representations of G are no longer such a natural class of objects to consider. This is first apparent in the work of Auslander and Kostant [3], who established a version of Theorem 1.1 for solvable Lie groups.

In light of all these complications, what one might hope for along the lines of Theorem 1.1 for general Lie groups is something like this.

Problem 1.2. Suppose G is a type I Lie group. Find a construction attaching to each pair (X, τ) a unitary representation $\pi(X, \tau)$ of G. Here X is an orbit of G on $\mathfrak{g}^*$ satisfying an appropriate integrality hypothesis, and τ is some appropriate additional structure. The representations $\pi(X, \tau)$ should be close to irreducible, and they should include most of the interesting irreducible unitary representations of G.

We will eventually refine the statement of this problem substantially (see Problem 6.3). But we can discuss ideas for the solution of the problem without a clearer statement, and this we do next. The classical strategy, known as *geometric quantization*, is this. A coadjoint orbit X carries a natural G-invariant symplectic structure. (The definition will be recalled in Corollary

2.13.) There are various standard constructions of unitary representations of G, beginning with some data D and leading to a unitary representation $\pi(D)$. Attached to such a construction it is often possible to find a symplectic manifold $Y(D)$ with a G action. The idea of geometric quantization can be phrased in this way: given a coadjoint orbit X, one tries to find data D so that $X \simeq Y(D)$ (as symplectic manifolds with G action). If this can be done, then one says that $\pi(D)$ is the unitary representation associated to X. To use this idea to solve Problem 1.2, one must show that every (appropriately integral) coadjoint orbit is isomorphic to some $Y(D)$, and that $\pi(D)$ is independent of the choice of D (subject to the condition $X \simeq Y(D)$).

Here is an example. One standard construction of a unitary representation begins with an action of G on a smooth manifold M. (A good example to keep in mind is the action of $G = SL(2,\mathbb{R})$ on the real projective space $M = \mathbb{RP}^1$ of lines through the origin in $\mathbb{R}^2$. Thus M is just a circle, but the action of G is interesting and complicated.) To get a unitary representation we need a Hilbert space, and it is natural to consider something like $L^2(M)$, the space of square-integrable functions on M. To define this space we must choose a measure on M. In order to get a natural action of G on $L^2(M)$ by unitary operators, the measure must be preserved by the action of G. In many examples (including the action of $SL(2,\mathbb{R})$ on the circle described above) there is no nice G-invariant measure. To circumvent this problem, one can introduce the real line bundle $\mathcal{D}^{1/2}(M)$ of *half-densities on* M. (The precise definition will be recalled in Definition 5.5.) For us the central fact is that the tensor product of this bundle with itself is the density bundle $\mathcal{D}^1(M)$, whose sections are the (signed) smooth measures on M:

$$\mathcal{D}^{1/2}(M) \otimes \mathcal{D}^{1/2}(M) \simeq \mathcal{D}^1(M) \tag{1.3(a)}$$

Here the tensor product is of line bundles on M.

Consider now the space S of compactly supported smooth sections of $\mathcal{D}_{\mathbb{C}}^{1/2}(M)$. (The subscript $\mathbb{C}$ denotes complexification.) If s_1 and s_2 belong to S, then it follows from (1.3)(a) that $s_1\overline{s_2}$ is a compactly supported section of $\mathcal{D}_{\mathbb{C}}^1(M)$; that is, it is a compactly supported complex-valued density on M. We may therefore define a pre-Hilbert space structure on S by

$$\langle s_1, s_2 \rangle = \int_M s_1\overline{s_2} \tag{1.3(b)}$$

The completion of this pre-Hilbert space is written $L^2(M,\mathcal{D}^{1/2})$, the space of *square-integrable half-densities on* M. Each element g of G (and indeed each diffeomorphism of M) acts on the space S and preserves the inner product; so we get a unitary representation

$$\pi(M)\colon G \to U(L^2(M,\mathcal{D}^{1/2})) \tag{1.3(c)}$$

of G. (We write $U(\mathcal{H})$ for the group of unitary operators on a Hilbert space $\mathcal{H}$.) Now geometric quantization asks that we find attached to M a symplectic manifold with a G action. The candidate we choose is the cotangent bundle:

$$Y(M) = T^*(M). \tag{1.3)(d}$$

Geometric quantization says that if a coadjoint orbit X is isomorphic to the cotangent bundle $T^*(M)$ of a G-manifold M, then we should attach to X the unitary representation of G on half-densities on M. This is reasonable statement as far as it goes, but it doesn't go very far. The action of G on $T^*(M)$ preserves the zero section $M \subset T^*(M)$, so it is never transitive unless M is discrete. Since a coadjoint orbit is by definition a homogeneous space, it can be isomorphic to a cotangent bundle only if it is discrete.

Fortunately it is possible to generalize this construction in many ways. The simplest is to consider in addition to M a Hermitian line bundle $\mathcal{L}$ over M, equipped with an action of G preserving the metric. Then $\mathcal{L}\otimes\mathcal{D}^{1/2}(M)$ is a complex line bundle on M; write $S(M,\mathcal{L})$ for the space of compactly supported smooth sections. If σ_1 and σ_2 belong to $S(M,\mathcal{L})$, then we can write $\sigma_i = l_i \otimes s_i$; here l_i is a compactly supported section of $\mathcal{L}$ and s_i a nowhere vanishing section of $\mathcal{D}^{1/2}(M)$. Then $\langle l_1, l_2\rangle_{\mathcal{L}}$ is a compactly supported complex-valued function on M and $s_1 s_2$ is a smooth density; so the product is a compactly supported complex-valued density. We may therefore define a pre-Hilbert space structure on $S(M,\mathcal{L})$ by

$$\langle \sigma_1, \sigma_2\rangle = \int_M \langle l_1, l_2\rangle_{\mathcal{L}} s_1 s_2. \tag{1.4)(a}$$

The completion of this pre-Hilbert space is written $L^2(M, \mathcal{L}\otimes\mathcal{D}^{1/2})$. We get a unitary representation

$$\pi(M,\mathcal{L})\colon G \to U(L^2(M,\mathcal{L}\otimes\mathcal{D}^{1/2})) \tag{1.4)(b}$$

The philosophy of geometric quantization requires also a symplectic manifold attached to M and $\mathcal{L}$. This is provided by a construction of Kostant (see [13] or [21], Proposition 4.6). From a Hermitian line bundle $\mathcal{L}$ on a real manifold M one can construct the *twisted cotangent bundle*

$$Y(M,\mathcal{L}) = T^*(M,\mathcal{L}). \tag{1.4)(c}$$

This is a fiber bundle over M with a natural symplectic structure $\omega(\mathcal{L})$; it is an affine bundle over the cotangent bundle. In particular, the fiber $T^*_m(M,\mathcal{L})$ over m in M is an affine space for the vector space $T^*_m(M)$; that is, it is a copy of $T^*_m(M)$ with the origin forgotten. Sections of $T^*(M,\mathcal{L})$ are certain special connections on $\mathcal{L}$. We can now formulate

Philosophy of Geometric Quantization (first form). Suppose that a coadjoint orbit X for a Lie group G is isomorphic to a twisted cotangent bundle $T^*(M,\mathcal{L})$ (with M a smooth G-manifold and $\mathcal{L}$ a Hermitian line bundle on M). Then the unitary representation $\pi(M,\mathcal{L})$ is attached to X.

This philosophy has some content: the proof of Kirillov's Theorem 1.1 shows that every coadjoint orbit for a connected nilpotent Lie group is a twisted cotangent bundle. For more complicated groups the full power of geometric quantization requires considering not just the real analysis construction of (1.4)(b) but also some complex-analytic analogues; but for the purposes of this introduction the present statement will suffice.

The geometric part of the geometric quantization approach to Problem 1.2 is therefore this: given a coadjoint orbit X, find a twisted cotangent bundle $T^*(M, \mathcal{L})$ to which X is isomorphic (as a symplectic G-space). To understand how to do that, we need to know a little more about the geometry of $T^*(M, \mathcal{L})$. The main point is that the fibers $T^*_m(M, \mathcal{L})$ are Lagrangian submanifolds; that is, the tangent space to a fiber is always a maximal isotropic subspace for the symplectic form. In this way the symplectic manifold $T^*(M, \mathcal{L})$ has a foliation with Lagrangian leaves; the base manifold M may be identified with the space of leaves.

The idea now is to find a parallel structure in our coadjoint orbit X. Let us fix a base point $f \in X$, with isotropy group G_f; then

$$X \simeq G/G_f. \tag{1.5)(a}$$

We would like to find a G-invariant Lagrangian foliation of X. Since X is homogeneous, the space of leaves must be homogeneous. The whole foliation will therefore be determined as soon as we know the leaf Λ_f through the base point f; the other leaves will just be the translates $g \cdot \Lambda_f$. The purely set-theoretic requirement that X be the disjoint union of these translates imposes a very strong constraint on Λ_f: it implies that there must be a subgroup $H \supset G_f$ with

$$\Lambda_f = H \cdot f \simeq H/G_f \subset G/G_f \simeq X. \tag{1.5)(b}$$

The requirement that Λ_f be an isotropic submanifold is a further condition on H: it must be a Lie group, and

$$f|_{[\mathfrak{h},\mathfrak{h}]} = 0. \tag{1.5)(c}$$

That is, f must define a one-dimensional representation of the Lie algebra of H. In the presence of (1.5)(c), the assumption that Λ_f is Lagrangian is equivalent to

$$\dim H/G_f = \frac{1}{2}\dim G/G_f. \tag{1.5)(d}$$

In order for the space of leaves to be a nice manifold M, we need H to be a closed subgroup of G; and in order for M to carry an appropriate G-equivariant Hermitian line bundle, we need a one-dimensional unitary representation

$$\tau \in \widehat{H}, \qquad d\tau = 2\pi i f. \tag{1.5)(e}$$

(Notice that if H is connected and simply connected, then (1.5)(c) guarantees the existence of a unique τ with differential $2\pi i f$.)

Conversely, suppose that $X \simeq G/G_f$ is given, and that we can find a closed subgroup $H \supset G_f$ together with a unitary character τ of H, satisfying (1.5)(c)–(e). Then τ defines a Hermitian line bundle $\mathcal{L}$ on $M = G/H$, and one can show that some open set in $T^*(M, \mathcal{L})$ is G-equivariantly symplectomorphic to a covering space of X. We will say that X is *locally isomorphic to* $T^*(M, \mathcal{L})$. In this way the geometric problem of relating coadjoint orbits to twisted cotangent bundles is reduced to the group-theoretic problem of finding appropriate subgroups H.

When G is a nilpotent group, the family of subgroups of G is rather rich; that is why one can find a group H making each coadjoint orbit a twisted cotangent bundle. As G becomes more complicated, the supply of subgroups dwindles. For reductive groups, one has the following remarkable result.

Theorem 1.6 (Ozeki and Wakimoto [16]) *Suppose G is a reductive Lie group and $f \in \mathfrak{g}^*$; write G_f for the isotropy group. Suppose $\mathfrak{h} \supset \mathfrak{g}_f$ is a Lie subalgebra of $\mathfrak{g}$ such that*

i) the linear functional f vanishes on $[\mathfrak{h}, \mathfrak{h}]$; and
ii) $\dim \mathfrak{h}/\mathfrak{g}(f) = \frac{1}{2} \dim \mathfrak{g}/\mathfrak{g}_f$.

Then $\mathfrak{h}$ must be a parabolic subalgebra of $\mathfrak{g}$. In particular, the dimension of the coadjoint orbit $G \cdot f$ must be exactly twice the codimension of a parabolic subalgebra.

Ozeki and Wakimoto are actually much more precise about the relationship between f and $\mathfrak{h}$.

Corollary 1.7. *Suppose G is a reductive Lie group, and X is a coadjoint orbit for G. Assume that the dimension of X is not equal to twice the codimension of any parabolic subalgebra of $\mathfrak{g}$. Then X is not locally isomorphic to a twisted cotangent bundle $T^*(M, \mathcal{L})$ for G.*

The dimensions appearing in Corollary 1.7 are easy to compute. One finds, for example

Corollary 1.8. *Suppose G is a split simple group over $\mathbb{R}$ or $\mathbb{C}$, not of type A; and suppose X is a coadjoint orbit of minimal non-zero dimension. Then X is not locally isomorphic to a twisted cotangent bundle for G.*

For these coadjoint orbits, the philosophy of geometric quantization as described above does not suggest a representation to attach to X. Our goal in this paper is to find an appropriate extension of that philosophy. The main idea, taken from [8] and [7], is to replace the Lagrangian foliation considered in (1.5) by a family of Lagarangian submanifolds which are allowed to overlap. Here is a formal definition.

Definition 1.9 (Guillemin-Sternberg and Ginsburg; see [8], Definition 2.1, and [7], A.1). Suppose X is a symplectic manifold. A *Lagrangian covering of*

X is a diagram of smooth manifolds and smooth maps

$$\begin{array}{ccc} & Z & \\ \pi \swarrow & & \searrow \rho \\ X & & M \end{array}$$

subject to the two conditions below.

a) The diagram is a double fibration ([9], page 340). That is, the maps π and ρ are fibrations, and $\pi \times \rho$ is an embedding of Z in $X \times M$.

This condition allows us to identify each fiber $L_m = \rho^{-1}(m)$ with a subset of X, and each fiber $M_x = \pi^{-1}(x)$ with a subset of M.

b) Each fiber L_m is a Lagrangian submanifold of X.

If X carries a symplectic action of G, then we say that the Lagrangian covering is *equivariant* if Z and M carry actions of G making π and ρ G-maps. It is called *homogeneous* if Z is a homogeneous space (in which case X and M must be as well).

Just as in the setting of (1.4) and (1.5), the manifold M is indexing a collection of Lagrangian submanifolds that cover X. The first observation, due essentially to Ginzburg, is that nice Lagrangian coverings often exist.

Theorem 1.10 (see [7], end of Appendix A). *Suppose G is a complex reductive Lie group, and $X = G{\cdot}f \subset \mathfrak{g}^*$ is a coadjoint orbit. Then there is an equivariant Lagrangian covering of X (Definition 1.9) with $M = G/Q$ a partial flag variety for G.*

A proof will be given in section 8; the main ingredient is a dimension estimate due to Spaltenstein in [19].

We turn now to a discussion of representation theory. In the setting of (1.5), the space of the representation was (roughly speaking) a space of sections of a line bundle $\mathcal{L}_M$ on $M = G/H$. We want to describe this space in terms of the symplectic manifold $X = G/G_f$. To do that, we first pull back the line bundle to a line bundle $\mathcal{L}_X$ on X; $\mathcal{L}_X$ is induced by the character $\tau_X = \tau|_{G_f}$ of G_f. Because $\mathcal{L}_X$ is pulled back from M by the projection $\rho\colon X \to M$, it makes sense to speak of sections of $\mathcal{L}_X$ that are "constant along the fibers of ρ." These fibers are just the leaves of the Lagrangian foliation of X. Sections of $\mathcal{L}_M$ may be identified with sections of $\mathcal{L}_X$ constant along the leaves of our specified Lagrangian foliation.

Suppose now that we are in the setting of Definition 1.9, and that G is acting compatibly on X, Z, and M. In order to have a parallel construction in the setting of Definition 1.9, we need first of all a (G-equivariant) line bundle

$$\mathcal{L}_M \to M. \tag{1.11)(a}$$

(Henceforth we will omit mention of the assumed G-equivariance of the structures introduced.) We can then define $\mathcal{L}_Z$ to be the pullback of $\mathcal{L}_M$ to Z by the fibration ρ:

$$\mathcal{L}_Z = \rho^*(\mathcal{L}_M). \tag{1.11)(b}$$

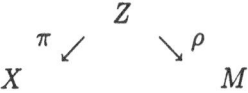

The representation we want will be on a space of sections of $\mathcal{L}_M$; equivalently, on a space of sections of $\mathcal{L}_Z$ that are constant along the fibers of ρ. Recall that these fibers may be identified with Lagrangian submanifolds of X; so already we have a construction reminiscent of (1.5).

The full space of sections of $\mathcal{L}_M$ is too large to carry the representation we want, however. In Definition 1.9, suppose that X has dimension $2n$, and that the fibers of π have dimension d. Then Z has dimension $2n+d$. Since the fibers of ρ are Lagrangian in X, they have dimension n; so M has dimension $n+d$. The philosophy of geometric quantization says that X should correspond to a representation of "functional dimension" n; that is, to something like a space of sections of a line bundle on an n-dimensional manifold. If d is not zero, M is too large. (If $d = 0$, then Z is a covering of X, and so inherits from it a symplectic structure. The map ρ provides a nice Lagrangian foliation of Z, and we are (at least in the homogeneous case) very close to the setting (1.5).) So we need a way to pick out a nice subspace of sections of $\mathcal{L}_M$. Carrying this out in detail will occupy most of this paper; for the moment we offer only a brief sketch.

There is a fiber bundle $\mathcal{B} = \mathcal{B}(X)$ over X for which the fiber $\mathcal{B}_x$ over x is the variety of Lagrangian subspaces of $T_x(X)$. (Thus $\mathcal{B}_x$ is a compact manifold of dimension $n(n+1)/2$.) We define a bundle map over X

$$\tau\colon Z \to \mathcal{B} \tag{1.11)(c}$$

as follows. Suppose $z \in Z$; write $x = \pi(z) \in X$ and $m = \rho(m) \in M$. Then the fiber L_m of ρ over m is a Lagrangian submanifold of X containing x; so its tangent space $T_x(L_m)$ is a Lagrangian subspace of $T_x(X)$. We set $\tau(z) = T_x(L_m)$.

Since each point of $\mathcal{B}$ is an n-dimensional real vector space, there is a tautological n-dimensional real vector bundle over $\mathcal{B}$. Taking its top exterior power and complexifying provides a line bundle $\mathcal{D}_{\mathcal{B}} \to \mathcal{B}$. (The $\mathcal{D}$ stands for determinant.) Roughly speaking, we need a square root $\mathcal{L}_{\mathcal{B}}$ of $\mathcal{D}_{\mathcal{B}}$. A little more precisely, we need a twisted version of Kostant's "symplectic spinors" on the base symplectic manifold X (see [12]). This is an infinite-dimensional vector bundle

$$\mathcal{S}_X \to X; \tag{1.11)(d}$$

the fiber over x may be identified with the smooth part of the metaplectic representation attached to the symplectic vector space $T_x(X)$, tensored with a one-dimensional twist. When X is a coadjoint orbit, a G-equivariant family of twisted symplectic spinors will be specified by an "admissible orbit datum" in the sense of Duflo (Definition 6.2). The reducibility of the metaplectic representation gives a natural decomposition $\mathcal{S} = \mathcal{S}^{even} \oplus \mathcal{S}^{odd}$. From these twisted spinors we will construct the line bundle $\mathcal{L}_{\mathcal{B}}$, and an embedding

$$\text{smooth sections of } \mathcal{S}^{even} \hookrightarrow \text{smooth sections of } \mathcal{L}_{\mathcal{B}} \tag{1.11)(e}$$

Now we can use the map $\tau: Z \to \mathcal{B}$ to pull $\mathcal{L}_{\mathcal{B}}$ back to Z. The last ingredient we will need is an isomorphism

$$\tau^*(\mathcal{L}_{\mathcal{B}}) \simeq \mathcal{L}_Z. \qquad (1.11)(f)$$

Using the embedding (1.11)(e) we can identify $\mathcal{S}^{even}$ with a space of sections of $\mathcal{L}_{\mathcal{B}}$. These sections may be pulled back using τ to sections of $\tau^*(\mathcal{L}_{\mathcal{B}})$ over Z, and then identified with sections of $\mathcal{L}_Z$ using (1.11)(f). Write $\mathcal{W}_Z^{even}$ for the resulting space of sections of $\mathcal{L}_Z$. Then the (smooth) representation of G we want to consider is on the space

$$\mathcal{W}_Z^{even} \cap (\text{smooth sections of } \mathcal{L}_M) \qquad (1.11)(g)$$

inside sections of $\mathcal{L}_Z$. The problem of constructing a unitary structure on this space we will leave to a future paper. (Even to guarantee the existence of an invariant Hermitian form requires an additional assumptions on $\mathcal{L}_M$.)

2. Symplectic and Poisson manifolds

We begin by recalling the definition of symplectic manifold.

Definition 2.1. A *symplectic manifold* is a smooth manifold X endowed with a 2-form ω_X, subject to the following conditions. Recall first of all that a 2-form may be regarded as a smoothly varying family of skew-symmetric bilinear forms ω_x on the various tangent spaces $T_x(X)$. We impose two conditions on ω_X.

a) Each form ω_x is non-degenerate; that is, for every non-zero tangent vector $v \in T_x(X)$ there is a vector $w \in T_x(X)$ so that $\omega_x(v,w) \neq 0$.
b) The 2-form ω_X is closed.

Condition (a) is equivalent to the assumption that $\dim X = 2n$ is even, and that the nth exterior power ω_X^n is a nowhere-vanishing volume form on X.

This definition may be made equally well in several other categories. We can define a *complex symplectic manifold*, which is a complex manifold endowed with a holomorphic 2-form satisfying analogues of (a) and (b); or a *complex symplectic algebraic variety*, which is a smooth complex algebraic variety endowed with an algebraic 2-form. We will invoke these generalizations as needed.

Any bilinear form on a vector space may be interpreted as a linear map from the vector space to its dual. On a symplectic manifold we therefore have maps from tangent spaces to cotangent spaces

$$\tau_x: T_x(X) \to T_x^*(X), \qquad \tau_x(v)(w) = \omega_x(w,v). \qquad (2.2)(a)$$

The non-degeneracy hypothesis on ω_X means that these maps are all isomorphisms, so they define a bundle isomorphism

$$\tau_X: T(X) \to T^*(X) \qquad (2.2)(b)$$

from the tangent bundle to the cotangent bundle. If $f \in C^\infty(X)$ is a (real-valued) smooth function, then df is a 1-form; that is, a smooth section of the cotangent bundle. We may therefore define

$$\xi_f = \tau^{-1}(df), \tag{2.2)(c}$$

a smooth section of the tangent bundle; that is, ξ_f is a vector field on X. It is called the *Hamiltonian vector field of* f. By inspection of the definitions, we see that ξ_f has the following characteristic property: if γ is any vector field on X, then

$$\omega_X(\gamma, \xi_f) = \gamma \cdot f; \tag{2.2)(d}$$

both sides are smooth functions on X.

Using the Hamiltonian vector fields, we can now define the *Poisson bracket* of smooth functions f and g on the symplectic manifold X:

$$\{f, g\} = \xi_f \cdot g = dg(\xi_f) = \omega_X(\xi_f, \xi_g) = -\xi_g \cdot f. \tag{2.2)(e}$$

Here the second expression may be taken as the definition; the remaining equalities are the definition of dg and (2.2)(d).

Proposition 2.3. *Suppose* (X, ω_X) *is a symplectic manifold.*

i) The Poisson bracket (2.2)(e) defines a Lie algebra structure on $C^\infty(X)$.

ii) The map $f \mapsto \xi_f$ *is a Lie algebra homomorphism from* $C^\infty(X)$ *to the Lie algebra of vector fields on* X. *Its kernel consists of the locally constant functions on* X.

iii) For each $f \in C^\infty(X)$, *the endomorphism* $g \mapsto \{f, g\}$ *is a derivation of* $C^\infty(X)$. *That is, it is linear in* g, *and*

$$\{f, gh\} = \{f, g\}h + g\{f, h\}.$$

iv) If we identify derivations of $C^\infty(X)$ *with vector fields on* X, *then bracket with* f *corresponds to the Hamiltonian vector field* ξ_f:

$$\{f, g\} = \xi_f \cdot g.$$

The elementary proof may be found in [2], Chapter 8 or [1], Chapter 3; both references use a different sign convention from ours. The Jacobi identity for the Lie algebra structure amounts to the fact that ω_X is closed. Of course (iv) is just the definition of the Poisson bracket given in (2.2)(e). One may also interpret it as defining the Hamiltonian vector field ξ_f in terms of the Poisson bracket.

For most of what we do, the Poisson bracket is more fundamental than the symplectic structure. At the same time, a Poisson bracket can be defined even on some singular spaces (like closures of nilpotent coadjoint orbits) where a symplectic structure does not make sense. We therefore recall a few highlights from the theory of Poisson spaces.

Definition 2.4. A *Poisson algebra A* over a field F is a commutative algebra (with 1) over F endowed with a Poisson bracket

$$\{,\}: A \times A \to A,$$

subject to the following conditions.

a) The Poisson bracket makes A a Lie algebra over F. That is, it is bilinear, skew-symmetric, and satisfies the Jacobi identity

$$\{f,\{g,h\}\} = \{\{f,g\},h\} + \{g,\{f,h\}\} \qquad (f,g,h \in A).$$

b) For each $f \in A$, the endomorphism ξ_f of A defined by

$$\xi_f \cdot g = \{f,g\}$$

is a derivation:

$$\{f,gh\} = \{f,g\}h + g\{f,h\}.$$

We call ξ_f the *Hamiltonian vector field of f*. The Jacobi identity for the Poisson bracket says that the map

$$A \to \operatorname{Der} A, \qquad f \mapsto \xi_f$$

is a Lie algebra homomorphism. (Recall that the commutator of two derivations of A is a derivation; this defines the Lie algebra structure on Der A.)

A *Poisson manifold* is a smooth manifold X endowed with a real Poisson algebra structure on $C^\infty(X)$. An *affine Poisson algebraic variety* (over F) is an affine algebraic variety X endowed with a Poisson algebra structure on the algebra $R(X)$ of regular functions on X. A *Poisson algebraic variety* is a an algebraic variety for which the sheaf of rings is a sheaf of Poisson algebras. More generally, a *Poisson space* is a ringed space in which the sheaf of rings is a sheaf of Poisson algebras.

There is a small consistency problem to worry about in the last definitions: whether an affine Poisson algebraic variety X is a Poisson algebraic variety. In fact it is, essentially because Poisson structures localize well: after inverting an element h, one can (and is forced to) define

$$\{f/h, g/h\} = \left(h\{f,g\} - f\{h,g\} - g\{f,h\}\right)/h^3.$$

In this way a Poisson structure on an algebra gives rise to Poisson structures on all localizations; so the Poisson structure on $R(X)$ makes the sheaf of rings on X into a sheaf of Poisson algebras. Similar remarks apply in the setting of manifolds: a Poisson structure on $C^\infty(X)$ gives rise to one on the sheaf of germs of smooth functions. In this case the key fact is that the Poisson bracket is local: that if f vanishes near x, then all Poisson brackets $\{f,h\}$ also vanish near x.

Let us see how to recover something close to a symplectic structure from a Poisson structure. To fix ideas we will concentrate on the case of algebraic varieties, but it is easy to give a parallel discussion for manifolds. What we will see is that a Poisson structure on X provides a foliation of "most" of X by symplectic manifolds; Poisson brackets are computed by restricting to the leaves and using the symplectic Poisson brackets there.

Suppose therefore that A is a Poisson algebra over F. Write $X = \operatorname{Spec} A$; then A can be thought of as an algebra of functions on X. A closed point x of X is a maximal ideal $\mathfrak{m}_x \subset A$. Write

$$F_x = A/\mathfrak{m}_x, \tag{2.6)(a}$$

an extension field of F. Recall that the *Zariski cotangent space of X at x* is the F_x vector space

$$T_x^*(X) = \mathfrak{m}_x/\mathfrak{m}_x^2. \tag{2.6)(b}$$

It is clear from (b) in Definition 2.4 that

$$\{\mathfrak{m}_x, \mathfrak{m}_x^2\} \subset \mathfrak{m}_x. \tag{2.6)(c}$$

The Poisson bracket therefore descends to a skew-symmetric F_x-bilinear pairing

$$\{,\}_x: T_x^*(X) \times T_x^*(X) \to F_x, \qquad \{f + \mathfrak{m}_x^2, g + \mathfrak{m}_x^2\} = \{f, g\} + \mathfrak{m}_x \tag{2.6)(d}$$

using the identifications (2.6)(a) and (2.6)(b). Write $\mathcal{R}_x$ for the radical of $\{,\}_x$; the form $\{,\}_x$ descends to a non-degenerate skew-symmetric form on $T_x^*/\mathcal{R}_x$. The Zariski tangent space at x is the dual space

$$T_x(X) = (T_x^*(X))^* \tag{2.6)(e}$$

Using the form $\{,\}_x$ each cotangent vector v defines a tangent vector $\delta(v)$:

$$\delta(v)(w) = \{w, v\}_x \qquad (v, w \in T_x^*(X)). \tag{2.6)(f}$$

Write $\mathcal{S}_x \subset T_x(X)$ for the image of δ, and $\mathcal{R}_x^\perp \subset T_x(X)$ for the annihilator of $\mathcal{R}_x$ in T_x. Then

$$\mathcal{S}_x \subset \mathcal{R}_x^\perp \simeq (T_x^*(X)/\mathcal{R}_x)^* \tag{2.6)(g}$$

On the other hand, the map δ factors to an isomorphism

$$\overline{\delta}: T_x^*(X)/\mathcal{R}_x \simeq \mathcal{S}_x. \tag{2.6)(h}$$

If $\mathcal{S}_x$ is finite-dimensional, it follows that $\mathcal{S}_x = \mathcal{R}_x^\perp$; so we can use $\overline{\delta}$ to transfer the skew-symmetric form $\{,\}_x$ to a non-degenerate skew-symmetric form ω_x on $\mathcal{S}_x$. (In the case of a symplectic manifold, the map $\overline{\delta}$ is just the inverse of the map τ of (2.2)(b).)

Suppose now that X is an irreducible complex affine algebraic variety, with A the algebra of regular functions. Then all the Zariski cotangent spaces are finite-dimensional, of dimension at least equal to the dimension n of X. Equality holds exactly on the smooth part X_s of X, which is a dense open subset. Write $2r(x)$ for the rank of the bilinear form $\{,\}_x$. It is easy to check that each point x has an open neighborhood U_x with $r(y) \geq r(x)$ for all $y \in U_x$. Consequently the rank function assumes its maximum value r on a dense open subset U_s of X_s. On U_s, the subspaces $\mathcal{R}_x$ define a subbundle of the cotangent bundle of rank $n - 2r$. The annihilators of these spaces define a subbundle $\mathcal{R}^\perp = \mathcal{S}$ of the tangent bundle TU_s, of rank $2r$. It is not difficult to see that the fiber $\mathcal{S}_x$ consists precisely of the values at x of all Hamiltonian vector fields ξ_f. Recall that we have defined non-degenerate skew-symmetric forms ω_x on each fiber $\mathcal{S}_x$.

We now do a little holomorphic differential geometry on X. Since the Hamiltonian vector fields are closed under Lie bracket, it follows easily that the distribution $\mathcal{S}$ is integrable; that is, that the family of smooth vector fields with values in $\mathcal{S}$ is closed under Lie bracket. The Frobenius theorem therefore provides a smooth foliation of U_s by complex submanifolds of dimension $2r$. Writing S for the submanifold through x, we have $T_xS = \mathcal{S}_x$. It is not difficult to check that the forms ω_x define a holomorphic symplectic structure on S, and that the corresponding Poisson brackets fit together to give the original Poisson bracket on U_s, as we wished to show.

Poisson manifolds have sometimes been defined in such a way that the nice open set U_s is all of X. This approach excludes interesting behavior that we want to consider, however.

We record one consequence of these ideas: a characterization of symplectic manifolds among Poisson manifolds.

Proposition 2.7. *Suppose X is a smooth Poisson manifold. The following conditions are equivalent.*

a) The Poisson bracket arises from a symplectic structure ω_X on X.
b) For every $x \in X$, the radical $\mathcal{R}_x \subset T_x^*(X)$ of $\{,\}_x$ is zero (cf. (2.6)(d)).
c) For every $x \in X$, the subspace $\mathcal{S}_x \subset T_x(X)$ is all of $T_x(X)$.
d) For every $x \in X$, the collection $\{\xi_f(x) | f \in C^\infty(X)\}$ of values at x of Hamiltonian vector fields is all of $T_x(X)$.

We leave the proof to the reader. Of course there is a parallel statement for algebraic varieties.

Here is a fundamental example. Suppose $\mathfrak{g}$ is a finite-dimensional real Lie algebra. Let $\{X_1, \ldots, X_n\}$ be a basis of $\mathfrak{g}$, and $\{\lambda_1, \ldots, \lambda_n\}$ the dual basis of the dual vector space $\mathfrak{g}^*$. Each λ_i gives rise to a vector field $\frac{\partial}{\partial \lambda_i}$ on $\mathfrak{g}^*$. We define a Poisson bracket on $C^\infty(\mathfrak{g}^*)$ by

$$\{f, g\}(\lambda) = \sum_{i,j} \lambda([X_i, X_j]) \frac{\partial f}{\partial \lambda_i} \frac{\partial g}{\partial \lambda_j}. \tag{2.8)(a}$$

(Notice that this bracket preserves the subspace $S(\mathfrak{g})$ of polynomial functions on $\mathfrak{g}^*$. It therefore makes $\mathfrak{g}^*$ into a real affine Poisson algebraic variety.) Among the axioms for a Poisson algebra, the only difficult one to check is the Jacobi identity. To compute the iterated Poisson brackets appearing there, we need to compute the derivative

$$\frac{\partial\lambda([X_i, X_j])}{\partial\lambda_q.}$$

The function being differentiated is linear, so the derivative is constant; it is $\lambda_q([X_i, X_j])$. Therefore

$$\{f, \{g, h\}\} = \sum_{p,q,i,j} \lambda([X_p, X_q])\cdot$$

$$\frac{\partial f}{\partial\lambda_p}\left(\lambda_q([X_i, X_j])\frac{\partial g}{\partial\lambda_i}\frac{\partial h}{\partial\lambda_j} + \lambda([X_i, X_j])\left(\frac{\partial^2 g}{\partial\lambda_q\partial\lambda_i}\frac{\partial h}{\partial\lambda_j} + \frac{\partial g}{\partial\lambda_i}\frac{\partial^2 h}{\partial\lambda_q\partial\lambda_j}\right)\right).$$

The other iterated brackets in the Jacobi identity are similar. The terms involving second derivatives cancel for easy reasons. The remaining term is

$$\sum_{p,q,i,j} \lambda([X_p, X_q])\lambda_q([X_i, X_j])\frac{\partial f}{\partial\lambda_p}\frac{\partial g}{\partial\lambda_i}\frac{\partial h}{\partial\lambda_j}.$$

Now for any $Y \in \mathfrak{g}$ we have

$$Y = \sum_q \lambda_q(Y)X_q.$$

Inserting this above gives

$$\sum_{p,i,j} \lambda([X_p, [X_i, X_j]])\frac{\partial f}{\partial\lambda_p}\frac{\partial g}{\partial\lambda_i}\frac{\partial h}{\partial\lambda_j}.$$

Now it is more or less obvious that the contribution of these terms to the Jacobi identity for the Poisson bracket vanishes because of the Jacobi identity for $\mathfrak{g}$.

Although the expression (2.8)(a) for the Poisson bracket is very attractive, there is a less symmetric formulation that offers more information. The Lie algebra $\mathfrak{g}$ acts on $\mathfrak{g}^*$ by the coadjoint action:

$$(\mathrm{ad}^*(Y)\cdot\tau)(X) = -\tau(\mathrm{ad}(Y)(X)) = \tau([X, Y]).$$

In this way every element Y of $\mathfrak{g}$ defines a vector field on $\mathfrak{g}^*$. By abuse of notation, we will still call this vector field Y. One computes easily that

$$Y(\lambda) = \sum_j \lambda([Y, X_j)])\frac{\partial}{\partial\lambda_j}.$$

Using this formula, we can rewrite (2.8)(a) as

$$\begin{aligned}\{f,g\} &= \sum_i \frac{\partial f}{\partial \lambda_i}(X_i \cdot g) \\ &= -\sum_i (X_i \cdot f)\frac{\partial g}{\partial \lambda_i}.\end{aligned} \tag{2.8)(b}$$

Here the action of X_i is the coadjoint action.

Suppose G is a Lie group with Lie algebra $\mathfrak{g}$. The conclusion we want to draw from (2.8)(b) is this. Suppose $S \subset \mathfrak{g}^*$ is preserved by the coadjoint action of G. Let g be any smooth function on $\mathfrak{g}^*$ vanishing on S, and f any smooth function. Then we claim that the Poisson bracket $\{f,g\}$ also vanishes on S. The reason is that $X_i \cdot g$ may be computed at λ by differentiating g along the path $\mathrm{Ad}^*(\exp(tX_i))\cdot\lambda$. If λ belongs to S, then this path is contained in S; so g vanishes on the path, and the derivative is zero. The first formula in (2.8)(b) therefore shows that $\{f,g\}$ vanishes at λ.

What follows is that the Poisson bracket on $C^\infty(\mathfrak{g}^*)$ descends to the algebra of restrictions to S of smooth functions on $\mathfrak{g}^*$. Here is a formal setting for this fact.

Definition 2.9. Suppose A is a Poisson algebra. An ideal $J \subset A$ is called a *Poisson ideal* if $\{A, J\} \subset J$. In this case the quotient algebra A/J inherits a Poisson algebra structure.

There is a minor technical point about restricting smooth functions that should be addressed as well.

Definition 2.10. Suppose M is a smooth manifold and $S \subset M$. A function f on S is said to be *smooth* if for every $s \in S$ there is a neighborhood U_s of s in M and a smooth function $f_s \in C^\infty(U_s)$ with the property that $f_s|_{U_s \cap S} = f|_{U_s \cap S}$. In this way S becomes a ringed space: the value of the sheaf on an open set $V \subset S$ is $C^\infty(V)$

If S is closed then a partition of unity argument shows that $C^\infty(S) = C^\infty(M)|_S$.

Proposition 2.11. *Suppose G is a Lie group, $S \subset \mathfrak{g}^*$ is an $\mathrm{Ad}^*(G)$-stable subset, and $U \subset \mathfrak{g}^*$ is open. Then the ideal $J_S(U)$ of smooth functions on U vanishing on $S \cap U$ is a Poisson ideal in $C^\infty(U)$. Consequently $C^\infty(S)$ is a Poisson algebra; and the ringed space (S, C^∞) is a Poisson space.*

Suppose now that G is an algebraic group (over $\mathbb{R}$ or $\mathbb{C}$) and $X \subset \mathfrak{g}^*$ is an $\mathrm{Ad}^*(G)$-stable subvariety. Then X is an affine Poisson algebraic variety, with algebra of functions

$$R(X) = S(\mathfrak{g})/(J_X \cap S(\mathfrak{g})).$$

This is clear from the discussion after (2.8)(b).

Fix $\lambda \in \mathfrak{g}^*$. Using (2.8)(b) it is easy to check that the space S_λ defined in (2.6)(g) is just the set of values at λ of the coadjoint action vector fields. This is precisely the tangent space to the orbit through λ:

$$S_\lambda = \{\mathrm{ad}^*(Y) \mid Y \in \mathfrak{g}\} = T_\lambda(G \cdot \lambda). \tag{2.12)(a}$$

The formula (2.8)(b) also shows that if Y is regarded as a (linear) function on $\mathfrak{g}^*$, then the derivation (Poisson bracket with Y) is just the coadjoint action of Y:

$$\xi_Y = \mathrm{ad}^*(Y). \tag{2.12)(b}$$

(This is clearest for the basis vectors X_i, but it follows at once for all $Y \in \mathfrak{g}$.)

Corollary 2.13. *Suppose G is a Lie group, and X is an orbit of the coadjoint action of G on $\mathfrak{g}^*$. Then the Poisson structure on X provided by Proposition 2.11 is actually a symplectic structure. At a point $\lambda \in X$, the symplectic form on $T_\lambda(X)$ is given by*

$$\omega_\lambda(\xi_Y, \xi_Z) = \lambda([Y, Z]) \qquad (Y, Z \in \mathfrak{g}).$$

Sketch of proof. We give X its topology as a homogeneous space for G, to make it a manifold. In order to apply Proposition 2.11 we need to assume that this agrees with the subspace topology from $\mathfrak{g}^*$. This is automatic if G is reductive, so we omit a discussion of the minor modification of Proposition 2.11 needed for the general case. At any rate, we find on X the structure of a Poisson manifold. By the construction in Proposition 2.11, the Hamiltonian vector fields on X are just the restrictions to X of Hamiltonian vector fields on $\mathfrak{g}^*$. By (2.12), these span all the tangent spaces to X. By Proposition 2.7, it follows that X is symplectic. To compute the form explicitly, we use the ideas in (2.6). They show that

$$\omega_x(\xi_f(x), \xi_g(x)) = \{f, g\}(x) \tag{2.14)(a}$$

whenever f and g belong to the maximal ideal $\mathfrak{m}_x$ of functions vanishing at x. Poisson bracket with a constant is zero; so both sides of (2.14) are unchanged if we add constants to f and g. Therefore (2.14) is true for all f and g. In the case of a coadjoint orbit, we take Y and Z to be linear functions on $\mathfrak{g}^*$ corresponding to elements of $\mathfrak{g}$; then the conclusion is that

$$\omega_\lambda(\xi_Y, \xi_Z) = \{Y, Z\}(\lambda). \tag{2.14)(b}$$

Finally, notice that the Poisson bracket of Y and Z is just the linear functional $[Y, Z]$:

$$\{Y, Z\} = [Y, Z] \tag{2.14)(c}$$

This is clear from (2.7)(a) if Y and Z belong to the basis of $\mathfrak{g}$, and the general case is immediate from bilinearity. Now the formula for ω_λ in the proposition follows from (2.14)(b) and (2.14)(c). Q.E.D.

3. Hamiltonian G-spaces and moment maps

In this section we recall a little of the general theory of Poisson spaces with Lie group actions. Since Lie groups are assembled from one-parameter subgroups, it is helpful to begin by examining one-parameter groups of automorphisms. We are mostly interested in motivating the definition of Hamiltonian actions, so there is no need to strive for maximum generality. We begin therefore with a Poisson manifold X and a one-parameter group $\{U_t | t \in \mathbb{R}\}$ of automorphisms of X. This means first of all that each U_t is a smooth automorphism of X, and that $U_t U_s = U_{t+s}$. We can make U_t an algebra automorphism of $C^\infty(X)$ by $(U_t f)(x) = f(U_{-t}x)$; then the last requirement is that these automorphisms preserve the Poisson bracket.

We now have a homomorphism $\mathbb{R} \to \mathrm{End}(C^\infty(X))$. A sensible requirement to impose on the original automorphisms U_t is that this should be differentiable. This is equivalent to the differentiability of the map

$$\mathbb{R} \times X \to X, \qquad (t, x) \mapsto U_t(x).$$

The derivative at 0 of U_t is an endomorphism $\xi(U)$ of $C^\infty(X)$: the defining relation is

$$\frac{d}{dt}(U_t f)|_{t=0} = \xi(U)(f). \tag{3.1)(a}$$

By differentiating the requirement that U_t is an algebra automorphism, we find that $\xi(U)$ is a derivation:

$$\xi(U)(fg) = \xi(U)(f)g + f\xi(U)(g). \tag{3.1)(b}$$

This means that $\xi(U)$ is a vector field on X. Differentiating the requirement that U_t preserve Poisson brackets shows that $\xi(U)$ is a derivation of the Poisson bracket:

$$\xi(U)(\{f, g\}) = \{\xi(U)(f), g\} + \{f, \xi(U)(g)\}. \tag{3.1)(c}$$

We can recover U_t from $\xi(U)$ by solving the differential equation

$$\frac{d}{dt}(U_t(f)) = \xi(U)(U_t(f)), \qquad U_0(f) = f. \tag{3.1)(d}$$

This follows immediately from (3.1)(a) and the composition law for U_t.) Conversely, suppose $\xi(U)$ is an endomorphism of $C^\infty(X)$ satisfying (3.1)(b) and (3.1)(c), and U_t is a family of endomorphisms of $C^\infty(X)$ satisfying (3.1)(d). Then U_t is a one-parameter group of Poisson algebra automorphisms of $C^\infty(X)$. This identifies smooth one-parameter groups U of automorphisms of the Poisson algebra $C^\infty(X)$ with certain vector fields $\xi(U)$ satisfying (3.1)(c). (Not all vector fields appear, because the differential equation (3.1)(d) may not have a solution for all t.) But it certainly suggests that we should look for interesting vector fields satisfying (3.1)(c).

If $f \in C^\infty(X)$, then the Hamiltonian vector field ξ_f of Definition 2.4 satisfies (3.1)(c). If there is a corresponding one-parameter group $U(f)_t$ of automorphisms of X, then the automorphisms must preserve each leaf of the symplectic foliation (of an open subset of X) discussed in (2.6). More general Poisson automorphisms may permute these leaves. It is natural to regard the automorphisms attached to Hamiltonian vector fields as "inner." Here is a definition.

Definition 3.2. Suppose X is a Poisson manifold (Definition 2.4) endowed with a smooth action of a Lie group G by Poisson automorphisms. We say that the action is *Hamiltonian* (or that X is a *Hamiltonian G-space*) if we are given a linear map

$$\tilde{\mu}: \mathfrak{g} \to C^\infty(X)$$

with the following properties.

a) The map $\tilde{\mu}$ intertwines the adjoint action of G with its action on $C^\infty(X)$.
b) For every $Y \in \mathfrak{g}$, write Y_t for the one-parameter group of automorphisms of X given by the action of $\exp(tY)$. Then the corresponding vector field $\xi(Y)$ on X (cf. (3.1)) is the Hamiltonian vector field attached to the function $\tilde{\mu}(Y)$ on X.

The assumption is slightly stronger than the requirement that each one-parameter subgroup in G be generated by a Hamiltonian vector field. By differentiating the requirement in (a) and using (b), we find that $\tilde{\mu}$ must be a Lie algebra homomorphism.

A linear map from a finite-dimensional vector space into $C^\infty(X)$ is precisely the same thing as a smooth map from X to the dual vector space. In the setting of Definition 3.2, we define the *moment map* for the Hamiltonian G-space to be

$$\mu: X \to \mathfrak{g}^*, \qquad \mu(x)(Y) = \tilde{\mu}(Y)(x) \qquad (Y \in \mathfrak{g}, x \in X). \tag{3.3)(a}$$

The same formula shows how to define $\tilde{\mu}$ in terms of μ. The first condition in Definition 3.2 may be immediately reformulated in terms of μ: it just says

$$\mu \text{ intertwines the action of } G \text{ on } X \text{ with the coadjoint action on } \mathfrak{g}^*. \tag{3.3)(b}$$

For the second condition, recall that the smooth map μ gives a pullback map on smooth functions:

$$\mu^*: C^\infty(\mathfrak{g}^*) \to C^\infty(X).$$

We know that $\tilde{\mu}$ is a Lie algebra homomorphism. This turns out to be equivalent to the requirement that μ^* preserve Poisson brackets:

$$\{\mu^* f, \mu^* g\} = \mu^*\{f, g\} \qquad (f, g \in C^\infty(\mathfrak{g}^*)) \tag{3.3)(c}$$

(That (3.3)(c) implies that $\tilde{\mu}$ is a Lie algebra homomorphism is obvious from (2.14)(c). For the other direction, we try to prove (3.3)(c) at a point $x \in X$, say with $\mu(x) = \lambda$. We can find "Taylor" expansions for f and g at λ, as

$$f = a + \sum_i a_i(X_i - \lambda(X_i)) + \sum_{i \leq j}(X_i - \lambda(X_i))(X_j - \lambda(X_j))f_{ij};$$

here a and a_i are constants, and f_{ij} are smooth functions. Of course there is a similar expansion for g, and we can pull them all back to X by μ^*. Now use these expansions to compute Poisson brackets at x. The constants like a and $\lambda(X_i)$ drop out. The linear terms satisfy

$$\{\mu^*(X_i), \mu^*(X_j)\} = \mu^*\{X_i, X_j\}$$

because $\mu^*(X_i) = \tilde{\mu}(X_i)$, and we are assuming that $\tilde{\mu}$ is a Lie algebra homomorphism. The quadratic terms belong to the square of the maximal ideal of functions vanishing at x, and so don't contribute to the Poisson bracket at x (compare (2.6)(c)). The conclusion is that (3.3)(c) holds at the point x.) Finally, we can obviously write condition (b) of Definition 3.2 as

$$\text{for every } Y \in \mathfrak{g}, \text{ the vector field on } X \text{ induced by } Y \text{ is the Hamiltonian vector field } xi_{\mu^*Y}. \tag{3.3)(d}$$

We have shown that Definition 3.2 is equivalent to

Definition 3.4. Suppose X is a Poisson manifold endowed with a smooth action of a Lie group G by Poisson automorphisms. We say that the action is *Hamiltonian* if we are given a smooth G-equivariant Poisson mapping (the *moment map*)

$$\mu: X \to \mathfrak{g}^*$$

with the following property: for every $Y \in \mathfrak{g}$, write $\xi(Y)$ for the vector field on X induced by the one-parameter group of automorphisms given by the action of $\exp(tY)$. Then

$$\xi(Y) = \xi_{\mu^*Y}.$$

Here on the right we regard Y as a linear functional on $\mathfrak{g}^*$, pull it back to a smooth function on X via μ, and then form the corresponding Hamiltonian vector field.

In light of (3.1)(d), this condition on μ may be reformulated as follows: for every smooth function f on X, we have

$$\frac{d}{dt}(f(\exp(-tY) \cdot x))|_{t=0} = \{\mu^*Y, f\}.$$

The notion of Hamiltonian G-space extends readily to some singular spaces. The best general theory of such spaces is in the algebraic setting, so we begin with that. Suppose X is a complex affine Poisson algebraic variety, with $R(X)$ its ring of regular functions. (This means that $R(X)$ is a complex Poisson algebra that is finitely generated as a commutative algebra.) If G is an algebraic group, then an algebraic action of G on X is just an algebraic action of G on $R(X)$ by Poisson algebra automorphisms:

$$G \times R(X) \to R(X). \tag{3.5)(a}$$

To say that this action is algebraic means that every element of $R(X)$ belongs to a finite-dimensional G-invariant subspace $V \subset R(X)$, and that the corresponding homomorphism from G to $GL(V)$ is algebraic. Such a homomorphism has a differential, which is a Lie algebra homomorphism from $\mathfrak{g}$ to $\mathrm{End}(V)$. (The Lie algebra structure on $\mathrm{End}(V)$ is commutator of linear transformations.) The action of G on $R(X)$ may therefore be differentiated to get a Lie algebra homomorphism

$$\xi: \mathfrak{g} \to \mathrm{End}(R(X)). \tag{3.5)(b}$$

Because G acts by algebra automorphisms, ξ acts by derivations:

$$\xi(Y)(fg) = (\xi(Y)f)g + f(\xi(Y)g) \tag{3.5)(c}$$

Similarly, ξ acts by Poisson algebra derivations:

$$\xi(Y)(\{f,g\}) = \{\xi(Y)f, g\} + \{f, \xi(Y)g\}. \tag{3.5)(d}$$

Definition 3.6. Suppose X is a complex affine Poisson algebraic variety, and G is an algebraic group acting algebraically on X by Poisson automorphisms. Write

$$\xi: \mathfrak{g} \to \mathrm{End}(R(X))$$

for the differential of this action. We say that the action is *Hamiltonian* if we are given a linear map

$$\tilde{\mu}: \mathfrak{g} \to R(X)$$

with the following properties.

a) The map $\tilde{\mu}$ intertwines the adjoint action of G with its action on $R(X)$.
b) For every $Y \in \mathfrak{g}$, the endomorphism $\xi(Y)$ of $R(X)$ is the Hamiltonian vector field attached to $\tilde{\mu}(Y)$:

$$\xi(Y) = \xi_{\tilde{\mu}(Y)}.$$

Just as in the setting of manifolds, this definition may conveniently be recast in terms of moment maps.

Definition 3.7. Suppose X is a complex affine Poisson algebraic variety, and G is an algebraic group acting algebraically on X by Poisson automorphisms. We say that the action is *Hamiltonian* if we are given an algebraic G-equivariant Poisson mapping (the *moment map*)

$$\mu: X \to \mathfrak{g}^*$$

with the following property: the differential ξ of the G action on X is given by

$$\xi(Y) = \xi_{\mu^* Y}.$$

Here on the right we regard Y as a linear function on $\mathfrak{g}^*$, pull it back to a regular function on X via μ, then form the corresponding Hamiltonian vector field.

Similarly we can define Hamiltonian actions on general algebraic varieties. To define a Hamiltonian action on a general Poisson space, we need to know how to differentiate the group action on the sheaf of rings, in order to write a condition like (b) in Definition 3.6. As usual we avoid trying to write the most general natural definition, since we do not know what that might be.

In the philosophy of geometric quantization, Hamiltonian G-spaces are the (crudest) classical analogues of unitary representations of G. Corresponding to irreducible representations are the homogeneous Hamiltonian G-spaces. These were classified by Kostant.

Proposition 3.8 ([11], Theorem 5.4.1.) *Suppose X is a homogeneous Hamiltonian Poisson G-manifold (Definition 3.4). Then the moment map μ exhibits X as a covering of a coadjoint orbit $G \cdot f \subset \mathfrak{g}^*$; so X is actually a symplectic manifold.*

Proof. Fix $x \in X$, and write G_x for the isotropy group at x; so $X \simeq G/G_x$. Define $f = \mu(x) \in \mathfrak{g}^*$, and write G_f for the isotropy group; then $G_x \subset G_f$ since μ is G-equivariant. It remains to show that G_x is open in G_f; that is, that the two groups have the same Lie algebra. For $Y \in \mathfrak{g}$, we regard Y as a linear function on $\mathfrak{g}^*$, and pull it back to a smooth function $\mu^* Y$ on X. By the definition of Hamiltonian action, this map from $\mathfrak{g}$ to $C^\infty(X)$ is a Lie algebra homomorphism. Exactly as in the proof of Corollary 2.13, we may therefore compute

$$\begin{aligned}\omega_x(\xi_{\mu^* Y}, \xi_{\mu^* Z}) &= \mu^*([Y, Z])(x)\\ &= [Y, Z](\mu(x)) = [Y, Z](f)\\ &= \omega_f(\xi_Y, \xi_Z).\end{aligned}$$

Suppose now that $Y \in \mathfrak{g}_f$. Then $\xi_Y(f)$ belongs to the radical of the form ω_f, so the preceding equation implies that $\xi_{\mu^* Y}(x)$ belongs to the radical of ω_x. Because ω_x is non-degenerate on the values at x of Hamiltonian vector fields (see the discussion after (2.6)), it follows that $\xi_{\mu^* Y}(x) = 0$. By Definition 3.2(b), it follows that the vector field $\xi(Y)$ vanishes at x; that is, that $Y \in \mathfrak{g}_x$. The other containment $\mathfrak{g}_x \subset \mathfrak{g}_f$ is clear. Q.E.D.

Here is a standard way to construct Hamiltonian G-spaces. Suppose M is a manifold with a smooth action of G. This means in particular that we are given a Lie algebra homomorphism ξ from $\mathfrak{g}$ to vector fields on M. Now let $X = T^*M$ be the cotangent bundle of M with its standard symplectic structure. The action of G by diffeomorphisms of M automatically lifts to an action of G by symplectomorphisms of X. Now each vector field τ on M

may be identified with a smooth function f_τ on X; the value of f_τ at the cotangent vector (m, v) is given by evaluating the tangent vector $\tau(m)$ on the covector v. This mapping sends Lie bracket of vector fields to Poisson bracket of functions on X. The map $\tilde{\mu}$ for the Hamiltonian G-space X (Definition 3.2) sends $Y \in \mathfrak{g}$ to the function $f_{\xi(Y)}$. One can check that this map makes X a Hamiltonian G-space; of course the difficult part is condition (b) of Definition 3.2. The corresponding moment map μ sends the cotangent vector (m, v) to the linear functional $Y \mapsto \langle \xi(Y)(m), v\rangle$ on $\mathfrak{g}$. Here the pairing on the right is between tangent vectors and covectors at m.

4. Lagrangian subspaces

A central notion in symplectic geometry is that of a Lagrangian submanifold. In this section we will consider how to extend that notion to Poisson spaces. We begin with the simplest linear setting, and gradually generalize it.

Definition 4.1. Suppose F is a field, W is a finite-dimensional F-vector space, and ω is a non-degenerate skew-symmetric bilinear form on W. (We say that the pair (W, ω) is a *symplectic vector space over* F.) The *symplectic group of* W is the group $Sp(W)$ of F-linear transformations of W preserving ω. In analogy with (2.2), we define a linear isomorphism

$$\tau: W \to W^*, \qquad \tau(v)(w) = \omega(w, v). \tag{4.1)(a}$$

Suppose $V \subset W$ is any subspace. Set

$$V^\perp = \{w \in W \mid \omega(w, V) = 0\}, \tag{4.1)(b}$$

the preimage under τ of the annihilator of V in W^*. If we need to emphasize the dependence on ω, we may write $V^{\perp,\omega}$. Then τ factors to a linear isomorphism

$$\tau_V: W/V^\perp \to V^*. \tag{4.1)(c}$$

From this it follows that

$$\dim V + \dim V^\perp = \dim W. \tag{4.1)(d}$$

Evidently $(V^\perp)^\perp \supset V$; because (4.1)(d) shows that these spaces have the same dimension, we get

$$(V^\perp)^\perp = V. \tag{4.1)(e}$$

Dually, the restriction of τ to V provides an isomorphism

$$\delta_V: V \to (W/V^\perp)^*. \tag{4.1)(f}$$

This map is the transpose of τ_V.

We say that a subspace $I \subset W$ is *isotropic* if $I \subset I^\perp$; that is, if $\omega|_I = 0$. We say C is *co-isotropic* if $C^\perp \subset C$; that is, if $C^\perp$ is isotropic. We say L is *Lagrangian* if it is both isotropic and co-isotropic; that is, if $L^\perp = L$.

Definition 4.2. Suppose (W, ω) is a symplectic vector space over F of dimension $2n$. The *Lagrangian Grassmannian* is the collection $\mathcal{B}(W)$ of Lagrangian subspaces of W; it is a subset of the Grassmannian of all n-dimensional subspaces L of W, defined by the algebraic condition $\omega|_L = 0$.

Fix a Lagrangian subspace L of W. The *Siegel parabolic subgroup defined by* L is the isotropy group

$$P(L) = \{g \in Sp(W) \mid gL \subset L\}. \tag{4.2)(a}$$

Its *unipotent radical* is the normal subgroup

$$U(L) = \{g \in P(L) \mid g|_L = \mathrm{Id}_L\}. \tag{4.2)(b}$$

Obviously restriction to L defines an inclusion

$$\rho(L)\colon P(L)/U(L) \hookrightarrow GL(L). \tag{4.2)(c}$$

We will see in Proposition 4.4 that $\rho(L)$ is an isomorphism. We define the *determinant character* of $P(L)$ by

$$\chi(L)\colon P(L) \to F^\times, \qquad \chi(L)(p) = \det(p|_L) = \det(\rho(L)(p)). \tag{4.2)(d}$$

Recall from (4.1)(f) the isomorphism $\delta_L\colon L \to (W/L)^*$. If $u \in U(L)$, $w \in W$, and $v \in L$, then $\omega((u-1)w, v) = \omega(w, (u^{-1}-1)v) = \omega(w, 0) = 0$. Consequently $(u-1)w \in L^\perp = L$, so $u - 1 \in \mathrm{Hom}_F(W/L, L)$. In light of the isomorphism δ_L, this gives

$$u - 1 \in \mathrm{Hom}_F(W/L, L) \simeq \mathrm{Hom}_F(W/L, (W/L)^*). \tag{4.2)(e}$$

That is, $u - 1$ gives rise to a bilinear form on the vector space W/L. By inspection of the definition of δ_L, the form is

$$B_u(v + L, w + L) = \omega(w, (u-1)v). \tag{4.2)(f}$$

Proposition 4.3. *Suppose (W, ω) is a finite-dimensional symplectic vector space over F, L is a Lagrangian subspace, and $u \in U(L)$ (cf. (4.2)(b)). Then the bilinear form B_u on W/L defined by (4.2)(f) is symmetric.*

Conversely, suppose B is any symmetric bilinear form on W/L. Identify B with a linear map

$$T_B \in \mathrm{Hom}_F(W/L, (W/L)^*), \qquad T_B(v+L)(w+L) = B(v+L, w+L);$$

and further identify T_B as an element of

$$\mathrm{Hom}_F(W/L, L) \subset \mathrm{Hom}_F(W, W)$$

using (4.2)(e). Explicitly, T_B is characterized as an endomorphism of W by the property

$$\omega(w, T_B v) = B(v + L, w + L) \qquad (v, w \in W).$$

Then $u_B = 1 + T_B$ belongs to the subgroup $U(L)$ of $Sp(W)$, and the corresponding bilinear form is B.

In this way $U(L)$ is naturally isomorphic with the (additive) group of symmetric bilinear forms on W/L.

Proof. To see that the form B_u is symmetric, we use (4.2)(f) and the fact that u preserves ω to compute

$$\begin{aligned} B_u(v + L, w + L) &= \omega(w, (u-1)v) = -\omega(w, v) + \omega(w, uv) \\ &= -\omega(uw, uv) + \omega(w, uv) \\ &= -\omega((u-1)w, uv) = -\omega(uv, (u-1)w). \end{aligned}$$

Now (4.2)(e) says that $(u-1)w \in L$, and that $uv - v \in L$. Since ω is zero on L, we can replace uv by v in the last formula without affecting the value. It is then precisely $B_u(w + L, v + L)$, as we wished to show.

For the converse, the last formula certainly defines the endomorphism T_B of W; T_B is zero on L and carries W into L. The symmetry of B implies that $\omega(w, T_B v) = \omega(v, T_B w) = -\omega(T_B w, v)$. Since T_B takes values in L, where ω vanishes, we also have $\omega(T_B w, T_B v) = 0$. Together these properties imply that $\omega((1 + T_B)w, (1 + T_B)v) = \omega(w, v)$, and therefore that $u = 1 + T_B \in Sp(W)$. Since T_B annihilates L, u acts trivially there; so $u \in U(L)$. The remaining assertions are easy. Q.E.D.

Proposition 4.4. *Suppose (W, ω) is a finite-dimensional symplectic vector space over F. Then the Lagrangian Grassmannian $\mathcal{B}(W)$ is a homogeneous space for $Sp(W)$. The isotropy group at a Lagrangian subspace L is the Siegel parabolic subgroup $P(L)$ (Definition 4.2). The map $\rho(L)$ of (4.2)(c) is surjective. More precisely, we can find a second Lagrangian subspace $L' \subset W$ so that $L \cap L' = 0$. In this case the intersection of the two Siegel parabolics is naturally isomorphic to $GL(L)$, and the isomorphism is implemented by restriction to L.*

Proof. It is convenient to begin near the end and work backwards. So fix L; we seek a second Lagrangian subspace L' with $L \cap L' = 0$. Here is a way to construct one. First choose a basis $\{p_1, \ldots, p_n\}$ of L. We want to choose elements $\{q_1, \ldots, q_n\}$ of W so that

$$\omega(p_i, q_j) = \delta_{i,j} \tag{4.5)(a}$$

and

$$\omega(q_i, q_j) = 0, \qquad i \leq j. \tag{4.5)(b}$$

We will do this by induction on j; that is, we suppose that $q_1, \ldots, q_{j-1}$ have been chosen satisfying (4.5), and we try to choose q_j. Now (4.1)(f) guarantees that (4.5)(a) will be satisfied for some element q_j' of W; in fact it says that q_j' is uniquely determined modulo L. That means that we may modify q_j' by an element of L without affecting (4.5)(a). We therefore define

$$q_j = q_j' + \sum_{i<j} \omega(q_i, q_j') p_i. \tag{4.5)(c}$$

Then it is easy to see that (4.5)(b) is satisfied for j.

Define L' to be the span of $\{q_1, \ldots, q_n\}$. By (4.5)(b), L' is isotropic; so it is Lagrangian by dimension. It is clear from (4.5)(a) that τ_L (Definition 4.1) maps L' isomorphically onto L^*; that is, that the restriction of ω to $L \times L'$ defines

$$L' \simeq L^* \tag{4.5)(d}$$

Because the pairing of L with itself is zero, it follows that $L \cap L' = 0$. By dimension counting, we get

$$W \simeq L \oplus L' \simeq L \oplus L^*. \tag{4.5)(e}$$

In this last picture, the symplectic form on W is given by

$$\omega((v, \lambda), (v', \lambda')) = \lambda'(v) - \lambda(v') \qquad (v, v' \in L, \lambda, \lambda' \in L^*) \tag{4.5)(f}$$

We want to compute $P(L) \cap P(L')$. The stabilizer of L and L' in $GL(W)$ is $GL(L) \times GL(L')$. Using (4.5)(d), this can be written as $GL(L) \times GL(L^*)$. The question is which elements of the product preserve the symplectic form ω. Using (4.5)(f), we see immediately that if $T \in GL(L)$ and $S \in GL(L^*)$, then (T, S) preserves ω if and only if

$$(S\lambda')(Tv) = \lambda'(v) \qquad (\lambda' \in L^*, v \in V).$$

This in turn means that S must be the inverse of the transpose of T. That is,

$$P(L) \cap P(L') = \{(T, {}^tT^{-1}) \mid T \in GL(L)\}. \tag{4.5)(g}$$

This proves the last claim in the proposition, and also the surjectivity of (4.2)(c).

Finally, we must show that $\mathcal{B}(W)$ is homogeneous for $Sp(W)$. Suppose L_1 and L_2 are any two Lagrangian subspaces. For each of them choose a Lagrangian complement L_i' as in (4.5), so that we have natural isomorphisms

$$L_i' \simeq L_i^*$$

and so on as above. Let T be any linear isomorphism from L_1 onto L_2; such a T exists since L_1 and L_2 have the same dimension. Let S be the induced (inverse transpose) isomorphism from L_1^* to L_2^*; equivalently, from L_1' to L_2'. Then (T, S) defines a linear isomorphism from $W = L_1 \oplus L_1'$ onto $W = L_2 \oplus L_2'$; that is, an element $g \in GL(W)$. Using the descriptions (4.5)(f) for ω, we see that $g \in Sp(W)$. By construction $g \cdot L_1 = L_2$, as we wished to show. Q.E.D.

Corollary 4.6. *Suppose (W, ω) is a $2n$-dimensional symplectic vector space over F, and $L \subset W$ is Lagrangian. Then the Siegel parabolic $P(L)$ acts transitively on the set*

$$\mathcal{B}(W)_{0,L} = \{L' \in \mathcal{B}(W) \mid L' \cap L = 0\}$$

of Lagrangian complements to L. The stabilizer in $P(L)$ of one such Lagrangian L' is a Levi subgroup $GL(L)$ of $P(L)$; this is a complement for the normal subgroup $U(L)$ of Definition 4.2. Consequently $U(L)$ acts simply transitively on $\mathcal{B}(W)_{0,L}$, which is therefore (algebraically) isomorphic to an F vector space of dimension $n(n+1)/2$.

Proof. Suppose L_1' and L_2' are Lagrangian complements to L. In the argument at the end of the proof of Proposition 4.4, take $L_1 = L_2 = L$; then the element $g \in Sp(W)$ constructed there preserves L, and carries L_1 to L_2. It is even clear that we can choose g to be the identity on L; that is, $g \in U(L)$. The remaining assertions are now clear from Propositions 4.4 and 4.3. Q.E.D.

Although we will make no use of it, we mention in passing a description of the other orbits of $P(L)$ on $\mathcal{B}(W)$.

Proposition 4.7. *In the setting of Corollary 4.6, define*

$$\mathcal{B}(W)_{r,L} = \{L' \in \mathcal{B}(W) \mid \dim(L \cap L') = r\}.$$

Similarly, for each subspace S of L, define

$$\mathcal{B}(W)_{S,L} = \{L' \in \mathcal{B}(W) \mid L \cap L' = S\}.$$

a) For S and L as above, the quotient $W_S = S^\perp/S$ inherits from W a natural non-degenerate symplectic form. The subspace $L_S = L/S$ of W_S is Lagrangian. If S has dimension r, then W_S has dimension $2(n-r)$.

b) Each $L' \in \mathcal{B}(W)_{S,L}$ defines a Lagrangian complement $L_S' = L'/S$ for L_S in W_S. This correspondence provides an algebraic isomorphism

$$\mathcal{B}(W)_{S,L} \simeq \mathcal{B}(W_S)_{0,L_S}.$$

c) The set $\mathcal{B}(W)_{S,L}$ is an orbit of $U(L)$. It is algebraically isomorphic to an F vector space of dimension $(n-r)(n-r+1)/2$.

d) The set $\mathcal{B}(W)_{r,L}$ is an orbit of $P(L)$. It is algebraically isomorphic to a vector bundle of dimension $(n-r)(n-r+1)/2$ over the Grassmannian of r-dimensional subspaces of L. The base space is a projective algebraic variety of dimension $r(n-r)$, so $\mathcal{B}(W)_{r,L}$ has dimension $(n-r)(n+r+1)/2$. It is of codimension $(r^2+r)/2$ in $\mathcal{B}(W)$.

Because we will not use the result, we leave the elementary proof to the reader.

The preceding linear algebra can be thought of as infinitesimal symplectic geometry; we will apply it to tangent and cotangent spaces of symplectic manifolds (or algebraic varieties). In order to treat simultaneously the case of Poisson manifolds, we need to weaken the assumptions in Definition 4.1. The most obvious way to do that is to drop the assumption that ω is non-degenerate. Perhaps surprisingly, it is also important to examine the case when ω is non-degenerate, but is defined only on a subspace. The two possibilities are related by duality (of vector spaces). We begin with the second.

Definition 4.8. Suppose F is a field. A *degenerate symplectic vector space over F* is a triple (V, W, ω) subject to the following conditions.

a) The space V is a finite-dimensional F-vector space, and W is a subspace.
b) The form ω is a non-degenerate skew-symmetric bilinear form on W.

The *symplectic group of V* is the group $Sp(V)$ of F-linear automorphisms of V preserving the subspace W and the form ω. It maps (by restriction of linear transformations from V to W) surjectively to $Sp(W)$.

Suppose $S \subset V$ is any subspace. Set

$$S^{\perp} = \{w \in W \mid \omega(w, S \cap W) = 0\}.$$

Notice that $S^{\perp} \subset W$. We say that a subspace $I \subset V$ is *isotropic* if $I \cap W \subset I^{\perp}$; that is, if $I \cap W$ is isotropic in W in the sense of Definition 4.1. We say that C is *co-isotropic* if $C^{\perp} \subset C$; that is, if $C \cap W$ is a co-isotropic subspace of W in the sense of Definition 4.1. We say L is *Lagrangian* if it is both isotropic and co-isotropic; that is, if $L \cap W$ is a Lagrangian subspace of W in the sense of Definition 4.1.

The definition of isotropic is not entirely an obvious one; it might seem natural to consider instead the condition $I \subset I^{\perp}$, meaning that I is an isotropic subspace of W. We might call this *strongly isotropic*, since it is more restrictive than the condition in Definition 4.8. The definition of Lagrangian would then change as well; a *strongly Lagrangian* subspace of V is a Lagrangian subspace of W. But the definition we have given seems well suited to representation theory.

Definition 4.9. Suppose F is a field. A *degenerate cosymplectic vector space over F* is a pair (U, η) subject to the following conditions.

a) The space U is a finite-dimensional F-vector space.
b) The form η is a (possibly degenerate) skew-symmetric bilinear form on U.

The *symplectic group of U* is the group $Sp(U)$ of F-linear automorphisms of U preserving the form η. Define $R \subset U$ to be the radical of η:

$$R = \{u \in U \mid \eta(u, U) = 0\}.$$

Then η defines a non-degenerate symplectic form on the quotient space $W = U/R$. Every element of $Sp(U)$ preserves the subspace R, and so we get a quotient map

$$Sp(U) \to Sp(W).$$

This map is surjective.

Suppose $T \subset U$ is any subspace. Set

$$T^\perp = \{u \in U \mid \eta(u, T) = 0\}.$$

Notice that $T^\perp \supset R$. We say that a subspace $I \subset U$ is *isotropic* if $I \subset I^\perp$; that is, if $I/(I \cap R)$ is isotropic in U/R in the sense of Definition 4.1. We say that C is *co-isotropic* if $C^\perp/R \subset C/(R \cap C)$; that is, if $C/(R \cap C)$ is a co-isotropic subspace of U/R in the sense of Definition 4.1. We say L is *Lagrangian* if it is both isotropic and co-isotropic; that is, if $L/(R \cap L)$ is a Lagrangian subspace of U/R in the sense of Definition 4.1.

Again we could have changed the definition of co-isotropic by requiring $C^\perp \subset C$; that is, that $C \supset R$, and C/R be co-isotropic in U/R. We could call this requirement *strongly co-isotropic*, and get a corresponding notion of *strongly Lagrangian*.

We have said that Definitions 4.8 and 4.9 differ by duality of vector spaces. Here is a more precise formulation. Suppose (V, W, ω) is a degenerate symplectic vector space. Define $U = V^*$ to be the dual vector space. There is an order-reversing bijection from subspaces of V to subspaces of U, sending a subspace $S \subset V$ to

$$S^\perp = \{\lambda \in V^* \mid \lambda(S) = 0\}. \tag{4.10)(a}$$

Then restriction of linear functionals to S defines an isomorphism

$$S^* \simeq U/S^\perp. \tag{4.10)(b}$$

Now let $R = W^\perp$ be the subspace of U corresponding to $W \subset V$, so that

$$W^* \simeq U/R. \tag{4.10)(c}$$

The non-degenerate symplectic form ω on W defines via the duality isomorphism $\tau_W : W \to W^*$ (cf. (4.1)(a)) a non-degenerate symplectic form η on U/R. We regard η as a symplectic form on U with radical R. Then (U, η) is a degenerate cosymplectic vector space. It is easy to construct the inverse correspondence, and we leave that to the reader. The duality correspondence $S \mapsto S^\perp$ carries isotropic to co-isotropic, co-isotropic to isotropic, and Lagrangian to Lagrangian. Taking inverse transpose defines a natural isomorphism $GL(V) \simeq GL(U)$; this isomorphism restricts to an isomorphism of $Sp(V)$ onto $Sp(U)$.

We can now explain the sense in which degenerate symplectic and cosymplectic vector spaces are infinitesimal versions of Poisson spaces. To fix ideas we discuss affine Poisson algebraic varieties; as usual it is a simple matter to modify the discussion for other nice Poisson spaces.

Proposition 4.11. *Suppose X is an affine Poisson algebraic variety over F, with ring of regular functions $R(X)$. Suppose $x \in X$ is a closed point; that is, a maximal ideal $\mathfrak{m}_x \subset R(X)$. Set $F_x = R(X)/\mathfrak{m}_x$, a finite extension field of F.*

a) The bilinear form $\{,\}_x$ of (2.6)(d) makes the Zariski cotangent space $T_x^*(X)$ into a degenerate cosymplectic vector space over F_x.

b) Define $S_x \subset T_x(X)$ to be the space of values at x of Hamiltonian vector fields defined by functions in $\mathfrak{m}_x$ (cf. (2.6)); and define a non-degenerate symplectic form ω_x on S_x as in (2.6). Then $(T_x(X), S_x, \omega_x)$ is a degenerate symplectic vector space over F_x. It is dual to the degenerate cosymplectic vector space $T_x^*(X)$ of (a).

This is more or less obvious from the definitions (see also the discussion at (2.6)).

We can now begin to consider non-linear versions of the notion of coisotropic subspace. An algebraic subvariety Y of X is specified by an ideal

$$J(Y) \subset R(X), \qquad R(Y) = R(X)/J(Y). \tag{4.12)(a}$$

(We are ignoring the Poisson structure for the moment.) Now suppose we are in the setting of Proposition 4.11, and that x is also a point of Y; that is,

$$J(Y) \subset \mathfrak{m}_x. \tag{4.12)(b}$$

Then $J(Y)$ defines a natural subspace of $T_x^*(X)$ by

$$T_{Y,x}^*(X) = J(Y)/(J(Y) \cap \mathfrak{m}_x^2) \hookrightarrow T_x^*(X). \tag{4.12)(c}$$

Its annihilator is a subspace of the tangent space at x:

$$T_x(Y) = \{\lambda \in T_x(X) \mid \lambda(T_{Y,x}^*(X)) = 0\}. \tag{4.12)(d}$$

As the notation indicates, this subspace may be naturally identified with the tangent space at x to Y. One way to think of the reason is in terms of the short exact sequence

$$0 \to T_{Y,x}^*(X) \to T_x^*(X) \to T_x^*(Y) \to 0. \tag{4.12)(e}$$

Proposition 4.13. *In the setting of Proposition 4.11, suppose Y is an algebraic subvariety of X defined by an ideal $J(Y) \subset R(X)$. Assume that $J(Y) \subset \mathfrak{m}_x$, and define subspaces of tangent and cotangent spaces as in (4.12) above. Then the following conditions are equivalent.*

a) $\{J(Y), J(Y)\} \subset \mathfrak{m}_x$.

b) $T^*_{Y,x}(X)$ is an isotropic subspace of the degenerate cosymplectic vector space $T^*_x(X)$ (Definition 4.9).

c) $T_x(Y)$ is a co-isotropic subspace of the degenerate symplectic vector space $T_x(X)$ (Definition 4.8).

Proof. The equivalence of (a) and (b) is clear from the definition of the degenerate symplectic form $\{,\}_x$ in (2.6)(d). The equivalence of (b) and (c) is a general feature of the duality relationship between Definitions 4.8 and 4.9, as explained in (4.10): a subspace S of a degenerate symplectic vector space V is co-isotropic if and only if its annihilator $S^\perp \subset V^*$ is isotropic in V^*. Q.E.D.

Definition 4.14. Suppose X is an affine Poisson algebraic variety over F, and Y is a subvariety corresponding to an ideal $J(Y) \subset R(X)$. We say that Y (or $J(Y)$) is *co-isotropic* if $\{J(Y), J(Y)\} \subset J(Y)$. (Similar definitions apply to Poisson structures on more general ringed spaces.)

The point of Proposition 4.13 is that this terminology is reasonable. Here is a precise statement.

Proposition 4.15. *Suppose X is an affine Poisson algebraic variety over F, and Y is a co-isotropic subvariety corresponding to an ideal $J(Y)$. If $\mathfrak{m}_x$ is any maximal ideal containing $J(Y)$, then $T_x(Y)$ is a co-isotropic subspace of the degenerate symplectic vector space $T_x(X)$.*

Conversely, suppose that Y is any subvariety of X, corresponding to an ideal $J(Y)$; and assume that $T_x(Y)$ is co-isotropic in $T_x(X)$ whenever $J(Y) \subset \mathfrak{m}_x$. If $J(Y)$ is the intersection of the maximal ideals containing it (that is, if $J(Y)$ is a radical ideal) then Y is co-isotropic.

Proof. This is immediate from Proposition 4.13 and Definition 4.14. Q.E.D.

Example 4.16. If r, s, and f belong to a Poisson algebra $R(X)$, then

$$\{rf, sf\} = (r\{f, s\} + s\{r, f\} + f\{r, s\})f.$$

From this formula it follows that the ideal $\langle f \rangle$ generated by f is always co-isotropic. A similar argument shows that if J has a collection of generators that is closed under Poisson bracket, then J is co-isotropic.

Suppose V is a degenerate symplectic vector space over F. We make V into an algebraic variety with algebra of functions $R(V) = S(V^*)$ (the symmetric algebra of V^*). The construction of (4.10) provides a degenerate symplectic form η on V^*. It turns out that there is a unique Poisson algebra structure on $S(V^*)$ characterized by

$$\{\lambda, \mu\} = \eta(\lambda, \mu);$$

the function on the right is the constant function. If $T \subset V$ is any linear subspace, then we can regard T as a subvariety defined by the ideal

$$J(T) = \langle \lambda \in V^* \mid \lambda(T) = 0 \rangle.$$

Then it is easy to check that T is a co-isotropic subvariety (Definition 4.14) if and only if it is a co-isotropic subspace (Definition 4.8).

Finally, we turn to the "dual" notion of isotropic subvariety. Here is the classical definition.

Definition 4.17. In the setting and notation of (4.12), assume that Y is reduced. Recall from (2.6) the symplectic vector space $\mathcal{S}_x \subset T_x(X)$ inside each tangent space to X. We say that Y is *isotropic* if at every smooth point x of Y, T_xY is an isotropic subspace of the degenerate symplectic vector space T_xX (Defintion 4.8). That is, we require that the intersection $T_x(Y) \cap \mathcal{S}_x$ is an isotropic subspace of $\mathcal{S}_x$.

We say that Y is *Lagrangian* if it is both isotropic and co-isotropic; that is, if $T_x(Y) \cap \mathcal{S}_x$ is Lagrangian in $\mathcal{S}_x$ at every smooth point x of Y.

Even for a symplectic manifold X, the tangent space condition of Definition 4.17 can fail at singular points of Y (where T_xY has larger dimension). For this reason, we would like to have a definition more along the lines of Definition 4.14. We have not done all the checking needed to verify the equivalence with Definition 4.17, but here is a possibility. In the setting and notation of (4.12), define

$$J(Y)^c = \{f \in R(X) \mid \{f, J(Y)\} \subset J(Y)\} \tag{4.18)(a}$$

If $f \in J(Y)^c$ and $x \in Y$, then the value $\xi_f(x)$ at x of the Hamiltonian vector field ξ_f belongs to $T_x(Y)$. (Here is the reason. By(4.12), we must show that $\xi_f(x)$ annihilates the subspace of $T_x^*(X)$ spanned by functions in J. If g is any function in $\mathfrak{m}_x$, then the value of $\xi_f(x)$ on the corresponding tangent vector is $\{f, g\} + \mathfrak{m}_x \in R(X)/\mathfrak{m}_x$. If $g \in J$, then the first bracket is in $J \subset \mathfrak{m}_x$, so we get zero as required.) Recalling from (2.6) that $\mathcal{S}_x$ denotes the space of Hamiltonian tangent vectors at x, we have therefore shown that

$$\{\xi_f(x) \mid f \in J(Y)^c\} \subset \mathcal{S}_x \cap T_x(Y). \tag{4.18)(b}$$

What we have not verified is a partial converse:

if Y is reduced, then equality holds in (4.18)(b) at every smooth point of Y. (4.18)(c)

Assuming this to be the case, an argument along the lines of Proposition 4.13 proves that Definition 4.17 is equivalent to

Definition 4.19. Suppose X is an affine Poisson algebraic variety over F, and Y is a subvariety corresponding to an ideal $J(Y)$. Define $J(Y)^c$ as in (4.18) above. We say that Y (or $J(Y)$) is *isotropic* if $\{J(Y)^c, J(Y)^c\} \subset J(Y)$.

In any case the requirement in Definition 4.17 certainly implies the one in Definition 4.19. In this paper we will use only Definition 4.17.

5. THE METAPLECTIC REPRESENTATION

In this section we recall the construction of the metaplectic or oscillator representation. A convenient reference for most of this material is [14]; original sources include [18] and [23].

Definition 5.1. Suppose F is a field of characteristic not equal to 2, and (W, ω) is a finite-dimensional symplectic vector space over F. The *Heisenberg group of* W is the set $H(W) = W \times F$ with group law

$$(w, s) \cdot (v, t) = (w + v, s + t + \omega(w, v)/2)) \qquad (5.1)(a)$$

The center of $H(W)$ is the subgroup F, and this is also the commutator subgroup; so $H(W)$ is a two-step unipotent algebraic group over F, with $H(W)/F \simeq W$. The group $Sp(W)$ (Definition 4.1) acts by automorphisms on $H(W)$, by acting trivially on F.

If V is any subspace of W, then $H(V) = V \times F$ is a subgroup of $H(W)$, and its centralizer is $H(V^{\perp})$. (Taking $V = 0$ or $V = W$, we get the previous claim that F is the center of $H(W)$.) If I is isotropic, then $H(I)$ is abelian; and if L is Lagrangian, then $H(L)$ is a maximal abelian subgroup of $H(W)$.

Assume now that F is a finite or local field. Then $H(W)$ has a natural locally compact topology. Fix once and for all a non-trivial additive unitary character χ of F. (It is traditional and convenient to assume that $\chi(t) = \exp(2\pi i t)$ if $F = \mathbb{R}$, but we will carry χ along in the notation as a reminder of the importance of this choice.) If L is any Lagrangian subspace of W, then χ extends to a character of the abelian group $H(L)$ by

$$\chi(L)(w, s) = \chi(s) \qquad (w \in L, s \in F). \qquad (5.1)(b)$$

The *Schrödinger representation of* $H(W)$ *attached to* L is

$$\sigma_{\chi,L} = \mathrm{Ind}_{H(L)}^{H(W)} \chi(L).$$

We may omit the subscript χ when no confusion can result.

We will return in a moment to a careful discussion of the space of $\sigma_{\chi,L}$. Essentially it is the Hilbert space of square-integrable sections of the line bundle on $H(W)/H(L) \simeq W/L$ induced by the character $\chi(L)$. The first thing to observe is that the center F of $H(W)$ acts in a Schrödinger representation by the scalar χ. This property turns out to be characteristic.

Theorem 5.2 (Stone and von Neumann; see [14], Theorem 1.3.3, or [15], Théorème 2.I.2 and Lemme 2.I.8). *Suppose W is a symplectic vector space over a finite or local field F of characteristic not 2, and χ is a non-trivial additive unitary character of F.*

a) *The representations $\sigma_{\chi,L}$ of $H(W)$ (Definition 5.1) are all irreducible and equivalent.*

b) Suppose σ_χ is any unitary representation of $H(W)$ with the property that $\sigma_\chi(t) = \chi(t)\mathrm{Id}$ for $t \in F$. Then σ_χ is equivalent to a multiple of $\sigma_{\chi,L}$.

Corollary 5.3 (Segal-Shale-Weil) *In the setting of Theorem 5.2, fix a Schrödinger representation $(\sigma_\chi, \mathcal{H})$ of $H(W)$ (associated to the character χ of F). Suppose $g \in Sp(W)$. Then there is a unitary automorphism $T(g)$ of $\mathcal{H}$ satisfying*

$$\sigma_\chi(g \cdot h) = T(g)\sigma_\chi(h)T(g)^{-1} \qquad (h \in H(W)).$$

Here g acts on $H(W)$ as in Definition 5.1. The operator $T(g)$ is determined by this condition up to multiplication by a scalar of absolute value one.

Define $GMp(W)$ to be the group of unitary operators on $\mathcal{H}$ generated by the various $T(g)$ and the scalar multiplications. Sending $T(g)$ to g defines an exact sequence of groups

$$1 \to \mathbb{T} \to GMp(W) \to Sp(W) \to 1.$$

The group $GMp(W)$ carries a natural locally compact topology making these maps continuous. There is a closed subgroup $Mp(W) \subset GMp(W)$ so that $Mp(W) \cap \mathbb{T} = \{\pm 1\}$, and we have an exact sequence

$$1 \to \{\pm 1\} \to Mp(W) \to Sp(W) \to 1.$$

Proof. The map $h \mapsto \sigma(g \cdot h)$ is an irreducible unitary representation of $H(W)$ on $\mathcal{H}$. By Theorem 5.2, it is unitarily equivalent to σ. The existence of the operator $T(g)$ follows, and the exact sequence for $GMp(W)$ follows easily. The reduction of the covering to $\{\pm 1\}$ was carried out by Shale in the real case (see [14]) and by Weil in general ([23], section 43). Q.E.D.

Definition 5.4. Suppose W is a symplectic vector space over a finite or local field F of characteristic not 2, and χ is a non-trivial character of F. Let $(\sigma, \mathcal{H})$ be a Schrödinger representation of $H(W)$. The group $Mp(W)$ of Corollary 5.3 is called the *metaplectic group of* W, and its (tautological) representation τ on $\mathcal{H}$ is the *metaplectic representation.*

We want to realize the metaplectic representation more concretely. Although much of the discussion below applies to any finite or local field, it is convenient from now on to assume that $F = \mathbb{R}$. We begin with a more precise description of the space of a Schrödinger representation; and for that we need half densities.

Definition 5.5. Suppose V is a finite-dimensional real vector space. The space $D(V)$ of *densities on* V consists of all real multiples of Lebesgue measure on V. This is a one-dimensional real vector space. If dv is a Lebesgue measure, then

$$D(V) = \{c\, dv \mid c \in \mathbb{R}\}. \qquad (5.5)(a)$$

If $d'v$ is another Lebesgue measure, then necessarily $d'v = a\,dv$ for some positive scalar a; and in the coordinates of (5.5)(a),

$$c\,dv = c'\,d'v \quad \text{if and only if} \quad c' = c/a. \tag{5.5)(b}$$

The space $D_{\mathbb{C}}(V)$ of *complex densities on* V consists of all complex multiples of Lebesgue measure on V; it is the complexification of $D(V)$.

Suppose now that t is any real number. A *t-density on* V is a formal symbol $c(dv)^t$, with dv a Lebesgue measure on V and c a real number. We define equality of such formal symbols by analogy with (5.5)(b): if $d'v = a\,dv$ is another Lebesgue measure on V, then

$$c(dv)^t = c'(d'v)^t \quad \text{if and only if} \quad c' = c/a^t. \tag{5.5)(c}$$

The space of t-densities on V is written $D^t(V)$; it is a one-dimensional real vector space. There is a well-defined multiplication

$$D^t(V) \otimes D^s(V) \to D^{t+s}(V), \qquad (c(dv)^t) \otimes (c'(dv)^s) \mapsto cc'(dv)^{t+s}. \tag{5.5)(d}$$

This defines an isomorphism $D^t(V) \otimes D^s(V) \simeq D^{t+s}(V)$. Similarly we can define a one-dimensional complex vector space $D^z_{\mathbb{C}}(V)$ of z-densities on V for any complex number z.

Suppose now that M is a real manifold. The real line bundle of *t-densities on* M is the bundle $\mathcal{D}^t(M)$ whose fiber at m is $D^t(T_mM)$, the t-densities on the tangent space at m. The isomorphisms of (5.5)(d) give bundle isomorphisms

$$\mathcal{D}^t(M) \otimes \mathcal{D}^s(M) \simeq \mathcal{D}^{t+s}(M). \tag{5.5)(e}$$

Similarly we can define $\mathcal{D}^z_{\mathbb{C}}(M)$, the complex line bundle of z-densities on M.

Proposition 5.6. *In the setting of Definition 5.5, the space of smooth sections of $\mathcal{D}^1(M)$ may be identified with the space of smooth densities on M; that is, with signed measures on M that are given by a smooth function times Lebesgue measure in every coordinate chart. In particular, every compactly supported smooth section of $\mathcal{D}^1(M)$ has a well-defined integral over M (a real number). Similarly, there is a complex-valued integral for sections of $\mathcal{D}^1_{\mathbb{C}}(M)$.*

Using Proposition 5.6, we can define on the space of compactly supported smooth sections of $\mathcal{D}^{1/2}_{\mathbb{C}}(M)$ a natural pre-Hilbert space structure, as follows. Suppose σ_1 and σ_2 are such sections. Because $\mathcal{D}^{1/2}_{\mathbb{C}}(M)$ is the complexification of $\mathcal{D}^{1/2}(M)$, the complex conjugate $\overline{\sigma_2}$ is a well-defined section of $\mathcal{D}^{1/2}_{\mathbb{C}}(M)$. Using the multiplication (5.5)(e), we can regard $\sigma_1\overline{\sigma_2}$ as a compactly supported smooth section of $\mathcal{D}^1_{\mathbb{C}}(M)$. Such a section has an integral, by Proposition 5.6; so we define

$$\langle \sigma_1, \sigma_2 \rangle = \int_M \sigma_1(m)\overline{\sigma_2(m)} \tag{5.7)(a}$$

It is easy to check that this is a pre-Hilbert space structure. The corresponding Hilbert space is called $L^2(M, \mathcal{D}_{\mathbb{C}}^{1/2})$, the space of *square-integrable half-densities on* M.

The great advantage of this Hilbert space over the (isomorphic) one $L^2(M, d\mu)$ of square-integrable functions on M with respect to a chosen measure $d\mu$ is this. Any diffeomorphism T of M acts on compactly supported smooth sections of $\mathcal{D}_{\mathbb{C}}^{1/2}(M)$, preserving the pre-Hilbert structure. Consequently T defines a unitary operator $\rho(T)$ on $L^2(M, \mathcal{D}_{\mathbb{C}}^{1/2})$, and ρ is a unitary representation of $\mathrm{Diff}(M)$.

A little more generally, suppose $\mathcal{H} \to M$ is a Hilbert space bundle over M. Then we can form the tensor product $\mathcal{H} \otimes \mathcal{D}_{\mathbb{C}}^{1/2}(M)$; topologically its fibers are Hilbert spaces, but there is no longer a distinguished inner product. More precisely, the fiber over m comes equipped with a positive sesquilinear pairing not into $\mathbb{C}$, but into the space $\mathcal{D}_{\mathbb{C},m}^{1}(M)$ of densities at m. In this way the space of compactly supported continuous sections of $\mathcal{H} \otimes \mathcal{D}_{\mathbb{C}}^{1/2}(M)$ acquires a pre-Hilbert space structure. Explicitly, suppose h_1 and h_2 are compactly supported continuous sections of $\mathcal{H}$, and σ_1 and σ_2 are compactly supported continuous sections of $\mathcal{D}_{\mathbb{C}}^{1/2}(M)$. Then we define

$$\langle h_1 \otimes \sigma_1, h_2 \otimes \sigma_2 \rangle = \int_M \langle h_1(m), h_2(m) \rangle_{\mathcal{H}_m} \sigma_1(m) \overline{\sigma_2(m)} \tag{5.7)(b}$$

The completion of this Hilbert space is called $L^2(M, \mathcal{H} \otimes \mathcal{D}_{\mathbb{C}}^{1/2})$, the space of *square-integrable half-density sections of* $\mathcal{H}$. (Of course the space M is implicit in the bundle $\mathcal{H}$, and we may sometimes omit it from the notation.) Sometimes it is helpful to think of this space in terms of real half-densities, using the isomorphism

$$\mathcal{H} \otimes_{\mathbb{C}} \mathcal{D}_{\mathbb{C}}^{1/2}(M) \simeq \mathcal{H} \otimes_{\mathbb{R}} \mathcal{D}^{1/2}(M). \tag{5.7)(c}$$

Example 5.8. Suppose G is a Lie group, H is a closed subgroup, and $(\tau, \mathcal{H}_\tau)$ is a unitary representation of H. Then τ defines a G-equivariant Hilbert bundle $\mathcal{H} = G \times_H \mathcal{H}_\tau$ over G/H. The total space of the bundle is $G \times \mathcal{H}_\tau$ modulo the equivalence relation

$$(gh, v) \sim (g, \tau(h)v) \qquad (g \in G, h \in H, v \in \mathcal{H}_\tau).$$

The action of G on the total space is by the left action in the first variable; this respects the equivalence relation. At the same time the action of G on G/H defines an action on half-densities on G/H; so G acts on the compactly supported continuous half-density sections of $\mathcal{H}$. This action respects the pre-Hilbert space structure, and so gives a unitary action of G on $L^2(G/H, \mathcal{H} \otimes \mathcal{D}_{\mathbb{C}}^{1/2})$. This is nothing but the induced representation:

$$\mathrm{Ind}_H^G(\mathcal{H}_\tau) = L^2((G \times_H \mathcal{H}_\tau) \otimes \mathcal{D}_{\mathbb{C}}^{1/2}(G/H)).$$

The twist by half-densities corresponds to the "ρ shift" appearing in more purely group-theoretic descriptions of induction. There the space of the representation is described as a space of functions f on G with values in $\mathcal{H}_\tau$, satisfying a transformation law

$$f(gh) = \rho_{G/H}(h^{-1})\tau(h^{-1})f(g) \qquad (g \in G, h \in H).$$

Here $\rho_{G/H}$ is a certain real-valued character of H. The point is that the half-density bundle on G/H is also an induced bundle, and so is characterized by the action of H on the fiber at the identity coset eH; that is, by the action of H on half-densities on $T_{eH}(G/H) = \mathfrak{g}/\mathfrak{h}$. This action is by one over the square root of the absolute value of the determinant of the adjoint action of H on $\mathfrak{g}/\mathfrak{h}$; and that is precisely the character $\rho_{G/H}$ of H.

This entire discussion depends only on the notion of Lebesgue measure on finite-dimensional vector spaces, and so works equally well over any local or finite field. (All t-density bundles are trivial in the case of a finite field.)

With this machinery in hand, we can return to the problem of describing the space of the Schrödinger representation. According to Example 5.8, we can take the Hilbert space of $\sigma_{\chi,L}$ to be the space of square-integrable half-density sections of the line bundle on $H(W)/H(L)$ induced by the character $\chi(L)$. In order to understand this space, we will trivialize the line bundle. For that, we need to pick a second Lagrangian subspace L' of W so that $L \cap L' = 0$; this is possible by Proposition 4.4. In that case L' is an abelian subgroup of $H(W)$, and

$$L' \simeq H(W)/H(L); \tag{5.9)(a}$$

this follows from (4.5)(e). Because the isotropy group of the L' action here is trivial, the line bundle is trivial (as an L'-equivariant bundle). The Hilbert space is therefore

$$\mathcal{H}(L) = L^2(L', D_{\mathbb{C}}^{1/2}), \tag{5.9)(b}$$

the space of square-integrable half-densities on the vector space L'. If dx' is a Lebesgue measure on L', then a typical element of this space is a symbol $\phi(x')(dx')^{1/2}$; here ϕ is an L^2 function on L'. This isomorphism respects the action of L'; that is, L' acts by the left regular representation (translation of half densities). Explicitly,

$$\sigma_{\chi,L}(l')[\phi(x')(dx')^{1/2}] = \phi(x' - l')(dx')^{1/2}. \tag{5.9)(c}$$

It is also fairly easy to calculate the action of the subgroup L in this picture. Recall from (4.5)(d) that each element $l \in L$ gives a well-defined linear functional $x' \mapsto \omega(l, x')$ on L', and therefore a well-defined unitary character $x' \mapsto \chi(\omega(l, x'))$ of L'. The action of $\sigma_{\chi,L}(l)$ is multiplication by this unitary character:

$$\sigma_{\chi,L}(l)[\phi(x')(dx')^{1/2}] = \chi(\omega(l, x'))\phi(x')(dx')^{1/2}. \tag{5.9)(d}$$

To see this, recall that the isomorphism (5.9)(b) identifies the half density $\phi(x')(dx')^{1/2}$ with a function Φ on $H(W)$ transforming according to the character $\chi(L)$ of $H(L)$. We now compute

$$\sigma_{\chi,L}(l)[\phi(x')(dx')^{1/2}] = \Phi((l,0)^{-1}(x',0));$$

here the multiplication on the right takes place in the group $H(W)$. Now

$$(l,0)^{-1}(x',0) = (-l,0)(x',0) = (x'-l,\omega(-l,x')/2) = (x',0)(-l,-\omega(l,x')).$$

By the transformation law for Φ under $H(L)$, we get

$$\Phi((l,0)^{-1}(x',0)) = \Phi(x',0)\chi(L)(-l,-\omega(l,x'))^{-1} = \phi(x')(dx')^{1/2}\chi(\omega(l,x')).$$

This is (5.9)(d).

So far this discussion applies equally well over any local or finite field. Now we turn to something special to $\mathbb{R}$: the differentiated representation. We begin with the Lie algebra $\mathfrak{h}(W) = W \times \mathbb{R}$ of $H(W)$. (The identification of $H(W)$ with the vector space $W \times \mathbb{R}$ naturally identifies the Lie algebra with this vector space as well.) The Lie bracket is

$$[(w,s),(v,t)] = (0,\omega(w,v)). \tag{5.10)(a}$$

Obviously the center of $\mathfrak{h}(W)$ is equal to the commutator subalgebra, which is $\mathbb{R}$. In the setting of (5.9), we get

$$\mathfrak{h}(W) = L \oplus L' \oplus \mathbb{R}. \tag{5.10)(b}$$

In order to differentiate the Schrödinger representation, we must first understand the differential of the character χ of $\mathbb{R}$. This is a map from the Lie algebra $\mathbb{R}$ of $\mathbb{R}$ to the Lie algebra $i\mathbb{R}$ of the unit circle; so $d\chi(1)$ is a purely imaginary number. With the standard choice $\chi(t) = \exp(2\pi it)$ made in Definition 5.1, we get $d\chi(1) = 2\pi i$.

Proposition 5.11. *Suppose W is a real symplectic vector space, and L and L' are Lagrangian subspaces with $L \cap L' = 0$; realize the Schrodinger representation $\sigma_{\chi,L}$ (Definition 5.1) as in (5.9). Then the differentiated representation of $\mathfrak{h}(W)$ (cf. (5.10)) may be calculated as follows.*

a) Suppose $t \in \mathbb{R}$. Then $d\sigma_{\chi,L}(0,t) = td\chi(1) = 2\pi it$.

b) Suppose $l \in L$. Identify l with the linear functional $\tau'_L(l)$ on L' (Definition 4.1), defined by $\tau_{L'}(l)(l') = \omega(l',l)$. Then

$$d\sigma_{\chi,L}(l,0) = \text{multiplication by } -d\chi(1)\tau_{L'}(l) = -2\pi i\tau_{L'}(l).$$

c) Suppose $l' \in L'$. Then $d\sigma_{\chi,L}(l',0)$ is the directional derivative in the direction $-l'$.

Proof. Part (a) follows from the fact that the center $\mathbb{R}$ of $H(W)$ acts in $\sigma_{\chi,L}$ by the character χ. Parts (b) and (c) are differentiated versions of (5.9)(d) and (c) respectively. For (b), for example, we calculate

$$\begin{aligned} d\sigma_{\chi,L}(l,0)[\phi(x')(dx')^{1/2}] &= d/dt(\sigma_{\chi,L}(tl,0)[\phi(x')(dx')^{1/2}])|_{t=0} \\ &= d/dt(\chi(\omega(tl,x'))\phi(x')(dx')^{1/2})|_{t=0} \\ &= d/dt(\chi(-t\tau_{L'}(l)(x')))|_{t=0}\phi(x')(dx')^{1/2} \\ &= -d\chi(1)\tau_{L'}(l)(x')\phi(x')(dx')^{1/2} \end{aligned}$$

This is (b). Q.E.D.

Corollary 5.12. *In the setting of Definition 5.1 and (5.9), the space $\mathcal{H}(L)^\infty$ of smooth vectors in the Schrödinger representation is the Schwartz space $\mathcal{S}(L', D_{\mathbb{C}}^{1/2})$: the space of all half-densities $\phi(x')(dx')^{1/2}$ such that any derivative of ϕ times any polynomial in x' is bounded.*

Proof. If we replace the phrase "is bounded" by "belongs to L^2," then the statement is immediate from Proposition 5.11 and the definition of the space of smooth vectors. So what we must show is that if ϕ is a function on $\mathbb{R}^n$ with the property that $x^\alpha \frac{\partial^\beta \phi}{\partial x^\beta}$ belongs to L^2 for all multiindices α and β, then ϕ belongs to the Schwartz space. We write $L^2\mathcal{S}(\mathbb{R}^n)$ temporarily to distinguish the function space defined by these L^2 conditions. Since polynomials times derivatives of ϕ again belong to $L^2\mathcal{S}$, it suffices to show that ϕ is bounded. This is a consequence of the Sobolev lemma. We sketch a direct argument. For compactly supported smooth ϕ, we have

$$\phi(a) = \int_{x_i \le a_i} \frac{\partial^n \phi}{\partial x_1 \cdots \partial x_n} dx_1 \cdots dx_n.$$

We need to estimate the right side using L^2 norms. For that, consider the polynomial $p(x) = (1+|x|^2)^n$. Then $1/p \in L^2(\mathbb{R}^n)$, so

$$\phi(a) = \int_{x_i \le a_i} (1/p(x))(p(x)\frac{\partial^n \phi}{\partial x_1 \cdots \partial x_n}) dx_1 \cdots dx_n.$$

This integral may be interpreted as the L^2 inner product of $p\frac{\partial^n \phi}{\partial x_1 \cdots \partial x_n}$ with a cutoff of $1/p$. By the Cauchy-Schwarz inequality, we get

$$|\phi(a)| \le \|1/p\|_2 \|p\frac{\partial^n \phi}{\partial x_1 \cdots \partial x_n}\|_2.$$

This gives the estimate we need:

$$\|\phi\|_\infty \le \|1/p\|_2 \|p\frac{\partial^n \phi}{\partial x_1 \cdots \partial x_n}\|_2.$$

Once it is established for compactly supported smooth functions, the estimate follows for general $\phi \in L^2\mathcal{S}$ by continuity (more precisely, by the density of C_c^∞ in $L^2\mathcal{S}$). Q.E.D.

Proposition 5.11 and Corollary 5.12 show clearly the relevance to the Schrödinger representation of the algebra of polynomial coefficient differential operators on L'. Here is an abstract definition of it.

Definition 5.13. Suppose W is a real symplectic vector space and χ is a non-trivial additive character of $\mathbb{R}$; as usual we will generally assume $\chi(t) = \exp(2\pi i t)$. The *Weyl algebra of* W is the complex associative algebra $A_\chi(W)$ with unit generated by W, subject to the relations

$$vw - wv = d\chi(1)\omega(v,w) = 2\pi i\omega(v,w) \qquad (v, w \in W).$$

That is, $A_\chi(W)$ is the complex tensor algebra $T(W_{\mathbb{C}})$ divided by the ideal generated by elements $v \otimes w - w \otimes v - 2\pi i\omega(v,w)$. Since we have specified a choice of χ, we may sometimes omit it from the notation. As a quotient of a graded algebra, $A_\chi(W)$ inherits an increasing filtration; $A^p_\chi(W)$ is spanned by the images of elements $w_1 \otimes \cdots \otimes w_q$ with $q \leq p$. Thus

$$A^0_\chi(W) = \mathbb{C}, \qquad A^1_\chi(W) = \mathbb{C} + W_{\mathbb{C}}, \qquad A^p A^q \subset A^{p+q}.$$

The generators by which we divide are all sums of tensors of even degree; so $A_\chi(W)$ inherits from $T(W_{\mathbb{C}})$ a $\mathbb{Z}/2\mathbb{Z}$ grading

$$A_\chi(W) = A^{even}_\chi(W) \oplus A^{odd}_\chi(W).$$

Finally, there is a complex conjugate-linear antiautomorphism $r \mapsto r^*$ of $A_\chi(W)$, characterized by the property $v^* = -v \quad (v \in W)$. (The existence of this map is a formal consequence of the definition of $A_\chi(W)$. One defines $A^*_\chi(W)$ to be the opposite algebra of $A_\chi(W)$, with complex multiplication by z given by the old multiplication by $\bar{z}$. Then the elements $\{-v \mid v \in W\}$ of $A^*_\chi(W)$ satisfy the same relations as the generators W of $A_\chi(W)$.)

Proposition 5.14. *In the setting of Proposition 5.11, the following three algebras are naturally isomorphic.*

a) *The Weyl algebra $A_\chi(W)$ of Definition 5.13.*
b) *The quotient $U(\mathfrak{h}(W))/I_\chi$ of the universal enveloping algebra of the Heisenberg Lie algebra (cf. (5.10)) by the ideal I_χ generated by the element $(0,1) - d\chi(1) = (0,1) - 2\pi i$. Here (0,1) is the central element of $\mathfrak{h}(W)$.*
c) *The algebra $\mathcal{D}(L')$ of polynomial coefficient differential operators on the Lagrangian subspace L' of W.*

The isomorphism of $A_\chi(W)$ with $\mathcal{D}(L')$ sends an element $l \in L$ to multiplication by the purely imaginary linear functional $-d\chi(1)\tau'_L(l) = -2\pi i\tau'_L(l)$; and it sends an element $l' \in L'$ to the directional derivative in the direction $-l'$. The involution $r \mapsto r^*$ is given in $\mathcal{D}(L')$ by the formal adjoint of differential operators.

Proof. By definition the universal enveloping algebra $U(\mathfrak{h}(W))$ is the complex associative algebra generated by $W + \mathbb{R}$, subject to certain relations. To describe them it is convenient to write z for the central element $(0,1) \in \mathfrak{h}(W)$. Then the relations are $vw - wv = \omega(v,w)z$ and $zv = vz$ for $v, w \in W$. Dividing by I_χ amounts to identifying z with $d\chi(1) = 2\pi i$. It follows that $U(\mathfrak{h})/I_\chi$ may be described as the complex associative algebra generated by W, subject to the relations $vw - wv = \omega(v,w) \cdot d\chi(1)$, or $vw - wv = 2\pi i\omega(v,w)$. This is just $A_\chi(W)$ by definition. The filtration defined on $A_\chi(W)$ in Definition 5.13 is clearly the one inherited from the standard filtration on $U(\mathfrak{h}(W))$. It follows easily from the Poincaré-Birkhoff-Witt theorem that the associated graded algebra is commutative; specifically, that

$$\operatorname{gr} A_\chi(W) \simeq S(W_{\mathbb{C}}), \tag{5.15}$$

the complexified symmetric algebra of W.

To get the isomorphism with $\mathcal{D}(L')$, we consider the action of $U(\mathfrak{h}(W))$ on $\mathcal{H}(L)^\infty$ by $d\sigma_{\chi,L}$. Clearly the element z acts by $d\chi(1)$, so the ideal I_χ acts by 0. Proposition 5.11 shows that the image of the enveloping algebra is precisely $\mathcal{D}(L')$; so we have a surjective homomorphism $U(\mathfrak{h}(W))/I_\chi \to \mathcal{D}(L')$. The Poincaré-Birkhoff-Witt theorem has already shown us how to find a basis in the domain. More explicitly, we can take a basis $\{x_i\}$ of L, followed by a dual basis $\{y_j\}$ of L'. (This means that $\omega(y_j, x_i) = \delta_{ij}$.) Then the elements $x^\alpha y^\beta$ (for multiindices α and β) form a basis of $U(\mathfrak{h}(W))/I_\chi$. Proposition 5.11 shows that their images in $\mathcal{D}(L')$ are (up to scale factors) the standard basis elements $x^\alpha \frac{\partial^\beta}{\partial x^\beta}$ of $\mathcal{D}(L')$. Our homomorphism is therefore an isomorphism. Q.E.D.

We return now to the metaplectic representation. Corollary 5.3 provides a global description of it; we want an infinitesimal description, giving the action of the Lie algebra of the symplectic group. This Lie algebra is

$$\mathfrak{sp}(W) = \{M \in \operatorname{End} W \mid \omega(Mv,w) + \omega(v,Mw) = 0\}. \tag{5.16}$$

We will realize this Lie algebra using the Weyl algebra $A_\chi(W)$. To do that, we want to take advantage of the non-commutative nature of the Weyl algebra. Suppose $r \in A_\chi^p(W)$ and $s \in A_\chi^q(W)$. Then the products rs and sr both belong to $A_\chi^{p+q}(W)$. In fact their images in $A_\chi^{p+q}(W)/A_\chi^{p+q-1}(W)$ are the same; this is the content of the assertion in (5.15) that the associated graded algebra is commutative. That is, $rs - sr \in A_\chi^{p+q-1}(W)$. But actually even more is true.

Proposition 5.17. *Suppose W is a real symplectic vector space, and $\chi(t) = \exp(2\pi it)$ is our standard non-trivial character of $\mathbb{R}$. Define the Weyl algebra $A_\chi(W)$ as in Definition 5.13, and use the isomorphism (5.15) of the associated graded algebra with the complex symmetric algebra on W.*

a) Suppose $r \in A_\chi^p(W)$ and $s \in A_\chi^q(W)$. Then $rs - sr \in A_\chi^{p+q-2}(W)$.

b) *There is a Poisson bracket* $\{,\}_\chi$ *on* $S(W_{\mathbb{C}})$, *homogeneous of degree* -2, *defined as follows. Suppose* $R \in S^p(W)$ *and* $S \in S^q(W)$ *are homogeneous polynomials of degrees* p *and* q. *Choose representatives* $r \in A_\chi^p(W)$ *and* $s \in A_\chi^q(W)$ *for* R *and* S, *under the isomorphisms* $S^m(W) \simeq A_\chi^m(W)/A_\chi^{m-1}(W)$. *Then* $rs - sr \in A_\chi^{p+q-2}(W)$; *and we define* $\{R, S\}_\chi$ *to be the polynomial represented by* $rs - sr$. *This bracket makes* $S(W_{\mathbb{C}})$ *into a complex Poisson algebra (Definition 2.4).*

c) *The Poisson bracket of (b) is characterized by the property* $\{v, w\}_\chi = d\chi(1)\omega(v, w) = 2\pi i\omega(v, w)$.

Proof. We use the $\mathbb{Z}/2\mathbb{Z}$ grading of $A_\chi(W)$ from Definition 5.13. It is clear that the image of this grading in $S(W_{\mathbb{C}})$ is the usual grading into even and odd polynomials. To prove the claim in (a), it is enough to replace r and s by any other representatives of their classes in gr $A_\chi(W)$. Say for definiteness that p is even and q is odd; then we may choose these new representatives so that $r \in A_\chi^{even}(W)$ and $s \in A_\chi^{odd}(W)$. Then rs and sr both belong to $A_\chi^{odd}(W)$, so $rs - sr$ does as well. But $p+q-1$ is even, so the class of $rs - sr$ in gr $A_\chi(W)$ is an odd polynomial of even degree; so it is zero.

For (b), that the bracket is well-defined is immediate from (a); and the axioms of Definition 2.4 follow easily. (For example, the Jacobi identity is a consequence of the identity $[r, [s, t]] = [[r, s], t] + [s, [r, t]]$ for commutators in an associative algebra.) Finally (c) is immediate from the defining relations of $A_\chi(W)$ in Definition 5.13. Q.E.D.

Of course essentially this same Poisson structure has appeared earlier. The symplectic structure on W defines a symplectic structure on the dual vector space W^*, by means of the isomorphism τ of W with W^* (Definition 4.1). We may regard $S(W_{\mathbb{C}})$ as the space of complex-valued polynomial functions on the symplectic manifold W^*. The Poisson structure of Proposition 2.3 on $C^\infty(W^*)$ preserves $S(W_{\mathbb{C}})$, and (after multiplication by $d\chi(1) = 2\pi i$) it is precisely the structure of Proposition 5.17. For a less obvious identification, one can look inside the dual $\mathfrak{h}(W)^*$ of the Heisenberg Lie algebra at the hyperplane

$$M_\chi = \{\lambda \in \mathfrak{h}(W)^* \mid \lambda(0, 1) = d\chi(1)/2\pi i = 1\}.$$

Because $(0, 1)$ is central in $\mathfrak{h}(W)$, M_χ is preserved by the coadjoint action of $H(W)$. (Actually it is a single coadjoint orbit.) By Proposition 2.11, it follows that M_χ is a Poisson manifold. Restriction of linear functionals to W identifies M_χ with W^*, and then τ^{-1} provides an identification with W. Again the Poisson structure of Proposition 2.11 differs from that of Proposition 5.17 by the factor $2\pi i$. We will make no use of these identifications, so we leave the verifications to the reader.

Corollary 5.18. *In the setting of Definition 5.13, the subspace* $A_\chi^2(W)$ *of the Weyl algebra is closed under commutator. It is therefore a finite-dimensional complex Lie algebra. There is a natural real form*

$$A_\chi^2(W)_{\mathbb{R}} = \{r \in A_\chi^2(W) \mid r^* = -r\}$$

(cf. Definition 5.13).

a) *The subspace $A^1_\chi(W)_{\mathbb{R}}$ is an ideal naturally isomorphic to the Heisenberg Lie algebra $\mathfrak{h}(W)$ (under the map $U(\mathfrak{h}(W)) \to A_\chi(W)$ of Proposition 5.14).*

b) *The subspace $A^{1,odd}_\chi(W)_{\mathbb{R}}$ is naturally isomorphic to W. Its Lie bracket into $A^0_\chi(W)_{\mathbb{R}} \simeq i\mathbb{R}$ is given by $d\chi(1)\omega = 2\pi i\omega$.*

c) *The adjoint action of $A^{2,even}_\chi(W)_{\mathbb{R}}$ on $A^{1,odd}_\chi(W)_{\mathbb{R}}$ defines a Lie algebra homomorphism*

$$\tau\colon A^{2,even}_\chi(W)_{\mathbb{R}} \to \mathfrak{sp}(W).$$

This map is surjective, and its kernel is $A^0_\chi(W)_{\mathbb{R}} \simeq i\mathbb{R}$.

d) *The short exact sequence of (c) splits uniquely. That is, there is a unique Lie subalgebra $\mathfrak{mp}(W) \subset A^{2,even}_\chi(W)_{\mathbb{R}}$ with the property that*

$$A^{2,even}_\chi(W)_{\mathbb{R}} \simeq i\mathbb{R} \oplus \mathfrak{mp}(W).$$

The subalgebra $\mathfrak{mp}(W)$ is mapped isomorphically onto $\mathfrak{sp}(W)$ by the adjoint action in (c).

Proof. The first claim is immediate from Proposition 5.17(a). That $A^2_\chi(W)_{\mathbb{R}}$ is a real Lie algebra follows from the formula $[r, s]^* = [s^*, r^*]$, which in turn follows from the fact that $r \mapsto r^*$ is an antiautomorphism. That it is a real form follows from the fact that $r \mapsto r^*$ is conjugate linear. The isomorphism in (b) comes from (5.15), and then the description of the bracket from Definition 5.13.

For (c), we must check that if $r \in A^{2,even}_\chi(W)_{\mathbb{R}}$ and $v, w \in W$, then $\tau(r)$ satisfies the condition in (5.16) to belong to the symplectic Lie algebra. Multiplying by $d\chi(1) = 2\pi i$ and using (b), this condition becomes $[[r, v], w] + [v, [r, w]] = 0$. By the Jacobi identity, the left side is $[r, [v, w]] = [r, d\chi(1)\omega(v, w)]$. The second term on the right is a scalar, so the bracket is indeed zero. Because W generates $A_\chi(W)$, the kernel of τ is the intersection of its domain with the center of $A_\chi(W)$, which is $A^0_\chi(W) = \mathbb{C}$. So the kernel of τ is indeed $A^0_\chi(W)_{\mathbb{R}}$. Now (5.15) shows that the image of τ is isomorphic to $A^{2,even}_\chi(W)_{\mathbb{R}}/A^0_\chi(W)_{\mathbb{R}} \simeq iS^2(W)$. To show that τ is an isomorphism, it remains only to show that $\mathfrak{sp}(W)$ has the same dimension as $S^2(W)$. Say W has dimension $2n$; then $S^2(W)$ has dimension $n(2n + 1)$. The dimension of $Sp(W)$, on the other hand, is equal to the dimension of the Lagrangian Grassmannian $\mathcal{B}(W) \simeq Sp(W)/P(L)$ plus the dimension of a Siegel parabolic $P(L)$. The first number is $n(n + 1)/2$ by Corollary 4.6, and the second is n^2 plus the dimension of $U(L)$ by Proposition 4.4. Finally, the dimension of $U(L)$ is $n(n + 1)/2$ by Proposition 4.3; so the total dimension of $Sp(W)$ is $n(n + 1)/2 + n^2 + n(n + 1)/2 = n(2n + 1)$, as we wished to show.

For (d), one knows that $\mathfrak{sp}(W)$ is a semisimple Lie algebra. The short exact sequence of (c) is a central extension of that Lie algebra. But every central extension of a semisimple Lie algebra is trivial. We can take for $\mathfrak{mp}(W)$ the commutator subalgebra of $A^{2,even}_\chi(W)_{\mathbb{R}}$. Q.E.D.

Theorem 5.19. *Suppose W is a real symplectic vector space, and L and L' are Lagrangian subspaces with $L \cap L' = 0$. Realize the Schrödinger representation $\sigma_{\chi,L}$ on $\mathcal{H}(L)$ as in (5.9), and define groups $GMp(W)$ and $Mp(W)$ of unitary operators as in Corollary 5.3.*

a) The space of smooth vectors for the representation of $GMp(W)$ coincides with the smooth vectors $\mathcal{H}(L)^\infty$ of the Schrödinger representation.

b) Use Proposition 5.14 to identify the Weyl algebra $A_\chi(W)$ with an algebra of operators on $\mathcal{H}(L)^\infty$. Then the Lie algebra of $GMp(W)$ is precisely

$$\mathfrak{gmp}(W) = A_\chi^{2,even}(W)_{\mathbb{R}}.$$

In terms of differential operators, these are the even skew-adjoint polynomial coefficient differential operators of total degree (both polynomial and differentiation) at most 2. The short exact sequence of groups

$$1 \to \mathbb{T} \to GMp(W) \to Sp(W) \to 1$$

in Corollary 5.3 gives rise to the short exact sequence of Lie algebras

$$1 \to i\mathbb{R} \to \mathfrak{gmp}(W) \to \mathfrak{sp}(W) \to 1$$

of Corollary 5.18(c).

c) The Lie algebra of the subgroup $Mp(W)$ is precisely the subalgebra $\mathfrak{mp}(W)$ of Corollary 5.18(d). The double covering $Mp(W) \to Sp(W)$ of Corollary 5.3 gives rise to the Lie algebra isomorphism $\mathfrak{mp}(W) \simeq \mathfrak{sp}(W)$ of Corollary 5.18(d).

Sketch of proof. Suppose $X \in \mathfrak{gmp}(W)$. This means that first of all that X is a densely defined self-adjoint operator on $\mathcal{H}(L)$, and that $\exp(tX)$ is a unitary operator in $GMp(W)$ for every real t. This means in turn that for every t there is an element $A(t) \in Sp(W)$ so that

$$\sigma_{\chi,L}(A(t) \cdot h) = \exp(tX)\sigma_{\chi,L}(h)\exp(-tX) \qquad (h \in H(W)). \qquad (5.20)(a)$$

It is not difficult to see that this condition determines $A(t)$ uniquely. Because the map $GMp(W) \to Sp(W)$ is continuous, it follows that $A(t) = \exp(tY)$ for some $Y \in \mathfrak{sp}(W)$.

Now $Sp(W)$ acts by algebra automorphisms on the Weyl algebra $A_\chi(W)$ of Definition 5.13, by its action on the generating subspace W. Recall that Proposition 5.14 allows us to identify $A_\chi(W)$ with certain densely defined operators on $\mathcal{H}(L)$. If we differentiate (5.20)(a) with respect to h and use Proposition 5.11, we find that the operators $\exp(tX)$ preserve $\mathcal{H}(L)^\infty$, and define (by conjugation) algebra automorphisms of $\mathcal{D}(L')$. More explicitly, for any $v \in A_\chi^{1,odd}(W)$, we have

$$\exp(tY) \cdot v = \exp(tX)v\exp(-tX). \qquad (5.20)(b)$$

If we now (formally) differentiate both sides with respect to t, we get

$$Y \cdot v = [X, v]. \tag{5.20)(c}$$

This equation is the infinitesimal version of the definition of the Weil representation in Corollary 5.3. It begins with the family of operators $v \in A_\chi^{1,odd}(W)$, and the linear transformation Y of that family; and it seeks a new operator X satisfying (5.20)(c). Corollary 5.18(c) provides solutions to (5.20)(c). The problem is essentially to show that they really arise by differentiating solutions to (5.20)(a). Perhaps the easiest way to do that is simply to exhibit some solutions to (5.20)(a).

For that, let $P(L) \subset Sp(W)$ be the Siegel parabolic subgroup preserving the Lagrangian subspace L (Definition 4.2). The action of $P(L)$ on the Heisenberg group $H(W)$ preserves $H(L)$, and so descends to an action of $P(L)$ by diffeomorphisms on the homogeneous space $H(W)/H(L) \simeq W/L$. The action of $P(L)$ on $H(L)$ fixes the character $\chi(L)$, so $P(L)$ acts by automorphisms on the Hermitian line bundle $\mathcal{L}_\chi \to H(W)/H(L)$ induced by $\chi(L)$. Recall from Example 5.8 that the Hilbert space $\mathcal{H}(\mathcal{L})$ of the Schrödinger representation may be identified with the space $L^2(H(W)/H(L), \mathcal{L}_\chi \otimes D_{\mathbb{C}}^{1/2})$ of square-integrable half density sections of $\mathcal{L}$. From what we have just said, there is a natural unitary representation $\tau(L)$ of $P(L)$ on $\mathcal{H}(L)$ compatible with its action by automorphisms of $H(W)$. Explicitly,

$$\tau_L(p)\sigma_{\chi,L}(h)\tau_L(p)^{-1} = \sigma_{\chi,L}(p \cdot h) \qquad (p \in P(L), h \in H(W)). \tag{5.21)(a}$$

This is the defining relation for the metaplectic representation (Corollary 5.3). Consequently τ_L may be regarded as a homomorphism from $P(L)$ into $GMp(W)$, a section (over $P(L)$) of the natural projection from $GMp(W)$ onto $Sp(W)$ (Corollary 5.3). That is, we have a commutative diagram

$$\begin{array}{ccc} & & GMp(W) \\ & {}^{\tau_L}\nearrow & \downarrow \\ P(L) & \longrightarrow & Sp(W) \end{array} \tag{5.21)(b}$$

It is now clear that any one-parameter subgroup $A(t)$ of $P(L)$ gives a solution of (5.20)(a): we take for X a generator of the one-parameter group $\tau_L(A(t))$ of unitary operators. So we would like to compute τ_L explicitly. As in (5.9) we begin by identifying the Hilbert space $\mathcal{H}(L)$ with square-integrable half-densities on L'. As in Proposition 4.4, identify $GL(L)$ with the subgroup of $P(L)$ preserving L'; recall that the symplectic form identifies L' with the dual of L, so $GL(L) \simeq GL(L')$ (the isomorphism sending g to ${}^tg^{-1}$). Now $GL(L')$ is a group of diffeomorphisms of L', so there is a natural action of $GL(L')$ on square integrable half-densities. This is precisely τ_L; so we get

$$\tau_L(g)[\phi(x')(dx')^{1/2}] = |\det g|^{1/2}\phi({}^tgx')(dx')^{1/2} \qquad (g \in GL(L) \subset P(L)). \tag{5.21)(c}$$

(The determinant factor arises from the action of g on the half-density $(dx')^{1/2}$.) It is a simple matter to differentiate this representation. If we choose coordinates $(x'_1,\ldots,x'_n)$ on L', then we get at the same time an identification of $GL(L)$ with $GL(n,\mathbb{R})$. The Lie algebra may be identified with $n\times n$ real matrices, and the standard basis matrices act by

$$d\tau_L(e_{pq}) = x_p\frac{\partial}{\partial x_q} + \frac{1}{2}\delta_{pq}. \qquad (5.21)(d)$$

Next we compute the action of the unipotent radical $U(L)$. Recall from Proposition 4.3 that each element $u\in U(L)$ corresponds to a symmetric bilinear form B on $W/L\simeq L'$. This correspondence uses the linear map $u-1=T$, which carries L' to L. To compute the action of u on $\phi(x')(dx')^{1/2}$, recall from (5.9) that ϕ corresponds to a function Φ on $H(W)$ transforming according to $\chi(L)$ under $H(L)$. We have

$$\tau_L(u)[\phi(x')(dx')^{1/2}] = \Phi(u^{-1}\cdot(x',0)) = \Phi((x'-Tx',0)).$$

Now we use the multiplication law in $H(W)$ given in (5.1)(a) to write

$$(x'-Tx',0) = (x',0)\cdot(-Tx',\omega(x',Tx')/2).$$

In light of the identifications in Proposition 4.3, the second factor is equal to $(-Tx',B(x',x')/2)$. This term belongs to $H(L)$, and the character χ_L takes the value $\chi(B(x',x')/2)$ on it. Because of the transformation property of Φ under $H(L)$, we get

$$\Phi(u^{-1}\cdot(x',0)) = \Phi((x',0)\chi(-B(x',x')/2).$$

That is,

$$\tau_L(u)[\phi(x')(dx')^{1/2}] = \phi(x')(dx')^{1/2}\chi(-B(x',x')/2) \qquad (u\in U(L)\subset P(L)). \qquad (5.21)(e)$$

The Lie algebra of $U(L)$ may also be identified with symmetric bilinear forms on L', and we compute

$$d\tau_L(B) \text{ is multiplication by } -d\chi(1)B/2. \qquad (5.21)(f)$$

That is, $d\tau_L$ carries the Lie algebra of $U(L)$ onto multiplication operators by purely imaginary quadratic polynomial functions on L'. In terms of the coordinates chosen in (5.21)(d), this is the span of the multiplication operators $ix'_px'_q$.

These calculations establish the isomorphism in (b) of the proposition for that part of $\mathfrak{gmp}(W)$ lying over the parabolic subalgebra $\mathfrak{p}(L)$ of $\mathfrak{sp}(W)$. We just sketch the rest of the proof. Because $\mathfrak{p}(L)$ and $\mathfrak{p}(L')$ together span $\mathfrak{sp}(W)$, part (b) follows. At the same time we see that the operators in the enveloping algebra of $\mathfrak{gmp}(W)$ are contained in the enveloping algebra of the Heisenberg

Lie algebra; so the smooth vectors for the Schrödinger representation are contained in the smooth vectors for the metaplectic representation. On the other hand, if $\phi \in L^2(L')$ is a smooth vector for the metaplectic representation, then it follows from (b) that $D\phi$ must belong to L^2 for every even polynomial coefficient differential operator D. This forces ϕ to belong to the Schwartz space, proving (a). The rest of the proposition is formal. Q.E.D.

Proposition 5.22. *In the setting of Theorem 5.19, write*

$$\mathcal{H}(L) = \mathcal{H}^{even}(L) \oplus \mathcal{H}^{odd}(L)$$

for the decomposition into even and odd half-densities on L'. Then these spaces are invariant under the metaplectic representation τ of Definition 5.4; they are inequivalent irreducible representations τ^{even} and τ^{odd} of $Mp(W)$, independent of the choice of Lagrangian subspaces L and L'.

Proof. Proposition 5.19 allows us to identify $\mathcal{H}(L)$ with $L^2(\mathbb{R}^n)$, in such a way that the smooth vectors of τ correspond to the Schwartz space. The Lie algebra of $\mathfrak{sp}(W)$ is spanned by the operators

$$ix_p x_q, \quad x_p\frac{\partial}{\partial x_q} + \frac{1}{2}\delta_{pq}, \quad i\frac{\partial^2}{\partial x_p \partial x_q} \qquad (1 \leq p, q \leq n) \tag{5.23)(a}$$

The invariance of the subspaces $\mathcal{H}^{even}(L)$ and $\mathcal{H}^{odd}(L)$ is immediate. For the irreducibility, we use the theory of Harish-Chandra modules. Write $\mathcal{F}$ for the standard Fourier transform on $L^2(\mathbb{R}^n)$:

$$(\mathcal{F}f)(\xi) = \int_{\mathbb{R}^n} f(x)e^{ix\cdot\xi}\,dx. \tag{5.23)(b}$$

The Fourier transform induces an automorphism θ of order 4 on the Weyl algebra (of polynomial coefficient differential operators) by the requirement

$$\mathcal{F}(Df) = (\theta D)\mathcal{F}(f). \tag{5.23)(c}$$

This is given on generators by the familiar formulas

$$\theta(x_p) = -i\frac{\partial}{\partial x_p}, \qquad \theta(\frac{\partial}{\partial x_p}) = -ix_p. \tag{5.23)(d}$$

Obviously θ restricts to $\mathfrak{sp}(W)$ as an automorphism of order 2. Explicitly,

$$\theta(ix_p x_q) = -i\frac{\partial^2}{\partial x_p \partial x_q}, \quad \theta(x_p\frac{\partial}{\partial x_q} + \frac{1}{2}\delta_{pq}) = x_q\frac{\partial}{\partial x_p} + \frac{1}{2}\delta_{pq}),$$
$$\theta(i\frac{\partial^2}{\partial x_p \partial x_q}) = -ix_p x_q. \tag{5.23)(e}$$

We want to check that θ is a Cartan involution of $\mathfrak{sp}(W)$. Because θ is an involution, it is equivalent to show that the fixed subalgebra $\mathfrak{k}$ is compact. Because $Mp(W)$ is by definition a group of unitary operators, this in turn is equivalent to showing that $\mathfrak{k}$ preserves a family of finite-dimensional subspaces of $\mathcal{H}(L)$ whose union is dense. (Then the group K will be embedded in the direct product of the corresponding finite-dimensional unitary groups.)

Consider now the function $e(x) = \exp(-\sum x_p^2/2)$. Then $\frac{\partial e}{\partial x_p} = -x_p e$. Consequently

$$i\frac{\partial^2 e}{\partial x_p \partial x_q} = i x_p x_q e - i\delta_{pq} e,$$

and therefore

$$i(\frac{\partial^2}{\partial x_p \partial x_q} - x_p x_q)e = -i\delta_{pq} e.$$

By (5.23)(e), the operator on the left side is of the form $X + \theta X$ for some $X \in \mathfrak{sp}(W)$. Combining this with similar formulas for $x_p \frac{\partial}{\partial x_q}$, we find a character $\lambda\colon \mathfrak{k} \to i\mathbb{R}$ so that

$$X \cdot e = \lambda(X) e \qquad (X \in \mathfrak{k} = \mathfrak{sp}(W)^\theta). \tag{5.23)(f}$$

Recall the filtration of the Weyl algebra introduced in Definition 5.13, and the action of the Weyl algebra on $\mathcal{H}(L)$ from Proposition 5.14. Define

$$\mathcal{H}^p(L) = A^p_\chi(W) \cdot e. \tag{5.23)(g}$$

This is a finite-dimensional subspace of $\mathcal{H}(L)$; clearly it consists of polynomials of degree at most p multiplied by e. By Proposition 5.17 and (5.23)(f), we have

$$\text{the action of } \mathfrak{k} \text{ preserves } \mathcal{H}^p(L). \tag{5.23)(h}$$

Now the remarks preceding the definition of e show that K is compact. At the same time we have computed the Harish-Chandra module of K-finite vectors in τ: it is

$$\mathcal{H}^K(L) = \bigcup_p \mathcal{H}^p(L),$$

the space of polynomials times the Gaussian e. The irreducibility we want is equivalent to the algebraic irreducibility of $\mathcal{H}^{K,even}(L)$ and $\mathcal{H}^{K,odd}(L)$ under the enveloping algebra of $\mathfrak{sp}(W)$; that is, under the even part $A^{even}(W)$ of the Weyl algebra.

Write $\mathcal{P}$ for the space of polynomial functions on $\mathbb{R}^n$. There is an obvious linear isomorphism m from $\mathcal{P}$ onto $\mathcal{H}^K(L)$, sending f to $f \cdot e$. Of course m does not respect the actions of the Weyl algebra: for the generators we have

$$\frac{\partial fe}{\partial x_p} = (\frac{\partial f}{\partial x_p} - x_p f)e, \qquad x_q \cdot fe = (x_q f) \cdot e. \tag{5.24)(a}$$

Now the linear transformation of the generators of the Weyl algebra defined by

$$j(\frac{\partial}{\partial x_p}) = \frac{\partial}{\partial x_p} - x_p, \qquad j(x_q) = x_q \tag{5.24)(b}$$

respects the defining relations; so it extends uniquely to an automorphism j of $A(W)$. Then (5.24)(a) gives

$$a \cdot m(f) = m(j(a)f) \qquad (a \in A(W), f \in \mathcal{P}) \tag{5.24)(c}$$

So our irreducibility problem for $\mathcal{H}^K(L)$ is equivalent to the irreducibility of $\mathcal{P}^{even}$ and $\mathcal{P}^{odd}$ under the even polynomial coefficient differential operators; and this is very easy to prove.

It remains to establish the inequivalence of τ^{even} and τ^{odd}. For that, we need to understand the Lie algebra $\mathfrak{k}$ and its representation in τ. We have described $\mathfrak{k}$ fairly explicitly in (5.23)(e), and the map j in (5.24)(b). A straightforward calculation shows that $j(\mathfrak{k})$ has a basis of elements

$$\begin{aligned} E^+_{pq} &= i\left(x_p \frac{\partial}{\partial x_q} + x_q\frac{\partial}{\partial x_p} + \delta_{pq} - \frac{\partial^2}{\partial x_p \partial x_q}\right) \qquad (1 \le p \le q \le n) \\ E^-_{pq} &= x_p\frac{\partial}{\partial x_q} - x_q \frac{\partial}{\partial x_p} \qquad (1 \le p < q \le n). \end{aligned} \tag{5.25}$$

Notice that the second derivative terms act to lower degree by two, and all other terms preserve degree. It follows that $j(\mathfrak{k})$ preserves the filtration of $\mathcal{P}$ by degree, and that the action in the associated graded space $\operatorname{gr}\mathcal{P}$ is given by the same formulas without the second derivative.

Lemma 5.26. *Consider the natural action π of $G = GL(n, \mathbb{C})$ on the space $\mathcal{P}$ of complex polynomials in n variables. Identify the Lie algebra $\mathfrak{g}$ of G with complex $n \times n$ matrices. Then on the standard basis matrices the differential of π is given by*

$$d\pi(e_{pq}) = x_p \frac{\partial}{\partial x_q}.$$

The Lie algebra of $U(n)$ (consisting of $n \times n$ skew-Hermitian matrices) has a basis consisting of elements $e^+_{pq} = i(e_{pq} + e_{qp})$ and $e^-_{pq} = e_{pq} - e_{qp}$. These act by the operators

$$d\pi(e^+_{pq}) = i\left(x_p\frac{\partial}{\partial x_q} + x_q \frac{\partial}{\partial x_p}\right), \quad d\pi(e^-_{pq}) = x_p\frac{\partial}{\partial x_q} - x_q\frac{\partial}{\partial x_p}.$$

Write $\det_{\mathbb{C}}$ *for the determinant character of $U(n)$. Then*

$$d\det_{\mathbb{C}}(e^+_{pq}) = 2i\delta_{pq}, \qquad d\det_{\mathbb{C}}(e^-_{pq}) = 0.$$

This is elementary and standard.

Corollary 5.27. *The Lie algebra $\mathfrak{k}$ is isomorphic to $\mathfrak{u}(n)$. Write $\mathbb{C}_{1/2}$ for the one-dimensional space on which $\mathfrak{u}(n)$ acts by half the differential of the determinant character. Then the action of $j(\mathfrak{k})$ (cf. (5.24)) on* $\operatorname{gr}\mathcal{P}$ *is naturally isomorphic to the action of $\mathfrak{u}(n)$ on $\mathcal{P}\otimes\mathbb{C}_{1/2}$ The isomorphism sends the basis elements $E^{\pm}_{pq}$ of $j(\mathfrak{k})$ to $e^{\pm}_{pq}$.*

In particular, the irreducible representations of $\mathfrak{k} \simeq \mathfrak{u}(n)$ appearing in τ^{even} (respectively τ^{odd}) are $S^k(\mathbb{C}^n)\otimes\mathbb{C}_{1/2}$ with k even (respectively odd).

Proof. The assertions in the first paragraph are clear from (5.25) and Lemma 5.26. Those in the second follow at once (since the natural representation of $U(n)$ on homogeneous polynomials of degree k is irreducible. Q.E.D.

It follows at once from Corollary 5.27 that τ^{even} and τ^{odd} are inequivalent as representations of the maximal compact subgroup K of $Mp(W)$, which completes the proof of Proposition 5.22. Q.E.D.

As a corollary of the proof, we get a description of the group K.

Proposition 5.28. *In the setting of Theorem 5.19, fix a positive-definite bilinear form B on L', and use it to introduce a Fourier transform $\mathcal{F}$ as a unitary operator on $\mathcal{H}(L) \simeq L^2(L', D_{\mathbb{C}}^{1/2})$ (cf. (5.23)(b)). As in Proposition 4.3, identify L' with W/L and L^*, and so identify B with an isomorphism $T_B\colon L' \to L$. Define a linear transformation on $W = L\oplus L'$ by*

$$\sigma_B = \begin{pmatrix} 0 & -T_B^{-1} \\ T_B & 0 \end{pmatrix}.$$

a) The element σ_B belongs to $Sp(W)$. We have $\sigma_B^2 = -1$, so σ_B is a complex structure on W.
b) Conjugation by σ_B defines an involutive automorphism θ_B of $Sp(W)$.
c) The symplectic form ω is the imaginary part of a unique positive definite Hermitian form h_B on the complex vector space (W, σ_B).
d) Write $U(W, h_B)$ for the unitary group of the Hermitian form. Then $U(W, h_B)$ is the group of fixed points of θ_B; it is a maximal compact subgroup of $Sp(W)$. The complex-valued determinant of an automorphism of (W, σ_B) defines a unitary character

$$\det{}_{\mathbb{C}}\colon U(W, h_B) \to \mathbb{C}^{\times}.$$

e) Define

$$\widetilde{U}(W, h_B) = \{(u, z) \in U(W, h_B)\times\mathbb{C}^{\times} \mid \det{}_{\mathbb{C}}(u) = z^2\},$$

the square root of the determinant cover of $U(W, h_B)$. Then projection on the first factor is a two-fold covering

$$\widetilde{U}(W, h_B) \to U(W, h_B),$$

and projection on the second factor is a unitary character

$$\det_{\mathbb{C}}^{1/2}: \widetilde{U}(W, h_B) \to \mathbb{C}^\times.$$

f) The Fourier transform $\mathcal{F}$ is a preimage of σ_B in the metaplectic group (Corollary 5.3). The corresponding maximal compact subgroup K of $Mp(W)$ is naturally isomorphic to $\widetilde{U}(W, h_B)$.

Proof. The proof that $\sigma_B \in Sp(W)$ is similar to the proof in Proposition 4.3 that $u_B \in Sp(W)$; we leave it to the reader. That $\sigma_B^2 = -1$ is obvious, and then (b) follows. For (c), the Hermitian form must be

$$h_B(v, w) = \omega(\sigma_B(v), w) + i\omega(v, w).$$

Conversely, this formula is easily seen to define a Hermitian form (compare the proof of Proposition 4.3). For (d), the fixed points of θ_B consists of the complex-linear elements of $Sp(W)$; and this in turn is obviously the unitary group of h_B. Since $Sp(W)$ is a noncompact simple real group, any compact group of fixed points of an involution must be a maximal compact subgroup. The assertions in (e) are elementary. Finally for (f), we have seen in Proposition 5.22 that a maximal compact subgroup K of $Mp(W)$ may be constructed as a double cover of the maximal compact subgroup of $U(W, h_B)$ of $Sp(W)$; and that K admits a one-dimensional character δ (its action on the Gaussian function e described before (5.23)) whose differential is one half the differential of the determinant character of $U(W, h_B)$. Write $\pi: K \to U(W, h_B)$ for the covering map; then $(\pi, \delta): K \to U(W, h_B) \times \mathbb{C}^\times$ is an isomorphism from K onto $\widetilde{U}(W, h_B)$, as we wished to show. Q.E.D.

6. ADMISSIBLE ORBIT DATA

Suppose G is a Lie group. As explained in the introduction, we need a little more than a coadjoint orbit to hope to construct a unitary representation. With the discussion of the metaplectic representation in the last section, we now have in place all the ideas needed to describe Duflo's version of what that "little more" should be. We begin with an element $f \in \mathfrak{g}^*$, and form the coadjoint orbit

$$X = G \cdot f \simeq G/G_f; \tag{6.1)(a}$$

here of course G_f is the isotropy group for the coadjoint action of G at f, a closed subgroup of G. From the formula for the differentiated coadjoint action given before (2.8)(b), we find

$$\mathfrak{g}_f = \{Y \in \mathfrak{g} \mid f([Y, \mathfrak{g}]) = 0\}. \tag{6.1)(b}$$

Recall from Corollary 2.13 that X carries a G-invariant symplectic structure. On the tangent space at f, the symplectic form is given by

$$T_f(X) \simeq \mathfrak{g}/\mathfrak{g}_f, \qquad \omega_f(Y + \mathfrak{g}_f, Z + \mathfrak{g}_f) = f([Y, Z]). \tag{6.1)(c}$$

Of course the tangent vector $Y+\mathfrak{g}_f$ is just the value at f of the vector field ξ_Y, the coadjoint action of Y (see (2.12)(b)). It is clear that the isotropy action of G_f on $T_f(X)$ preserves the form ω_f; so we get a Lie group homomorphism

$$j_f: G_f \to Sp(\mathfrak{g}/\mathfrak{g}_f, \omega_f). \tag{6.1)(d}$$

In Corollary 5.3 we constructed a natural double covering

$$p: Mp(\mathfrak{g}/\mathfrak{g}_f, \omega_f) \to Sp(\mathfrak{g}/\mathfrak{g}_f, \omega_f), \qquad \ker p = \{1, \epsilon\}.$$

of the symplectic group. We can use the homomorphism j_f to pull this back to a double cover of G_f. Explicitly, we define

$$\widetilde{G}_f = \{(g,m) \in G_f \times Mp(\mathfrak{g}/\mathfrak{g}_f, \omega_f) \mid j_f(g) = p(m)\}. \tag{6.1)(e}$$

Then projection on the first factor defines a double covering

$$p_f: \widetilde{G}_f \to G_f, \qquad \ker p_f = \{1, \epsilon\}. \tag{6.1)(f}$$

That is, $p_f(g,m) = g$. Similarly, projection on the second factor defines a Lie group homomorphism

$$\widetilde{j}_f: \widetilde{G}_f \to Mp(\mathfrak{g}/\mathfrak{g}_f, \omega_f). \tag{6.1)(g}$$

Finally, recall that the metaplectic group was defined as a group of unitary operators on a Hilbert space. The homomorphism $\widetilde{j}_f$ therefore gives rise to (or may be interpreted as) a unitary representation

$$\tau_f: \widetilde{G}_f \to U(\mathcal{H}_f), \qquad \tau_f(\epsilon) = -1. \tag{6.1)(h}$$

We call τ_f the *metaplectic representation of* $\widetilde{G}_f$. By Proposition 5.22, this representation decomposes as $\tau_f = \tau_f^{even} \oplus \tau_f^{odd}$. Various descriptions of the Hilbert space $\mathcal{H}_f$ are given in section 5. (For example, if we write $\mathfrak{g}/\mathfrak{g}_f$ as a direct sum of Lagrangians L and L', then $\mathcal{H}_f$ may be identified with square-integrable half-densities on L'. This is most useful if G_f preserves L and L'; but we will not always be able to arrange that.)

Definition 6.2 (see [6].) Suppose G is a Lie group and $f \in \mathfrak{g}^*$. Use the notation of (6.1); recall also that we have fixed a non-trivial character χ of $\mathbb{R}$ (see (5.1)). An *admissible orbit datum at* f is an irreducible unitary representation $(\pi, \mathcal{H}_\pi)$ of $\widetilde{G}_f$ with the following two properties:

$$\pi(\epsilon) = -1 \tag{6.2)(a}$$

(cf. (6.1)(f)), and

$$d\pi(Y) = d\chi(1) f(Y) \qquad (Y \in \mathfrak{g}_f). \tag{6.2)(b}$$

(Here the scalars all mean the corresponding multiples of the identity operator on $\mathcal{H}_\pi$.) An equivalent formulation is

$$\pi(\exp Y) = \chi(f(Y)) \qquad (Y \in \mathfrak{g}_f). \tag{6.2)(b'}$$

The pair (f, π) will be called an *admissible orbit datum.* If there is an admissible orbit datum at f, we say that the orbit $G \cdot f$ is *admissible.*

The first of the defining properties says that π should be a "genuine" representation, not descending to G_f. To understand the second, notice that (6.1)(b) implies that $f: \mathfrak{g}_f \to \mathbb{R}$ is a Lie algebra homomorphism. Therefore $d\chi(1) \cdot f$ is a Lie algebra homomorphism to $i\mathbb{R}$, the Lie algebra of the unit circle in $\mathbb{C}^\times$. The second condition therefore says that the restriction of π to the identity component $\widetilde{G}_{f,0}$ should be a multiple of a (specified) unitary character.

Notice that the group G acts on admissible orbit data, as follows. Suppose (f, π) is an admissible orbit datum. If $g \in G$, then $g \cdot f \in \mathfrak{g}^*$. Conjugation by g defines an isomorphism c_g from G_f to $G_{g \cdot f}$. The covering $\widetilde{G}_f$ pushes forward under this isomorphism to a double cover of $G_{g \cdot f}$. A little more explicitly, the covering group is just $\widetilde{G}_f$, and the covering map is $c_g \circ p_f$ (notation as in (6.1)(f)). It is easy to check that this covering is naturally isomorphic to the one $\widetilde{G}_{g \cdot f}$ defined in (6.1)(e). Under this isomorphism, the representation π of $\widetilde{G}_f$ is identified with a representation that we call $g \cdot \pi$ of $\widetilde{G}_{g \cdot f}$. Again it is easy to check that $g \cdot \pi$ is an admissible orbit datum at $g \cdot f$; so it makes sense to define $g \cdot (f, \pi) = (g \cdot f, g \cdot \pi)$. Because inner automorphisms act trivially on representations (up to equivalence), the stabilizer of (f, π) is precisely G_f.

For our purposes Duflo's definition of admissible orbit datum is always exactly the right "integrality hypothesis" required in Problem 1.2. Here is our promised refinement of that problem.

Problem 6.3. Suppose G is a type I Lie group, and (f, π) is an admissible orbit datum. Find a construction attaching to (f, π) a unitary representation $\gamma(f, \pi)$ of G. This representation should be close to irreducible, and should depend only on the G orbit of (f, π): that is, $\gamma(f, \pi)$ should be unitarily equivalent to $\gamma(g \cdot (f, \pi))$ for every g in G.

Even in this form the problem is still not perfectly formulated. If G is the double cover of $SL(3, \mathbb{R})$ and f is a nilpotent element with Jordan blocks of sizes 2 and 1, then there are exactly four admissible orbit data at f. Only three of these have associated unitary representations (see [20] and [22], Example 12.4). The example of [17] mentioned in the introduction is also not completely explained. We refer to [22] for a more extensive discussion of the shortcomings of Problem 6.3.

The work of Kirillov and Kostant emphasizes a condition different from admissibility, which is still widely used in work on geometric quantization. We recall this condition, partly for the light it sheds on Definition 6.2.

Definition 6.4 (see [11] or [9], Chapter V.) Suppose G is a Lie group and $f \in \mathfrak{g}^*$. Use the notation of (6.1); recall also that we have fixed a non-trivial character χ of $\mathbb{R}$ (see (5.1)). An *integral orbit datum at* f is an irreducible unitary representation (p, V_p) of G_f with the following property:

$$dp(Y) = d\chi(1)f(Y) \qquad (Y \in \mathfrak{g}_f). \tag{6.4)(a}$$

An equivalent formulation is

$$p(\exp Y) = \chi(f(Y)) \qquad (Y \in \mathfrak{g}_f). \tag{6.4)(a'}$$

(As in Definition 6.2, the scalars mean multiples of the identity operator on V_p.) If there is an integral orbit datum at f, we say that the orbit $G \cdot f$ is *integral.*

The unitary representation p is very often forced by (6.4)(a) to be one-dimensional; obviously this is true if G_f is connected, for example. In any case we can use it to define an equivariant Hermitian vector bundle $G \times_{G_f} V_p$ over the orbit G/G_f. Many descriptions of geometric quantization appear to depend heavily on this vector bundle, and the lack of any obvious analogue of it in the admissible case is at first disconcerting. Ultimately we will argue that the structure provided by Definition 6.2 is more natural. For the moment, we can at least explain why the notions of integral and admissible orbit sometimes coincide.

Proposition 6.5. *In the setting of (6.1), suppose $L \subset \mathfrak{g}/\mathfrak{g}_f$ is a Lagrangian subspace. Define H to be the subgroup of G_f preserving L. Each $h \in H$ defines a linear transformation of L, which has a non-zero determinant* $\det_L(h)$ *Taking the sign of this determinant, we get a character*

$$\mathrm{sgn}_L \colon H \to \{\pm 1\}.$$

Taking the square root of this character defines a double cover

$$1 \to \{1, \epsilon\} \to \widetilde{H} \to H \to 1$$

(cf. Proposition 5.28(e)); it is equipped with a character $\mathrm{sgn}_L^{1/2} \colon \widetilde{H} \to \{\pm 1, \pm i\}$ *which acts by -1 on ϵ.*

This covering of H is naturally isomorphic to the metaplectic covering of H induced by $\widetilde{G}_f$. If $H = G_f$—that is, if G_f preserves the Lagrangian subspace L—then tensoring with $\mathrm{sgn}_L^{1/2}$ *defines a bijection from admissible orbit data at f to integral orbit data at f. In particular, the orbit $G \cdot f$ is admissible if and only if it is integral in this case.*

If G is nilpotent, then G_f always preserves some Lagrangian subspace of $\mathfrak{g}/\mathfrak{g}_f$; so the notions of integral and admissible coincide.

We postpone the proof of Proposition 6.5 to section 7 (see the remarks after the proof of Proposition 7.2).

Here is the first geometric structure we can get from an admissible orbit datum.

Definition 6.6. Suppose G is a Lie group, $f \in \mathfrak{g}^*$, and $(\pi, \mathcal{H}_\pi)$ is an admissible orbit datum at f. Recall from (6.1) the metaplectic representation $(\tau_f, \mathcal{H}_f)$ of $\widetilde{G}_f$. Form the tensor product representation $(\pi \otimes \tau_f, \mathcal{H}_\pi \otimes \mathcal{H}_f)$. By (6.2)(a) and (6.1)(h), this representation is trivial on the kernel $\{1, \epsilon\}$ of the covering; so we may regard it as a representation of G_f. We may therefore define

$$\mathcal{S}_\pi = G \times_{G_f} \mathcal{H}_\pi \otimes \mathcal{H}_f, \tag{6.6)(a}$$

a Hilbert bundle over the orbit $G \cdot f$. One might call this the bundle of *twisted symplectic spinors on* $G \cdot f$. The decomposition of $\mathcal{H}_f$ as a direct sum of even and odd parts passes to the bundle:

$$\mathcal{S}_\pi = \mathcal{S}_\pi^{even} \oplus \mathcal{S}_\pi^{odd}. \tag{6.6)(b}$$

We will also want to consider the (Fréchet) subbundles corresponding to the smooth vectors in the metaplectic representation, such as

$$\mathcal{S}_\pi^{even,\infty} = G \times_{G_f} \mathcal{H}_\pi \otimes \mathcal{H}_f^{even,\infty}. \tag{6.6)(c}$$

Here we mean vectors smooth with respect to the action of the full metaplectic group $Mp(\mathfrak{g}/\mathfrak{g}_f)$. This is the Schwartz space described in Corollary 5.12 and Theorem 5.19(a). (In every case we consider seriously, the orbit datum π will be finite-dimensional; so $\mathcal{H}_\pi$ consists entirely of smooth vectors.) Similarly, we can enlarge $\mathcal{H}_f$ to the corresponding space of distribution vectors (the continuous dual of the smooth vectors, obtaining bundles like

$$\mathcal{S}_\pi^{even,-\infty} = G \times_{G_f} \mathcal{H}_\pi \otimes \mathcal{H}_f^{even,-\infty}. \tag{6.6)(d}$$

The space $\mathcal{H}_f^{-\infty}$ may be identified as in Corollary 5.12 with a space of tempered distributions on $\mathbb{R}^n$.

The representations we want to associate to $G \cdot f$ will be related to spaces of sections of these symplectic spinor bundles. It will be convenient to interpret these sections (of infinite-dimensional bundles) as sections of finite-dimensional bundles over a larger space. We conclude this section by introducing this larger space.

Definition 6.7. Suppose (X, ω_X) is a symplectic manifold of dimension $2n$ (Definition 2.1). The *bundle of infinitesimal Lagrangians on* X is a fiber bundle $\mathcal{B}(X)$ over X. The fiber over a point $x \in X$ is $\mathcal{B}(T_x(X))$ (Definition 4.2), the Lagrangian Grassmannian of Lagrangian subspaces of the tangent space at x to X.

7. Symplectic spinors and the Lagrangian Grassmannian

In this section we will describe a realization of the even half τ^{even} of the metaplectic representation as a space of sections of a line bundle on the Lagrangian Grassmannian. In more traditional representation-theoretic language, we are realizing τ^{even} as a subrepresentation of a degenerate principal series representation, induced from a non-unitary one-dimensional character of a Siegel parabolic subgroup. These results are known to many people; one reference is [12], section 5.

We begin as in Definition 4.2 with a finite-dimensional real symplectic vector space (W, ω) and a Lagrangian subspace L. Write $P(L)$ for the stabilizer of L in $Sp(W)$, and $\chi(L): P(L) \to \mathbb{R}^\times$ for the determinant character (the determinant of the action of $P(L)$ on L). In analogy with Proposition 5.28(e), we define the *square root of the determinant cover of* $P(L)$ by

$$\widetilde{P}(L) = \{(p, z) \in P(L) \times \mathbb{C}^\times \mid \chi(L)(p) = z^2\}. \tag{7.1)(a}$$

Just as in Proposition 5.28(e), projection on the second factor defines a character

$$\chi(L)^{1/2}: \widetilde{P}(L) \to \mathbb{C}^\times; \tag{7.1)(b}$$

it takes values in $\mathbb{R}^\times \cup i\mathbb{R}^\times$.

Proposition 7.2. *The covering $\widetilde{P}(L)$ defined by (7.1) is naturally isomorphic to the covering of $P(L)$ induced by the double cover $Mp(W)$ of $Sp(W)$ (Corollary 5.3). A little more precisely, let $\mathcal{H}(L)$ be the realization of the Schrödinger representation of the Heisenberg group in (5.9), and let τ_L be the representation of $P(L)$ on $\mathcal{H}(L)$ constructed in (5.21). Then the metaplectic representation τ of $Mp(W)$ on $\mathcal{H}(L)$ is given by*

$$\tau(x) = (\chi(L)^{1/2}(x)/|\chi(L)(\overline{x})|^{1/2})\tau_L(\overline{x}).$$

Here $x \in \widetilde{P}(L)$, and we write $\overline{x}$ for its image in $P(L)$.

The factor in front on the right is a character of $\widetilde{P}(L)$ taking values in $\{\pm 1, \pm i\}$. It is trivial on the identity component of $\widetilde{P}(L)$.

Proof. Write $MP(L)$ for the preimage of $P(L)$ in $Mp(W)$. This is a double cover of $P(L)$. By the definition of the metaplectic representation in Corollary 5.3, and the construction of τ_L in (5.21), we find that there is a genuine character ϕ of $MP(L)$ (that is, $\phi(\epsilon) = -1$ for ϵ the non-trivial element of the kernel of the covering map) with the property that $\tau(y) = \phi(y)\tau_L(\overline{y})$. (Here $y \in MP(L)$ and $\overline{y}$ is its image in $P(L)$.) What we propose to show is

$$\phi(y)^2 = \operatorname{sgn} \chi(L)(\overline{y}) \qquad (y \in MP(L)) \tag{7.3)(a}$$

It follows that

$$(\phi(y)|\chi(L)(\overline{y})|^{1/2})^2 = \chi(L)(\overline{y}) \qquad (y \in MP(L)) \tag{7.3)(b}$$

The character $\phi(y)|\chi(L)(\overline{y})|^{1/2}$ therefore provides the isomorphism we want from $MP(L)$ to $\widetilde{P}(L)$: explicitly, it sends y to the pair $(\overline{y}, \phi(y)|\chi(L)(\overline{y})|^{1/2})$. The formula for τ in the proposition also follows immediately.

So we need only prove (7.3)(a). The differential of τ_L is computed explicitly in (5.21). It takes values in the commutator subalgebra of $A^{2,even}_{\chi}(W)_{\mathbb{R}}$, just as the differential of τ does. Consequently the differential of ϕ is zero; that is, ϕ is trivial on the identity component of $MP(L)$. Now (7.3)(a) follows for y in the identity component. Since ϕ^2 does descend to a character of $P(L)$, it remains only to prove (7.3)(a) for some element y such that $\chi(L)(\overline{y}) < 0$.

Now suppose we are in the setting of Theorem 5.19 and Proposition 5.28. The representation space $\mathcal{H}(L)$ for τ is the space of square-integrable half-densities on L', which is an inner product space. We therefore have a well-defined orthogonal group $O(L') \subset GL(L') \simeq GL(L) \subset P(L)$. Choose an element $\overline{y} \in O(L')$ of determinant -1, and a preimage y in $MP(L)$. Recall from (5.23) the Gaussian $e \in \mathcal{H}(L)$. By (5.21)(c), $\tau_L(\overline{y})(e) = e$. On the other hand, $\overline{y}$ belongs to the unitary group $U(W, h_B)$ described in Proposition 5.28; so Corollary 5.27 implies that $\tau(y)(e) = ae$, for a scalar a which is a square root of $\det_{\mathbb{C}}(\overline{y})$. Now this complex determinant character on the unitary group restricts to the real determinant on the orthogonal subgroup; so $\det_{\mathbb{C}}(\overline{y}) = -1$, and $a = \pm i$. Consequently $\phi(y) = \pm i$, and (7.3)(a) follows. Q.E.D.

Proposition 6.4 is an immediate corollary of Proposition 7.2: the subgroup H of G_f is just the preimage in G_f of the parabolic subgroup $P(L) \subset Sp(\mathfrak{g}/\mathfrak{g}_f)$.

Suppose $(\tau, \mathcal{H})$ is a unitary representation of a Lie group G, and $\mathcal{H}^\infty$ is the subspace of smooth vectors; this is a Fréchet subrepresentation of τ. We want to define a corresponding "superrepresentation" $\mathcal{H}^{-\infty}$ of distribution vectors. Roughly speaking this should be the dual space of $\mathcal{H}^\infty$. The difficulty is that the dual space doesn't contain $\mathcal{H}$. If $w \in \mathcal{H}$, then the linear functional λ_w on $\mathcal{H}^\infty$ defined by

$$\lambda_w(v) = \langle v, w \rangle$$

does indeed belong to the dual space $(\mathcal{H}^\infty)^*$, but the map sending w to λ_w is conjugate-linear. We therefore define $\mathcal{H}^{-\infty}$ to be the Hermitian dual of $\mathcal{H}^\infty$. This means that as a real vector space, $\mathcal{H}^{-\infty} = (\mathcal{H}^\infty)^*$, but complex multiplication is defined by

$$(z \cdot \lambda)(v) = \lambda(\overline{z} \cdot v).$$

With this definition the map $w \mapsto \lambda_w$ above provides an inclusion of $\mathcal{H}$ in $\mathcal{H}^{-\infty}$. The transpose of the representation τ^∞ defines an algebraic representation $\tau^{-\infty}$ of G on $\mathcal{H}^{-\infty}$ by continuous operators. Without further assumptions on τ, it need not be a continuous representation, however.

Proposition 7.4. *Suppose $(\tau, \mathcal{H})$ is a metaplectic representation of $Mp(W)$ (Corollary 5.3). Write $\mathcal{H} = \mathcal{H}^{even} \oplus \mathcal{H}^{odd}$ as in Proposition 5.22, and $\mathcal{H}^\infty$ for the subspace of smooth vectors. Finally write $\mathcal{H}^{-\infty}$ for the Hermitian dual of*

$\mathcal{H}^\infty$, the space of distribution vectors of τ. Suppose $L \subset W$ is a Lagrangian subspace, $P(L)$ is its stabilizer in $Sp(W)$, and $U(L)$ is the unipotent radical of $P(L)$. Write also $U(L)$ for the identity component of the inverse image of $U(L)$ in $Mp(W)$. Identify the preimage $\widetilde{P}(L)$ of $P(L)$ in $Mp(W)$ as in Proposition 7.2.

a) *For each Lagrangian subspace L, the space $\mathcal{L}^*(L)$ of $U(L)$-fixed even distribution vectors has dimension 1:*

$$\mathcal{L}^*(L) = \{\lambda \in \mathcal{H}^{even,-\infty} \mid \tau(u)(\lambda) = \lambda \quad (u \in U(L)\}.$$

b) *The representation of $\widetilde{P}(L)$ on $\mathcal{H}^{-\infty}$ preserves $\mathcal{L}^*(L)$, and acts there by the character $(\overline{\chi(L)^{1/2}})^{-1}$.*

c) *Write $\mathcal{L}(L)$ for the Hermitian dual space of $\mathcal{L}^*(L)$. These lines may be assembled into a smooth $Mp(W)$-equivariant line bundle $\mathcal{L}$ on the Lagrangian Grassmannian*

$$\mathcal{B}(W) \simeq Sp(W)/P(L) \simeq Mp(W)/\widetilde{P}(L).$$

It is isomorphic to the line bundle induced by the character $\chi(L)^{1/2}$ of $\widetilde{P}(L)$.

d) *There is a canonical $Mp(W)$-equivariant embedding*

$$\gamma: \mathcal{H}^{even,\infty} \to C^\infty(\mathcal{B}(W), \mathcal{L}),$$

defined as follows. Suppose $v \in \mathcal{H}^{even,\infty}$ and $L \in \mathcal{B}(W)$. We need to specify the value $\gamma(v)(L)$ of the section $\gamma(v)$ at the point L. This is an element of $\mathcal{L}(L)$, and therefore a linear functional on $\mathcal{L}^*(L)$. Its value at $\lambda \in \mathcal{L}^*(L)$ is

$$\gamma(v)(L)(\lambda) = \lambda(v).$$

Here we are using the description of λ from (a) as a distribution vector in $\mathcal{H}^{even,-\infty}$.

Proof. For (a), we use the realization of $\mathcal{H}^\infty$ in Corollary 5.12, as the Schwartz space $\mathcal{S}(L', D_{\mathbb{C}}^{1/2})$ of rapidly decreasing half-densities on L'. The space $\mathcal{H}^{-\infty}$ of is then identified with tempered distributions. The action of $\widetilde{P}(L)$ is computed in (5.21) and Proposition 7.2. Since the Lie algebra of $U(L)$ acts by multiplication by purely imaginary quadratic polynomials, it is easy to see that the $U(L)$-fixed distribution vectors are spanned by evaluation and first derivatives at the origin. Only evaluation is even; so $\mathcal{L}^*(L)$ is spanned by the linear functional λ defined by

$$\lambda(\phi(x')(dx')^{1/2}) = \phi(0).$$

(In this definition we have implicitly chosen a half-density $(dx')^{1/2}$.) Because $\widetilde{P}(L)$ normalizes $U(L)$, it automatically preserves $\mathcal{L}^*(L)$; the formula for the action on λ follows from (5.21)(c) and Proposition 7.2. (The complex conjugate arises because of the twist in the complex structure on $\mathcal{H}^{-\infty}$.) Part (c) is immediate from (b). The mapping defined in (d) is non-zero and $Mp(W)$-equivariant to the space of arbitrary sections of $\mathcal{L}$. The first thing that requires proof is that $\gamma(v)$ is a smooth section, and that γ is a continuous map. To see this, choose a coordinate neighborhood X of L in $\mathcal{B}(W) \simeq Mp(W)/\widetilde{P}(L)$ that lifts to $t(X) \subset Mp(W)$. This means that $t(X)$ is a smooth submanifold of $Mp(W)$, and that group multiplication identifies $t(X) \times \widetilde{P}(L)$ with an open subset of $Mp(W)$. Now the smooth structure on $\mathcal{L}^*$ arises from its identification with an induced bundle. It follows that the section $x \mapsto t(x) \cdot \lambda$ is a smooth local trivialization of $\mathcal{L}^*$ over X. To say that $\gamma(v)$ is smooth at L therefore means precisely that the function $x \mapsto (t(x) \cdot \lambda)(v)$ is smooth in x. But this may be written as $\lambda(\tau^\infty(t(x)^{-1})(v))$, which is smooth in x because τ^∞ is a smooth representation. This argument also shows that the map γ is continuous (from the Fréchet space of smooth vectors to that of smooth sections of $\mathcal{L}$). Q.E.D.

Proposition 7.4 is our promised realization of τ^{even} in a degenerate principal series representation. Here are some useful technical facts about it.

Proposition 7.5. *Suppose we are in the setting of Proposition 7.4.*

a) The map γ is an isomorphism from $\mathcal{H}^{even,\infty}$ onto a closed subspace of $C^\infty(\mathcal{B}(W), \mathcal{L})$.

b) Write $D_{\mathbb{C}}$ for the complex line bundle of densities on $\mathcal{B}(W)$. Then $D_{\mathbb{C}}$ is isomorphic to the line bundle induced by the character $|\chi(L)|^{n+1}$ of $P(L)$.

c) There is a natural Hermitian pairing between smooth sections of $\mathcal{L}$ and smooth sections of $\mathcal{L}^* \otimes D_{\mathbb{C}}$ (compare (5.7)). For this reason we can define $C^{-\infty}(\mathcal{B}(W), \mathcal{L}^* \otimes D_{\mathbb{C}})$ (the space of distribution sections) to be the continuous Hermitian dual of $C^\infty(\mathcal{B}(W), \mathcal{L})$.

d) The transpose of γ is an $Mp(W)$-equivariant continuous surjection

$$\gamma^*: \to C^{-\infty}(\mathcal{B}(W), \mathcal{L}^* \otimes D_{\mathbb{C}}) \to \mathcal{H}^{even,-\infty}.$$

e) The map γ^* of (d) restricts to a continuous surjection

$$\gamma^{*,\infty}: \to C^{\infty}(\mathcal{B}(W), \mathcal{L}^* \otimes D_{\mathbb{C}}) \to \mathcal{H}^{even,\infty}.$$

f) The map γ of Proposition 7.4 extends to a continuous embedding with closed range

$$\gamma^{-\infty}: \mathcal{H}^{even,-\infty} \to C^{-\infty}(\mathcal{B}(W), \mathcal{L}).$$

Sketch of proof. Part (a) follows from the general theory of smooth globalizations of Casselman and Wallach (see [5]): the smooth globalization of a finite length Harish-Chandra module may be realized as the space of smooth vectors in any reflexive Banach space globalization. In this case we compare the two globalizations of the metaplectic Harish-Chandra module given by the metaplectic Hilbert space and by the degenerate principal series. The description of the density bundle amounts to a calculation of the character by which $P(L)$ acts on the top exterior power of $\mathfrak{g}/\mathfrak{p}(L)$. Part (c) is essentially a definition, and (d) follows from (a) and the Hahn-Banach theorem. Finally (e) and (f) follow from (d) and Proposition 7.4 (respectively), again by the general results in [5] on uniqueness of distribution globalizations. Q.E.D.

Definition 7.6. Suppose we are in the setting of Definition 6.6. Write $X = G \cdot f$ for the coadjoint orbit of f, a symplectic manifold. Write $\mathcal{B}(X)$ for the bundle of infinitesimal Lagrangians (Definition 6.6). Then the admissible orbit datum gives rise to a G-equivariant vector bundle on $\mathcal{B}(X)$, as follows. The fiber at a Lagrangian L in $\mathfrak{g}/\mathfrak{g}_f$ is by definition $\mathcal{H}_\pi \otimes \mathcal{L}(L)$. Here $\mathcal{H}_\pi$ is the representation space of the admissible orbit datum, and $\mathcal{L}(L)$ is the line defined in Proposition 7.4(c) (using the metaplectic representation $\mathcal{H}_f$). We write this vector bundle as $\mathcal{V}_\pi$.

8. Existence of Lagrangian coverings

In this section we will prove Theorem 1.10. We therefore fix a complex reductive algebraic group G, and a coadjoint orbit

$$X = G \cdot f \subset \mathfrak{g}^*. \tag{8.1)(a}$$

We fix also a Borel subgroup B of G, with unipotent radical N. This defines an $\mathrm{Ad}^*(B)$-invariant linear subspace

$$\mathfrak{n}^\perp = \{\phi \in \mathfrak{g}^* \mid \phi|_\mathfrak{n} = 0\} \subset \mathfrak{g}^*. \tag{8.1)(b}$$

Because G is reductive, we may use an invariant symmetric form to identify $\mathfrak{g}^*$ with $\mathfrak{g}$. Under such an identification, $\mathfrak{n}^\perp$ is sent to $\mathfrak{b}$. We will also mention the B-invariant subspace

$$\mathfrak{b}^\perp = \{\phi \in \mathfrak{g}^* \mid \phi|_\mathfrak{b} = 0\} \subset \mathfrak{n}^\perp \subset \mathfrak{g}^*. \tag{8.1)(c}$$

In the identification of $\mathfrak{g}^*$ with $\mathfrak{g}$, $\mathfrak{b}^\perp$ corresponds to $\mathfrak{n}$. Here is the easy part of what we want to prove.

Lemma 8.2. *In the setting (8.1), the intersection*

$$X_\mathfrak{b} = X \cap \mathfrak{n}^\perp$$

is a non-empty B-invariant closed subset of X. If X is semisimple, then $X_\mathfrak{b}$ is the union of finitely many closed B orbits. If X is nilpotent, then $X_\mathfrak{b} \subset \mathfrak{b}^\perp$.

Proof. We identify $\mathfrak{g}^*$ with $\mathfrak{g}$ using an invariant symmetric bilinear form as above. For the first assertion, only the non-emptiness requires proof. The element f of X corresponds to an element $Z \in \mathfrak{g}$. The subspace $\mathbb{C}Z$ is a solvable subalgebra of $\mathfrak{g}$, and is therefore contained in a maximal solvable subalgebra $\mathfrak{b}'$. By definition $\mathfrak{b}'$ is a Borel subalgebra, so it is conjugate by G

to $\mathfrak{b}$: $\mathrm{Ad}(g)(\mathfrak{b}') = \mathfrak{b}$ for some $g \in G$. Since $Z \in \mathfrak{b}'$, it follows that $\mathrm{Ad}(g)(Z) \in \mathfrak{b}$. Expressed in terms of X, this says that $\mathrm{Ad}^*(g)(f) \in \mathfrak{n}^\perp$, and therefore that $X_\mathfrak{b}$ is non-empty.

For the second assertion, fix a maximal torus $H \subset B$; this is a Cartan subgroup of G. Each semisimple conjugacy class for B in $\mathfrak{b}$ meets $\mathfrak{h}$ exactly once (see for example [4], Theorem III.10.6). By [4], Theorem III.9.2, each B orbit on $X_\mathfrak{b}$ is closed; so we need only show that these orbits are finite in number. This amounts to the fact that a semisimple orbit in $\mathfrak{g}$ meets $\mathfrak{h}$ finitely often (in fact in a single orbit of the Weyl group of H). This is well known.

The last assertion says that the nilpotent elements in $\mathfrak{b}$ are exactly those in $\mathfrak{n}$. This is [4], Theorem III.10.6(4). Q.E.D.

As an orbit for an algebraic group action, X is a locally closed algebraic subvariety of $\mathfrak{g}^*$. Consequently $X_\mathfrak{b}$ is a locally closed algebraic subvariety of the vector space $\mathfrak{n}^\perp$. We may therefore write $X_\mathfrak{b}$ as the union of irreducible components:

$$X_\mathfrak{b} = X_\mathfrak{b}^1 \cup \cdots \cup X_\mathfrak{b}^r. \tag{8.3}$$

Here each $X_\mathfrak{b}^i$ is an irreducible locally closed B-stable algebraic subvariety of $\mathfrak{n}^\perp$. The intersection of any two components is a proper subvariety of each, and hence of lower dimension than either.

Lemma 8.4. *In the setting (8.3), suppose $\phi \in X_\mathfrak{b}$. Then the linear functional $\psi = \phi|_\mathfrak{b} \in \mathfrak{b}^*$ vanishes on $\mathfrak{n} = [\mathfrak{b}, \mathfrak{b}]$. It takes a constant value ψ^i on each component $X_\mathfrak{b}^i$.*

Proof. As in the proof of Lemma 8.1, we use an invariant bilinear form to translate into statements about conjugacy classes of G in $\mathfrak{g}$. So we are fixing a conjugacy class X, and $X_\mathfrak{b}$ is its intersection with $\mathfrak{b}$. That each ϕ vanishes on $\mathfrak{n}$ follows from the remark after (8.1)(b). The operation of restricting linear functionals to $\mathfrak{b}$ amounts to projecting from $\mathfrak{b}$ to $\mathfrak{b}/\mathfrak{n} \simeq \mathfrak{h}$. For Z in $\mathfrak{b}$, the element of $\mathfrak{h}$ obtained in this way represents the conjugacy class of the semisimple part of Z; so there are only finitely many possibilities for ψ as ϕ varies over $X_\mathfrak{b}$. It follows at once that ψ is constant on components of $X_\mathfrak{b}$. Q.E.D.

Proposition 8.5 (Spaltenstein [19]). *In the setting (8.3), each component $X_\mathfrak{b}^i$ has dimension equal to half the dimension of X.*

Sketch of proof. For X nilpotent, this is the main theorem in [19]. The general case may easily be reduced to that, using the Jordan decomposition. We omit the details. Q.E.D.

Theorem 8.6 (Ginsburg [7], Theorem 4.1). *Suppose X is a complex Poisson algebraic variety endowed with a Hamiltonian action of the solvable algebraic group B (Definition 3.7) with moment map $\mu_B\colon X \to \mathfrak{b}^*$, and $\Omega \subset \mathfrak{b}^*$ is a coadjoint orbit. Then $\mu_B^{-1}(\Omega)$ is a co-isotropic subvariety of X (Definition 4.14).*

Corollary 8.7 (Ginsburg [7], Proposition 4.3). *In the setting of (8.1) and (8.3), each component $X^i_{\mathfrak{b}}$ is Lagrangian in X.*

Proof. We will apply Ginsburg's theorem to the symplectic variety $X = G \cdot f$ of (8.1)(a), and the solvable group B. The action of G on X is Hamiltonian, with tautological moment map μ_G the inclusion of X in $\mathfrak{g}^*$. It follows that the action of B on X is Hamiltonian, with moment map μ_B given by μ_G composed with the projection $\mathfrak{g}^* \to \mathfrak{b}^*$ (restriction of linear functionals). Fix a component $X^i_{\mathfrak{b}}$ of $X_{\mathfrak{b}}$ (cf. (8.3)), and define $\psi^i \in \mathfrak{b}^*$ to be the constant value of μ_B on $X^i_{\mathfrak{b}}$ (Lemma 8.4). Because $X^i_{\mathfrak{b}}$ is B-stable, $\Omega^i = \{\psi^i\}$ is an orbit of B. (This is also a consequence of the fact that ψ^i vanishes on $[\mathfrak{b}, \mathfrak{b}] = \mathfrak{n}$.) The inverse image of Ω^i under μ_B is contained in $X_{\mathfrak{b}}$ (by Lemma 8.2), and so must be a union of components:

$$\mu_B^{-1}(\Omega^i) = \bigcup_{\{j \mid \psi^j = \psi^i\}} X^j_{\mathfrak{b}}.$$

By Theorem 8.6, this union, and in particular its irreducible component $X^i_{\mathfrak{b}}$, is a co-isotropic subvariety of X. By Proposition 4.13, every tangent space $T_x(X^i_{\mathfrak{b}})$ is a co-isotropic subspace of the symplectic vector space $T_x(X)$. If x is a smooth point of $X^i_{\mathfrak{b}}$, this tangent space has dimension exactly half the dimension of $T_x(X)$ (Proposition 8.5), and is therefore Lagrangian (cf. (4.1)). By Definition 4.17, $X^i_{\mathfrak{b}}$ is a Lagrangian subvariety, as we wished to show. Q.E.D.

(The proof in [7] that $X^i_{\mathfrak{b}}$ is isotropic requires a little elucidation; we prefer to deduce it from Spaltenstein's Proposition 8.5, which Ginsburg claims as a corollary.)

Proof of Theorem 1.10. In the setting of Corollary 8.7, fix a component $X^i_{\mathfrak{b}}$. Write L for the smooth part of this component, a smooth Lagrangian subvariety of X. (We have dropped the superscript i since it will be fixed henceforth.) Because $X^i_{\mathfrak{b}}$ is B-stable, L must be as well. Define

$$Q = \{q \in G \mid q \cdot L = L\}, \tag{8.8)(a}$$

a subgroup of G containing B. Any subgroup containing B is parabolic, so

$$M = \{g \cdot L \mid g \in G\} \simeq G/Q \tag{8.8)(b}$$

is a partial flag variety for G. Now L is a smooth algebraic variety with an algebraic action of Q; so we can form a fiber bundle

$$Z = G \times_Q L \xrightarrow{\rho} G/Q \simeq M. \tag{8.8)(c}$$

A point of Z is an equivalence class in $G \times L$, with (gq, l) equivalent to $(g, q \cdot l)$ whenever $g \in G$, $q \in Q$, and $l \in L$. The action of G on X gives a natural map

$G \times L \to X$, $(g,l) \mapsto g \cdot l$. It is now clear that this map is constant on the equivalence classes defining Z; so it descends to an algebraic map

$$\pi: Z \to X, \qquad \pi(g,l) = g \cdot l. \tag{8.8)(d}$$

We have now constructed all the spaces and maps required for a Lagrangian covering (Definition 1.9). By construction ρ is a fibration, and the fact that $\pi \times \rho$ is injective is trivial. Define M_f to be the fiber of π over the base point f of X. (The notation is chosen because M is the space of translates of L in X, and M_f may be identified with the subvariety of Lagrangians in M containing f.) Because π is a submersion, M_f is smooth. The isotropy group G_f acts on M_f, so we can form the (smooth) fiber product $G \times_{G_f} M_f$. For formal reasons there is a G-equivariant algebraic map

$$i: G \times_{G_f} M_f \to Z, \qquad i(g,m) = g \cdot m. \tag{8.8)(e}$$

It is easy to check that i is a bijection on points. The tangent space to X at a point $z \in M_f$ fits in a short exact sequence

$$0 \to T_z(M_f) \to T_z(Z) \to T_{eG_f}(G/G_f) \to 0.$$

Here the second map is $d\pi$, and the first is the differential of the inclusion of M_f in Z. There is a similar exact sequence for $T_z(G \times_{G_f} M_f)$, and di provides a map from the first exact sequence to the second. The five lemma then guarantees that di is an isomorphism, and, it follows that i must be an isomorphism. Therefore π is a fibration. (It also follows that M_f is smooth.) The last requirement in Definition 1.9 (that $\pi \times \rho$ embeds Z in $X \times M$) follows similarly by inspecting tangent spaces; we omit the details. Q.E.D.

9. Construction of representations

In this section we will fill in some details in the construction of representations outlined at the end of the introduction. We work with a complex reductive group G, and a coadjoint orbit $X = G \cdot f \simeq G/G_f$ (cf. (8.1)). Fix a metaplectic representation $(\tau_f, \mathcal{H}_f)$ of the metaplectic cover $Mp(\mathfrak{g}/\mathfrak{g}_f)$, and the corresponding cover $\widetilde{G}_f$ of G_f as in (6.1). We fix also an admissible orbit datum $(\pi, \mathcal{H}_\pi)$ at f (Definition 6.2). Because the group of connected components of $\widetilde{G}_f$ is finite, the representation π is necessarily finite-dimensional. As in Definition 6.6, this gives rise to a Hilbert bundle

$$\mathcal{S}_\pi = G \times_{\widetilde{G}_f} \mathcal{H}_\pi \otimes \mathcal{H}_f, \tag{9.1)(a}$$

and to various Fréchet subbundles like $\mathcal{S}_\pi^{even,\infty}$ (cf. (6.6)(c)).

We recall from Definition 6.7 the bundle $\mathcal{B}(X)$ of infinitesimal Lagrangians in X, and from Definition 7.6 the finite-dimensional vector bundle

$$\mathcal{V}_\pi \to \mathcal{B}(X) \tag{9.1)(b}$$

over $\mathcal{B}(X)$. We write $C^\infty(\mathcal{B}(X), \mathcal{V}_\pi)$ for its space of smooth sections. Using Proposition 7.4, we find a natural inclusion

$$\gamma_X: C^\infty(X, \mathcal{S}_\pi^{even,\infty}) \hookrightarrow C^\infty(\mathcal{B}(X), \mathcal{V}_\pi). \tag{9.1)(c}$$

The image of γ_X consists of those smooth sections of $\mathcal{V}_\pi$ whose restriction to each fiber of $\mathcal{B}(X)$ belongs to the image of the corresponding map γ in Proposition 7.4. (Recall that the fiber over f of $\mathcal{B}(X)$ is the Lagrangian Grassmannian of the symplectic vector space $\mathfrak{g}/\mathfrak{g}_f$.)

As in (8.8), we fix a component $X_\mathfrak{b}^i$, and write Q for its stabilizer in G (a parabolic subgroup) and L for its smooth locus (a locally closed smooth Lagrangian subvariety of X). As in (8.8), we write $M \simeq G/Q$ for the family of translates of L, and $Z = G \times_Q L$. The construction of (1.11)(c) provides a map of bundles over X

$$\tau: Z \to \mathcal{B}(X). \tag{9.2)(a}$$

Using τ, we can pull the bundle $\mathcal{V}_\pi$ back to a G-equivariant vector bundle

$$\tau^*(\mathcal{V}_\pi) \to Z. \tag{9.2)(b}$$

(The rank of this vector bundle is the dimension of the admissible orbit datum π.) Smooth sections of $\mathcal{V}_\pi$ pull back to smooth sections of $\tau^*(\mathcal{V}_\pi)$:

$$\tau^* : C^\infty(\mathcal{B}(X), \mathcal{V}_\pi)) \to C^\infty(Z, \tau^*(\mathcal{V}_\pi). \tag{9.2)(c}$$

Composing the maps of (9.1)(c) and (9.2)(c) gives a map

$$\tau^* \circ \gamma_X: C^\infty(X, \mathcal{S}_\pi^{even,\infty}) \to C^\infty(Z, \tau^*(\mathcal{V}_\pi)). \tag{9.2)(d}$$

The next ingredient we need is a finite-dimensional smooth representation (γ, W_γ) of the parabolic subgroup Q. Such a representation gives a vector bundle

$$\mathcal{W}_\gamma \to G/Q \simeq M, \tag{9.3)(a}$$

which pulls back by the fibration ρ to a vector bundle

$$\rho^*(\mathcal{W}_\gamma) \to Z. \tag{9.3)(b}$$

In this way the space of smooth sections of $\mathcal{W}_\gamma$ may be identified with a space of sections of $\rho^*(\mathcal{W}_\gamma)$:

$$\rho^*: C^\infty(M, \mathcal{W}_\gamma) \hookrightarrow C^\infty(Z, \rho^*(\mathcal{W}_\gamma)). \tag{9.3)(c}$$

Last but not least, we need a G-equivariant isomorphism of vector bundles

$$j_{\gamma,\pi}: \tau^*(\mathcal{V}_\pi) \xrightarrow{\sim} \rho^*(\mathcal{W}_\gamma). \tag{9.3)(d}$$

The existence of this isomorphism is of course not automatic; it imposes a strong constraint on γ, which may be impossible to satisfy. Now (9.3)(c) and (9.2)(d) define G-invariant spaces of sections of the same vector bundle over Z; so it makes sense to consider their intersection. This intersection is the representation we want. Here is a precise statement.

Definition 9.4. Suppose we are in the setting of (9.1)–(9.3). That is, we fix a coadjoint orbit $X = G\cdot f$ for a complex reductive group G, an admissible orbit datum $(\pi, \mathcal{H}_\pi)$ at f (Definition 6.2), a smooth Lagrangian L (constructed as in (8.8)) and stabilized by a parabolic Q. Fix also a finite-dimensional smooth representation (γ, W_γ) of Q, and a G-equivariant isomorphism of vector bundles $j_{\gamma,\pi}$ as in (9.3)(d). (Recall that γ and $j_{\gamma,\pi}$ need not exist.) Then the smooth representation of G attached to $(f,\pi,L,\gamma,j_{\gamma,\pi})$ is by definition

$$V(f,\pi,L,\gamma,j_{\gamma,\pi}) = \rho^*(C^\infty(M,\mathcal{W}_\gamma)) \cap j_{\gamma,\pi}(\tau^* \circ \gamma_X(C^\infty(X,\mathcal{S}_\pi^{even,\infty}))),$$

a space of sections of $\rho^*(\mathcal{W}_\gamma)$ over Z. Thus $V(f,\pi,L,\gamma,j_{\gamma,\pi})$ may be identified with a G-invariant subspace of $C^\infty(M,\mathcal{W}_\gamma)$, which in turn is the space of smooth vectors in the degenerate principal series representation induced from γ on Q (non-normalized induction). In terms of the normalized induction of Example 5.8, this is

$$V(f,\pi,L,\gamma,j_{\gamma,\pi}) \subset \mathrm{Ind}_Q^G(\gamma\otimes\rho_{G/Q}^{-1}).$$

Example 9.5. Suppose $G = GL(4,\mathbb{C})$. We identify $\mathfrak{g}^*$ with the Lie algebra $M(4,\mathbb{C})$ (consisting of all four by four complex matrices), sending a matrix T to the linear functional f_T defined by $f_T(S) = \mathrm{tr}\, TS$. We consider the coadjoint orbit X consisting of all rank two matrices f with $f^2 = 0$. These are the nilpotent matrices corresponding to the partition $2+2$ of 4; the orbit has dimension 8. We can take for a representative the matrix (written with two by two blocks)

$$f = \begin{pmatrix} 0 & I \\ 0 & 0 \end{pmatrix}.$$

The isotropy group is the centralizer in G of the matrix f, namely

$$G_f = \left\{ \begin{pmatrix} A & B \\ 0 & A \end{pmatrix} \mid A \in GL(2,\mathbb{C}), B \in M(2,\mathbb{C}) \right\}. \qquad (9.5)(a)$$

Because G is complex, the metaplectic cover $\widetilde{G}_f$ is trivial (isomorphic to $G_f \times \mathbb{Z}/2\mathbb{Z}$). There is only one admissible orbit datum π: it is trivial on G_f, and acts by the non-trivial character on $\mathbb{Z}/2\mathbb{Z}$.

The variety $X_\mathfrak{b}$ is easily calculated by writing down the condition for an upper triangular matrix to have square zero; we find

$$X_\mathfrak{b} = \left\{ T = \begin{pmatrix} 0 & a & b & c \\ 0 & 0 & d & e \\ 0 & 0 & 0 & f \\ 0 & 0 & 0 & 0 \end{pmatrix} \mid ad = df = ae + bf = 0, \mathrm{rank}\, T = 2 \right\} \qquad (9.5)(b)$$

The rank condition picks out an open subset of the four-dimensional variety determined by the three equations. There are exactly two irreducible components:

$$X_{\mathfrak{b}}^1 = \left\{ T = \begin{pmatrix} 0 & a & b & c \\ 0 & 0 & 0 & e \\ 0 & 0 & 0 & f \\ 0 & 0 & 0 & 0 \end{pmatrix} \mid ae + bf = 0, (a, b) \neq 0, (e, f) \neq 0 \right\} \quad (9.5)(c)$$

and

$$X_{\mathfrak{b}}^2 = \left\{ T = \begin{pmatrix} 0 & 0 & b & c \\ 0 & 0 & d & e \\ 0 & 0 & 0 & 0 \\ 0 & 0 & 0 & 0 \end{pmatrix} \mid be - dc \neq 0 \right\}. \quad (9.5)(d)$$

The second is preserved by G_f, and therefore leads to a polarization of X. We therefore concentrate on the first. It is smooth, and so equal to L; the stabilizer of L is the standard parabolic subgroup Q with Levi factor $GL(1) \times GL(2) \times GL(1)$. Notice that L contains the base element f. Calculations are simplified by the fact that Q acts transitively on L; so

$$L \simeq Q/Q_f, \quad (9.6)(a)$$

$$Q_f = Q \cap G_f = \left\{ \begin{pmatrix} A & B \\ 0 & A \end{pmatrix} \mid A \in GL(2) \text{ upper triangular}, B \in M(2, \mathbb{C}) \right\}.$$

It follows that $Z \simeq G/Q_f$. The equivariant line bundle $\tau^* \mathcal{V}_\pi$ is necessarily induced by a character α of Q_f. Proposition 7.4 implies that α is given by the square root of the absolute value of the (real) determinant of Q_f acting on the tangent space $\mathfrak{q}/\mathfrak{q}_f$ of L at f. This is

$$\alpha \begin{pmatrix} A & B \\ 0 & A \end{pmatrix} = |xz^{-1}|^2, \qquad A = \begin{pmatrix} x & y \\ 0 & z \end{pmatrix}. \quad (9.6)(b)$$

The condition (9.3)(d) on the character γ of Q is simply $\gamma|_{Q_f} = \alpha$. Such a character γ is determined by an arbitrary complex character β of $\mathbb{C}^\times$, by the formula

$$\gamma \begin{pmatrix} r & * & * \\ 0 & S & * \\ 0 & 0 & t \end{pmatrix} = \beta((\det S)(rt)^{-1})|rt^{-1}|^2 \qquad (r, t \in \mathbb{C}^\times, S \in GL(2)). \quad (9.6)(c)$$

The half-density bundle on G/Q is given by the character

$$\rho_{G/Q} \begin{pmatrix} r & * & * \\ 0 & S & * \\ 0 & 0 & t \end{pmatrix} = |rt^{-1}|^3. \quad (9.7)(a)$$

Define $\gamma' = \gamma \otimes \rho_{G/Q}^{-1}$; then

$$\gamma' \begin{pmatrix} r & * & * \\ 0 & S & * \\ 0 & 0 & t \end{pmatrix} = \beta((\det S)(rt)^{-1})|rt^{-1}|^{-1}, \tag{9.7)(b}$$

and

$$V(f, \pi, L, \gamma, j_{\gamma,\pi}) \subset \mathrm{Ind}_Q^G(\gamma'). \tag{9.7)(c}$$

We wish to replace Q by the associate standard parabolic subgroup Q_1 with Levi subgroup $GL(2) \times GL(1) \times GL(1)$. Define a character γ_1 of Q_1 by

$$\gamma_1 \begin{pmatrix} S & * & * \\ 0 & r & * \\ 0 & 0 & t \end{pmatrix} = \beta((\det S)(rt)^{-1})|rt^{-1}|^{-1}. \tag{9.8)(a}$$

By standard results about parabolic induction, $\mathrm{Ind}_Q^G(\gamma')$ and $\mathrm{Ind}_{Q_1}^G(\gamma_1)$ have exactly the same composition factors; and in fact they are isomorphic if β is a unitary character. Roughly speaking, therefore

$$V(f, \pi, L, \gamma, j_{\gamma,\pi}) \subset \mathrm{Ind}_{Q_1}^G(\gamma_1); \tag{9.8)(b}$$

the containment must be interpreted in terms of composition series if β is non-unitary. Now define Q_2 to be the standard parabolic subgroup with Levi factor $GL(2) \times GL(2)$. Define a character γ_2 of Q_2 by

$$\gamma_2 \begin{pmatrix} A & * \\ 0 & B \end{pmatrix} = \beta((\det A)(\det B)^{-1}). \tag{9.8)(c}$$

By calculation in $GL(2)$ and induction by stages, we have a containment

$$\mathrm{Ind}_{Q_2}^G(\gamma_2) \subset \mathrm{Ind}_{Q_1}^G(\gamma_1). \tag{9.8)(d}$$

The containments (9.8)(b) and (9.8)(d) suggest (but do not prove) that (at least if β is unitary)

$$V(f, \pi, L, \gamma, j_{\gamma,\pi}) = \mathrm{Ind}_{Q_2}^G(\gamma_2). \tag{9.8)(e}$$

The representations on the right are obviously unitary whenever β is; so we may hope that Definition 9.4 is actually producing unitary representations in that case.

Suppose now that we repeat the entire calculation using the other Lagrangian L^2 (from (9.5)(d)). This time the stabilizer is Q_2, and it turns out that the characters of Q_2 allowed by the condition (9.3)(d) are precisely those given by (9.8)(c).

We want to draw two conclusions from the example. First, the geometric considerations of this paper (involving symplectic spinors and so on) led to certain non-unitary degenerate series representations; yet these non-unitary representations very often had interesting unitary components. Second, the non-canonical choice of Lagrangian L in Definition 9.4 may not affect the representations finally constructed as much as one might first guess.

We conclude with a few general remarks about Definition 9.4. Each element $f \in X$ defines a subvariety M_f of M, the collection of all Lagrangians in M containing f. This is just the fiber of π over f (see (8.8)(d)). The map τ of (1.11)(c) carries M_f into $\mathcal{B}(\mathfrak{g}/\mathfrak{g}_f)$, the Lagrangian Grassmannian for the tangent space to X at f. Over $\mathcal{B}(\mathfrak{g}/\mathfrak{g}_f)$ we have the vector bundle $\mathcal{V}_\pi$; Proposition 7.4 embeds $\mathcal{H}_\pi \otimes \mathcal{H}_f^{even,\infty}$ as a space of smooth sections of $\mathcal{V}_\pi$. Let us call these sections *metaplectic.*

Suppose ϕ is a section of $\mathcal{W}_\gamma$ on M belonging to our representation space $V(f, \pi, L, \gamma, j_{\gamma,\pi})$. Then the restriction of ϕ to M_f must be equal to the pullback (via τ and the isomorphism $j_{\gamma,\pi}$) of a metaplectic section of $\mathcal{V}_\pi$. Said more loosely, ϕ must be metaplectic on each subvariety M_f of M. This condition is probably not sufficient for belonging to $V(f, \pi, L, \gamma, j_{\gamma,\pi})$, but it is certainly necessary; and in some sense it seems to be the main requirement.

We have concentrated almost exclusively on complex groups. For nilpotent orbits in real reductive groups, Corollary 8.7 is almost certainly still true; so most of the formalism of sections 8 and 9 can be set up. This leads to subrepresentations of degenerate principal series again. It is not entirely clear that this is the best or only way to proceed, however. Lemma 8.2 fails for elliptic semisimple orbits in the real case, and one is forced to introduce complex polarizations (and the machinery of cohomological parabolic induction) to construct unitary representations. It may be that nilpotent orbits in the real case should be treated using ideas from cohomological induction, and that at least some of the associated representations should appear inside cohomologically induced representations. We hope to return to these questions in a future paper.

References

[1] R. Abraham and J. Marsden, *Foundations of Mechanics.* Benjamin/Cummings, Reading, Massachusetts, 1978.

[2] V. Arnold, *Mathematical Methods of Classical Mechanics.* Springer-Verlag, New York-Heidelberg-Berlin, 1978.

[3] L. Auslander and B. Kostant, "Polarization and unitary representations of solvable Lie groups," Invent. Math. **14** (1971), 255–354.

[4] A. Borel, *Linear algebraic groups.* Springer-Verlag, New York, 1991.

[5] W. Casselman, "Canonical extensions of Harish-Chandra modules to representations of G," Can. J. Math. **41** (1989), 385–438.

[6] M. Duflo, "Théorie de Mackey pour les groupes de Lie algébriques," Acta Math. **149** (1982), 153–213.

[7] V. Ginsburg, "𝔊-modules, Springer's representations and bivariant Chern classes," Adv. in Math. **61** (1986), 1–48.

[8] V. Guillemin and S. Sternberg, "A generalization of the notion of polarization," Ann. Glob. Analysis and Geometry **4** (1986), 327–347.

[9] V. Guillemin and S. Sternberg, *Geometric Asymptotics*, revised edition. Mathematical Surveys and Monographs **14**. American Mathematical Society, Providence, Rhode Island (1990).

[10] A. Kirillov, "Unitary representations of nilpotent Lie groups," Uspehi Mat. Nauk. **17** (1962), 57–110.

[11] B. Kostant, "Quantization and unitary representations," in *Lectures in Modern Analysis and Applications*, (C. Taam, ed.), Lecture Notes in Mathematics **170**. Springer-Verlag, Berlin-Heidelberg-New York, 1970.

[12] B. Kostant, "Symplectic spinors," 139–152 in *Geometria Simplettica e Fisica Matematica.* Symposia Mathematica XIV, Istituto Nazionale di Alta Matematica. Academic Press, London and New York, 1974.

[13] B. Kostant, "Coadjoint orbits and a new symbol calculus for line bundles," in *Conference on Differential Geometric Methods in Theoretical Physics* (G. Denardo and H.D. Doebner, eds.). World Scientific, Singapore, 1983.

[14] G. Lion and M. Vergne, *The Weil representation, Maslov index, and theta series.* Birkhäuser, Boston-Basel-Berlin, 1980.

[15] C. Moeglin, M.-F. Vigneras, and J.-L. Waldspurger, *Correspondances de Howe sur un corps p-adique,* Lecture Notes in Mathematics **1291**. Springer-Verlag, Berlin-Heidelberg-New York, 1987.

[16] H. Ozeki and M. Wakimoto, "On polarizations of certain homogeneous spaces," Hiroshima Math. J. **2** (1972), 445–482.

[17] L. Rothschild and J. Wolf, "Representations of semisimple Lie groups associated to nilpotent coadjoint orbits," Ann. Sci. École Norm. Sup. (4) **7** (1974), 155–173.

[18] D. Shale, "Linear symmetries of free boson fields," Trans. Amer. Math. Soc. **103** (1962), 149–167.

[19] N. Spaltenstein, "On the fixed point set of a unipotent element on the variety of Borel subgroups," Topology **16** (1977), 203–204.

[20] P. Torasso, "Quantification geometrique, operateurs d'entrelacement et représentations unitaires de $SL_3(\mathbb{R})$," Acta Math. **150** (1983), 153–242.

[21] D. Vogan, "Noncommutative algebras and unitary representations," in *The Mathematical Heritage of Hermann Weyl*, R. O. Wells, Jr., ed., Proceedings of Symposia in Pure Mathematics, **48**. American Mathematical Society, Providence, Rhode Island, 1988.

[22] D. Vogan, "Associated varieties and unipotent representations," in *Harmonic Analysis on Reductive Groups*, W. Barker and P. Sally, eds. Birkhäuser, Boston-Basel-Berlin, 1991.

[23] A. Weil, "Sur certaines groupes d'opérateurs unitaires," Acta Math. **111** (1964), 143–211.

William Graham
Department of Mathematics
University of Georgia
Athens, Georgia 30602

and

David A. Vogan, Jr.
Department of Mathematics
Massachusetts Institute of Technology
Cambridge, Massachusetts 02139

A REMARK ON CASSELMAN'S COMPARISON THEOREM

HENRYK HECHT AND JOSEPH L. TAYLOR

This paper is an outgrowth of our attempt to understand a comparison theorem of Casselman which asserts that Lie algebra homology groups, with respect to certain nilpotent algebras of a Harish-Chandra module and its C^∞ completion, coincide. Let us start with a precise statement of this theorem.

Let G_0 be a connected semisimple real Lie group with finite center. We fix a maximal compact subgroup K_0 of G_0. Recall that a continuous topological representation (π, M_π) of G_0 is called *admissible* if the subspace M of M_π, consisting of K_0-finite vectors splits under the action of K_0 with finite multiplicities. Although G_0 does not act on M, its Lie algebra $\mathfrak{g}_0$ does, as does the complexification $\mathfrak{g}$ of $\mathfrak{g}_0$. By definition, K_0 acts on M, and this action extends to the action of the complexification K of K_0. As usual, we refer to M as the Harish-Chandra module of M_π. We say also that M_π is a globalization, or completion of M.

Among many possible globalizations of a Harish-Chandra module M there are canonical ones which behave in a functorial way. We will consider two of them in this paper. By $\tilde{M}$ we denote the *minimal globalization* of M: It is isomorphic to the space of analytic vectors in any Banach space globalization of M. This globalization was constructed and investigated by Schmid ([7], [5]). We also consider the C^∞, or *smooth globalization* M^∞ of M, which is isomorphic to the space of C^∞ vectors in any Banach globalization of M.

This globalization has been defined and studied by Casselman and Wallach ([2], [8]). The correspondences $M \to \tilde{M}$, and $M \to M^\infty$ both constitute exact functors from the category of Harish-Chandra modules to the appropriate categories of G_0-modules. Let P_0 be a minimal parabolic subgroup of G_0 and $P_0 = M_0A_0U_0$, its Langlands decomposition. We get corresponding decompositions on the level of real and complexified Lie algebras: $\mathfrak{p}_0 = \mathfrak{m}_0 \oplus \mathfrak{a}_0 \oplus \mathfrak{u}_0$, and $\mathfrak{p} = \mathfrak{m} \oplus \mathfrak{a} \oplus \mathfrak{u}$. It is a well-known fact that the $\mathfrak{u}$-homology groups $H_p(\mathfrak{u}, M)$ are finite dimensional modules for the group M_0A_0. On the other hand, $H_p(\mathfrak{u}, M^\infty)$ also have natural structures of M_0A_0-modules. This is evident from looking at the Koszul complex $\wedge^*\mathfrak{u} \otimes M^\infty$ which computes the $\mathfrak{u}$-homology. Clearly, the natural embedding $M \to M^\infty$ induces M_0A_0-invariant maps

$$H_p(\mathfrak{u}, M) \to H_p(\mathfrak{u}, M^\infty).$$

1. Theorem. *(Casselman, unpublished) For each p, the map $H_p(\mathfrak{u}, M) \to H_p(\mathfrak{u}, M^\infty)$ is an isomorphism.*

In particular, it follows that the homology groups $H_p(\mathfrak{u}, M^\infty)$ are finite dimensional.

We want to emphasize that the theorem stated above is a special case of a more general result of Casselman, which holds for arbitrary parabolic subgroups of G_0, not necessarily minimal. In this paper, however, we are interested only in the special case described above.

We want to give a proof of Casselman's theorem that is based on a different circle of ideas.* Rather than proving it directly, we utilize another comparison theorem.

2. Proposition. *The map $H_p(\mathfrak{u}, M) \to H_p(\mathfrak{u}, \tilde{M})$ is an isomorphism.*

This result can be easily deduced from the main result of [4].

We want to point out that Proposition 2 is a special case of a *much more general* comparison theorem proved by Bratten ([1]). It was also recently obtained by U. Bunke and M. Olbrich by applying results of W. Schmid and M. Kashiwara ([5]). Therefore to prove the theorem it is enough to show that the natural embedding $\tilde{M} \to M^\infty$ induces an isomorphism

$$H_p(\mathfrak{u}, \tilde{M}) \cong H_p(\mathfrak{u}, M^\infty) \tag{1}$$

for each Harish-Chandra module M and each p.

The first step is to reduce the proof to the case of principal series representations. We keep assuming that P_0 is a minimal parabolic subgroup of G_0. By a *generalized principal series representation* we mean a representation of G_0 obtained by induction from a finite dimensional representation of P_0. If the inducing representation of P_0 is in addition irreducible, we call the resulting induced representation a *principal series representation.* If I is a Harish-Chandra module of a generalized principal series representation, then $\tilde{I}$ and I^∞ denote the corresponding analytic and smooth completions.

3. Proposition. *It is enough to prove (1) in the case of a principal series representation.*

Proof. Each Harish-Chandra module M can be embedded into a generalised principal series Harish-Chandra module I^0. Continuing, we get the resolution of M,

$$M \to I^0 \to I^1 \to \cdots I^n \to \cdots$$

by generalized principal series Harish-Chandra modules. By exactness of the minimal and smooth globalization functors, we get resolutions

$$\tilde{M} \to \tilde{I}^0 \to \tilde{I}^1 \to \cdots \tilde{I}^n \to \cdots$$

*W. Schmid informed us that Casselman's Theorem also follows from his joint work with Kashiwara.

and

$$M^\infty \to (I^0)^\infty \to (I^1)^\infty \to \cdots (I^n)^\infty \to \cdots$$

The natural map of double complexes

$$\wedge^p \mathfrak{u} \otimes \tilde{I}^q \to \wedge^p \mathfrak{u} \otimes (I^q)^\infty$$

can be analyzed by means of the (convergent) spectral sequences associated to the two standard filtrations. The spectral sequence corresponding to the second filtration degenerates at E_2: $E_2^{q,-p}$-terms are zero except when $q = 0$, and the resulting map on the $(-p)$-cohomology of the total complex is nothing but

$$H_p(\mathfrak{u}, \tilde{M}) \to H_p(\mathfrak{u}, M^\infty) \tag{2}$$

On the other hand, the map between the $E_1^{-p,q}$ terms of the spectral sequence associated to the first filtration is

$$H_p(\mathfrak{u} \otimes \tilde{I}^q) \to H_p(\mathfrak{u} \otimes (I^q)^\infty) \tag{3}$$

Thus, if the maps in (3) are isomorphisms, then so is the map in (2). This shows that to prove (1) it is enough to consider the case of a Harish-Chandra module in the generalized principal series. The proposition follows as such a module has a finite composition series with principal series modules as associated quotients. □

Let M be a Harish-Chandra module of a principal series representation.

For expository reasons we assume that the group G_0 is split. This is not an essential restriction: It simplifies the notation and makes the idea of the proof more transparent. The assumption means that $\mathfrak{p}$ is a Borel subalgebra split over $\mathbb{R}$. Let X be the flag variety of $\mathfrak{g}$, i.e., the space of all Borel subalgebras of $\mathfrak{g}$. For $x \in X$ let $\mathfrak{b}_x$ denote the corresponding Cartan subalgebra and $\mathfrak{n}_x$ its nilpotent radical.

The group G_0 acts on X. Under this action there is precisely one closed orbit S characterized by the following property: $x \in S$ if and only if $\mathfrak{b}_x$ is split over $\mathbb{R}$. Thus we assume that $\mathfrak{p} = \mathfrak{b}_x$ and $\mathfrak{u} = \mathfrak{n}_x$ for some $x \in S$. In this notation the proof of the theorem is reduced to the following proposition:

4. Proposition. *Let M be a principal series module. Then*

$$H_p(\mathfrak{n}_x, \tilde{M}) \to H_p(\mathfrak{n}_x, M^\infty)$$

is an isomorphism for $x \in S$.

Fix $y \in S$. The stabilizer $B_{y,0}$ of y in G_0 is a Borel subgroup of G_0. The representation M may be regarded as induced from an irreducible representation σ of $B_{y,0}$. The (complexified) differential of σ is a linear form $\nu \in \mathfrak{b}_y^*$. This representation induces a G_0-homogeneous real analytic line bundle $\mathbb{L}$ on

S. Let $\mathcal{O}(\mathbb{L})$ be the sheaf of germs of (complex valued) real analytic sections of $\mathbb{L}$. Similarly, let $C^\infty(\mathbb{L})$ be the sheaf on germs of (complex valued) C^∞ sections of $\mathbb{L}$. Then

$$\tilde{M} = \Gamma(S, \mathcal{O}(\mathbb{L})) = H^0(S, \mathcal{O}(\mathbb{L}))$$

and

$$M^\infty = \Gamma(S, C^\infty(\mathbb{L})) = H^0(S, C^\infty(\mathbb{L}))$$

Thus we need to prove that the map

$$(4) \qquad H_p(\mathfrak{n}_x, H^0(S, \mathcal{O}(\mathbb{L}))) \to H_p(\mathfrak{n}_x, H^0(S, C^\infty(\mathbb{L})))$$

is an isomorphism. Now, the modules in (4) are E_2 terms of a convergent spectral sequence associated to a double complex (for details see [4]). Since higher cohomology of both sheaves $\mathcal{O}(\mathbb{L}))$ and $C^\infty(\mathbb{L})$ vanish, this spectral sequence degenerates and the terms in (4) are the $(-p)$-cohomology groups of the total complex of the double complex. On the other hand, this double complex can be analyzed by the second spectral sequence (which is also convergent). This results in the maps

$$(5) \qquad H^q(S, H_p(\mathfrak{n}_x, \mathcal{O}(\mathbb{L}))) \to H^q(S, H_p(\mathfrak{n}_x, C^\infty(\mathbb{L})))$$

We remark that, for each $y \in S$, the stalks $\mathcal{O}(\mathbb{L})_y$ and $C^\infty(\mathbb{L})_y$ are naturally $\mathcal{U}(\mathfrak{g})$ modules, and the embedding

$$i : \mathcal{O}(\mathbb{L})_y \to C^\infty(\mathbb{L})_y$$

is a map of $\mathcal{U}(\mathfrak{g})$ modules.

Suppose we can show that for each $y \in S$ the map

$$(6) \qquad H_p(\mathfrak{n}_x, \mathcal{O}(\mathbb{L})_y) \to H_p(\mathfrak{n}_x, C^\infty(\mathbb{L})_y)$$

is an isomorphism. This would imply that (5) is an isomorphism, and, consequently, (4) is an isomorphism. This will prove Proposition 4.

The reason we prefer to prove (6), rather than (4) directly, lies in the fact that $\mathcal{O}(\mathbb{L})_y$ and $C^\infty(\mathbb{L})_y$ have much simpler structures than $H^0(S, \mathcal{O}(\mathbb{L}))$ and $H^0(S, C^\infty(\mathbb{L}))$.

Let $\hat{\mathcal{O}}(\mathbb{L})_y$ be the completion of $\mathcal{O}(\mathbb{L})_y$ with respect to the ideal of germs, at y, of real analytic functions on S which vanish at y. This is the space of "formal power series" at y with values in $\mathbb{L}$. It inherits the $\mathcal{U}(\mathfrak{g})$ structure from $\mathcal{O}(\mathbb{L})_y$ and the embedding

$$j : \mathcal{O}(\mathbb{L})_y \to \hat{\mathcal{O}}(\mathbb{L})_y$$

is $\mathcal{U}(\mathfrak{g})$ invariant. Let

$$k : C^\infty(\mathbb{L})_y \to \hat{\mathcal{O}}(\mathbb{L})_y$$

be the map which assigns to an element of $C^\infty(\mathbb{L})_y$ its Taylor expansion at y. This is also a $\mathcal{U}(\mathfrak{g})$ map. We note that $j = k \circ i$. Therefore the proof of (6) follows from the following proposition:

6. Proposition. *The maps j and k induce isomorphisms*

(a) $$H_p(\mathfrak{n}_x, \mathcal{O}(\mathbb{L})_y) \to H_p(\mathfrak{n}_x, \hat{\mathcal{O}}(\mathbb{L})_y)$$

(b) $$H_p(\mathfrak{n}_x, C^\infty(\mathbb{L})_y) \to H_p(\mathfrak{n}_x, \hat{\mathcal{O}}(\mathbb{L})_y)$$

Proof. We choose, as we may, a Cartan subalgebra $\mathfrak{c}_0$ contained in $(\mathfrak{b}_x)_0 \cap (\mathfrak{b}_y)_0$ (the subscript "0" means "taking the real points"). The choice of $\mathfrak{c}_0$ results in a root space decomposition

$$\mathfrak{g}_0 = (\mathfrak{n}_y)_0 \oplus \mathfrak{c}_0 \oplus \mathfrak{q}_0$$

Let $\alpha_1, \cdots, \alpha_n$ be the roots of $(\mathfrak{c}_0, \mathfrak{q}_0)$. For $i = 1, \cdots, n$ fix a root vector u_i of α_i. The polynomial maps

$$u \to \exp(u)y$$

$$(x_1, \cdots, x_n) \to \exp(\sum_{1=1}^{n} x_i u_i)y$$

identify $\mathfrak{q}_0$, and, respectively, $\mathbb{R}^n$, with an open dense neighborhood (Bruhat cell) of y in S.

On this neighborhood the bundle $\mathbb{L}$ trivializes in the standard way. In particular,

(7a) $$\mathcal{O}(\mathbb{L})_y \cong \mathcal{O}_0 \otimes \mathbb{L}_y$$

(7b) $$\hat{\mathcal{O}}(\mathbb{L})_y \cong \hat{\mathcal{O}}_0 \otimes \mathbb{L}_y$$

(7c) $$C^\infty(\mathbb{L})_y \cong C_0^\infty \otimes \mathbb{L}_y$$

where $\mathbb{L}_y$ is the fiber of $\mathbb{L}$ at y, $\mathcal{O}_0, C_0^\infty$ are the spaces of germs at $0 \in \mathfrak{q}_0$, of real analytic and C_0^∞-functions on S, respectively, and $\hat{\mathcal{O}}_0$ is the space of formal power series at $0 \in \mathfrak{q}_0$. Under this identification, the Cartan subalgebra $\mathfrak{c}_0$ acts on (7) via the tensor product action. It is important to understand how $\mathfrak{c}_0$ acts on the first factor in (7). For our purpose it suffices to fix $v \in \mathfrak{c}_0$ such that, for $i = 1, \cdots, n$, $\mu_i = -\alpha_i(v)$ is a positive integer. Then

(8) $$v(f) = \sum_{i=1}^{n} x_i \mu_i \frac{\partial}{\partial x_i}(f)$$

For the remainder of the paper we fix such a v. Whenever v acts as an endomorphism of a vector space U, and $\mu \in \mathbb{C}$, we denote by U_μ the maximal v invariant subspace of U on which $v - \mu \cdot 1$ is locally nilpotent. It follows from (8) that monomials in $x_1, \cdots, x_n$ are eigenvectors of v acting on $\mathcal{O}_0$, each $(\mathcal{O}_0)_\mu$ is finite dimensional and is contained in the space of polynomials. Moreover, $(\mathcal{O}_0)_\mu$ has a unique v invariant complement $(\mathcal{O}_0)_\mu^\perp$ in $\mathcal{O}_0$, which can be described as follows: Identify $\mathcal{O}_0$ with the space of convergent power series at 0. Then $(\mathcal{O}_0)_\mu^\perp$ consists of those convergent power series which do not involve monomials in $(\mathcal{O}_0)_\mu$. Alternately, $(\mathcal{O}_0)_\mu^\perp$ is the closure, in the natural topology of $\mathcal{O}_0$, of $\oplus_{\nu\neq\mu}(\mathcal{O}_0)_\mu$.

The same statement, properly reinterpreted, holds with $\mathcal{O}_0$ replaced by terms of the Koszul complex $\wedge^p \mathfrak{n}_x \otimes \mathcal{O}(\mathbb{L})_y$. Because the differentials in the Koszul complex are continuous v invariant maps, we conclude that the complex $\wedge \mathfrak{n}_x \otimes \mathcal{O}(\mathbb{L})_y$ has a unique v invariant splitting as a direct sum of two complexes:

(9a) $$\wedge \mathfrak{n}_x \otimes \mathcal{O}(\mathbb{L})_y = (\wedge \mathfrak{n}_x \otimes \mathcal{O}(\mathbb{L})_y)_\mu \oplus (\wedge \mathfrak{n}_x \otimes \mathcal{O}(\mathbb{L})_y)^\perp$$

and $v - \mu \cdot 1$ is invertible on $(\wedge \mathfrak{n}_x \otimes \mathcal{O}(\mathbb{L})_y)^\perp$. From this we conclude

(9b) $$H_p(\mathfrak{n}_x, \mathcal{O}_y)_\mu \cong H_p(\wedge \mathfrak{n}_x \otimes \mathcal{O}(\mathbb{L})_y)_\mu$$

Word by word, (9a) and (9b) hold with $\mathcal{O}_y$ replaced by $\hat{\mathcal{O}}_y$. As $(\wedge \mathfrak{n}_x \otimes \mathcal{O}(\mathbb{L})_y)_\mu \cong (\wedge \mathfrak{n}_x \otimes \hat{\mathcal{O}}(\mathbb{L})_y)_\mu$, we conclude

$$H_p(\mathfrak{n}_x, \mathcal{O}_y)_\mu \cong H_p(\mathfrak{n}_x, \hat{\mathcal{O}}_y)_\mu$$

for each $\mu \in \mathbb{C}$. As both $\mathcal{O}_y$ and $\hat{\mathcal{O}}_y$ are finite over the center of $\mathcal{U}(\mathfrak{g})$, the Casselman–Osborne Lemma ([3]) implies $H_p(\mathfrak{n}_x, \mathcal{O}_y) \cong H_p(\mathfrak{n}_x, \hat{\mathcal{O}}_y)$. This proves part (a) of the proposition.

Let $\mathcal{C}^{\infty,0}(\mathbb{L})_y$ be the space of germs of sections in $\mathcal{C}^\infty(\mathbb{L})_y$ vanishing to infinite order at y. Then, by a result of Emile Borel, we have an exact sequence

(10) $$0 \to \mathcal{C}^{\infty,0}(\mathbb{L})_y \to \mathcal{C}^\infty(\mathbb{L})_y \to \hat{\mathcal{O}}(\mathbb{L})_y \to 0$$

Clearly the maps in (10) are maps of $\mathfrak{g}$-modules. The proof of part (b) of the proposition is therefore equivalent to showing the following lemma.

6. Lemma.
$$H_p(\mathfrak{n}_x, \mathcal{C}^{\infty,0}(\mathbb{L})_y) = 0$$
for all p.

The proof of Lemma 6 is *the* crucial point of our argument: $\mathcal{C}^{\infty,0}(\mathbb{L})_y$ encapsulates the essential difference between the smooth and analytic cases. The lemma asserts that this difference does not affect the $\mathfrak{n}_x$-homology.

To prove this it is enough to show that for each $\mu \in \mathbb{C}$

(11) $v - \mu \cdot 1$ is invertible on the Koszul complex $\wedge\, \mathfrak{n}_x \otimes \mathcal{C}^{\infty,0}(\mathbb{L})_y$

In fact, (11) implies that $v-\mu\cdot 1$ is invertible when acting on $H_p(\mathfrak{n}_x, \mathcal{C}^{\infty,0}(\mathbb{L})_y)$. Thus $H_p(\mathfrak{n}_x, \mathcal{C}^{\infty,0}(\mathbb{L})_y)_\mu = 0$. Therefore by the Casselman–Osborne Lemma, $H_p(\mathfrak{n}_x, \mathcal{C}^{\infty,0}(\mathbb{L})_y) = 0$. We note that $\mathcal{C}^{\infty,0}(\mathbb{L})_y \cong \mathcal{C}_0^{\infty,0} \otimes \mathbb{L}$, where $\mathcal{C}_0^{\infty,0}$ denotes the subspace of $\mathcal{C}_0^{\infty}$ which consists of elements which vanish to infinite order at 0. We have

$$\wedge\mathfrak{n}_x \otimes \mathcal{C}^{\infty,0}(\mathbb{L})_y \cong \wedge\mathfrak{n}_x \otimes \mathcal{C}_0^{\infty,0} \otimes \mathbb{L}_y$$

As v is semisimple on $\mathfrak{n}_x$ and on $\mathbb{L}_y$, we see that (11) is equivalent to

$$\text{(12)} \qquad v - \mu \cdot 1 \text{ is invertible on } \mathcal{C}_0^{0,\infty} \text{ for each } \mu \in \mathbb{C}$$

Choose the standard inner product on $\mathbb{R}^n \cong \mathfrak{q}_0$. Let S^{n-1} be the unit sphere in $\mathfrak{q}_0$. Define

$$\varphi : \mathbb{R}^n - \{0\} \simeq \mathbb{R}^+ \times S^{n-1} \to \mathfrak{q}_0 - \{0\}$$

by $\varphi(r,u) = \mathrm{Ad}_{\exp(-\ln(r)v)}\, u$. In Cartesian coordinates

$$\varphi(t_1, t_2 \cdots t_n) = \sum_{i-1}^{n} r^{u_i - 1} t_i u_i$$

where r is the length of $(t_1, t_2 \cdots t_n)$. Then φ is a diffeomorphism which extends to a homeomorphism of $\mathbb{R}^n$ onto $\mathfrak{q}_0$. In what follows we replace $f \in \mathcal{C}_0^{\infty,0}$ by $f \circ \varphi$. We denote the pullback of $\mathcal{C}_0^{\infty,0}$, via φ, by the same symbol. This is permitted since φ, as well as its inverse, preserve the germs of smooth functions vanishing to infinite order at the origin. A simple computation shows that

$$\text{(13)} \qquad v(f) = r\frac{\partial}{\partial r} \text{ on } \mathbb{R}^n - \{0\}$$

where r is the radial coordinate on $\mathbb{R}^n$. Define a complex valued function χ_μ on $\mathbb{R}^n - \{0\}$ by $\chi_\mu(u) = r(u)^\mu$. We note that multiplication by χ_μ defines a linear bijection of $\mathcal{C}_0^{0,\infty}$. Let us observe that

$$v - \mu \cdot 1 = \chi_{\mu+1} \circ \frac{\partial}{\partial r} \circ \chi_{-\mu}.$$

This reduces the proof of Lemma 6 to showing that $\frac{\partial}{\partial r} : \mathcal{C}_0^{0,\infty} \to \mathcal{C}_0^{0,\infty}$ is invertible. This is clear, as the inverse is given by integration along r-rays. This completes the proof of Theorem 1. □

REFERENCES

1. Bratten, T., *A Comparison Theorem for Lie Algebra Homology Groups*, preprint, FaMAF UNC (1995).
2. Casselman, W., *Canonical extensions of Harish-Chandra modules to representations of G*, Can.Jour. Math. (1989), 385–438.

3. Casselman, W., Osborne, M. S., *The $\mathfrak{n}$-homology of representations with infinitesimal character*, Compositio Math. (1975), 219–227.
4. Hecht, H., Taylor, J. L., *A comparison theorem for $\mathfrak{n}$-homology*, Compositio Mathematica **86** (1993), 189–207.
5. Kashiwara, M., Schmid, W., *Quasi-Equivariant $\mathcal{D}$-modules, Equivariant Derived Category, and Representations of Reductive Lie Groups*, RIMS-980, Research Institute for Mathematical Sciences (1994).
6. Matsuki, *The orbits of affine symmetric spaces under the action of minimal parabolic subgroups*, J. Math. Soc. Japan **31** (1979), 332–357.
7. Schmid, W., *Boundary value problems for group invariant differential equations*, Astérisque, 1985. Proc. Cartan Symposium (Lyon, 1984).
8. Wallach, N., *Asymptotic expansion of generalized matrix entries of representations of real reductive groups*, Lie group representations I, (Proceedings, University of Maryland 1982-1983), Lecture Notes in Mathematics 1024, Springer Verlag, New York, 1983.

Department of Mathematics
University of Utah
Salt Lake City, UT 84112

PRINCIPAL COVARIANTS, MULTIPLICITY-FREE ACTIONS, AND THE K-TYPES OF HOLOMORPHIC DISCRETE SERIES

ROGER HOWE AND HANSPETER KRAFT

ABSTRACT. We prove a result on the structure of the K-types for holomorphic discrete series of $\mathrm{Sp}_{2n}(\mathbb{R})$. The proof applies the theory of multiplicity-free actions to the realization of holomorphic discrete series by means of the dual pair $(\mathrm{Sp}_{2n}, \mathrm{O}_m)$.

INTRODUCTION

A major method for studying representations of a semisimple Lie group G has been to consider the decomposition of such a representation under restriction to the maximal compact subgroup K of G. One of HARISH-CHANDRA's early basic results was that the multiplicity of an irreducible representation of K (a K-*type*) in the restriction of an irreducible representation of G is finite, in fact bounded by the dimension of the K-type ([Kna], [Wal], [War]). Formulas for the multiplicities of K-types are known in many cases. In particular, the method of cohomological induction produces attractive formulas for multiplicities of K-types, expressed as alternating sums over some Weyl group ([Kna], [KnVo], [Wal]). A well-known problem with this kind of formula is that it is hard to use to answer practical questions, for example, "Is the multiplicity zero or positive?" Thus for understanding K-types, it is desirable to supplement the alternating sum formulas with other information. The main goal of this paper is to provide such supplementary information for holomorphic discrete series representations.

The holomorphic discrete series were first constructed by HARISH-CHANDRA in an early essay toward his Plancherel Formula [HC IV-VI]. An alternating sum formula for K-types of holomorphic discrete series was established in [Sch1], [Sch2]. Alternatively, one has a description of a holomorphic discrete series representation as a generalized Verma module ([Kna], [Sch1,2]). This description reveals a good deal of information about the K-structure of such representations. We recall it.

Let $\mathfrak{g}$ and $\mathfrak{k}$ denote the Lie algebras of G and K respectively, and let $\mathfrak{g}_{\mathbb{C}}$ and $\mathfrak{k}_{\mathbb{C}}$ denote their complexifications. We have the Cartan decomposition

$$\mathfrak{g} = \mathfrak{k} + \mathfrak{p}, \tag{1}$$

The first author was partially supported by NSF Grant DMS-9224358, the second by SNF (Schweizerischer Nationalfonds).

where $\mathfrak{p}$ is the orthogonal complement to $\mathfrak{k}$ with respect to the Killing form on $\mathfrak{g}$. When G allows holomorphic discrete series, $\mathfrak{k}$ has a non-trivial center, and, on complexification, we get

$$\mathfrak{g}_{\mathbb{C}} = \mathfrak{k}_{\mathbb{C}} + \mathfrak{p}^+ + \mathfrak{p}^-, \tag{2}$$

where $\mathfrak{p}^+$ and $\mathfrak{p}^-$ are the eigenspaces for the center of $\mathfrak{k}$ acting on $\mathfrak{p}_{\mathbb{C}}$ by the adjoint representation. Each of $\mathfrak{p}^+$ and $\mathfrak{p}^-$ is an abelian Lie algebra. Hence the universal enveloping algebra of $\mathfrak{p}^+$ is isomorphic to the symmetric algebra $\mathcal{S}(\mathfrak{p}^+)$. The adjoint action of K on $\mathfrak{p}^+$ makes $\mathcal{S}(\mathfrak{p}^+)$ into a K-module. The K-module structure of $\mathcal{S}(\mathfrak{p}^+)$ is very well understood [Jo],[Sch1], [GW]. It is multiplicity-free, that is, each irreducible representation of K occurs in $\mathcal{S}(\mathfrak{p}^+)$ at most once. Furthermore (and as a consequence of multiplicity-freeness; see, for example, [Ho]), if one fixes a maximal unipotent subalgebra $\mathfrak{u}_K$ of $\mathfrak{k}_{\mathbb{C}}$, then the algebra $\mathcal{S}(\mathfrak{p}^+))^{\mathfrak{u}_K}$ of $\mathfrak{u}_K$ invariant vectors in $\mathcal{S}(\mathfrak{p}^+)$ (the highest weight vectors with respect to $\mathfrak{u}_K$), is a polynomial algebra on certain canonical and explicitly known generators.

Let ρ be a holomorphic discrete series representation of G, and let V be the associated $(\mathfrak{g}, K)$-module of K-finite vectors [Kna], [Wal]. The first key property of holomorphic discrete series is that the subspace of V consisting of vectors annihilated by the elements of $\mathfrak{p}^-$ defines an irreducible representation for K ([HC IV], [Kna], [Sch1,2]). This is called the *lowest K-type* for ρ. Denote the lowest K-type by μ_ρ. Then the structure of all of V as a K-module is given by taking a tensor product with $\mathcal{S}(\mathfrak{p}^+)$:

$$V_{|K} \simeq \mu_\rho \otimes \mathcal{S}(\mathfrak{p}^+). \tag{3}$$

The space $V^{\mathfrak{u}_K}$ of K-highest weight vectors in V will be a module for $\mathcal{S}(\mathfrak{p}^+)^{\mathfrak{u}_K}$, and knowledge of $V^{\mathfrak{u}_K}$ as an $\mathcal{S}(\mathfrak{p}^+)^{\mathfrak{u}_K}$-module implies an understanding of V as a K-module. Thus it is nice to know that the $\mathcal{S}(\mathfrak{p}^+)^{\mathfrak{u}_K}$-module structure of $V^{\mathfrak{u}_K}$ is as simple as it could be.

Theorem 1. *For $G = \mathrm{Sp}_{2n}(\mathbb{R})$, the real symplectic group in $2n$ variables, the space $V^{\mathfrak{u}_K}$ of K-highest weight vectors in the holomorphic discrete series representation V is a free module for $\mathcal{S}(\mathfrak{p}^+)^{\mathfrak{u}_K}$.*

Remark. It can be shown by a counting argument that the rank of $V^{\mathfrak{u}_K}$ over $\mathcal{S}(\mathfrak{p}^+)^{\mathfrak{u}_K}$ must be $\dim \mu_\rho$.

There is no reason to expect this result to fail for other groups that have holomorphic discrete series. However, our method of proof does not apply to the general case. It relies on the realization of the holomorphic discrete series in the local theta correspondence for the dual pair $(\mathrm{O}_m, \mathrm{Sp}_{2n}(\mathbb{R}))$ [Ge], [KaVe], [Sa]. A similar proof works for the groups O^*_{2m} and $\mathrm{U}_{p,q}$. It would be interesting to construct an argument valid for all groups whose associated symmetric space is hermitian.

Our approach to Theorem 1 uses the realization of holomorphic discrete series associated to the reductive dual pair $(\mathrm{O}_m, \mathrm{Sp}_{2n}(\mathbb{R}))$, with $m \geq 2n$.

(See [Ge], [GK], [KaVe], and [Sa]; we note that these correspondences were among the earliest examples of dual pair correspondences.) This gives a coherent realization of all holomorphic discrete series in the same vector space, which may be realized as a Fock space of holomorphic functions on $\mathbb{C}^{mn}$. The $\mathrm{O}_m \times \mathrm{U}_n(\mathbb{C})$-finite functions then appear as polynomial functions, and we can reduce the statement of the theorem to a result in invariant theory. We can then use the theory of multiplicity-free actions [Ho], [Krf], [Krm], [Kno] and the geometry of the action of $K_{\mathbb{C}}$ on complex flag manifolds to establish the desired result.

Let O_m denote the orthogonal group in m variables. Denote by $\mathbb{C}^{mn}$ the space of $m \times n$ matrices, and let $\mathrm{O}_m(\mathbb{C})$ act on $\mathbb{C}^{mn}$ by matrix multiplication on the left. Let $\mathrm{GL}_n(\mathbb{C})$ act on $\mathbb{C}^{mn}$ by multiplication on the right. These actions commute with each other. We can extend them to actions on the polynomial functions $\mathcal{P}(\mathbb{C}^{mn})$ in the usual way. $\mathrm{O}_m(\mathbb{C})$ is the complexification of $\mathrm{O}_m = \mathrm{O}_m(\mathbb{R})$ and the natural action of $\mathrm{O}_m(\mathbb{C})$ on $\mathcal{P}(\mathbb{C}^{mn})$ may be identified to the holomorphic extension to $\mathrm{O}_m(\mathbb{C})$ of the action of O_m on Fock space. Similarly, $\mathrm{GL}_n(\mathbb{C})$ is the complexification of U_n. The natural action of $\mathrm{GL}_n(\mathbb{C})$ on $\mathcal{P}(\mathbb{C}^{mn})$ is not quite, but almost the holomorphic extension to $\mathrm{GL}_n(\mathbb{C})$ of the action of U_n on Fock space. To get the holomorphic extension exactly, we should twist the natural action of $\mathrm{GL}_n(\mathbb{C})$ by $(\det)^{m/2}$. (See formulas (4b) below for the infinitesimal action.)

Denote by r_{ij}^2 the inner product of the i-th and j-th column of an element of $\mathbb{C}^{mn}$. The r_{ij}^2 are quadratic polynomials on $\mathbb{C}^{mn}$ and they are invariant under the action of O_m. The First Fundamental Theorem of Classical Invariant Theory (see [Ho], [We]) for O_m says that the r_{ij}^2 generate the full algebra $\mathcal{P}(\mathbb{C}^{mn})^{\mathrm{O}_m}$ of polynomials invariant under O_m. Denote by Δ_{ij} the "partial Laplacian" corresponding to the i-th and j-th rows of $\mathbb{C}^{mn}$. Explicitly, if $\{z_{ij} \mid 1 \leq i \leq m\}$ are the coordinates on the j-th copy of $\mathbb{C}^m$ with respect to a fixed orthonormal basis, then

$$\text{(4a)} \qquad r_{ij}^2 = \sum_{a=1}^{m} z_{ai} z_{aj} \quad \text{and} \quad \Delta_{ij} = \sum_{a=1}^{m} \frac{\partial^2}{\partial z_{ai} \partial z_{aj}}, \qquad i, j = 1, \dots, n.$$

Denote by $\mathfrak{sp}^{(2,0)}$ the linear span of the r_{ij}^2 considered as operators on $\mathcal{P}(\mathbb{C}^{mn})$, and denote by $\mathfrak{sp}^{(0,2)}$ the span of the Δ_{ij}. Denote by $\mathfrak{sp}^{(1,1)}$ the span of the commutators $[\Delta_{ij}, r_{k\ell}^2]$. Then, $\mathfrak{sp}^{(1,1)}$ is a Lie algebra—it is a very mild perturbation of the Lie algebra defined by the infinitesimal action of the Lie algebra $\mathfrak{gl}_n$ of $\mathrm{GL}_n(\mathbb{C})$ acting on $\mathbb{C}^{mn}$ by multiplication on the right. Precisely, it is spanned by operators

$$\text{(4b)} \qquad \tilde{E}_{ij} = \left(\sum_{a=1}^{m} z_{ai} \frac{\partial}{\partial z_{aj}} \right) + \delta_{ij} \frac{m}{2} = E_{ij} + \delta_{ij} \frac{m}{2},$$

where the E_{ij} are the standard "polarization operators," which are the infinitesimal generators of the standard action of GL_n. Then

$$\text{(4c)} \qquad \mathfrak{sp}_{\mathbb{C}} = \mathfrak{sp}^{(1,1)} \oplus \mathfrak{sp}^{(0,2)} \oplus \mathfrak{sp}^{(2,0)}$$

is a Lie algebra isomorphic to $\mathfrak{sp}(2n, \mathbb{C})$, and decomposition (4c) is an instance of the complexified Cartan decomposition (2), with $\mathfrak{sp}^{(1,1)}$ playing the role of $\mathfrak{k}_{\mathbb{C}}$, and $\mathfrak{sp}^{(2,0)}$ being $\mathfrak{p}^+$, and $\mathfrak{sp}^{(0,2)}$ being $\mathfrak{p}^-$. The operators spanning $\mathfrak{sp}_{\mathbb{C}}$ commute with the action of O_m on $\mathcal{P}(\mathbb{C}^{mn})$. The joint action of O_m and $\mathfrak{sp}_{\mathbb{C}}$ on $\mathcal{P}(\mathbb{C}^{mn})$ gives rise to a decomposition ([KaVe], [Ho])

$$\mathcal{P}(\mathbb{C}^{mn}) = \sum_{\tau} \mathcal{P}(\mathbb{C}^{mn})^{\mathrm{O}_m,\tau} = \sum_{\tau} H_\tau \cdot \mathbb{C}[r_{ij}^2] = \sum_{\tau} (\tau \otimes \tilde{\tau}) \cdot \mathbb{C}[r_{ij}^2] = \sum_{\tau} \tau \otimes V_\tau. \tag{5}$$

The variable of summation in formulas (5) runs over the set of irreducible representations τ of O_m, and each summand $\mathcal{P}(\mathbb{C}^{mn})^{\mathrm{O}_m,\tau}$ is the τ-isotypic component for the action of O_m on $\mathcal{P}(\mathbb{C}^{mn})$. (Some summands may be trivial.) The three last summations express three different aspects of the structure of these isotypic components. The space H_τ is the space of τ-isotypic harmonics, consisting of elements in $\mathcal{P}(\mathbb{C}^{mn})^{\mathrm{O}_m,\tau}$ that are annihilated by $\mathfrak{sp}^{(0,2)}$, i.e., by all the partial Laplacians Δ_{ij}. The second summation expresses the well-known general fact that $\mathcal{P}(\mathbb{C}^{mn})^{\mathrm{O}_m,\tau}$ is generated as a $\mathbb{C}(r_{ij}^2)$-module by H_τ. The third summation tells us that the space H_τ, which has the structure of $\mathrm{O}_m \times \mathrm{GL}_n$-module, is isomorphic to a tensor product $\tau \otimes \tilde{\tau}$ where $\tilde{\tau}$ is a GL_n-module determined by τ. The fourth summation further indicates that we may combine $\tilde{\tau} \cdot \mathbb{C}[r_{ij}^2]$ into a module V_τ for $\mathfrak{sp}_{\mathbb{C}}$. This module V_τ is irreducible and is determined by τ.

The modules V_τ will belong to the holomorphic discrete series if $m > 2n$, and will include all holomorphic discrete series when $m = 2n$ [Ge]. As long as $m \geq 2n$, the multiplications $\tilde{\tau} \cdot \mathbb{C}[r_{ij}^2]$ by $\mathbb{C}[r_{ij}^2]$ yield tensor product decompositions:

$$V_\tau = \tilde{\tau} \cdot \mathbb{C}[r_{ij}^2] \simeq \tilde{\tau} \otimes \mathbb{C}[r_{ij}^2]. \tag{6}$$

This enables us to prove Theorem 1 by establishing an appropriate result about the structure of the polynomial ring $\mathcal{P}(\mathbb{C}^{mn})$.

Let U_n denote the standard maximal unipotent subgroup of $\mathrm{GL}_n(\mathbb{C})$ of upper triangular matrices. The highest weight vectors for $K \subseteq \mathrm{Sp}(2n, \mathbb{R})$ acting on $\mathcal{P}(\mathbb{C}^{mn})$ are the invariant polynomials for U_n. As was remarked in the discussion leading up to the statement of Theorem 1, the algebra $\mathbb{C}[r_{ij}^2]^{U_n}$ of highest weight vectors in $\mathbb{C}[r_{ij}^2]$ is a polynomial algebra. We may describe its generators as follows. Consider the r_{ij}^2 to be the entries of a symmetric $k \times k$-matrix. Define δ_k to be the determinant of the leading $k \times k$ submatrix:

$$\delta_1 := r_{11}^2 \qquad \delta_2 := \begin{vmatrix} r_{11}^2 & r_{12}^2 \\ r_{12}^2 & r_{22}^2 \end{vmatrix} \qquad \delta_3 := \begin{vmatrix} r_{11}^2 & r_{12}^2 & r_{13}^2 \\ r_{12}^2 & r_{22}^2 & r_{23}^2 \\ r_{13}^2 & r_{23}^2 & r_{33}^2 \end{vmatrix}, \tag{7}$$

and so forth. Thus

$$\mathbb{C}[r_{ij}^2]^{U_n} = \mathcal{P}(\mathbb{C}^{mn})^{\mathrm{O}_m \times U_n} = \mathbb{C}[\delta_1, \delta_2, \dots, \delta_n].$$

From the description of $\mathcal{P}(\mathbb{C}^{mn})$ given in formulas (5), we can see that Theorem 1 will follow from the following result.

Theorem 2. *Under the assumption that $m \geq 2n$, the algebra $\mathcal{P}(\mathbb{C}^{mn})^{U_n}$ of $\mathrm{GL}_n(\mathbb{C})$ highest weight vectors in $\mathcal{P}(\mathbb{C}^{mn})$ is a free module for the subalgebra $\mathbb{C}[r_{ij}^2]^{U_n} = \mathbb{C}[\delta_1, \delta_2, \dots, \delta_n]$.*

This theorem can be regarded as a first step toward understanding the structure of $\mathcal{P}(\mathbb{C}^{mn})$ as an $\mathrm{O}_m \times \mathrm{GL}_n$-algebra. A final goal is to give an explicit description of the algebra of $\mathrm{O}_m \times \mathrm{GL}_n$ highest weight vectors $\mathcal{P}(\mathbb{C}^{mn})^{U \times U_n}$ where $U \subset \mathrm{O}_m$ is a maximal unipotent subgroup.

Example. To simplify the notation we will work with the orthogonal group O_m with respect to the quadratic form $\delta = \sum_1^m x_i x_{m-i+1}$. Then the lower triangular matrices in O_m form a maximal unipotent subgroup $U \subset \mathrm{O}_m$, and the diagonal matrices a maximal torus.

The description of $\mathcal{P}(\mathbb{C}^{mn})^{U \times U_n}$ in case $n = 1$ is easy:

$$\mathcal{P}(\mathbb{C}^m)^U = \mathbb{C}[x_1, \delta] \supset \mathcal{P}(\mathbb{C}^m)^{\mathrm{O}_m} = \mathbb{C}[\delta].$$

This is also expressed by the well-known decomposition formula of the space $\mathcal{P}_d(\mathbb{C}^m)$ of homogeneous functions of degree d as an O_m-module:

$$\mathcal{P}_d(\mathbb{C}^m) = \langle x_1^d \rangle_{\mathrm{O}_m} \oplus \delta \cdot \mathcal{P}_{d-2}(\mathbb{C}^m).$$

Here $\langle f \rangle_G$ denotes the G-module linearly spanned by the G-orbit of the element f. In particular, $\langle x_1^d \rangle_{\mathrm{O}_m}$ is the irreducible O_m-module of highest weight $d\omega_1$.

The case $n = 2$ is more complicated. We first introduce the following elements:

$$\alpha_1 := x_{11} \quad \text{and} \quad \alpha_2 := \begin{vmatrix} x_{11} & x_{12} \\ x_{21} & x_{22} \end{vmatrix}.$$

Clearly, $\alpha_1^{a_1} \alpha_2^{a_2}$ is a $\mathrm{O}_m \times \mathrm{GL}_2$ highest weight vector of weight $a_1\omega_1 + a_2\omega_2$ (with respect to both groups) corresponding to the diagram $(a_1 + a_2, a_2)$. In fact, these are exactly the $\mathrm{O}_m \times \mathrm{GL}_2$ harmonic highest weight vectors, i.e. $\langle \alpha_1^{a_1} \alpha_2^{a_2} \rangle_{\mathrm{O}_m \times \mathrm{GL}_2} = H_\tau$ in the notation of formula (5) where τ is the irreducible O_m-module of highest weight $a_1\omega_1 + a_2\omega_2$. One can easily verify the following explicit CLEBSCH-GORDAN decomposition in $\mathcal{P}(\mathbb{C}^{m2})$:

$$\langle \alpha_1 \rangle_{\mathrm{GL}_2} \cdot \langle \delta_1 \rangle_{\mathrm{GL}_2} = \langle \alpha_1 \delta_1 \rangle_{\mathrm{GL}_2} \oplus \langle \beta_1 \rangle_{\mathrm{GL}_2}$$
$$\langle \alpha_1^2 \rangle_{\mathrm{GL}_2} \cdot \langle \delta_1 \rangle_{\mathrm{GL}_2} = \langle \alpha_1^2 \delta_1 \rangle_{\mathrm{GL}_2} \oplus \langle \alpha_1 \beta_1 \rangle_{\mathrm{GL}_2} \oplus \langle \beta_2 \rangle_{\mathrm{GL}_2}$$

where

$$\beta_1 := \begin{vmatrix} x_{11} & r_{11}^2 \\ x_{12} & r_{21}^2 \end{vmatrix} \quad \text{and} \quad \beta_2 := \begin{vmatrix} 0 & x_{11} & x_{12} \\ x_{11} & r_{11}^2 & r_{12}^2 \\ x_{12} & r_{21}^2 & r_{22}^2 \end{vmatrix}$$

are $\mathrm{O}_m \times \mathrm{GL}_2$ highest weight vectors of GL_2-weight $(2,1)$ and $(2,2)$ and O_m-weight $(1,0)$ and $(2,0)$, respectively. (δ_1 and δ_2 have been defined earlier.)

Moreover, we have the following LEWIS CARROLL identity. (For the origin of the terminology, see [Do], [RR].)

$$\beta_2\delta_1 + \alpha_1^2\delta_2 + \beta_1^2 = 0.$$

Now one can show (see [ATZ]) that

$$\mathcal{P}(\mathbb{C}^{m2})^{U\times U_2} = \mathbb{C}[\alpha_1, \alpha_2, \beta_1, \beta_2, \delta_1, \delta_2]/(\beta_2\delta_1 + \alpha_1^2\delta_2 + \beta_1^2).$$

It follows that the monomials $\alpha_1^{a_1}\alpha_2^{a_2}\beta_1^{\varepsilon}\beta_2^{b_2}$ with $a_1, a_2, b_2 \geq 0$ and $\varepsilon = 0$ or 1, form a basis for $\mathcal{P}(\mathbb{C}^{m2})^{U\times U_2}$ as a $\mathbb{C}[\delta_1, \delta_2]$-module. The monomials with a fixed O_m-highest weight $\lambda = a_1\omega_1 + a_2\omega_2$ have the form $\alpha_1^{a_1-\varepsilon-2b_2}\alpha_2^{a_2}\beta_1^{\varepsilon}\beta_2^{b_2}$. From this description it is easy to explicitly verify the statements made so far.

From the above presentation, we can easily calculate the multiplicities for representations of K_2 (= the unitary group of 2×2 matrices) in holomorphic discrete series of $\mathrm{Sp}_4(\mathbb{R})$. As above, we indicate a dominant weight λ by a pair of integers (a_1, a_2), $a_1 \geq a_2$ where $\lambda = (a_1 - a_2)\omega_1 + a_2\omega_2$. Let $\lambda = (a_1, a_2)$ be the lowest highest weight of the holomorphic discrete series representation V_λ and let (b_1, b_2) denote a typical highest weight. Then the multiplicity of the representation of K_2 with highest weight (b_1, b_2) is described as follows. First, the multiplicity is zero unless the following conditions are satisfied:

$$b_1 \geq a_1 \quad b_2 \geq a_2 \quad \text{and} \quad b_1 + b_2 \equiv a_1 + a_2 \bmod 2.$$

We assume this from now on. Then

(i) $\mathrm{mult}_{(b_1,b_2)} V_\lambda = \min\{\frac{a_1-a_2+1}{2}, \lfloor\frac{b_1-a_1}{2}\rfloor + 1, \lfloor\frac{b_2-a_2}{2}\rfloor + 1, \frac{b_1-b_2+1}{2}\}$ if $a_1 - a_2$ is odd.
(Here $\lfloor x\rfloor$ denotes the greatest integer not exceeding x.)

(ii) $\mathrm{mult}_{(b_1,b_2)} V_\lambda = \min\{\frac{a_1-a_2}{2} + 1, \frac{b_1-a_1}{2} + 1, \frac{b_2-a_2}{2} + 1, \frac{b_1-b_2}{2} + 1\}$ if $a_1 - a_2$ and both $b_j - a_j$ are even.

(iii) $\mathrm{mult}_{(b_1,b_2)} V_\lambda = \min\{\frac{a_1-a_2}{2}, \frac{b_1-a_1+1}{2}, \frac{b_2-a_2+1}{2}, \frac{b_1-b_2}{2}\}$ if $a_1 - a_2$ is even and both $b_j - a_j$ are odd.

Remark. The phenomenon described in Theorem 2 seems to be quite special. First of all it is already very rare that the ring of U-invariants of a representation of a reductive group is a polynomial ring. (See [Br] where Brion classifies all irreducible representations of simple groups with this property. In our situation above, the representation is $\wedge^2\mathbb{C}^n$.) Moreover, consider the following general question which illustrates the phenomenon. Let G be a reductive group with maximal unipotent subgroup U and let $\pi: X \to Y$ be an equivariant morphism of G-varieties such that $\mathcal{O}(X)$, the ring of regular functions on X, is a free module over $\mathcal{O}(Y)$. *Then when is $\mathcal{O}(X)^U$ a free module over $\mathcal{O}(Y)^U$?*

This certainly fails in general. For example, let $G = \mathrm{SL}_2$, $X = Y \times \mathbb{C}^2$ and π the projection onto the first factor. If $\mathcal{O}(X)^U$ is free over $\mathcal{O}(Y)^U$ then $\mathcal{O}(X)^G \simeq \mathcal{O}(Y)^U$ is free over $\mathcal{O}(Y)^G$. Hence, the covariants for Y should be a free module over the invariants. This typically does not happen. A representation V with this property is called *cofree* and there are only finitely many for a given group G [Pop]. For simple groups G they have been classified by Schwarz [Schw], and the irreducible cofree representations V for semisimple groups by Littelmann [Li].

Proof of Theorem 2. The algebra $R := \mathcal{P}(\mathbb{C}^{mn})^{U_n}$ of U_n-invariants is finitely generated and Cohen-Macaulay ([KrP] section 18.7; cf. [Pop2]). Denote by X the corresponding affine variety spec R, the maximal spectrum of R, and by $\delta: X \to \mathbb{C}^n$ the morphism $\delta(x) := (\delta_1(x), \delta_2(x), \dots, \delta_n(x))$. Since $\mathbb{C}[\delta_1, \delta_2, \dots, \delta_n]$ is the O_m-invariant ring R^{O_m} this means that the map δ is the quotient morphism by O_m ([Krf, II.3]). The essential step in the proof is to show that the zero fiber

$$\delta^{-1}(0) = \{x \in X \mid \delta_1(x) = \delta_2(x) = \cdots = \delta_n(x) = 0\}$$

has codimension n in X (Proposition 1 below). Since δ is homogeneous this implies that all fibers of δ have codimension n, i.e., δ is equidimensional. It follows that δ is flat because X is Cohen-Macaulay and $\mathbb{C}^n$ is smooth ([Mat, Theorem 23.1 and Corollary]). Decompose R into a direct sum of isotypic components with respect to O_m:

$$R = \bigoplus_{\tau} R^{\mathrm{O}_m, \tau}.$$

Then each $R^{\mathrm{O}_m, \tau}$ is a finitely generated graded module over R^{O_m} ([Krf, II.3.2]). Because of flatness, every summand $R^{\mathrm{O}_m, \tau}$ is therefore a free R^{O_m}-module, and Theorem 2 follows. □

As in the proof above we denote by X the affine variety spec R, the maximal spectrum of the U_n-invariant functions $R = \mathcal{P}(\mathbb{C}^{mn})^{U_n}$ on $\mathbb{C}^{mn}$. Let us call a sequence $f_1, f_2, \dots, f_r$ of regular functions on an irreducible variety Z *good* if the locus of common zeros has codimension r in Z.

Proposition 1. *Under the assumption $m \geq 2n$, the functions $\delta_1, \delta_2, \dots, \delta_n$ form a good sequence on X.*

Proof. By classical invariant theory the ring R of U_n-invariant functions on $\mathbb{C}^{mn}$, the so-called *primary covariants*, is generated by the $k \times k$-minors extracted from the first k columns of $\mathbb{C}^{mn}$ for $k = 1, 2, \dots, n$. (This follows immediately from the well-known fact that the $U_m \times U_n$-invariants are generated by the principal minors, see [Krf, III.3.7] or [Ho]). Thus we obtain the following description of the variety X as a subvariety of

$$V \times \bigwedge^2 V \times \bigwedge^3 V \times \cdots \times \bigwedge^n V$$

where $V := \mathbb{C}^m$. Let $w := (e_1, e_1 \wedge e_2, e_1 \wedge e_2 \wedge e_3, \dots, e_1 \wedge \cdots \wedge e_n) \in W$. Then

$$X = \overline{\mathrm{GL}_m \cdot w},$$

the closure of the GL_m orbit of the point w, because $\mathrm{GL}_m \times U_n$ has a dense orbit in $\mathbb{C}^{mn}$. In fact, even $B_m \times U_n$ has a dense orbit where B_m are the upper triangular matrices, and thus X is multiplicity-free and contains only finitely many GL_m-orbits. We can describe them in the following way. For any subset $J \subset N := \{1, 2, \ldots, n\}$, $J = \{j_1 < j_2 < \cdots < j_s\}$ we define $w_J \in W$ by

$$(w_J)_j := \begin{cases} 0 & \text{if } j \notin J \\ e_1 \wedge e_2 \wedge \cdots \wedge e_j & \text{if } j \in J \end{cases}$$

Then

$$X = \overline{\mathrm{GL}_m \cdot w} = \bigcup_{J \subset N} \mathrm{GL}_m \cdot w_J$$

Let P_J be the parabolic subgroup stabilizing the flag $\mathbb{C}^{j_1} \subset \mathbb{C}^{j_2} \subset \cdots \subset \mathbb{C}^{j_s} \subset \mathbb{C}^m$. Then the stabilizer of w_J is the subgroup Q_J of P_J subject to the condition that the automorphism induced on $\mathbb{C}^{j_k}/\mathbb{C}^{j_{k-1}}$ has determinant 1 for $k = 1, \ldots, s$.

Next we remark that the functions δ_k are defined on all of W:

$$\delta_k(\omega_1, \omega_2, \ldots, \omega_n) = (\omega_k, \omega_k)$$

where $(\ ,\ )$ is the symmetric bilinear form on $\bigwedge^k V$ induced by the standard form on V, i.e., the k-fold wedge products from an orthonormal basis for V form an orthonormal basis for $\bigwedge^k V$. If $\omega_k = v_1 \wedge \cdots \wedge v_k$ then $(\omega_k, \omega_k) = 0$ if and only if the subspace of V spanned by $v_1, \ldots, v_k$ is degenerate with respect to the standard form on V. In particular we see that δ_j vanishes on the orbit $\mathrm{GL}_m \cdot w_J$ if $j \notin J$. The following Main Lemma implies that the locus of common zeroes of $\delta_1, \ldots, \delta_n$ has codimension $\#J$ in the orbit $\mathrm{GL}_m \cdot w_J$. Now we remark that $\dim Q_J - \dim Q_N \geq n - \#J$ which means that the orbit $\mathrm{GL}_m \cdot w_J$ has codimension at least $n - \#J$ in X. This proves Proposition 1. □

Main Lemma. *The functions $\{\delta_j \mid j \in J\}$ form a good sequence on the orbit $\mathrm{GL}_m \cdot w_J$.*

Proof. Consider the orbit map $p_J\colon \mathrm{GL}_m \to \mathrm{GL}_m \cdot w_J$. For any $j \in J$ the pull-back $\tilde{\delta}_j = p_J \circ \delta_j$ has the following interpretation: For any matrix $g \in \mathrm{GL}_m$ with column vectors $v_1, v_2, \ldots, v_m$ we have $\tilde{\delta}_j(g) = 0$ if and only if the span of $v_1, \ldots, v_j$ in $V (= \mathbb{C}^m)$ is degenerate with respect to the form $(\ ,\ )$. If the sequence $(\tilde{\delta}_j \mid j \in J)$ is good then its zero locus $Z_J \subset \mathrm{GL}_m$ has codimension $\#J$ in GL_m. Moreover, Z_J is stable under right multiplication by Q_J. It follows that the image $\overline{Z_J} := p_J(Z_J)$ is the zero set of $\{\delta_j \mid j \in J\}$ and has codimension $\#J$ in the orbit $\mathrm{GL}_m \cdot w_J$. Thus it remains to prove that the sequence $\tilde{\delta}_1, \tilde{\delta}_2, \ldots, \tilde{\delta}_n$ of functions on GL_m, and, more generally, that every subsequence of the $\tilde{\delta}_j$ is good.

For this purpose consider the map $\tilde{\delta} := (\tilde{\delta}_1, \tilde{\delta}_2, \ldots, \tilde{\delta}_n)\colon \mathrm{GL}_m \to \mathbb{C}^n$. By definition, it can be decomposed in the form

$$\mathrm{GL}_m \xrightarrow{p} \mathrm{Sym}_m \xrightarrow{\mu} \mathbb{C}^n,$$
$$g \overset{p}{\mapsto} g^t g = A \overset{\mu}{\mapsto} (\mu_1(A), \mu_2(A), \ldots, \mu_n(A))$$

where Sym_m are the symmetric $m \times m$-matrices and $\mu_1(A), \ldots, \mu_n(A)$ are the first n principal minors of A. The first map is a principal O_m-bundle over its image and therefore equidimensional (i.e., the inverse image under p of a closed subset of codimension d has codimension d). Thus, it remains to show that $\mu_1, \mu_2, \ldots, \mu_n$ form a good sequence on Sym_m, i.e., that the zero fiber $\mu^{-1}(0)$ has codimension n in Sym_m. In fact, since the functions μ_i are all homogeneous this implies that $\mu_1, \mu_2, \ldots, \mu_n$ form a *regular sequence* and so every subsequence $(\mu_j \mid j \in J)$ is good, too (cf. [Mat, Theorem 17.4]). It follows that the locus of common zeroes in Sym_m has codimension $\#J$ and hence the same holds for the locus of common zeroes of $(\tilde{\delta}_j \mid j \in J)$ in GL_m. So everything is reduced to the following proposition. □

Proposition 2. *The morphism*

$$\mu\colon \mathrm{Sym}_m \to \mathbb{C}^m,\ A \mapsto (\mu_1(A), \mu_2(A), \ldots, \mu_m(A)),$$

where $\mu_i(A)$ is the i-th principal minor of A, is equidimensional.

The following proof was communicated to us by GERRY SCHWARZ. It replaces our more complicated argument using induction on m.

Proof. It is enough to find a regular sequence for $\mathcal{P}(\mathrm{Sym}_m)$ that contains the μ_i. First choose linear functions $l_1, \ldots, l_r$, $r := \dim \mathrm{Sym}_m - m = \binom{m}{2}$, such that their zeros consist of matrices of the form:

$$A = \begin{pmatrix} a_1 & a_2 & a_3 & \ldots & a_m \\ a_2 & a_3 & \ldots & a_m & 0 \\ a_3 & \vdots & & 0 & 0 \\ \vdots & a_m & 0 & 0 & \vdots \\ a_m & 0 & 0 & \ldots & 0 \end{pmatrix}.$$

Then the functions μ_i send the matrix A above to

$$\mu_1(A) = a_1,\ \mu_2(A) = a_1 a_3 - a_2^2,$$
$$\mu_3(A) = -a_3^3 - a_2^2 a_5 - a_1 a_4^2 + a_1 a_3 a_5 + 2a_2 a_3 a_4,$$

etc. We see that $\mu_i(A)$ contains the variable a_i to the pure power i, and the rest of the expression consists of monomials involving at least one of the previous variables $a_1, a_2, \ldots, a_{i-1}$. Thus the $\binom{m+1}{2}$ equations

$$l_1(A) = l_2(A) = \cdots = l_r(A) = \mu_1(A) = \mu_2(A) = \cdots = \mu_m(A) = 0$$

force $A = 0$. It follows that the l_j followed by the μ_i form a homogeneous system of parameters in the polynomial ring $\mathcal{P}(\mathrm{Sym}_m)$, hence a regular sequence (cf. [Mat, Theorem 17.4]). □

APPENDIX

The essential point in the proof is the Main Lemma where we show that the functions $\{\delta_j \mid j \in J\}$ form a good sequence on the orbit $\mathrm{GL}_m \cdot w_J \simeq \mathrm{GL}_m / Q_J$ (see the proof of Proposition 1). In this appendix we make a more careful analysis of the situation and give a very precise description of the zero locus of the collection $\{\delta_b\}_{1 \le b \le \ell}$. The discussion could easily be extended to describe the zero locus of any collection of the δ_b.

We can regard functions on any one of the orbits GL_m / Q_J as functions on GL_m which happen to be invariant under right translations by elements of Q_J. Since the δ_b are invariant under left translation by the orthogonal groups O_m, their zero sets will evidently also be invariant under left translation by O_m. Further, the fact that the δ_b are also homogeneous means that they are, not invariant, but eigenfunctions under right translations by the full parabolic P_J containing Q_J. Thus, although the δ_b are not P_J invariant, their zero sets *will* be P_J invariant. In sum, then, when we ask about the zero set of the functions δ_b on the variety GL_m / Q_J, we are talking about a union of (O_m, P_J) double cosets, that is, a subset of the double coset space $\mathrm{O}_m \setminus \mathrm{GL}_m / P_J$.

If $B \subset P_J$ is a Borel subgroup, then the (O_m, P_J) double cosets are unions of (O_m, B) double cosets. It turns out that there are only a finite number of these, and there is a strong analogy between the description of $\mathrm{O}_m \setminus \mathrm{GL}_m / B$ and the Bruhat decomposition that describes $B \setminus \mathrm{GL}_m / B$. (The finiteness of these double coset spaces illustrates a result of BRION and VINBERG (cf. [Br2], [Vi]), which asserts the finiteness of the B-orbits in any multiplicity-free variety.)

The detailed description of $H \setminus G / B$ where G is a semisimple group, B a Borel subgroup and H a symmetric subgroup has been developed in [Ma], [Sp]. We will describe the specialization of their general results to the case at hand. We may think of (O_m, B) double cosets as GL_m-orbits of pairs $(\beta, \mathcal{F})$, where β is an inner product on $\mathbb{C}^m$, and $\mathcal{F} = \{\{0\} \subset V_1 \subset V_2 \subset V_3 \subset \cdots \subset V_m = \mathbb{C}^m\}$ is a complete flag in $\mathbb{C}^m$. Here GL_m acts on inner products and on flags in the usual ways, and acts on pairs by the product action. Somewhat more precisely, we have a diagram

$$\text{(A1)} \qquad \begin{array}{ccc} GL_m \times \mathrm{GL}_m & \longrightarrow & \mathrm{GL}_m / \mathrm{O}_m \times \mathrm{GL}_m / B \\ \big\downarrow & & \big\downarrow \\ \mathrm{GL}_m & \longrightarrow & \mathrm{O}_m \setminus \mathrm{GL}_m / B \end{array}$$

where the left vertical arrow is the principal GL_m-fibration $(g_1, g_2) \mapsto g_1^{-1} g_2$, and the upper horizontal arrow is the quotient mapping which is a principal $\mathrm{O}_m \times B$-fibration. It follows that the codimension of a GL_m-orbit in $\mathrm{GL}_m / \mathrm{O}_m \times \mathrm{GL}_m / B$ is equal to the codimension of a (O_m, B) double coset to which it is mapped by the right hand side vertical arrow.

Claim A2. *Given a pair $(\beta, \mathcal{F})$, we can find a basis $\{v_j\}_{1 \le j \le m}$ for $\mathbb{C}^m$ such* *that*

(i) $\{v_j\}_{1\le j\le k}$ *is a basis for* V_k, *for each* $k; 1 \le k \le m$.
(ii) *there is a permutation* $\sigma(\beta,\mathcal{F}) = \sigma$ *of* $\{1,2,3,\ldots,m\}$, *of order 2, such that* $\beta(v_i,v_j) = \delta_{i\sigma(j)} = \delta_{\sigma(i)j}$.

Indeed, consider V_1. If V_1 is not isotropic for the inner product β, we can find and element v_1 in V_1, such that $\beta(v_1,v_1) = 1$. (Such an element is unique up to multiplication by ±1.) Then let $Y = V_1^\perp$, where the orthogonal complement is taken with respect to β. Define a complete flag $\mathcal{F}' = \{V_j'\}_{1\le j\le m-1}$ in Y by setting

$$V_j' = V_{j+1}' \cap Y.$$

By induction on dimension, we can assume the claim is true for the pair $(\beta\mid_Y,\mathcal{F}')$. Let $\{v_j'\}$ be the basis of Y as specified in the claim. Then setting $v_j = v_{j-1}'$ for $j\ge 2$, we see that $\{v_j\}_{1\le j\le m}$ will satisfy the claim for $(\beta,\mathcal{F})$.

Consider, on the other hand, the possibility that V_1 is isotropic for β. Let a be the smallest index such that $V_a \not\subset V_1^\perp$. Choose v_1 arbitrarily in V_1, and choose v_a in V_a such that $\beta(v_1,v_a) = 1$, and $\beta(v_a,v_a) = 0$. Note that, if we have u in V_a such that $\beta(v_1,u) = 1$, then we can set $v_a = u - (\beta(u,u)/2)v_1$. Thus we see that, if we have a satisfactory v_a, we can add to it any element in V_{a-1}, then further add an appropriate multiple of v_1, to obtain another suitable candidate for v_a. Together with the fact that v_1 is only determined up to multiples, this lets us see that there is an $(a-1)$-dimensional family of satisfactory pairs $\{v_1,v_a\}$.

Let P be the plane spanned by v_1 and v_a, and set $Y = P^\perp$. Define a complete flag $\mathcal{F}' = \{V_j'\}_{1\le j\le m-2}$ in Y by the recipe

$$\begin{aligned} V_j' &= V_{j+1}\cap Y && \text{for } 1\le j\le a-2;\\ V_j' &= V_{j+2}\cap Y && \text{for } a-1\le j\le m-2. \end{aligned}$$

By induction on dimension we may assume we have a basis $\{v_j'\}$ for Y, which basis satisfies the claim for the pair $(\beta\mid_Y,\mathcal{F}')$. Now define

$$\begin{aligned} v_j &= v_{j-1}' && \text{for } 2\le j\le a-1,\\ v_j &= v_{j-2}' && \text{for } a+1\le j\le m. \end{aligned}$$

It is easy to check that the basis $\{v_j\}$ satisfies the claim for the pair $(\beta,\mathcal{F})$. □

The above procedure also allows one to compute $\dim(B_{\mathcal{F}}\cap O_\beta)$, where $B_{\mathcal{F}}$ is the stabilizer of $\mathcal{F}$ in GL_m, and O_β is the isometry group of β. This will also be the codimension of the GL_m-orbit of $(B,\mathcal{F})$, since $\dim B_V + \dim O_\beta = \dim \mathrm{GL}_m = m^2$. As mentioned above, this is also the codimension of the corresponding double coset. Since we had an $a-1$ dimensional set from which to choose the pair (v_1,v_a), we have the relation

(A3) $$\dim(B_{\mathcal{F}}\cap O_\beta) = a - 1 + \dim(B_{\mathcal{F}'}\cap(O_\beta\cap \mathrm{GL}(Y'))).$$

Note that this is also valid in the case $a = 1$.

We observe that $a - 1$ is the distance between 1 and $a = \sigma(1)$. This gives us the main term in the expression for $\dim(B_{\mathcal{F}} \cap O_\beta)$. For a pair $(c < d)$ of integers, let $\sigma(c, d)$ denote the transposition which exchanges c and d, and leaves all other integers fixed. Given pairs $(c < d)$ and $(c' < d')$ of integers, we say that they are *linked* if $c < c' < d < d'$, or if $c' < c < d' < d$. We say the permutations $\sigma(c, d)$ and $\sigma(c', d')$ are linked if the pairs (c, d) and (c', d') are.

Given an element σ of order two in the symmetric group on m letters, we can factor it into a product of transpositions:

$$\sigma = \prod_i \sigma(c_i, d_i). \tag{A4}$$

Lemma A5. *If the involution σ attached to the pair $(\beta, \mathcal{F})$ is factored as in formula* (A4), *then*

$$\dim(B_{\mathcal{F}} \cap O_\beta) = \sum_i (d_i - c_i) - \#\{(i, j) \mid \sigma(c_i, d_i) \text{ and } \sigma(c_j, d_j) \text{ are linked}\}.$$

Remark. We note that if we multiply σ by another factor $\sigma(c, d)$, the right hand side of this formula increases, since the number of pairs (c_i, d_i) with which (c, d) can be linked is at most $c - d - 1$.

Proof. If we assume the formula is true for Y, then it remains true for $\mathbb{C}^m$, by formula (A3), since if (c_i, d_i) is linked to $(1, a)$, it becomes $(c_i - 1, d_i - 2)$ in Y, while if it is not linked, it becomes $(c_i - e, d_i - e)$, where $e = 1$ or 2, as the case may be. □

Proof of the Main Lemma. Now let β_0 denote our fixed standard inner product on $\mathbb{C}^m$, and let δ_j be the O_m-invariant primary covariants of formula (7). Consider what (O_m, B) double cosets can be in the zero-locus of $\{\delta_j\}_{1 \le j \le \ell}$. For the flag $\mathcal{F}$ to be in the zero locus of δ_j, the space V_j should be singular for β_0. This follows by combining the diagram (A1) with the description of the zero locus of δ_j as given in the proof of Proposition 1.

This means, for the basis $\{v_j\}$ adapted to the pair $(\beta_0, \mathcal{F})$ as per Claim A2, that not every v_c with $c \le j$ is paired with a v_d with $d \le j$. In other words, some v_c with $c \le j$ is paired with v_d with $d > j$, or, the set $\{1, 2, 3, \ldots, j\}$ is not invariant under the involution $\sigma(\beta_0, \mathcal{F})$. That is, there is some factor $\sigma(c_i, d_i)$ of σ, with $c_i \le j < d_i$.

Suppose this holds for all j up to $j = \ell$. Taking $j = 1$, we see that 1 must not be fixed by σ. Hence in the factorization (A4) of σ, there is a factor $\sigma(1, d_1)$. If $d_1 > \ell$, then $\mathcal{F}$ is guaranteed to be in the zero locus of $\{\delta_j\}_{1 \le j \le \ell}$. However, if $d_1 \le \ell$, there must be another factor $\sigma(c_2, d_2)$ of σ, with $c_2 < d_1 < d_2$. If there is more than one such, we may chose the one with d_2 as large as possible. If $d_2 > \ell$, then we now are assured that $\mathcal{F}$ is in the zero locus of $\{\delta_j\}_{1 \le j \le \ell}$.

If, however, we still have $d_2 \leq \ell$, then there must be a third factor $\sigma(c_3, d_3)$ of σ, with $c_3 < d_2 < d_3$. Since we chose d_2 as large as possible at the previous stage, we know that $c_3 > d_1$. We may now select d_3 to be as large as possible. Continuing in this fashion, we conclude that σ has a set of factors $\sigma(c_i, d_i)$ for $1 \leq i \leq r$, such that

$$1 = c_1 < c_2 < d_1 < c_3 < d_2 < c_4 < d_3 < \cdots < c_r < d_{r-1} \leq \ell < d_r.$$

From Lemma A5 and the Remark following it, we see that the codimension of the coset containing $\mathcal{F}$ is at least equal to

$$\sum_{i=1}^{r}(d_i - c_i) - (r-1) \geq (d_1 - 1) + \sum_{i=2}^{r}(d_i - d_{i-1} + 1) - (r-1) \geq d_r - 1 \geq \ell,$$

as desired.

Remark. Using the above inequalities, we can be precise about which cosets have codimension exactly ℓ. They would be the ones for which $c_i = d_{i-1} - 1$, and $d_r = \ell + 1$. Thus they correspond to all possible sequences

$$1 < d_1 < d_2 < d_3 < \cdots < d_r = \ell + 1,$$

where $d_i - d_{i-1} \geq 2$ if $d_i \leq \ell$ (to allow for the insertion of c_{i+1}). The number of such sequences is the ℓ-th Fibonacci number.

References

[ATZ] Aslaksen, H., Tan, E.-C., Zhu, C.-B., *On certain rings of highest weight vectors*, J. Algebra **174** (1995), 159–186.

[Br] Brion, M., *Invariants d'un sous-groupe unipotent maximal d'un groupe semi-simple*, Ann. Inst. Fourier **33** (1983), 1–27.

[Br2] Brion, M., *Quelques propriétés des espaces homogènes sphériques*, manuscripta math. **55** (1986), 191–198.

[Do] Dodgson, C., *Condensation of determinants*, Proc. Royal Soc. London **15** (1866), 50-55.

[Ge] Gelbart, S., *Holomorphic discrete series for the real symplectic group*, Invent. Math. **19** (1973), 49–58.

[GK] Gross, K., Kunze, R., *Bessel functions and representation theory II*, J. Functional Anal. **25** (1977), 1–49.

[GW] Gross, B. and Wallach, N., *On quaternionic, discrete series representations and their continuation*, J. f. d. Reine u. Angew. Math. **481** (1996), 73– 123.

[HC-IV] Harish-Chandra, *Representations of semisimple Lie groups, IV*, Amer. J. Math. **77** (1955), 743-777.

[HC-V] ———, *Representations of semisimple Lie groups, V*, Amer. J. Math. **78** (1956), 1-41.

[HC-VI] ———, *Representations of semisimple Lie groups, VI*, Amer. J. Math. **78** (1956), 564–628.

[Ho] Howe, R., *Perspectives on Invariant Theory*, Israel Math. Conf. Proc. vol. 8 (I. Piatetski-Shapiro and S. Gelbart, eds.), The Schur Lectures (1992), American Math. Soc., Providence, RI, 1995, pp. 1–182.

[HSch] Hecht, H., Schmid, W., *A Proof of Blattner's Conjecture*, Invent. Math. **31** (1975), 129–154.

[Jo] Johnson, K., *On a ring of invariant polynomials on a Hermitian symmetric space*, J. Alg. **67** (1980), 72–81.

[KaVe] Kashiwara, M., and Vergne, M., *On the Segal-Shale-Weil representations and harmonic polynomials*, Invent. Math. **44** (1978), 1–47.

[Kna] Knapp, A., *Representation Theory of Semisimple Groups, An Overview Based on Examples*, Princeton University Press, Princeton, NJ, 1986.

[KnVo] Knapp, A., Vogan, D., *Cohomological Induction and Unitary Representations*, Princeton University Press, Princeton, NJ, 1995.

[Kno] Knop, F., *The Luna-Vust Theory of Spherical Embeddings*, Proc. Hyderabad Conf. on Algebraic Groups, National Boards for Higher Mathematics, 1989.

[Krf] Kraft, H., *Geometrische Methoden in der Invariantentheorie*, Aspekte der Mathematik, vol. D1, Vieweg, Braunschweig/Wiesbaden, 1984.

[KrP] Kraft, H.; Procesi, C., *On the geometry of conjugacy classes in classical groups*, Comment. Math. Helv. **57** (1982), 539–602.

[Krm] Krämer, M., *Sphärische Untergruppen in kompakten zusammenhängenden Lie Gruppen*, Compositio Math. **38** (1979), 129–153.

[Li] Littelmann, P., *Koreguläre und äquidimensionale Darstellungen*, J. Algebra **123** (1989), 193–222.

[Ma] Matsuki, T., *The orbits of affine symmetric spaces under the action of minimal parabolic subgroups*, J. Math. Soc. Japan **31** (1979), 331–357.

[Mat] Matsumura, H., *Commutative Ring Theory*, Cambridge Studies in Advanced Mathematics, vol. 8, Cambridge Univ. Press, Cambridge, 1989.

[Pop] Popov, V.L., *Syzygies in the theory of invariants*, Math. USSR-Izv. **22** (1984), 507–585.

[Pop2] Popov, V.L., *Contraction of the action of reductive algebraic groups*, Math. USSR-Sbornik **58** (1987), 311–335.

[PV] Popov, V.L., Vinberg, E., *On a class of quasihomogeneous affine varieties*, Math. USSR-Izv. **6** (1972), 743–758.

[RR] Robbins, D., Ramsey, H. Jr., *Determinants and alternating sign matrices*, Adv. Math. **62** (1986), 169-184.

[Sa] Saito, M., *Représentations unitaires des groupes symplectiques*, J. Math. Soc. Japan **24** (1972), 232–251.

[Sch1] Schmid, W., *Die Randwerte holomorpher Funktionen auf hermitesch symmetrischen Räumen*, Invent. Math. **9** (1969), 61–80.

[Sch2] ———, *On the characters of the discrete series (the Hermitian symmetric case)*, Invent. Math. **30** (1975), 47–144.

[Schw] Schwarz, G. W., *Representations of simple Lie groups with a free module of covariants*, Invent. Math. **50** (1978), 1–12.

[Sp] Springer, T.A., *Some results on algebraic groups with involutions*, Algebraic Groups and Related Topics, Advanced Studies in Pure Math. **6** (1985), North Holland, Amsterdam-New York-Oxford, 525–543.

[Vi] Vinberg, E. B., *Complexity of action of reductive groups*, Functional Anal. Appl. **20** (1986), 1–11.

[Vo] Vogan, D., *Irreducible characters of semisimple algebraic groups II, Proof of the Kazhdan-Lusztig conjecture in the integral case*, Invent. Math. **71** (1983), 381–417.

[Wal] Wallach, N., *Representations of Real Reductive Groups I*, Pure and App. Math 132, Academic Press, San Diego, CA, 1988.

[War] Warner, G., *Harmonic Analysis on Semi-Simple Lie Groups*, Grundlehren Math. Wiss., vol. 188, Springer Verlag, Berlin-Heidelberg-New York, 1972.

[We] Weyl, H., *The Classical Groups*, Princeton University Press, Princeton, NJ, 1940.

Roger Howe
Department of Mathematics, Yale University,
New Haven, CT 06520-8283, USA
email: howe-roger@math.yale.edu
and
Hanspeter Kraft
Mathematisches Institut, Universität Basel
Rheinsprung 21,
CH-4051 BASEL, Switzerland
email: kraft@math.unibas.ch

WHITTAKER MODELS FOR CARAYOL REPRESENTATIONS OF $GL_N(F)$

ROBERTO JOHNSON*

INTRODUCTION

Let F be a local non-Archimedean field with residual characteristic p and let G be the group of all invertible $N \times N$ matrices over F.

In this work we construct explicitly the Whittaker models for a special class of supercuspidal representation of G first obtained by Carayol [Ca]. To know these Whittaker models explicitly may be useful to compute the L and ε factors [Bo] for representations of p-adic group using the method of Shahidi [Sh]. To compute these factors, the representation must be generic, that is, it must have a Whittaker model.

Roughly described: Given a generic charater τ of the maximal unipotent radical subgroup U of G, let (ρ_τ, W_τ) be the induced representation of τ from U to G. A Whittaker model of a representation (Π, V) of G is a subrepresentation M of (ρ_τ, W_τ) which is isomorphic to (Π, V). I.M. Gelfand and D.A. Kazhdan proved [G-K] that irreducible admissible supercuspidal representations of $GL_N(F)$ admit Whittaker models. On the other hand, by a result of F. Rodier [R] we know that an irreducible admissible supercuspidal representation (Π, V) of G admits at most one Whittaker model with respect to τ. We then construct a convenient hereditary order $\mathcal{A}$ which allows us to prove, using Mackey's theorem for nonunitary representations [K], that $\mathrm{Hom}_G(\Pi, \rho_\tau)$ is isomorphic to $\mathrm{Hom}_{U \cap F[b]^\times \mathcal{U}^{[\frac{n+1}{2}]}}(\lambda, \tau)$ where λ and $F[b]^\times \mathcal{U}^{[\frac{n+1}{2}]}$ are such that $\Pi = \mathrm{Ind}^G_{F[b]^\times \mathcal{U}^{[\frac{n+1}{2}]}} \lambda$.

The plan of this paper is as follows. Sections 1 and 2 are essentially devoted to gathering together some facts about hereditary orders in the ring A of all $N \times N$ matrices over F and some preliminaries. In Section 3 we construct the hereditary order $\mathcal{A}$, which we use in Section 4 to prove the main result.

1. PRELIMINARIES

We denote the ring of integers in F by O_F, we fix a generator π_F of the maximal ideal P_F of O_F and we write $q_F = [O_F : P_F]$. For an element x in F, we denote the valuation of x by $\nu_F(x)$. Furthermore, we set $U_F = O_F^\times$ and

*Partially supported by FONDECYT grant 1950895, Chile, and NSF-Andes-Conicyt International Program.

write U_F^n for $1+P_F^n$, $n \geq 1$. If E is an extension field of F, $[E:F]$ denotes the degree of the extension, $e(E/F)$ the ramification degree and $f(E/F)$ the inertia degree. For details about local fields, see [Se].

By a representation (π, V) of G we mean a C-vector space V (C being the complex numbers) together with a homomorphism π of G into $\mathrm{Aut}_C(V)$. If (π, V) is such a representation, then a vector v in V is said to be smooth if its stability subgroup in G (via the action given by π) is open; the set V^∞ of smooth vectors in V is in fact a G-subspace of V; if $V = V^\infty$ then we say that (π, V) is a smooth representation of G. A smooth representation (π, V) of G is called supercuspidal if each of its coefficients is supported on a set that is compact-modulo-center; that is, given a coefficient c of π, there is a compact subset K of G such that the support of c is contained in the set $Z(G) \cdot K$, where $Z(G)$ is the center of G. For more details about smooth representations, see [Cr] or [Cs].

In what follows, $V = F^N$ and $\mathcal{B} = \{e_1, \ldots, e_N\}$ is the standard basis of V. Also, $\mathcal{A}$ is always the principal order in $A = M_N(F)$ with period $e(\mathcal{A}) = e$ associated to the O_F–lattice chain $\mathcal{L} = \{L_i \mid i \in Z\}$ in V where

$$L_i = P_F e_1 \oplus \cdots \oplus P_F e_{fi} \oplus O_F e_{fi+1} \oplus \cdots \oplus O_F e_N$$

with $f = N/e$, $i = 0, \ldots, e-1$ and $P_F L_i = L_{i+e}$ for every integer i. Thus, the Jacobson radical $\mathcal{P}$ of $\mathcal{A}$ is principal as both a left and a right ideal and

$$\mathcal{P} = \Pi_{\mathcal{L}} \cdot \mathcal{A} = \mathcal{A} \cdot \Pi_{\mathcal{L}}$$

where

$$\Pi_{\mathcal{L}} = \begin{bmatrix} 0_f & 0_f & \cdots & 0_f & \pi_F 1_f \\ 1_f & 0_f & & & 0_f \\ 0_f & \ddots & \ddots & & \vdots \\ \vdots & \ddots & 1_f & 0_f & 0_f \\ 0_f & \cdots & 0_f & 1_f & 0_f \end{bmatrix}$$

with 1_f the identity matrix in $M_f(O_F)$ and 0_f the zero matrix in $M_f(O_F)$.

Also, we can define a sequence of compact open subgroups of G by

$$\mathcal{U} = \mathcal{U}^0 = \mathcal{A}^\times$$

($\mathcal{A}^\times$ is the set of invertible elements of $\mathcal{A}$) and

1.1. $\mathcal{U}^n = 1 + \mathcal{P}^n$, for $n \geq 1$.

The set

1.2. $\mathcal{K}(\mathcal{A}) = \{x \in G \mid x\mathcal{U}x^{-1} = \mathcal{U}\}$

is a maximal open compact-modulo-center subgroup of G. Furthermore, the group $\mathcal{K}(\mathcal{A})$ is its own normalizer in G.

The $\mathcal{U}^n$, for $n \geq 0$, are normal subgroups of $\mathcal{K}(\mathcal{A})$, and in particular, $\mathcal{U}$ is the unique maximal compact subgroup of $\mathcal{K}(\mathcal{A})$.

For details and a more general treatment of principal orders and lattice chains, see 1.3 in [B-F].

2. The general setting

We denote by B the standard Borel subgroup of upper triangular matrices, U the maximal unipotent radical subgroup of G, U' the commutator subgroup of U and T the maximal torus of diagonal matrices in G.

A character of U is a smooth homomorphism τ from U into the group of complex numbers of absolute value 1.

Now, let Φ be the map from U into F^{N-1} defined by

$$\Phi\left(\begin{bmatrix} 1 & x_{1\,2} & x_{1\,3} & \cdots & x_{1\,N-1} & x_{1\,N} \\ 0 & 1 & \ddots & \ddots & \ddots & x_{2\,N} \\ 0 & \ddots & \ddots & x_{i\,i+1} & \ddots & \vdots \\ \vdots & \ddots & 0 & \ddots & \ddots & x_{N-2\,N} \\ 0 & \ddots & \ddots & \ddots & 1 & x_{N-1\,N} \\ 0 & 0 & \cdots & 0 & 0 & 1 \end{bmatrix}\right) = (x_{1\,2}, \dots, x_{N-1\,N}).$$

Then Φ is a group epimorphism with $\mathrm{Ker}\Phi = U'$. Thus, U/U' is isomorphic to F^{N-1}. If we fix a smooth character ψ of the additive group of F, with conductor P_F, (i.e. ψ is trivial on P_F but not on O_F) then any character τ on F^{N-1} is obtained as $\tau = \psi \circ L$, where L is an F-linear map from F^{N-1} into F. Since U/U' is isomorphic to F^{N-1} and any character τ on U factors through a character of U/U', then there are $\alpha_1, \dots, \alpha_{N-1} \in F$ such that

2.1. $$\tau(u) = \psi\left(\textstyle\sum_{i=1}^{N-1} \alpha_i x_{i\,i+1}\right) \qquad (u \in U).$$

The group T acts on the set of all characters of U by conjugation. That is, given a character τ of U we have that τ^t is also a character of U defined, as usual, by

$$\tau^t(u) = \tau(tut^{-1}) \qquad (t \in T, u \in U)$$

Definition 2.2. The character τ of U is said to be *generic* if the set of all t in T such that $\tau^t = \tau$ is the center $Z(G)$ of G.

Note that the character τ in 2.1 is generic if and only if $\alpha_i \neq 0$ for every $i = 1, \dots, N-1$, and any such character is conjugate, under the action of T, to the character τ defined by

$$\textbf{2.3.}\quad \tau\left(\begin{bmatrix} 1 & x_{1\,2} & x_{1\,3} & \cdots & x_{1\,N-1} & x_{1\,N} \\ 0 & 1 & \ddots & \ddots & \ddots & x_{2\,N} \\ 0 & \ddots & \ddots & x_{i\,i+1} & \ddots & \vdots \\ \vdots & \ddots & 0 & \ddots & \ddots & x_{N-2\,N} \\ 0 & \ddots & \ddots & \ddots & 1 & x_{N-1\,N} \\ 0 & 0 & \cdots & 0 & 0 & 1 \end{bmatrix}\right) = \psi\left(\textstyle\sum_{i=1}^{N-1} x_{i\,i+1}\right)$$

Now let τ be a generic character of U. Since U is a closed subgroup of G, we can construct the smooth induced representation (ρ_τ, W_τ) of the one-dimensional representation τ. That is, W_τ is the space of all functions ϕ from G to W satisfying $\phi(ug) = \tau(u)\phi(g)$, for every $u \in U$ and $g \in G$, such that ϕ is also smooth under the natural action ρ_τ of G, by right translation, on W_τ. This means

$$\textbf{2.4.}\qquad \rho_\tau = \mathrm{Ind}_U^G \tau$$

is defined by

$$[\rho_\tau(g)f](g') = f(g'g) \qquad (g, g' \in G; f \in W_\tau).$$

Definition 2.5. Let (Π, V) be a representation of G. A subspace M of W_τ is called a *Whittaker model of* (Π, V) *with respect to* τ if M is invariant under ρ_τ and the restriction of ρ_τ to M is equivalent to Π.

The results of Rodier, and Gelfand and Kazhdan mentioned in the introduction imply that

$$\textbf{2.6}\qquad \mathrm{Hom}_G(\Pi, \rho_\tau) \simeq C.$$

Thus we can think of a Whittaker model as an intertwining operator from Π to ρ_τ.

3. Main Lemmas

In this section we construct a specific hereditary order $\mathcal{A}$ in A which verifies some properties that allow us to proving Lemma 3.5. This lemma is the key to prove Proposition 4.5 which in turn we use to prove the main result, which is Theorem 4.6.

Let b be an element in $A = M_N(F)$ with irreducible characteristic polynomial $p(x) = a_0 + a_1 x + \cdots a_{N-1}x^{N-1} + x^N$ over F. In what follows, we pick b as the companion matrix of the monic polynomial $p(x) \in F[x]$.

3.1.
$$b = \begin{bmatrix} 0 & \cdots & \cdots & 0 & 0 & -a_0 \\ 1 & \ddots & & & 0 & -a_1 \\ 0 & 1 & 0 & & \vdots & \vdots \\ \vdots & \ddots & \ddots & \ddots & \vdots & \vdots \\ \vdots & & \ddots & 1 & 0 & -a_{N-2} \\ 0 & \cdots & \cdots & 0 & 1 & -a_{N-1} \end{bmatrix}.$$

So the ring $E = F[b]$, which is a subring of A, is a field extension of F of degree N. Also, we assume that $\nu_E(b) = -n$ and that b is E/F-minimal, i.e. $(\nu_E(b), e(E/F)) = 1$, and the element $\overline{\beta} = \pi_F^{-\nu_E(b)} b^{e(E/F)} + P_E$ generates the residue class field extension k_E/k_F. See, for example, [K-M].

We can take $A = \mathrm{End}_F(E)$ and $G = A^\times$. Identifying any element $\alpha \in E^\times$ with the endomorphism m_α defined by

$$m_\alpha(x) = \alpha x$$

for every $x \in E$, we can say that $E^\times \subset G$. We denote the stabilizer group of 1_E in G by Q, i.e.,

$$Q = \{g \in G \mid g \cdot 1_E = 1_E\}.$$

Proofs of the following lemma and proposition can be found in [J].

Lemma 3.2. *With notations as above, the following statements hold:*

(1) $E^\times \cap Q = \{1\}$.

(2) $G = E^\times Q$.

(3) Let $\mathcal{A}_o$ *be the principal order in* A *associated to the* O_F*-lattice chain* $\mathcal{L}_o = \{P_E^i \mid i \in Z\}$ *in* E *and* $\mathcal{U}_o^n = 1 + \mathcal{P}_o^n$*, for any* $n \geq 1$*, with* $\mathcal{P}_o$=*rad*$(\mathcal{A}_o)$*. Then*

$$\mathcal{U}_o^n = (E^\times \cap \mathcal{U}_o^n)(Q \cap \mathcal{U}_o^n).$$

(4) $E^\times \mathcal{U}_o^n = E^\times (Q \cap \mathcal{U}_o^n)$ *with* $E^\times \cap (Q \cap \mathcal{U}_o^n) = \{1\}$.

Recall that $\mathcal{B} = \{e_1, \ldots, e_N\}$ is the standard basis of $V = F^N$ and pick the basis $\widetilde{\mathcal{B}} = \{1, b, \ldots, b^{N-1}\}$ of E as an $F-$vector space. Let $\varphi : E \longrightarrow V$ be the coordinate map. Then, φ induces an algebra isomorphism

$$\Phi : \mathrm{End}_F(E) \longrightarrow A$$

defined by $\Phi(T) = \varphi T \varphi^{-1}$ for every $T \in \mathrm{End}_F(E)$.

Furthermore, since we know that $m_b \in \mathrm{End}_F(E)$ and b is the companion matrix of the polynomial $p(x)$ given above, then we have $\Phi(m_b) = \varphi m_b \varphi^{-1} = b \in A$. On the other hand, since Φ is a ring isomorphism and $E = F[b]$, we get $\Phi(E) = E$.

Proposition 3.3. *With $\mathcal{A}_o$ as above, put $\mathcal{A} = \Phi(\mathcal{A}_o)$ and $\mathcal{L} = \varphi(\mathcal{L}_o)$. Then*

(1) $\mathcal{L}$ is an O_F-lattice chain in V and $\mathcal{A}$ is the principal order in A associated to $\mathcal{L}$.

(2) $E^\times \subset \mathcal{K}(\mathcal{A})$, where $\mathcal{K}(\mathcal{A})$ is defined as in 1.2.

Definition 3.4. Let V be an F-vector space of dimension N. A basis $\{v_1, \ldots, v_N\}$ of V is called a *basis for an uniform O_F-lattice chain* $\mathcal{L} = \{L_i \mid i \in Z\}$ if

$$L_i = P_F v_1 + \cdots + P_F v_{if} + O_F v_{if+1} + \cdots + O_F v_N$$

for $i = 0, \ldots, e-1$ with $e = e(\mathcal{L})$ and $f = N/e$.

Since b is E/F-minimal we have that there exists a basis $\mathcal{B}' = \{\alpha_t = \pi_F^{l_t} b^{k_t} \mid t = 0, \ldots, N-1\}$ (see Proposition 2.2.2. in [J]) for the F-vector space E which is a basis for the uniform O_F-lattice chain $\mathcal{L}_0 = \{P_E^i \mid i \in Z\}$ such that

$$P_E^i = P_F 1_E + \cdots + P_F \alpha_{if-1} + O_F \alpha_i + \cdots + O_F \alpha_{N-1}$$

for $i = 1, \ldots, e$ where $e = e(E/F)$ and $f = f(E/F)$.

So $\varphi(\mathcal{B}')$ is a basis for the O_F-lattice chain $\mathcal{L} = \varphi(\mathcal{L}_o)$. Since $\alpha_t = \pi_F^{l_t} b^{k_t}$, then $\varphi(\alpha_t) = \pi_F^{l_t} \varphi(b^{k_t}) = \pi_F^{l_t} e_{k_t}$, which shows that φ is a monomial matrix and also that $\varphi(1_E) = e_0$.

Lemma 3.5. *Let $\mathcal{A}$ be the principal order defined in Proposition 3.3, $\mathcal{P} = rad(\mathcal{A})$ and $\mathcal{U}^n = 1 + \mathcal{P}^n$ for $n \geq 1$. Let U be the unipotent radical of the standard Borel subgroup of G. Then $U \cap E^\times \mathcal{U}^n = U \cap \mathcal{U}^n$.*

Proof: We know, from above, that $\varphi(\alpha_0) = \varphi(1_E) = e_0$; then $\Phi(Q) = Q = \{g \in G \mid g \cdot e_0 = e_0\}$ where $Q = \{T \in \mathrm{End}_F(E)^\times \mid T \cdot 1_E = 1_E\}$.

By (4) in Lemma 3.2 we have that

$$E^\times \mathcal{U}_0^n = E^\times (Q \cap \mathcal{U}_0^n).$$

Since $\Phi(Q) = Q$ is in $GL_N(F)$, $\Phi(E) = E$ and $\mathcal{A} = \Phi(\mathcal{A}_0)$ is in $M_N(F)$, then, using the fact Φ is an algebra isomorphism, we have

$$E^\times \mathcal{U}^n = E^\times (Q \cap \mathcal{U}^n).$$

So,

$$U \cap E^\times \mathcal{U}^n = U \cap E^\times (Q \cap \mathcal{U}^n)$$

and given $x \in U \cap E^\times (Q \cap \mathcal{U}^n)$, we have that there exist unique $\alpha \in E^\times$ and $q \in Q$ such that $x = \alpha q$. Hence $\alpha = q^{-1}x \in Q$, since $U \subset Q$ (U stabilizes e_0); from this follows $\alpha \in E^\times \cap Q = \{1\}$ and, finally, we can conclude that

$$U \cap E^\times \mathcal{U}^n = U \cap \mathcal{U}^n.$$

4. Construction of Whittaker Models

We now construct the Whittaker model for the irreducible admissible supercuspidal representations of G of level n, for any integer number $n > 1$. These admissible supercuspidal representations of G, obtained by Carayol [Ca], are constructed explicitly in [J] following the language of [B-K].

Let $\mathcal{A}$ be any principal order in A and $\mathcal{P}$ the radical of $\mathcal{A}$. Let ψ be a smooth character of F with conductor P_F (as in **2**). If n, r are integers satisfying $n \geq r$ then the map $b \longmapsto \widetilde{\psi}_b$, where $\widetilde{\psi}_b(x) = \psi(tr(bx))$ for $x \in \mathcal{P}^{r+1}$, induces an isomorphism of $\mathcal{P}^{-n}/\mathcal{P}^{-r}$ onto $(\mathcal{P}^{r+1}/\mathcal{P}^{n+1})^\wedge$, the smooth dual of $\mathcal{P}^{r+1}/\mathcal{P}^{n+1}$. See, for example, [Bu].

If r, n are integers satisfying $n \geq r \geq [\frac{n}{2}] > 0$ (where $[x]$ denotes the greatest integer less or equal than x), then we have a canonical isomorphism $\phi : \mathcal{U}^r/\mathcal{U}^n \longmapsto \mathcal{P}^r/\mathcal{P}^n$ given by $\phi(x) = x - 1$

So, $(\mathcal{U}^{r+1}/\mathcal{U}^{n+1})^\wedge$ is isomorphic to $\mathcal{P}^{-n}/\mathcal{P}^{-r}$ by $b + \mathcal{P}^{-r} \longmapsto \psi_b$ ($b \in \mathcal{P}^{-n}$) where,

$$\psi_b(x) = \widetilde{\psi}_b(x-1) = \psi(\mathrm{tr}(b(x-1))), \qquad (x \in \mathcal{U}^{r+1}).$$

Definition 4.1. An element b in $\mathcal{P}^{-n}$ is called $e(\mathcal{A})$-*cuspidal* if $F[b]$ is an extension field of F of degree N, b is a $F[b]/F$-minimal element and $F[b]^\times \subset \mathcal{K}(\mathcal{A})$. A character ψ_b associated with such a b is called $e(\mathcal{A})$-*cuspidal.* Further, a representation σ of $\mathcal{K}(\mathcal{A})$ is called *very cuspidal of level* n of $\mathcal{K}(\mathcal{A})$ if $\mathcal{U}^{n+1}$ is a subgroup of Kerσ and $\sigma_{|\mathcal{U}^n} = \bigoplus d(b)\psi_b$, where $d(b)$ is the multiplicity of ψ_b in σ and b ranges over $e(\mathcal{A})$-cuspidal elements of $\mathcal{A}$.

Carayol proved (see Theorem 4.2 in [Ca]) that $\Pi = \mathrm{Ind}_{\mathcal{K}(\mathcal{A})}^G \sigma$ is an irreducible admissible supercuspidal representation of G for any irreducible very cuspidal representation σ of $\mathcal{K}(\mathcal{A})$.

We can see that such a σ in fact contains a $\mathcal{U}^{[\frac{n}{2}]+1}$ subrepresentation ψ_b, b as above. Now the normalizer of ψ_b in $\mathcal{K}(\mathcal{A})$ is just $F[b]^\times \mathcal{U}^{[\frac{n+1}{2}]}$, by Proposition 3.6 in [Ca].

If n is odd, then it follows from Clifford theory, see [Cl], that σ contains a representation λ of the form $\theta\psi_b$ on $F[b]^\times\mathcal{U}^{\frac{n+1}{2}}$ where θ is a character of $F[b]^\times$ with $\theta = \psi_b$ on $F[b]^\times \bigcap \mathcal{U}^{\frac{n+1}{2}}$. So $\sigma = \mathrm{Ind}_{F[b]^\times\mathcal{U}^{\frac{n+1}{2}}}^{\mathcal{K}(\mathcal{A})} \lambda$ and therefore $\Pi = \mathrm{Ind}_{F[b]^\times\mathcal{U}^{\frac{n+1}{2}}}^{G} \lambda$.

If n is even we can still associate a representation λ of $F[b]^\times\mathcal{U}^{\frac{n}{2}}$ such that $\sigma = \mathrm{Ind}_{F[b]^\times\mathcal{U}^{\frac{n}{2}}}^{\mathcal{K}(\mathcal{A})} \lambda$ and so also $\Pi = \mathrm{Ind}_{F[b]^\times\mathcal{U}^{\frac{n}{2}}}^{G} \lambda$.

In fact, λ is an extension to $F[b]^\times\mathcal{U}^{\frac{n}{2}}$ of an irreducible representation κ of $U^1_{F[b]}\mathcal{U}^{\frac{n}{2}}$; see 5.4 and 5.7 in [Ca]. This κ is obtained inducing a character χ from a subgroup $U^1_{F[b]}H$ of $U^1_{F[b]}\mathcal{U}^{\frac{n}{2}}$. We get χ as an extension of the character $\theta\psi_b$ on $U^1_{F[b]}\mathcal{U}^{\frac{n}{2}+1}$ to $U^1_{F[b]}H$ where H is a subgroup of $U^{\frac{n}{2}}$ with the property that $(H-1)/(P^{\frac{n}{2}}_{F[b]} + \mathcal{P}^{\frac{n}{2}+1})$ is a maximal isotropic subspace of the O_F/P_F-vector space $\mathcal{P}^{\frac{n}{2}}/(P^{\frac{n}{2}}_{F[b]} + \mathcal{P}^{\frac{n}{2}+1})$ with respect to the O_F/P_F-non-degenerate alternating form $\langle x, y\rangle = tr(b(xy - yx))$. See Proposition 3.5 in [Ca].

An important fact that we will need later is that the character χ is a constituent of $\lambda_{|U^1_{F[b]}H}$. This follows just from the construction of λ and Frobenius reciprocity.

In conclusion we have, from the Carayol construction, that for every $n > 1$ we can produce a representation λ of $F[b]^\times\mathcal{U}^{[\frac{n+1}{2}]}$ such that

4.2. $$\Pi = \mathrm{Ind}_{F[b]^\times\mathcal{U}^{[\frac{n+1}{2}]}}^{G} \lambda.$$

Proposition 4.3. *Let $\mathcal{A}$ be the principal order in A defined in Proposition 3.3 and $\mathcal{P}$ its Jacobson radical. Assume that $n > r \geq [\frac{n}{2}]$ and ψ_b is the character of $\mathcal{U}^{r+1}/\mathcal{U}^{n+1}$ with b as in 3.1. Let U be the unipotent radical of the standard Borel subgroup of G. Then*

$$\psi_{b|\mathcal{U}^{r+1}\cap U} = \tau_{|\mathcal{U}^{r+1}\cap U}$$

where τ is the character of U defined in 2.3.

Proof: Straightforward.

Note that, without loss of generality, we can choose the principal order obtained in Proposition 3.3. In fact, given any b' in A such that $E = F[b']$ is an extension field of F of degree N, $\nu_E(b') = -n$, b' is E/F-minimal and $\mathcal{A}'$ is a principal order such that $E \subset \mathcal{K}(\mathcal{A}')$, then we can conjugate b' by some appropriate element $g \in G$ such that $gb'g^{-1} = b$ is the companion matrix of the polynomial $p(x)$ given in Section 3 and so $\Pi = \mathrm{Ind}_{F[b]^\times\mathcal{U}^{[\frac{n+1}{2}]}}^{G} \lambda$ is an irreducible admissible supercuspidal representation of G isomorphic to

$\Pi' = \mathrm{Ind}^G_{F[b']^\times \mathcal{U}'^{[\frac{n+1}{2}]}} \lambda'$ where $\mathcal{U}'^{[\frac{n+1}{2}]}$ is as in 1.1 and λ' is obtained similarly as in 4.2.

Proposition 4.4. *Let Π be an irreducible admissible supercuspidal representation of G as above. Then there exists a generic character τ of the standard unipotent subgroup U of G such that*

$$Hom_{U\cap\mathcal{U}^{[\frac{n+1}{2}]}}(\lambda,\tau) \neq 0$$

where λ is the representation of $F[b]^\times \mathcal{U}^{[\frac{n+1}{2}]}$ such that $\Pi = Ind^G_{F[b]^\times \mathcal{U}^{[\frac{n+1}{2}]}} \lambda$.

Proof: (Sketch) First of all we fix the principal order $\mathcal{A}$ defined in 3.3 and b as in 3.1.

(i) If n is odd, then $\lambda_{|\mathcal{U}^{\frac{n+1}{2}}}$ is ψ_b, and from Proposition 4.3 we have $\psi_{b|U\cap\mathcal{U}^{\frac{n+1}{2}}} = \tau_{|U\cap\mathcal{U}^{\frac{n+1}{2}}}$.

(ii) If n is even we can choose H containing $U\cap\mathcal{U}^{\frac{n}{2}}$ because $\left[(U\cap\mathcal{U}^{\frac{n}{2}}) - 1/(P^{\frac{n}{2}}_{F[b]} + \mathcal{P}^{\frac{n}{2}+1})\right]$ is an isotropic subspace of $\mathcal{P}^{\frac{n}{2}}/(P^{\frac{n}{2}}_{F[b]} + \mathcal{P}^{\frac{n}{2}+1})$. So, $\psi_b = \tau$ on $\mathcal{U}^{\frac{n}{2}+1}\cap(U\cap\mathcal{U}^{\frac{n}{2}})$ and we can extend ψ_b from $\mathcal{U}^{\frac{n}{2}+1}$ to $\psi_b\tau$ on $\mathcal{U}^{\frac{n}{2}+1}(U\cap\mathcal{U}^{\frac{n}{2}})$. Now we can take χ on $U^1_{F[b]}H$ being an extension of $\psi_b\tau$ from $\mathcal{U}^{\frac{n+1}{2}}(U\cap\mathcal{U}^{\frac{n}{2}})$. Hence $\chi_{|U\cap\mathcal{U}^{\frac{n}{2}}} = \tau_{|U\cap\mathcal{U}^{\frac{n}{2}}}$ and since χ is in $\lambda_{|U^1_{F[b]}H}$ then $\mathrm{Hom}_{U\cap\mathcal{U}^{\frac{n}{2}}}(\lambda,\tau) \neq 0$.

Proposition 4.5. *Let Π be an irreducible admissible supercuspidal representation of G as above. Let τ be the generic character of U defined in 2.3. Let ρ_τ be the smooth representation given in 2.4. Then*

$$Hom_G(\Pi,\rho_\tau) \simeq Hom_{U\cap F[b]^\times \mathcal{U}^{[\frac{n+1}{2}]}}(\lambda,\tau)$$

with λ as in the above proposition. *Proof:* From Lemma 3.5 we know that $U\cap F[b]^\times \mathcal{U}^{[\frac{n+1}{2}]} = U\cap\mathcal{U}^{[\frac{n+1}{2}]}$.

Using Mackey's theorem, see [K], we have that

$$\mathrm{Hom}_G(\Pi,\rho_\tau) \simeq \mathrm{Hom}_G(\mathrm{Ind}^G_{F[b]^\times \mathcal{U}^{[\frac{n+1}{2}]}}\lambda, \tag{1}$$

$$\mathrm{Ind}^G_U\tau) \simeq \prod_{x\in U\backslash G/F[b]^\times \mathcal{U}^{[\frac{n+1}{2}]}} \mathrm{Hom}_{x^{-1}Ux\cap F[b]^\times \mathcal{U}^{[\frac{n+1}{2}]}}(\lambda,\tau^x). \tag{2}$$

By 2.6, $\mathrm{Hom}_G(\Pi,\rho_\tau) \simeq C$. But by 4.4, $\mathrm{Hom}_{U\cap\mathcal{U}^{[\frac{n+1}{2}]}}(\lambda,\tau) \neq 0$ and we are done.

Now we are ready to give the Whittaker models for these irreducible admissible supercuspidal representations of G.

Theorem 4.6. *Let Π be an irreducible admissible supercuspidal representation of G and λ the representation of $F[b]^\times \mathcal{U}^{[\frac{n+1}{2}]}$ as above. Let τ be the generic character of U such that τ is a constituent of $\lambda_{|U\cap\mathcal{U}^{[\frac{n+1}{2}]}}$. Then, the Whittaker vector of Π for an element f in $Ind^G_{F[b]^\times\mathcal{U}^{[\frac{n+1}{2}]}}\lambda$ is given by*

$$[A_\Pi(f)](x) = \sum_{z\in F[b]^\times\mathcal{U}^{[\frac{n+1}{2}]}\backslash G} S_\lambda(xz^{-1})f(z) \qquad (x\in G)$$

where

$$S_\lambda(x) = \begin{cases} \tau(u)\lambda(h), & \text{if } x = uh \in UF[b]^\times\mathcal{U}^{[\frac{n+1}{2}]} \\ 0, & \text{if } x \notin UF[b]^\times\mathcal{U}^{[\frac{n+1}{2}]} \end{cases}$$

Proof: Straightforward, following the proof in [K] and using the above results.

Note that clearly A_Π defines an injective intertwining operator from $\mathrm{Ind}^G_{F[b]^\times\mathcal{U}^{[\frac{n+1}{2}]}}\lambda$ into $\mathrm{Ind}^G_U\tau$ that is called the *Whittaker function*, so the Whittaker model of Π with respect to τ is the subspace of $W_\tau = \mathrm{Ind}^G_U\tau$ given by $M = A_\Pi(\Pi)$.

Finally, we construct the Whittaker model for the representation $\Pi = \mathrm{Ind}^G_{\mathcal{K}(\mathcal{A})}\sigma$ of level 0; that is, $\mathcal{A}$ is the principal order in A associated to the O_F-lattice chain $\mathcal{L}$ with period $e(\mathcal{L}) = 1$ and σ is an irreducible very cuspidal representation of $\mathcal{K}(\mathcal{A}) = F^\times GL_N(O_F)$. This means that σ on $GL_N(O_F)/\mathcal{U}^1 \simeq GL_N(k_F)$ is cuspidal. (Recall that $\tilde{\tau}$ is a cuspidal representation of $GL_N(k_F)$ if $\mathrm{Hom}_{\tilde{U}}(\tilde{\tau}, 1_{\tilde{U}}) = 0$ for each non-trivial unipotent radical $\tilde{U}$ of $GL_N(k_F)$. See, for example, [G]).

So, such a representation σ is precisely of the form $\theta\xi$ where ξ is a representation of $GL_N(O_F)$ lifted from an irreducible cuspidal representation $\tilde{\xi}$ of $GL_N(k_F)$ and θ is a character of $F^\times$ with $\theta = \overline{\xi}$ on $F^\times\cap\mathcal{U}$ where $\overline{\xi}\cdot 1_\xi = \xi_{|F^\times\cap\mathcal{U}}$.

From the representation theory of $GL_N(k_F)$, see [G], we know that there is a generic character $\tilde{\tau}$ of the standard unipotent radical $\tilde{U}$ of $GL_N(k_F)$ such that $\mathrm{Hom}_{\tilde{U}}(\tilde{\xi}, \tilde{\tau}) \neq (0)$. So, if we choose $\tilde{\tau}$ as above, then we can lift it to τ on U and thus, $\mathrm{Hom}_{U\cap\mathcal{U}}(\theta\xi, \tau) \neq (0)$. Likewise, as before, we have that

$$\mathrm{Hom}_G(\Pi, \rho_\tau) \simeq \mathrm{Hom}_{U\cap\mathcal{U}}(\theta\xi, \tau)$$

and then, similar to Theorem 4.6, the Whittaker function is given, for every $f \in \mathrm{Ind}^G_{F^\times\mathcal{U}}\theta\xi$, by

$$[A_\Pi(f)](x) = \sum_{z\in F^\times\mathcal{U}\backslash G} S_{\theta,\xi}(xz^{-1})f(z) \qquad (x\in G),$$

where

$$S_{\theta,\xi}(x) = \begin{cases} \tau(u)\theta\xi(h), & \text{if } x = uh \in UF^{\times}\mathcal{U} \\ 0, & \text{if } x \notin UF^{\times}\mathcal{U} \end{cases}$$

All omitted details can be found in [J].

References

[Bo] A. Borel, *Automorphic L-functions*, Automorphic Forms, Representations and L-functions, Proc. Symps., Pure Math., Vol. **33,** part 2, Amer. Math. Soc., 1979, 27-33.

[Bu] C.J. Bushnell, *Hereditary orders, Gauss sums and supercuspidal representations of $GL_N(F)$*, J. Fur Reine und Angew. Math. **375/376**, (1987), 184-210.

[B-F] C.J. Bushnell and A. Fröhlich, *Non abelian congruence, Gauss sums and p-adic simple algebras*, Proc. London Math. Soc., **50**, (1985), 207-264.

[B-K] C.J. Bushnell and Philip C. Kutzko, *The admissible dual of $GL_N(F)$ via compact open subgroups,* Annals of Mathematics Studies, Number **129**, Princeton University Press, (1993).

[Ca] H. Carayol, *Représentations cuspidales du groupe linnéaire,* Ann. Sci. École Norm. Sup., Vol. **17**, (1984), 191-225.

[Cr] P. Cartier, *Représentations of p-adic groups: a survey,* Proc. of Symposia in Pure Math., Amer. Math. Soc., **23**, (1979), 111-155.

[Cs] W. Casselman, *Introduction to the theory of admissible representations of p-adic reductive groups,* Preprint.

[Cl] A. B. Clifford, *Representations induced in an invariant subgroup*, Ann. of Math. **38** (1937), 533-550.

[G] S. I. Gelfand, *Representations of the full linear group over a finite field,* (Russian) Mat. Sbornik, **83** (1970), 15-41.

[G-K] I.M. Gelfand and D.A. Kazhdan, *Representations of the group $GL_N(F)$ where F is a local field,* Proc. of the Summer School on Group Representations, Bolyai János Mathematical Society, Budapest, (1971).

[J] R. Johnson, *Some Whittaker Models for $GL_N(F)$*, To Appear.

[K] Philip C. Kutzko, *Mackey's theorem for nonunitary representations*, Proc. Amer. Math. Soc., Vol **64**, (1977), 173-175.

[K-M] Philip C. Kutzko and David C. Manderscheid, *On intertwining operators for $GL_N(F)$, F a non archimedean local field*, Duke Math. J., **57**, (1988), 275-293.

[R] F. Rodier, *Whittaker models for admissible representations of reductive p-adic split groups*, Proc. of Symposia in Pure Math., Amer. Math. Soc., **26**, (1973), 425-430.

[Se] J. P. Serre, *Corps Locaux*, Hermann, Paris, (1962).

[Sh] F. Shahidi, *Langlands conjecture on Plancherel measures for p-adic groups*, Harmonic Analysis on Reductive Groups, Progress in Math. **101**, Birkhäuser, 1991, 277-295.

Roberto Johnson
Instituto de Matemáticas
Universidad Católica de Valparaíso
Casilla 4059
Valparaíso
CHILE.
email: rjohnson@aix1.ucv.cl

SMOOTH REPRESENTATIONS OF REDUCTIVE p-ADIC GROUPS: An introduction to the theory of types

PHILIP C. KUTZKO[1]

The purpose of this paper is to report on recent joint work with C. J. Bushnell which forms part of our ongoing program of understanding the category of smooth (complex) representations of a p-adic group in terms of certain irreducible representations of compact, open subgroups. Motivation for this program comes from two special cases which may be viewed as extreme examples of what one hopes is a general phenomenon.

Let me begin with a brief description. If G is the group of F-points of a connected, reductive group over a p-adic field F, then it is natural to try to obtain irreducible representations of G by decomposing representations that are induced from proper parabolic subgroups. Most of the time this only partially succeeds. The remaining irreducible representations, those which are not composition factors of parabolically induced representations, are called *supercuspidal*. What has gradually become clear is that supercuspidal representations may be constructed in a very different way, namely, as representations induced from open, compact modulo center, subgroups of G. Further, the inducing data often contains the sort of arithmetic information necessary to study the role that supercuspidal representations play in the local (non-archimedean) version of what has come to be called "functoriality." (See, e.g. [KM]). In particular, all irreducible supercuspidal representations of the groups $GL(N, F)$, $SL(N, F)$ are now known to be induced [BK1-3] and there is substantial evidence that the same might be true for other groups.

At the other extreme are those representations of G which are composition factors of representations which are induced from unramified characters of a minimal parabolic subgroup. Here it is a result of Borel [Bo] and Casselman [Ca2] that the irreducible representations obtained in this way are in bijection with the simple unitary left modules of an algebra $\mathcal{H}(G, J)$ of compactly supported complex functions on G which are bi-invariant with respect to an explicitly given compact, open subgroup J of G. The algebra $\mathcal{H}(G, J)$ and its module category have been the object of much study (see, e.g.,[KL]; see also Barbasch's article in this volume) and, again, this approach has played a central role in the understanding of functoriality in this context.

Further one now knows, thanks to results of Bernstein [BD], that the irreducible representations described above are the irreducible objects in a natu-

[1] The research for this paper was partially supported, at various times, by SERC grant GR/H26901, EPSRC grant GR/K81584, NSF grant DMS-9003213, a University of Iowa Faculty Scholarship, and by the hospitality of IHES and the Universities of Paris VII and XI.

ral subcategory of the category of all smooth representations of G. One may then deduce that this category is equivalent to the category $\mathcal{H}(G,J)$-Mod of all left unitary $\mathcal{H}(G,J)$-modules, this equivalence being constructed in such a way that one may use it to study the decomposition of representations of the form $\mathrm{Ind}_P^G\chi$ where P is a minimal parabolic and χ is an unramified character of a Levi factor of P.

It is a fundamental observation of Howe [H] that the two examples above should be special cases of something more general. In particular, he proposed that the irreducible representations of G could be partitioned into subsets, each of which "contains" a particular irreducible representation ρ of a compact, open subgroup K and that information about the representations in a particular subset (such as, e.g., formal degree) could be obtained by studying the simple modules over an algebra constructed from (K,ρ). If one puts this proposal together with the categorical approach taken by Bernstein, one arrives at the point of view taken here.

These lectures are organized as follows. The first two sections give an elementary exposition of the smooth complex representation theory of a totally disconnected, locally compact topological group. I have limited myself to complex representations largely because the Bernstein theory is so far only available in that case. One may none-the-less carry out a large part of the program sketched out here for any algebraically closed field; notable success here has been achieved by Vigneras [V]. I have tried to stress here the more algebraic aspects of the theory and have largely followed [BD],[Ca1],[Car]. Section three gives a brief account of the Bernstein decomposition of the category of smooth representations of a connected, reductive p-adic group. (See [BD]). Section four explains the main concepts involved here; they are illustrated at some length in section five using $GL(N,F)$ as an example. Sections six and seven explain our new results [BK4].

It only remains for me to acknowledge with gratitude the generosity and skill of the organizers of this conference, the kindness and hospitality of the faculty and students of the Universidad de Cordoba, and the presentations of the other speakers from which I learned a great deal.

1. The category of smooth representations of a locally profinite group

Let G be a group. Then by a *representation* of G (or a *G-representation*) we mean a pair (π, V) consisting of a complex vector space V and a homomorphism $\pi : G \to Aut_{\mathbb{C}}(V)$. (We will often just write π, or V, in place of (π, V) when no confusion can arise.) Given G-representations (π_i, V_i), $i = 1,2$, a *G-map* from V_1 to V_2 is a linear transformation $T : V_1 \to V_2$ which satisfies $T\pi_1(g) = \pi_2(g)T$, $g \in G$; we denote the set of G-maps from V_1 to V_2 by $Hom_G(V_1, V_2)$. The class of all G-representations together with the sets $Hom_G(V_1, V_2)$ form a category which we denote here by *Rep*(G). *Rep*(G) is clearly an abelian category and so we may talk about G-subspaces, G-quotients, and so on.

Suppose now that G is a topological group. Given a representation (π, V) of G, we say that a vector $v \in V$ is *G-smooth* (or just *smooth*) if the subgroup of elements $g \in G$ for which $\pi(g)v = v$ is open. We denote the G-subspace of smooth vectors in V by V^∞. If $V = V^\infty$ then we say that V is *smooth.* We denote by $\mathfrak{R}(G)$ the full subcategory of $Rep(G)$ whose objects are the smooth representations of G. (We note that if V is a smooth G-representation, then so is any G-subquotient.)

A topological group need not, in general, have many smooth representations. For example, if (π, V) is a smooth representation of the additive group $\mathbb{R}$ of real numbers then we must have $\pi(x)v = v$ for all $v \in V$, $x \in \mathbb{R}$. This is because $\mathbb{R}$ is connected; it has no proper open subgroups. On the other hand, the category $\mathfrak{R}(G)$ is a natural category to consider when G is totally disconnected. Such groups arise in the study of local number theory and, in that case, they tend to be locally compact as well. Thus, for the rest of this paper, all groups under consideration will be taken as locally compact and totally disconnected. It is worth noting that this is equivalent to asserting that the group is *locally profinite*; that is, that the identity element has a neighborhood base consisting of compact, open subgroups. (This terminology stems from the fact that a compact, totally disconnected group is *profinite*; that is, it is the inverse limit of its finite quotients.)

Any locally compact group G has a left Haar measure; that is, a left-invariant positive Borel measure. We fix one such measure μ_G and denote by $\mathcal{H}(G)$ the convolution algebra of locally constant, compactly supported, complex functions on G. That is, multiplication in $\mathcal{H}(G)$ is given by

$$f * g(x) = \int_G f(y)g(y^{-1}x)d\mu(y).$$

Note that $\mathcal{H}(G)$ has no identity unless G is discrete; the wealth of compact, open subgroups of G ensures, however, that $\mathcal{H}(G)$ has lots of idempotents. In particular, if K is any such subgroup then $e_K = \frac{1}{\mu(K)}\chi_K$ is an idempotent. (Here, χ_K denotes the characteristic function of K.) Indeed, $\mathcal{H}(G)$ has the property that, given any pair of elements $f, g \in \mathcal{H}(G)$, there is an idempotent $e \in \mathcal{H}(G)$ such that $f, g \in e\mathcal{H}(G)e$. Rings with this property are called *idempotented rings* or *rings with local units*; such rings have been studied in contexts far removed from the one under consideration here[AM].

Given such a ring R, there is a natural module category associated to it: the category R-Mod of left *non-degenerate* (or *unitary*) R-modules. The objects in this category are just the left R-modules M satisfying $RM = M$ and the morphisms are the usual ones. The reason for introducing the algebra $\mathcal{H}(G)$ may now be described as follows:

1.1 Proposition. . *Let (π, V) be a smooth G-representation and for any element $h \in \mathcal{H}(G)$ and any vector $v \in V$, define the product hv by*

$$hv = \int_G h(x)\pi(x)v d\mu(x).$$

Then V is a non-degenerate $\mathcal{H}(G)$-module under this product and the map $(\pi, V) \to V$ identifies the categories $\mathfrak{R}(G)$ and $\mathcal{H}(G)$-Mod.

(It is worth noting that the integrals in this section both reduce to finite sums.)

2. Contragredient, change of group

We give a brief review of the smooth representation theory of a locally profinite group; proofs of any results stated here may be found in [Car]. Let G be a locally profinite group and let (π, V) be a smooth G-representation. Then there is a natural G-representation $\check{\pi}$ on the space $\hat{V} = Hom_{\mathbb{C}}(V, \mathbb{C})$ of functionals on V: $(\check{\pi}(x)(\psi))(v) = \psi(\pi(x^{-1})v)$, $\psi \in \hat{V}$, $v \in V$, $x \in G$. The representation $(\check{\pi}, \hat{V})$ is not necessarily smooth; we may, however, restrict $\check{\pi}$ to the space $\tilde{V} = (\hat{V})^{\infty}$ of smooth vectors in $\hat{V}$. We denote the restriction of $\check{\pi}$ to $\tilde{V}$ by $\tilde{\pi}$ and refer to $(\tilde{\pi}, \tilde{V})$ as the *contragredient* representation to (π, V). By a *matrix coefficient* of (π, V) we then mean a complex function on G of the form $x \to \psi(\pi(x)v)$ for fixed $\psi \in \tilde{V}$, $v \in V$.

Now suppose that H is a locally profinite group and that $\phi : H \to G$ is a homomorphism of topological groups. Then there is associated to ϕ a functor $\phi^* : \mathfrak{R}(G) \to \mathfrak{R}(H)$ defined on objects in $\mathfrak{R}(G)$ as follows: If (π, V) is a smooth G-representation then $\phi^*(V) = V$ and $\phi^*(\pi)(h) = \pi(\phi(h))$, $h \in H$. (ϕ^* is defined on morphisms in the obvious way.) Two special cases will interest us here: If H is a closed subgroup of G and ϕ is the canonical injection, then we write res_H^G in place of ϕ^*. If N is a closed normal subgroup of H, $G = H/N$ and ϕ is the canonical surjection, then we write infl_G^H in place of ϕ^*.

It is a fundamental fact that both res_H^G and infl_G^H have (easily described) *adjoints*. Let us start with res_H^G. What we are looking for is a functor, F, say, from $\mathfrak{R}(H)$ to $\mathfrak{R}(G)$ such that, for each smooth G-representation V and each smooth H-representation W, there is an isomorphism of vector spaces

$$\beta = \beta(V, W) : Hom_H(\mathrm{res}_H^G(V), W) \cong Hom_G(V, F(W)).$$

Furthermore, β is to be *natural* in both V and W. (See,e.g., [Borc]). These conditions determine the G-representation $F(W)$ up to a natural isomorphism. We now give a specific realization of an adjoint for res_H^G. If (σ, W) is a smooth H-representation, then G acts by right translation on the set of functions $f : G \to W$ which satisfy $f(hx) = \sigma(h)(f(x))$, $h \in H$, $x \in G$. The implied G- representation is not necessarily smooth but we may consider the space, $\mathrm{Ind}_H^G(W)$, of smooth vectors. We denote by $\mathrm{Ind}_H^G(\sigma)$ the G-representation on $\mathrm{Ind}_H^G(W)$ given by right translation; that is:

$$(\mathrm{Ind}_H^G(\sigma)(x)f)(y) = f(yx), \; f \in \mathrm{Ind}_H^G(W), \; x, y \in G.$$

2.1 Proposition. *Let H be a closed subgroup of G, and define the G- representation $\mathrm{Ind}_H^G(\sigma, W) = (\mathrm{Ind}_H^G(\sigma), \mathrm{Ind}_H^G(W))$ for any smooth H-representation (σ, W) as above. For smooth H-representations (σ_i, W_i), $i = 1,2$ and $\phi \in Hom_H(W_1, W_2)$, let us define $\mathrm{Ind}_H^G(\phi) : \mathrm{Ind}_H^G(W_1) \to \mathrm{Ind}_H^G(W_2)$ by $(Ind_H^G(\phi)(f))(x) = \phi(f(x))$. Then Ind_H^G is an adjoint to res_H^G.*

Remark. It is of some interest to consider res_H^G as a functor from $\mathcal{H}(G)$-Mod to $\mathcal{H}(H)$-Mod. Since $\mathcal{H}(H)$ is not in general a subalgebra of $\mathcal{H}(G)$, it is not immediately clear how to describe res_H^G in this context. However, one may show that there is a unique (and obvious) $(\mathcal{H}(H), \mathcal{H}(G))$-bimodule structure on $\mathcal{H}(G)$ so that res_H^G corresponds to the functor $\mathcal{H}(G)\otimes_{\mathcal{H}(G)} : \mathcal{H}(G)$-Mod $\to \mathcal{H}(H)$-Mod. It is then straightforward that the "smooth hom functor" $W \to \mathcal{H}(G) \cdot Hom_{\mathcal{H}(H)}(\mathcal{H}(G), W)$ furnishes an adjoint to res_H^G.

In this context, it is tempting to believe that the functor $\mathcal{H}(G)\otimes_{\mathcal{H}(H)} :$ $\mathcal{H}(H)$-Mod$\to$ $\mathcal{H}(G)$-Mod might furnish a *co-adjoint* to res_H^G; that is, that res_H^G might be an adjoint to $\mathcal{H}(G)\otimes_{\mathcal{H}(H)}$. Of course, to make sense of this, we need a $(\mathcal{H}(G), \mathcal{H}(H))$- bimodule structure on $\mathcal{H}(G)$ and here, we run into the fact that neither G nor H need be *unimodular*. Taking this into account, there is an appropriate $(\mathcal{H}(G), \mathcal{H}(H))$- bimodule structure on $\mathcal{H}(G)$ but it is not so obvious. With this structure, the functor $\mathcal{H}(G)\otimes_{\mathcal{H}(H)}$ is not in general a co-adjoint to res_H^G; it is however a co-adjoint in the important case that H is *open*. In addition, given any non-degenerate left H-module W, there is the usual isomorphism,

$$(\mathcal{H}(G) \otimes_{\mathcal{H}(H)} W)^{\sim} \cong \mathcal{H}(G) \cdot Hom_{\mathcal{H}(H)}(\mathcal{H}(G), \tilde{W}),$$

which relates the contragredient to induction. All of this will serve perhaps to provide some motivation for the normalization of induction. We turn to this now.

2.2 Definition. *With notation as in Proposition 2.1, we denote by c-$\mathrm{Ind}_H^G(W)$ the subspace of $\mathrm{Ind}_H^G(W)$ consisting of those functions which are supported on subsets of G of the form HC where C is compact. Then c-$\mathrm{Ind}_H^G(W)$ is a G-subspace of $\mathrm{Ind}_H^G(W)$; we denote the restriction of $\mathrm{Ind}_H^G(\sigma)$ to this subspace by c-$\mathrm{Ind}_H^G(\sigma)$.*

2.3 Proposition. *Let Δ_G be the* modulus character *on G; that is, Δ_G satisfies*

$$\int_G f(yx) d\mu_G(y) = \Delta_G(x) \int_G f(y) d\mu_G(y), \quad x \in G;$$

and let $\delta = \delta_{G/H} = \frac{\Delta_H}{\Delta_G}$, viewed as a character on H. Then given any smooth H-representation (σ, W), there is (with a slight abuse of notation) a natural G-isomorphism:

$$\mathcal{H}(G) \otimes_{\mathcal{H}(H)} W \cong \text{c-}\mathrm{Ind}_H^G\,(\sigma \otimes \delta, W)\,.$$

2.4 Definition. *We define the functors*

$$u\text{-Ind}_H^G,\, uc\text{-Ind}_H^G : \mathfrak{R}(H) \to \mathfrak{R}(G),$$

$$u\text{-res}_H^G : \mathfrak{R}(G) \to \mathfrak{R}(H)$$

of normalized *induction and restriction by*

$$u\text{-Ind}_H^G \sigma = \text{Ind}_H^G(\sigma \otimes \delta^{\frac{1}{2}}),$$

$$uc\text{-Ind}_H^G \sigma = c\text{-Ind}_H^G(\sigma \otimes \delta^{\frac{1}{2}}),$$

$$u\text{-res}_H^G \pi = (\text{res}_H^G \pi) \otimes \delta^{-\frac{1}{2}}.$$

2.5 Corollary. *With notation as in Proposition 2.3, we have*

(1) $(uc\text{-Ind}_H^G \sigma)^\sim \cong u\text{-Ind}_H^G \tilde{\sigma}$.
(2) *The functor* $u\text{-Ind}_H^G$ *is an adjoint to* $u\text{-res}_H^G$

We now turn our attention to infl_G^H;so we now take $G = H/N$ where N is a closed normal subgroup of H. Given a smooth H-representation (σ, W) we denote by $\text{inv}_G^H(W)$ the subspace of vectors $w \in W$ which satisfy $\sigma(n)w = w$, $n \in N$ and we define a G-representation $\text{inv}_G^H(\sigma)$ on $\text{inv}_G^H(W)$ by $\text{inv}_G^H(\sigma)(xH)w = \sigma(x)w$, $x \in G$, $w \in W$. Then $(\text{inv}_G^H(W), \text{inv}_G^H(\sigma))$ is a well-defined smooth G-representation.

Dually, we may consider the space $\text{coinv}_G^H(W) = W/W(N)$ where $W(N)$ is the subspace of W spanned by vectors of the form $\sigma(n)w - w$, $n \in N$, $w \in W$ and we may define a G-representation $\text{coinv}_G^H(\sigma)$ on $\text{coinv}_G^H(W)$ by $\text{coinv}_G^H(\sigma)(xH)(w + W(N)) = \sigma(x)w + W(N)$, $x \in g, w \in W$. Then $(\text{coinv}_G^H(W), \text{coinv}_G^H(\sigma))$ is also a well-defined smooth G-representation. Defining inv_G^H, coinv_G^H on morphisms in the obvious way, we obtain

2.6 Proposition. *The functor* inv_G^H *is an adjoint to* infl_G^H. *The functor* coinv_G^H *is a* co-adjoint *to* infl_G^H.

3. The Bernstein decomposition of $\mathfrak{R}(G)$

We now fix a p-adic field F and take G to be (the F-points of) a connected, reductive group over F. Then G comes equipped with a set of subgroups - the (F-) parabolic subgroups - which facilitates the analysis of the category $\mathfrak{R}(G)$. Among other things, a parabolic subgroup P of G has a normal subgroup, $N = N(P)$, its *unipotent radical*, such that P/N is again a connected reductive group. Furthermore, the sequence $1 \to N \to P \to P/N \to 1$ splits; any subgroup M of P such that $MN = P$, $M \cap N = \{1\}$ is called a *Levi factor* of P. Finally, given a parabolic subgroup P with Levi factor M, there is a unique parabolic subgroup $\bar{P}$ such that $\bar{P} = MN(\bar{P})$ and $P \cap N(\bar{P}) = \{1\}$. $\bar{P}$ is called the *opposite parabolic* to P with respect to M.

Given a parabolic subgroup P with Levi factor M it is natural to consider the functor $\iota_P = \iota_P^G = u\text{-Ind}_P^G \circ \text{infl}_M^P : \mathfrak{R}(M) \to \mathfrak{R}(G)$. We refer to this functor as the functor of *parabolic induction*. By Corollary 2.5 and Proposition 2.6, ι_P is an adjoint to the functor $r_P = r_P^G = \text{coinv}_M^P \circ u\text{-res}_P^G : \mathfrak{R}(G) \to \mathfrak{R}(M)$. We refer to the functor r_P as a *Jacquet* functor.

We are now in a position to list some of the basic theorems in this area [Ca1].

3.1 Theorem. *Let (π, V) be an irreducible representation of G. Then the following are equivalent:*

(1) $r_P(V) = \{0\}$ *for every proper parabolic subgroup $P < G$.*
(2) *Any matrix coefficient of (π, V) is compactly supported modulo the center of G.*

3.2 Definition. *An irreducible representation of G which satisfies the conditions of Theorem 3.1. is called* supercuspidal.

The first condition in Theorem 3.1 leads to the following classification of the irreducible representations of G:

3.3 Theorem.

(1) *Let M be a Levi subgroup of G (that is, a Levi factor of some parabolic subgroup) and let σ be an irreducible supercuspidal representation of M. Then $\iota_P\sigma$ has finite length for every parabolic subgroup P with Levi factor M. Further, the set of composition factors of $\iota_P^G\sigma$ is independent of P.*
(2) *Let M_i, $i = 1, 2$ be Levi subgroups of G and let σ_i, $i = 1, 2$ be irreducible supercuspidal representations of M_i. Then for any parabolic subgroups $P_i, i = 1, 2$ with Levi factors M_i we have that the representations $\iota_{P_i}\sigma_i$ either have the same set of composition factors or have no composition factor in common. The first of these possibilities occurs if and only if the pairs (M_i, σ_i) are* conjugate*; that is, there is an element $x \in G$ such that $M_2 = xM_1x^{-1}$ and the representations σ_2, σ_1^x are M_2-isomorphic.*
(3) *Given an irreducible supercuspidal representation π of G, there is a Levi subgroup M of G and an irreducible supercuspidal representation σ of M such that π is isomorphic to a composition factor of $\iota_P\sigma$ for any parabolic subgroup P with Levi factor M.*

3.4 Definition.

(1) *We refer to a pair (M, σ) consisting of a Levi subgroup M and an irreducible supercuspidal representation σ of M as a* cuspidal pair.
(2) *Let π be an irreducible representation of G. Then by Theorem 3.3, there is a unique conjugacy class of cuspidal pairs (M, σ) with the property that π is isomorphic to a composition factor of $\iota_P\sigma$ for some parabolic subgroup P. We call this conjugacy class of cuspidal pairs the* support *of π.*

The above classification of the set of irreducible smooth representations of G leads naturally to the following questions:

(1) What are the irreducible supercuspidal representations of a reductive group G?
(2) Given a cuspidal pair (M, σ) in G, how does one determine the composition factors of $\iota_P \sigma$?
(3) To what extent does the classification of irreducible representations of G by support yield information about the classification of more general representations?

We begin with a discussion of the third question and, here, we begin to see the significance of Theorem 3.1 above.

3.5 Proposition. *Suppose that (π, V) is an irreducible representation of G whose matrix coefficients are compactly supported. Then (π, V) is both a projective and an injective object in $\mathfrak{R}(G)$. In particular, if (ρ, X) is any smooth representation of G then we may write X uniquely as direct sum of two G-subspaces X_π and X'_π such that X_π is a direct sum of G-subspaces each one of which is isomorphic to V while X'_π has no G-subquotient isomorphic to V.*

Unfortunately, matrix coefficients of supercuspidal representations are only compactly supported mod center. One may get around this problem by fixing a central character but one runs into the problem again when dealing, as one must, with representations of Levi subgroups. Bernstein's approach to this problem may be described as follows. First of all, G has a very special subgroup, $^\circ G$:

3.6 Proposition.
Let $^\circ G$ be the smallest subgroup of G which contains all compact subgroups of G. Then

(1) *$^\circ G$ is an open normal subgroup of G and $G/{}^\circ G$ is a free abelian group of finite rank.*
(2) *The center of $^\circ G$ is compact.*
(3) *If π is an irreducible representation of G then $\mathrm{res}^G_{^\circ G}\pi$ is a finite direct sum of irreducible $^\circ G$-representations.*
(4) *The characters of G which factor through $G/{}^\circ G$ are precisely the* unramified *characters of G.*

(The group of unramified characters of G is, by definition, the group generated by characters of the form $\chi_0 \circ \chi_1$ where χ_1 is a rational character of G and χ_0 is an unramified character of $F^\times$; that is, a character which factors through $F^\times/O_F^\times$ where O_F is the ring of integers in F.)

In particular, if π is an irreducible supercuspidal representation, then we have $\mathrm{res}^G_{^\circ G}\pi \cong \oplus_{i=1}^t \pi_i$ where the representations π_i are irreducible and supercuspidal. Further, if π' is another irreducible supercuspidal representation of G and if we write $\mathrm{res}^G_{^\circ G}\pi' \cong \oplus \pi'_i$, then it follows from 3.6 (4) that the sets $\{\pi_i\}$, $\{\pi'_i\}$ are either the same (up to isomorphism) or disjoint depending on whether on not there is an unramified character χ of G such that $\pi' \cong \pi \otimes \chi$. If such a character χ exists then we say that π and π' are *inertially equivalent*; we write $[\pi]$ for the inertial equivalence class of π.

We now may consider the full subcategory $\mathfrak{R}^{[\pi]}(G)$ of $\mathfrak{R}(G)$ whose objects have the property that they are, upon restriction to ${}^{\circ}G$, sums of representations each of which is ${}^{\circ}G$-isomorphic to some π_i as above. Then one has the following generalization of Proposition 3.5:

3.7 Proposition.

(1) *A representation ρ of G is an object in $\mathfrak{R}^{[\pi]}(G)$ if and only if each irreducible subquotient of ρ lies in $[\pi]$.*

(2) *Let $\mathfrak{R}^{[\pi]}(G)'$ be the full subcategory of $\mathfrak{R}(G)$ whose objects have the property that none of their irreducible subquotients lies in $[\pi]$. Then*

$$\mathfrak{R}(G) = \mathfrak{R}^{[\pi]}(G) \times \mathfrak{R}^{[\pi]}(G)'$$

(This last statement means, first, that any object X in $\mathfrak{R}(G)$ may be written in the form $X = Y \oplus Y'$ with Y an object in $\mathfrak{R}^{[\pi]}(G)$ and Y' an object in $\mathfrak{R}^{[\pi]}(G)'$ and, second, that $Hom_G(W, W') = \{0\}$ for objects W in $\mathfrak{R}^{[\pi]}(G)$ and W' in $\mathfrak{R}^{[\pi]}(G)'$.)

An example:

To get some feeling for the structure of the subcategory $\mathfrak{R}^{[\pi]}(G)$, it is instructive to consider the very simple example of $G = GL_1(F) = F^\times$. Here, the irreducible representations are one-dimensional and all such representations are supercuspidal. Further, ${}^{\circ}G$ is just the group of units, $O_F^\times$, of the ring of integers, O_F, in F. Thus, if χ is any one-dimensional representation of $F^\times$, we may think of the map $\chi \longrightarrow [\chi]$ as being given by $\mathrm{res}_{O_F^\times}^{F^\times}$. As is well-known, the usual valuation, ν_F, gives rise to the following exact sequence:

$$1 \longrightarrow O_F^\times \longrightarrow F^\times \xrightarrow{\nu_F} \mathbb{Z} \longrightarrow 0.$$

It is then easy to see that the objects in $\mathfrak{R}^{[\chi]}(G)$ are simply those representations of the form $\chi \otimes \mathrm{infl}_{\mathbb{Z}}^{F^\times} \pi$ where π is an arbitrary representation of the discrete group $\mathbb{Z}$. In other words, the functor $\pi \longrightarrow \chi \otimes \mathrm{infl}_{\mathbb{Z}}^{F^\times} \pi$ serves to identify the categories $Rep(\mathbb{Z})$ and $\mathfrak{R}^{[\chi]}(G)$. Now $Rep(\mathbb{Z})$ is certainly a well-known category and may in turn be identified with the category of all modules over the group algebra, $\mathbb{C}(\mathbb{Z})$, of $\mathbb{Z}$. Since this algebra is naturally isomorphic to the polynomial ring $\mathbb{C}[x, x^{-1}]$, methods from algebraic geometry apply.

It is interesting to note that this simple example is actually quite general. Indeed, $G/{}^{\circ}G$ is always free abelian of finite rank; the only real complication which arises in the general case is that $\mathrm{res}_{{}^{\circ}G}^{G}\pi$ may be reducible. (See [BD] for details.)

Proposition 3.5 shows that we may, as it were, split off the subcategories of $\mathfrak{R}(G)$ which correspond to inertial equivalence classes of supercuspidal representations. To describe the rest of the category, we must have a corresponding notion of inertial equivalence for cuspidal pairs:

3.8 Definition. *Two cuspidal pairs (M,σ), (M',σ') will be called* G-inertially equivalent *(or just* inertially equivalent*) if there is an element $x \in G$ such that $M' = xMx^{-1}$ and $\sigma' \in [\sigma^x]$. We denote the G-inertial equivalence class of (M,σ) by $[M,\sigma]_G$ and the set of G-inertial equivalence classes of cuspidal pairs by $\mathfrak{B}(G)$.*

3.9 Definition. *Given an element $\mathfrak{s} = [M,\sigma] \in \mathfrak{B}(G)$ we define the category $\mathfrak{R}^{\mathfrak{s}}(G)$ to be the full subcategory of $\mathfrak{R}(G)$ whose objects π satisfy the following property:*

There exist parabolic subgroups P_i, $i = 1,\dots,t$ with Levi factor M and representations $\tau_i \in \mathfrak{R}^{[\sigma]}(M)$ $i = 1,\dots,t$ such that π is isomorphic to a subrepresentation of $\oplus_{i=1}^{t}\iota_{P_i}\tau_i$.

We may now state Bernstein's main result (2.8-2.10 of [BD]):

3.10 Theorem.

(1) *Let $\mathfrak{s} \in \mathfrak{B}(G)$. Then a representation π of G is an object in $\mathfrak{R}^{\mathfrak{s}}(G)$ if and only if the support of every irreducible subquotient of π lies in $\mathfrak{s}$.*

(2) *We have*

$$\mathfrak{R}(G) = \prod_{\mathfrak{s}\in\mathfrak{B}(G)} \mathfrak{R}^{\mathfrak{s}}(G)$$

(The above product of subcategories is defined just as in the parenthetical remark following Proposition 3.7)

4. $\mathfrak{s}$-IDEMPOTENTS AND $\mathfrak{s}$-TYPES

Given the decomposition of $\mathfrak{R}(G)$ provided by Theorem 3.10, it is natural that we now fix an inertial equivalence class $\mathfrak{s} = [M,\sigma]_G \in \mathfrak{B}(G)$ and turn our attention to the structure of the category $\mathfrak{R}^{\mathfrak{s}}(G)$. In view of the definition of $\mathfrak{R}^{\mathfrak{s}}(G)$ given in Definition 3.9, it is clear that this structure is intimately connected up with the second question we raised in the preceding section; namely: What are the composition factors of $\iota_P\sigma$?

It will be useful here to work in the category of smooth $\mathcal{H}(G)$-modules; we begin with the observation that Theorem 3.10 implies a decomposition $\mathcal{H}(G) = \oplus_{\mathfrak{s}\in\mathfrak{B}(G)}\mathcal{H}^{\mathfrak{s}}(G)$ of left $\mathcal{H}(G)$-modules. In fact, more is true:

4.1 Lemma.

(1) *The left ideals $\mathcal{H}^{\mathfrak{s}}(G)$ given above are in fact indecomposable two-sided ideals and they are maximal with this property.*

(2) *Let V be a smooth left $\mathcal{H}(G)$-module and set $V^{\mathfrak{s}} = \mathcal{H}^{\mathfrak{s}}(G)V$. Then $V^{\mathfrak{s}}$ is an object in $\mathfrak{R}^{\mathfrak{s}}(G)$ and we have $V = \oplus_{\mathfrak{s}\in\mathfrak{B}(G)} V^{\mathfrak{s}}$, this decomposition being the one given by Theorem 3.10 (2). In particular, V is an object in $\mathfrak{R}^{\mathfrak{s}}(G)$ if and only if $V = \mathcal{H}^{\mathfrak{s}}(G)V$.*

Proof. This is Proposition 3.1 of [BK4].

We now turn to a fundamental example, an example in which almost all of the structure with which we will be dealing in this section is present. Namely, we take M to be a *minimal* Levi subgroup and we take σ to be the trivial character, $\mathbf{1}_M$ of M. (We note that $\mathbf{1}_M$ is supercuspidal since M is compact modulo center.) Then the irreducible objects of $\mathfrak{R}^{\mathfrak{s}}(G)$ are just the composition factors of representations of the form $\iota_P\chi$ where χ is an unramified character of M; these composition factors are sometimes referred to as *unramified principal series* - at least when χ is unitary - and they have been the object of much study. This study has been greatly facilitated by the following result, for which see [Bo],[Ca2]:

4.2 Proposition. *There is a compact, open subgroup J of G with the following property:*

An irreducible representation (π, V) of G is an object in $\mathfrak{R}^{\mathfrak{s}}(G)$, $\mathfrak{s} = [M, \mathbf{1}_M]_G$ as above, if and only if there is a non-zero vector $v \in V$ such that $\pi(h)v = v$ for all $h \in J$.

To understand the significance of this result, it is helpful to recast it in the language of smooth $\mathcal{H}(G)$-modules. Let $e_J = \frac{1}{\mu(J)}\chi_J$ be the idempotent in $\mathcal{H}(G)$ associated to J as in §1. Then given a smooth representation (π, V) of G, we have that $e_J V = \{v \in V \mid \pi(h)v = v, \quad h \in J\}$. Thus, Proposition 4.2 says that (π, V) is an object in $\mathfrak{R}^{\mathfrak{s}}(G)$ if and only if $e_J V \neq \{0\}$. This formulation turns out to be the tip of a rather large iceberg! To be precise, given (π, V) as above, the space $e_J V$ is naturally a left module for the algebra $e_J\mathcal{H}(G)e_J$; in fact, the process $V \to e_J V$ induces a functor from $\mathfrak{R}^{\mathfrak{s}}(G)$ to $e_J\mathcal{H}(G)e_J$- Mod. (Note that $e_J\mathcal{H}(G)e_J$ is an algebra with identity - namely, e_J - so that $e_J\mathcal{H}(G)e_J$- Mod is the category of all left (unital) modules.) A simple consequence of Proposition 4.2 is then

4.3 Corollary. *The functor $V \to e_J V$ given above restricts to a bijection of the classes $Irr(\mathfrak{R}^{\mathfrak{s}}(G))$, $Irr(e_J\mathcal{H}(G)e_J$- Mod) of irreducible objects of $\mathfrak{R}^{\mathfrak{s}}(G)$, $e_J\mathcal{H}(G)e_J$- Mod respectively.*

Of course, this last result will not be of much use unless the algebra $e_J\mathcal{H}(G)e_J$ is susceptible to analysis and, here, we are in luck. The reason for this is that the algebra $e_J\mathcal{H}(G)e_J$ is just the subalgebra of $\mathcal{H}(G)$ consisting of functions f which are bi-invariant with respect to J; that is, functions f for which $f(hxk) = f(x)$, $x \in G$, $h, k \in J$. And this algebra is known to be isomorphic to an *affine Hecke algebra*, an algebra whose set of irreducible modules is well-understood due to the work of Kazhdan-Lusztig [KL].

There is an obvious question provoked by Corollary 4.3; namely: Is the functor $V \to eV$ an equivalence of categories? The answer here is 'yes' but this is by no means obvious. Indeed, this is a special case of Proposition 4.5 below. A second question is whether, given a parabolic subgroup with Levi factor M as above, we can somehow 'transfer' the functor ι_P to a functor involving module categories and thus transfer our question concerning the composition factors of $\iota_P\sigma$ to a question about modules. These questions may serve as motivation for all of what follows here.

We now return to our general setup. That is, G is a connected reductive group, (M, σ) is a cuspidal pair in G and $\mathfrak{s} = [M, \sigma]_G$ is an element in $\mathfrak{B}(G)$.

4.4 Definition. *We say that an idempotent $e \in \mathcal{H}(G)$ is an* $\mathfrak{s}$-idempotent *if the following are equivalent for any irreducible representation (π, V) of G.*

(1) *V is an object in $\mathfrak{R}^{\mathfrak{s}}(G)$*
(2) *$eV \neq \{0\}$.*

We now may state one of our main results. For any idempotent $e \in \mathcal{H}(G)$, let $\mathfrak{R}_e(G)$ be the full subcategory of $\mathfrak{R}(G)$ whose objects V satisfy $\mathcal{H}(G)eV = V$. Then

4.5 Proposition. *The following are equivalent for an idempotent $e \in \mathcal{H}(G)$:*

(1) *e is an $\mathfrak{s}$-idempotent,*
(2) *$\mathfrak{R}_e(G) = \mathfrak{R}^{\mathfrak{s}}(G)$ as subcategories of $\mathfrak{R}(G)$.*
(3) *$\mathcal{H}^{\mathfrak{s}}(G) = \mathcal{H}(G)e\mathcal{H}(G)$.*
(4) *The functor $\mathbf{m}_e : \mathfrak{R}^{\mathfrak{s}}(G) \to e\mathcal{H}(G)e$-Mod induced by $m_e(V) = eV$, $V \in \mathfrak{R}^{\mathfrak{s}}(G)$ is an equivalence of categories.*

Proof. This follows from [BK4] Propositions 3.3, 3.5.

The next result deals with the existence and uniqueness of $\mathfrak{s}$-idempotents.

4.6 Definition. *Let e, $e' \in \mathcal{H}(G)$ be idempotents. Then we say that e and e' are* associate *if there exist positive integers n, n' and surjections of left $\mathcal{H}(G)$-modules $\phi : (\mathcal{H}(G)e)^n \twoheadrightarrow \mathcal{H}(G)e'$, $\phi' : (\mathcal{H}(G)e')^{n'} \twoheadrightarrow \mathcal{H}(G)e$.*

4.7 Proposition. *Given any $\mathfrak{s} \in \mathfrak{B}(G)$, there exists an $\mathfrak{s}$-idempotent. If e is an $\mathfrak{s}$-idempotent and if e' is any other idempotent then e' is an $\mathfrak{s}$-idempotent if and only if e, e' are associate. If e, e' are $\mathfrak{s}$-idempotents, then the rings $e\mathcal{H}(G)e$, $e'\mathcal{H}(G)e'$ are canonically Morita equivalent.*

Proof. This is Propositions 3.13, 3.14 of [BK4].

Thus, for any $\mathfrak{s} \in \mathfrak{B}(G)$, we may identify the category $\mathfrak{R}^{\mathfrak{s}}(G)$ with the category of all left unital modules over an algebra with identity; namely, $e\mathcal{H}(G)e$. This will not be of much use, though, unless we can pick an $\mathfrak{s}$-idempotent e whose associated algebra $e\mathcal{H}(G)e$ is amenable to computation. It turns out that the example we gave above is illustrative here; it is helpful if e is associated to a representation of a compact, open subgroup. To be precise, let ρ be an irreducible representation of a compact, open subgroup K of G and let e_ρ be the idempotent in $\mathcal{H}(G)$ defined by

$$e_\rho(x) = \begin{cases} \dfrac{dim\rho}{\mu(K)} tr\rho(x^{-1}), & x \in K, \\ 0 & \text{otherwise.} \end{cases}$$

4.8. Definition. *Let* $\mathfrak{s} \in \mathfrak{B}(G)$ *and let* ρ *be an irreducible representation of a compact, open subgroup* K *of* G. *Then we say that the pair* (K, ρ) *is an* $\mathfrak{s}$-type *if* e_ρ *is an* $\mathfrak{s}$*-idempotent.*

(We give a somewhat broader definition in [BK4], $\mathfrak{s}$ above being replaced by a finite subset of $\mathfrak{B}(G)$; we will not discuss this here.)

This definition may be motivated by the following observation. For the moment, let K be any compact, open subgroup of G and let (ρ, W) be an irreducible representation of K. Then [BK4], (2.11) there is a natural isomorphism of G-spaces

$$c\text{-Ind}_K^G(\rho) \otimes_{\mathbb{C}} \tilde{W} \cong \mathcal{H}(G)e_\rho,$$

where, as in §2, $\tilde{W} = \text{Hom}_{\mathbb{C}}(W, \mathbb{C})$.

Computing the G-endomorphism ring of each side, (viewing endomorphisms as acting on the *right*) we get [BK4], (2.12):

$$\mathcal{H}(G, \rho) \otimes_{\mathbb{C}} \text{End}_{\mathbb{C}}(W) \cong e_\rho \mathcal{H}(G) e_\rho,$$

where $\mathcal{H}(G, \rho)$ is the usual Hecke (convolution) algebra of compactly supported functions for the *contragredient* representation, $\tilde{\rho}$, of ρ:

$$\mathcal{H}(G, \rho) = \left\{ h : G \to \text{End}_{\mathbb{C}}(\tilde{W}) \mid h(k_1 x k_2) = \tilde{\rho}(k_1) h(x) \tilde{\rho}(k_2),\ k_1, k_2 \in K, x \in G \right\}.$$

Thus, the algebras $e\mathcal{H}(G)e$ and $\mathcal{H}(G, \rho)$ are Morita equivalent and, combining this with Proposition 4.5 (3), we get an equivalence of categories $\mathbf{M}_\rho : \mathfrak{R}^{\mathfrak{s}}(G) \to \mathcal{H}(G, \rho)$-Mod. It is not hard to describe $\mathbf{M}_\rho$ explicitly. Given any smooth G-representation (π, V), we set $V_\rho = Hom_K(W, \text{res}_K^G V)$. Since K is open in G, $c\text{-Ind}_K^G$ is a co-adjoint to res_K^G. Thus we have a natural isomorphism of vector spaces $V_\rho \cong Hom_G(c\text{-Ind}_K^G W, V)$. Now as remarked above, $c\text{-Ind}_K^G W$ is a right $\mathcal{H}(G, \rho)$-module; we may thus view $Hom_G(c\text{-Ind}_K^G W, V)$ and hence V_ρ as a left-$\mathcal{H}(G, \rho)$-module. The functor $\mathbf{M}_\rho$ is then given on objects by $\mathbf{M}_\rho(V) = V_\rho$ and on morphisms $\phi \in Hom_G(V_1, V_2)$ by $\mathbf{M}_\rho(\phi)(f) = \phi \circ f,\ f \in (V_1)_\rho$.

The point of all this is that the structure of $\mathcal{H}(G, \rho)$ may often be elucidated; we illustrate this now using the example of $GL(N, F)$.

5. The case $G = GL(N, F)$

(Note: What follows here is a summary of some of the results in [BK1]. Following the notation there, we fix an F-vector space V of dimension N and set $A = \mathrm{End}_F(V)$. We then identify G with $A^\times$.)

5.1. Supercuspidal representations.

In [BK1], (5.5.10) an explicit description is given of a compact open subgroup J together with an irreducible representation λ of J; the pair (J, λ) is referred to as a *simple type*. Implicit in the description of (J, λ) is a subfield E of A such that $E^\times \subset N_G(J)$([BK1], (5.5.14)). (This field is labelled 'K' in [BK1].) When $[E : F] = N$, (J, λ) is called *maximal*; in that case, $N_G(J) = E^\times J$.

In (8.4.1) of [BK1] it is shown that every irreducible supercuspidal representation of G contains a maximal simple type. (That is, if π is an irreducible supercuspidal representation and $\mathfrak{s} = [G, \pi]_G$, then there exists a maximal simple type which is an $\mathfrak{s}$-type.) In (6.2.2) of [BK1] it is shown that if (J, λ) is a maximal simple type and if π is any irreducible supercuspidal representation of G which contains (J, λ) then there exists an extension, Λ, of λ to $N_G(J)$ such that:

$$\pi \cong \mathrm{Ind}_{N_G(J)}^G \Lambda \cong c\text{-}\mathrm{Ind}_{N_G(J)}^G \Lambda.$$

Note that in this case, we have

$$\mathcal{H}(G, \lambda) = \mathcal{H}(N_G(J), \lambda) \cong \mathbb{C}(N_G(J)/J) \cong \mathbb{C}(E^\times/O_E^\times) \cong \mathbb{C}(\mathbb{Z}).$$

5.2 The case that $\mathfrak{s}$ is simple.

Write $N = RS$, set $G_0 = GL(R, F)$, and pick some Levi subgroup L of G such that $L \cong G_0 \times G_0 \times \cdots \times G_0$. (So there are S factors.) Let σ_0 be an irreducible supercuspidal representation of G_0 and set $\sigma = \sigma_0 \otimes \sigma_0 \otimes \cdots \otimes \sigma_0$ as a representation of L. We obtain a point $\mathfrak{s} = [L, \sigma]_G \in \mathfrak{B}(G)$; any point $\mathfrak{s}$ constructed in this way will be called *simple*.

In some sense, the goal of [BK1] is the construction of $\mathfrak{s}$-types for simple points $\mathfrak{s}$ as above, these types being as useful as possible for the analysis of the subcategories $\mathfrak{R}^{\mathfrak{s}}(G)$. A first approximation to this is an $\mathfrak{s}$-type (J, λ) which is referred to in [BK1] as *simple*. This is the type referred to in §5.1 above. If $R < N$, however, this type will not be maximal; indeed, one always has $[E : F] = R$.

A simple type (J, λ) has the virtue that its Hecke algebra $\mathcal{H}(G, \lambda)$ is isomorphic in a natural way to an affine Hecke algebra of type A. To make this precise, we note that V is naturally an E-vector space since $E \subset \mathrm{End}_F(V)$.

Thus we may consider the subgroup $G' = Aut_E(V)$ of G. Now the pair (J, λ) has been constructed to have the following properties:

(1) $J' = J \cap G'$ is an Iwahori subgroup of G'.
(2) The *support* of $\mathcal{H}(G, \lambda)$ is $JG'J$; that is, the support of any function $h \in \mathcal{H}(G, \lambda)$ lies in $JG'J$.

One then shows (Theorem 5.6.6 of [BK1]) that:

There is a natural family of support preserving algebra isomorphisms:

$$\mathcal{H}(G, \lambda) \cong \mathcal{H}(G', \mathbf{1}_{J'}) \tag{5-1}$$

where $\mathbf{1}_{J'}$ is the trivial representation of J'.

Of course, (L, σ) also determines a point $\mathfrak{s}_L = [L, \sigma]_L \in \mathfrak{B}(L)$. And, in light of our results on supercuspidal representations, $\mathfrak{s}_L$ has a type; namely, (J_L, λ_L) where we obtain (J_L, λ_L) by starting with a $[G_0, \sigma_0]_{G_0}$-type (J_0, λ_0) and setting $J_L = J_0 \times J_0 \times \cdots \times J_0$, $\lambda_L = \lambda_0 \otimes \lambda_0 \otimes \cdots \otimes \lambda_0$. Also, given any parabolic subgroup P with Levi component L, the functors ι_P, r_P induce functors $\iota_P^{\mathfrak{s}}$, $r_P^{\mathfrak{s}}$ between the categories $\mathfrak{R}^{\mathfrak{s}_L}(L)$, $\mathfrak{R}^{\mathfrak{s}}(G)$. ($\iota_P^{\mathfrak{s}}$ is just the restriction of ι_P to $\mathfrak{R}^{\mathfrak{s}_L}(L)$; $r_P^{\mathfrak{s}}$ is obtained by composing the restriction of r_P to $\mathfrak{R}^{\mathfrak{s}}(G)$ with the projection of $\mathfrak{R}(L)$ on $\mathfrak{R}^{\mathfrak{s}_L}(L)$.) The simple type (J, λ) as described above does not reflect this extra structure. It is, however, possible to modify (J, λ) (chapter 7 of [BK1]) so as to obtain an $\mathfrak{s}$-type (which we again denote by (J, λ)) for which **(5-1)** continues to hold and one also has:

Let P be any parabolic subgroup with Levi component L. Then there is an embedding of algebras $t_P : \mathcal{H}(L, \lambda_L) \to \mathcal{H}(G, \lambda)$ so that the following diagrams commute:

$$\begin{array}{ccc} \mathfrak{R}^{\mathfrak{s}}(G) & \xrightarrow{\mathbf{M}_\lambda} & \mathcal{H}(G, \lambda)\text{-Mod} \\ {\scriptstyle r_P^{\mathfrak{s}}}\downarrow & & \downarrow{\scriptstyle t_P^*} \\ \mathfrak{R}^{\mathfrak{s}_L}(L) & \xrightarrow{\mathbf{M}_{\lambda_L}} & \mathcal{H}(L, \lambda_L)\text{-Mod}, \end{array}$$

(5-2)

$$\begin{array}{ccc} \mathfrak{R}^{\mathfrak{s}}(G) & \xrightarrow{\mathbf{M}_\lambda} & \mathcal{H}(G, \lambda)\text{-Mod} \\ {\scriptstyle \iota_P^{\mathfrak{s}}}\uparrow & & \uparrow{\scriptstyle (t_P)_*} \\ \mathfrak{R}^{\mathfrak{s}_L}(L) & \xrightarrow{\mathbf{M}_{\lambda_L}} & \mathcal{H}(L, \lambda_L)\text{-Mod}. \end{array}$$

(Here, $(t_P)_*$ is just the functor of "hom induction":

$$(t_P)_*(X) = Hom_{t_P(\mathcal{H}(L,\lambda_L))}(\mathcal{H}(G, \lambda), X)$$

for any left $\mathcal{H}(L,\lambda_L)$-module X.)

Properties **(5-1)**, **(5-2)** are compatible. That is, given any parabolic subgroup P as above, one has that $P' = P \cap G'$ is a Borel subgroup of G' and, with the obvious notation, one may choose isomorphisms $\mathcal{H}(G,\lambda) \cong \mathcal{H}(G',\mathbf{1}_{J'})$, $\mathcal{H}(L,\lambda_L) \cong \mathcal{H}(L',\mathbf{1}_{J'_L})$ so that the following diagram commutes for all P:

$$\begin{array}{ccc} \mathcal{H}(G,\lambda) & \xrightarrow{\cong} & \mathcal{H}(G,\mathbf{1}_{J'}) \\ {\scriptstyle t_P}\uparrow & & \uparrow{\scriptstyle t_{P'}} \\ \mathcal{H}(L,\lambda_L) & \xrightarrow{\cong} & \mathcal{H}(L',\mathbf{1}_{J'_L}) \end{array} \tag{5-3}$$

5.3. The general case.
[BK1] is limited to the case that $\mathfrak{s}$ is simple. Indeed, one of the purposes of [BK4] is to remove this limitation. We will return to this below.

6. Covers: a Hecke algebra approach to parabolic induction

We now turn to the question of whether it is possible to extend some of the results in §5 to the context of general reductive groups. We, and others, have some progress in the direction of **(5-1)** above for specific cases. Such isomorphisms appear to be connected with what has come to be called 'functoriality' and one may expect them to be difficult to come by in general. On the other hand, one can say a fair amount about the existence of $\mathfrak{s}$-types which satisfy **(5-2)**; we explain this now.

The set-up is as follows. We take G as in §3 and let M be a Levi subgroup of G. We need

6.1 Definition. *Let $P = MN$ be a parabolic subgroup with Levi component M and let $\bar{P} = M\bar{N}$ be its opposite. Let J be a compact open subgroup of G and let (τ, W) be an irreducible representation of J. Then the pair (J,τ) is said to be* decomposed *with respect to (M,P) if the following conditions hold:*

(1) $J = J \cap \bar{N} \cdot J \cap M \cdot J \cap N$;
(2) *the groups $J \cap \bar{N}$, $J \cap N$ are both contained in the kernel of τ.*

(We note that the pair $(B, \mathbf{1}_B)$ given in the example in §3 is decomposed with respect to any minimal parabolic. The $GL_N(F)$-types described above are also decomposed with respect to any parabolic subgroup with Levi factor L.)

Let (J,τ) be decomposed with respect to (M,P) as above, set $J_M = J \cap M$,and set $\tau_M = \tau \mid_M$. Then τ_M is irreducible and there is an obvious embedding of vector spaces:

$$T : \mathcal{H}(M, \tau_M) \hookrightarrow \mathcal{H}(G, \tau).$$

(See Proposition (6.3) of [BK4].)

Further, T restricts to an embedding of algebras on the subalgebra $\mathcal{H}^+(M, \tau_M)$ of $\mathcal{H}(M, \tau_M)$ whose elements are supported on "(J, P)-positive" elements; that is, those elements $y \in M$ with the property that $y(J \cap N)y^{-1} \subset J \cap N$, $y^{-1}(J \cap \bar{N})y \subset J \cap \bar{N}$. Our first goal is to give a necessary and sufficient condition that the map $T\,|_{\mathcal{H}^+(M,\tau_M)}$ extend to an embedding of **algebras**: $\mathcal{H}(M, \tau_M) \hookrightarrow \mathcal{H}(G, \tau)$.

To give this condition we need

6.2 Definition. *An element $\zeta \in M$ will be said to be* strongly (J, P)-positive *if it is (J, P) positive and*

(1) *ζ lies in the center $\mathcal{Z}(M)$ of M.*
(2) *For any compact open subgroups H_1, H_2 of N, there exists an integer $m \geq 0$ such that $\zeta^m H_1 \zeta^{-m} \subset H_2$.*
(3) *For any compact open subgroups K_1, K_2 of $\bar{N}$, there exists an integer $m \geq 0$ such that $\zeta^{-m} K_1 \zeta^m \subset K_2$.*

Strongly positive elements exist [BK4], (6.14). Further, given such an element ζ, there is a unique function $\phi_\zeta \in \mathcal{H}^+(M, \tau_M)$ which is supported on $J_M \zeta J_M$ and for which $\phi_\zeta(\zeta)$ is the identity endomorphism on $\tilde{W}$ [BK4], (7.1).

We may now state the first main result of this section:

6.3 Theorem. *The map $T : \mathcal{H}^+(M, \tau_M) \hookrightarrow \mathcal{H}(G, \tau)$ extends to an embedding of algebras $t' : \mathcal{H}(M, \tau_M) \hookrightarrow \mathcal{H}(G, \tau)$ if and only if $T(\phi_\zeta)$ is invertible, for some (hence any) strongly (J, P)-positive element ζ. If this extension exists, it is unique.*

(This is Theorem 7.2.i of [BK4].)

This result is neither surprising nor particularly difficult to prove. Our next result is more striking:

6.4 Theorem.

Let $\mathcal{H}(G, \tau)_M$ be the subspace of functions in $\mathcal{H}(G, \tau)$ which are supported on JMJ and suppose that the space $\mathcal{H}(G, \tau)_M$ is a subalgebra of $\mathcal{H}(G, \tau)$. Then $T(\phi_\zeta)$ is invertible for any strongly (J, P)-positive element ζ. Further, the resulting extension t' of T induces an isomorphism of algebras:

$$t' : \mathcal{H}(M, \tau_M) \xrightarrow{\cong} \mathcal{H}(G, \tau)_M$$

which is support preserving in the sense that

$$supp(t'(h)) = J \cdot supp(h) \cdot J, \quad h \in \mathcal{H}(M, \tau_M).$$

(This is Theorem 7.2.ii of [BK4].)

This result has a significant consequence:

6.5 Corollary.

Let P be a parabolic subgroup of G with Levi component M and let (J,τ) be decomposed with respect to (P,M). Suppose that the support of $\mathcal{H}(G,\tau)$ is contained in JMJ. Then with notation as above, there is a support preserving isomorphism of algebras:

$$t' : \mathcal{H}(M,\tau_M) \cong \mathcal{H}(G,\tau)$$

(We note that this result was known in special cases; see, in particular [HM].)

In order to apply the above results to the construction of $\mathfrak{s}$-types we need to consider the connection between the algebra map t' given above and the Jacquet map r_P. So suppose that we are in the situation of Definition 4.1; i.e., that we have a pair (J,τ) which is decomposed with respect to (M,P). Then for any smooth G-representation (π,V), the map r_P induces a map $q_P : V_\tau \to (r_P(V))_{\tau_M}$, q_P being given by $q_P(f) = r_P \circ f$, $f \in V_\tau$. The key result ([BK4], (7.9),(7.14)) is then:

6.6. Proposition. *Let P be a parabolic subgroup with Levi component M and let (J,τ) be decomposed with respect to (M,P) Then the following are equivalent:*

(1) *The map q_P is an injection for all irreducible representations (π,V).*
(2) *The map q_P is an isomorphism for all smooth representations (π,V).*
(3) *The map $T : \mathcal{H}^+(M,\tau_M) \hookrightarrow \mathcal{H}(G,\tau)$ extends to an embedding of algebras $t' : \mathcal{H}(M,\tau_M) \hookrightarrow \mathcal{H}(G,\tau)$*

If these conditions hold, then there exists an embedding of algebras $t : \mathcal{H}(M,\tau_M) \to \mathcal{H}(G,\tau)$ which is unique with the following properties:

(1) *$supp(t(h)) = supp(t'(h))$ for all $h \in \mathcal{H}(M,\tau_M)$*
(2) *$q_P(t(h)\cdot f) = h \cdot q_P(f)$, $\quad f \in V_\tau$, $h \in \mathcal{H}(M,\tau_M)$.*

We need one last definition.

6.7 Definition. *Let M be a proper F-Levi subgroup of G, let J_M be a compact open subgroup of M and let (τ_M, W) be an irreducible smooth representation of J_M. Let J be a compact open subgroup of G and let τ be an irreducible smooth representation of J. Then we say that the pair (J,τ) is a G-*cover *of (J_M,τ_M) if*

(1) *the pair (J,τ) is decomposed with respect to (M,P) for every F-parabolic subgroup P of G with Levi component M;*
(2) *$J \cap M = J_M$, $\quad \tau\,|_M \cong \tau_M$;*
(3) *the equivalent conditions of Proposition 6.6 (hence of Theorem 6.3) hold for every parabolic subgroup P with Levi component M.*

If (J,τ) is a G-cover of (J_M,τ_M) as above and if P is a parabolic subgroup with Levi factor M, then we write t_P for the algebra map t given in Proposition 6.6 above. We may now state

6.8 Theorem. *Let M be an F-Levi subgroup of G, let L be an F-Levi subgroup of M and let σ be an irreducible supercuspidal representation of L. Set $\mathfrak{s}_M = [L,\sigma]_M$, $\mathfrak{s} = [L,\sigma]_G$. Suppose that (J_M,τ_M) is an $\mathfrak{s}_M$-type in M and that (J,τ) is a G-cover of (J_M,τ_M). Then (J,τ) is an $\mathfrak{s}$-type in G.*

Further, if P is an F-parabolic subgroup with Levi component M then the following diagrams commute:

$$\begin{array}{ccc} \mathfrak{R}^{\mathfrak{s}}(G) & \xrightarrow{\mathbf{M}_\tau} & \mathcal{H}(G,\tau)\text{-Mod} \\ {\scriptstyle r_P^{\mathfrak{s}}}\downarrow & & \downarrow{\scriptstyle t_P^*} \\ \mathfrak{R}^{\mathfrak{s}_M}(M) & \xrightarrow{\mathbf{M}_{\tau_M}} & \mathcal{H}(M,\tau_M)\text{-Mod}, \end{array}$$

$$\begin{array}{ccc} \mathfrak{R}^{\mathfrak{s}}(G) & \xrightarrow{\mathbf{M}_\tau} & \mathcal{H}(G,\tau)\text{-Mod} \\ {\scriptstyle \iota_P^{\mathfrak{s}}}\uparrow & & \uparrow{\scriptstyle (t_P)_*} \\ \mathfrak{R}^{\mathfrak{s}_M}(M) & \xrightarrow{\mathbf{M}_{\tau_M}} & \mathcal{H}(M,\tau_M)\text{-Mod}. \end{array}$$

(This is Theorem 8.3 and Corollary 8.4 of [BK4].)

We close this section with the remark that the property of being a cover is transitive in an obvious sense. (Proposition 8.5 of [BK4].)

7. The structure of the Hecke algebra of a cover

Throughout this section, we fix an F- Levi subgroup L of G and an irreducible supercuspidal representation σ of L. We set $\mathfrak{s}_L = [L,\sigma]_L$, $\mathfrak{s} = [L,\sigma]_G$ and we assume that $\mathfrak{s}_L$ has a type (J_L,τ_L). In fact, we assume something stronger:

(1) There is an open, compact modulo center subgroup $\tilde{J}_L$ which contains J_L and an extension $\tilde{\tau}_L$ of τ_L to $\tilde{J}_L$ such that

$$\sigma \cong c\text{-Ind}_{\tilde{J}_L}^{L}\, \tilde{\tau}_L.$$

(2) The support of $\mathcal{H}(L,\tau)$ lies in $\tilde{J}_L$.

We remark that these assumptions are valid for all σ whenever L is isomorphic, for example, to a product of $GL(N)$'s. (See §5.1 above.)

Suppose now that there is a G-cover (J,τ) for (J_L,τ_L) and set $A = \mathcal{H}(G,\tau)$. For a fixed parabolic subgroup P with Levi component L, we have an embedding $t_P : \mathcal{H}(L,\tau_L) \to \mathcal{H}(G,\tau)$; set $B = t_P(\mathcal{H}(L,\tau_L))$. We are interested here in the structure of A as a left B-module. To see the sort of result which is to be expected, it might be useful to turn once again to the case that $G = GL(N, F)$, taking L to be the standard minimal Levi subgroup of diagonal matrices and $\sigma = \mathbf{1}_L$. In this simple case, $J_L = {}^\circ L$ - the subgroup of diagonal matrices with entries in $0_F^\times$, $\tau_L = \mathbf{1}_{{}^\circ L}$ and we may take J to be the standard (upper triangular) Iwahori subgroup and $\tau = \mathbf{1}_J$. Let P be the standard (upper triangular) Borel subgroup and let $K = GL(N, O_F)$. Then A is just the usual affine Hecke algebra of type A and parameter $q = [O_F : P_F]$ and t_P is the usual embedding of $\mathbb{C}(\mathbb{Z}^N)$ into A. Further, we have that the map $b \otimes k \to bk$ induces an isomorphism of left B-modules:

$$A \cong B \otimes_{\mathbb{C}} \mathcal{H}(K, \tau) \tag{7-1}$$

where $\mathcal{H}(K,\tau)$ is viewed as a subalgebra of A by extending functions in $\mathcal{H}(K,\tau)$ to be 0 away from K. $\mathcal{H}(K,\tau)$ is, of course, just the usual ordinary Hecke algebra of type A and parameter q and **(7-1)** may thus be thought of as describing part of what might be called the Bernstein-Lusztig presentation of the affine Hecke algebra [Lu].

Our main result here generalizes this. (We revert now to the notation at the beginning of the section.) Let $N(\mathfrak{s}_L)$ be the stabilizer in $N_G(L)$ of the point $\mathfrak{s}_L$, let $W = W(\mathfrak{s}_L) = N_G(\mathfrak{s}_L)/L$, and suppose that there is a compact, open subgroup K of G which contains J and which also contains a complete set of representatives, $\tilde{W}$, of W. Suppose also that $PK = G$ and that $K \cap L$ is a maximal compact subgroup of L. (These assumptions will hold, for example, if J is contained in a special maximal compact subgroup.) Suppose finally that $\tilde{W}$ can be chosen so that for all $w \in \tilde{W}$ we have $J_L^w = J_L$, $\tau_L^w = \tau_L$. (See the comments in 11.3 of [BK4] for the plausibility of these conditions.) Then we have

7.1 Theorem.

(1) *$dim_{\mathbb{C}}(\mathcal{H}(K,\tau)) = |W|$.*

(2) *The map $b \otimes k \to bk$ induces an isomorphism of left B-modules:*

$$A \cong B \otimes_{\mathbb{C}} \mathcal{H}(K, \tau).$$

(This is Theorem 11.4 of [BK4])

This result may be combined with the results in §**6** to give a complement to Theorem 6.4 as follows. With the above notation, let M be an F-Levi subgroup of G which contains L and let $J_M = J \cap M$, $\tau_M = \tau\mid_M$. Then ([BK4], 8.5) (J_M, τ_M) is a cover of (J_L, τ_L) and (J, τ) is a cover of (J_M, τ_M); in particular, (J_M, τ_M) is an $\mathfrak{s}_M$-type and we have

7.2 Theorem. *With the notation above, suppose that* $N_G(\mathfrak{s}_L) \subset M$. *Then for any parabolic subgroup* P *of* G *with Levi component* M *the map* $t_P : \mathcal{H}(M, \tau_M) \to \mathcal{H}(G, \tau)$ *is a support-preserving isomorphism.*

(This is 12.1 of [BK4])

This last result has an interesting corollary:

7.3 Corollary. *Under the above assumptions, the functor* $\iota_P : \mathfrak{R}^{\mathfrak{s}_M}(M) \to \mathfrak{R}^{\mathfrak{s}}(G)$ *is an equivalence of categories. In particular, if* W *is an irreducible object in* $\mathfrak{R}^{\mathfrak{s}_M}(M)$ *then* $\iota_P W$ *is irreducible.*

Note that although Theorem 7.2 and its corollary are proved under the hypothesis that $\mathfrak{s}_L$ has a type of a very special sort and that this type has a G-cover, the statement of Corollary 7.3 makes sense without this hypothesis. It would be interesting to know whether Corollary 7.3 remains valid without the hypothesis on types.

We remark in conclusion that the results here provide the technical tools to extend the results of [BK1] to the case of an arbitary point $\mathfrak{s} \in \mathfrak{B}(G)$. We will report on this elsewhere.

References

[AM] P.N. Anh and L. Marki, *Morita equivalence for rings without identity.* Tsukuba J. Math. **11** (1987), 1-16.

[BD] J.-N. Bernstein (rédigé par P. Deligne), *Le "centre" de Bernstein. Représentations des groupes réductifs sur un corps local.* Paris, 1984, pp. 1–32.

[Borc] F. Borceux, *Handbook of categorical algebra 1: Basic category theory, Cambridge University Press 1994.*

[Bo] *A. Borel, Admissible representations of a semisimple group with vectors fixed under an Iwahori subgroup.* Invent. Math. **35** (1976), 233-259.

[BK1] C.J. Bushnell and P. C. Kutzko, *The admissible dual of* $GL(N)$ *via compact open subgroups.* Annals of Math. Studies **129**, Princeton University Press 1993.

[BK2] C.J. Bushnell and P.C. Kutzko, *The admissible dual of* $SL(N)$ *I.* Ann. Scient. Éc. Norm. Sup. (4) **26** (1993), 261–279.

[BK3] C.J. Bushnell and P.C. Kutzko, *The admissible dual of* $SL(N)$ *II.* Proc. London Math. Soc. (3) **68** (1992), 317-379.

[BK4] C.J. Bushnell and P.C. Kutzko, *Smooth representations of reductive p-adic groups.* Preprint, 1995.

[Car] P. Cartier, *Representations of* $\mathfrak{p}$*-adic groups: a survey.* Automorphic forms, representations and L-functions (A. Borel & W. Casselman edd.), Proc. Symp. in Pure Math. **XXXIII** (AMS, Providence, 1979), 111-156.

[Ca1] W. Casselman, *Introduction to the theory of admissible representations of* $\mathfrak{p}$*-adic reductive groups.* Preprint 1974.

[**Ca2**] W. Casselman, *The unramified principal series of* $\mathfrak{p}$*-adic groups I.* Compositio Math. **40** (1980), 387–406.

[**H**] R.E. Howe, *Some qualitative results on the representation theory of* GL_n *over a p-adic field.* Pacific J. Math. **73** (1977), 479-538.

[**HM**] R.E. Howe and A. Moy, *Hecke algebra isomorphisms for* $GL(n)$ *over a p-adic field.* J. Alg. **131** (1990), 388–424.

[**KL**] D. Kazhdan and G. Lusztig, *Proof of the Deligne-Langlands conjecture for Hecke algebras.* Invent. Math. **87** (1987), 153-215.

[**KM**] P.C. Kutzko and A. Moy, *On the local Langlands conjecture in prime dimension.* Ann. Math. **121** (1985), 495–517.

[**Lu**] G. Lusztig, *Classification of unipotent representations of simple p-adic groups.* Preprint.

[**V**] M.F. Vigneras *Représentations l-modulaires d'un groupe réductif p-adique avec* $l \neq p$, Birkhäuser, 1997.

Department of Mathematics,
University of Iowa,
Iowa City,
Iowa 52242.
pkutzko@ blue.weeg.uiowa.edu

REGULAR METABELIAN LIE ALGEBRAS

FERNANDO LEVSTEIN AND ALEJANDRO TIRABOSCHI

ABSTRACT. We use a new characterization of the Lie algebras of H-type and the classification of regular 2-step nilpotent Lie algebras of center 2 to produce examples of regular Lie algebras that are not H-type.

§0. INTRODUCTION

The most elementary non-abelian Lie algebras are the so-called metabelian or two-step nilpotent Lie algebras. They can be defined by two vector spaces $\mathfrak{v}$ and $\mathfrak{z}$ and a skew symmetric bilinear form $B : \mathfrak{v} \to \mathfrak{z}$. An attempt at a classification in the complex case was made by Gauger in [Gau]. We are interested in a subclass that occurs only in the real case. These are the regular metabelian Lie algebras.

The defining condition is that for any non-zero λ in the dual of the center $\mathfrak{z}^*$, the bilinear form $\lambda \circ B$ is non-degenerate. From the work of C. Moore and J. Wolf [MW] it is known that these algebras have all their unitary representations, except the trivial one, square integrable. It is not easy to show that a Lie algebra is regular since one needs to check that a certain polynomial in several variables has no non-zero roots. One way to construct regular algebras is due to Kaplan [Ka1] (see Definition 1.3 below). The resulting algebras are called Lie algebras of the Heisenberg type or H-type algebras and give a generalization of the Heisenberg algebra that has proven to be useful in the construction of interesting examples in geometry and analysis; see [C], [Ka2], [KaR], [R], [B], [W], [TV]. For some time those were the only known examples of regular Lie algebras. Recently, Kaplan and Saal [KaS] found an example in dimension 10 that was not of H-type.

In this paper we give a characterization of H-type algebras that allows one to decide whether a given metabelian Lie algebra is of H-type or not. The procedure leads naturally to the definition of a new class of regular algebras that we called $\tilde{H}$-type. Furthermore, applying this method to a classification of metabelian Lie algebras with center of dimension 2, it was possible to give a classification of regular Lie algebras of dimension 10. This allowed us to produce an example of $\tilde{H}$-type not of H-type.

The paper is organized as follows: in §1 we give the basic definitions and the construction of H-type algebras. In §2 we characterize these algebras. In §3 we summarize some results about real skew pencils. In §4 we produce the classification of regular two-step nilpotent algebras with center of dimension 2, and we use it to show realizations of different kinds of regular algebras.

This work was partially supported by CONICET, CONICOR and UNC (Argentina).

This paper is an expanded version of a talk given by the first author at the Fifth Workshop on Representation Theory of Lie Groups and Its Applications. We want to thank the Scientific Organizing Committee of this workshop for the invitation to give this talk. We are indebted to Professor A. Kaplan for suggesting the problem and introducing us into the subject and we wish to thank Professors L. Saal, J. Vargas, and J. Wolf for helpful discussions.

§1. Two-step nilpotent Lie algebras: regular and H-type algebras

Let $\mathfrak{n}$ be a Lie algebra over a field F. Then we will say that $\mathfrak{n}$ is *two-step nilpotent* or *metabelian* if $[\mathfrak{n}, \mathfrak{n}]$ is contained in the center of $\mathfrak{n}$.

Definition 1.1. Let $\mathfrak{n}$ be a two-step nilpotent Lie algebra. We will say that $\mathfrak{n}$ is *regular* if for each $\lambda \neq 0$ in $\mathfrak{z}^*$ the dual of $\mathfrak{z}$, the bilinear form $< , >_\lambda$: $\mathfrak{n}/\mathfrak{z} \times \mathfrak{n}/\mathfrak{z} \longrightarrow F$ defined by $< x, y >_\lambda = \lambda([x, y])$ is not degenerated.

It follows from the definition that for $\dim(\mathfrak{z}) > 1$ there are no regular algebras over $\mathbb{C}$. We will consider only the case $F = \mathbb{R}$. Notice also that if $\mathfrak{n}$ is regular then $[\mathfrak{n}, \mathfrak{n}] = \mathfrak{z}$.

Let $\mathfrak{n}$ be a real two-step nilpotent Lie algebra, equipped with a scalar product $(,)$. Let $\mathfrak{v}$ be the orthogonal complement of $\mathfrak{z}$. We define the linear mapping $J : \mathfrak{z} \longrightarrow \mathrm{End}(\mathfrak{v})$ by the formula

$$(J_z x, x') = (z, [x, x']), \qquad \forall\, x, x' \in \mathfrak{v},\ \forall\, z \in \mathfrak{z},$$

whence $J_z^t = -J_z$ for all $z \in \mathfrak{z}$.

Lemma 1.2. *Let $\mathfrak{n}$ be a two-step nilpotent algebra. Then we have the following equivalences:*

(*i*) *$\mathfrak{n}$ is regular.*
(*ii*) *$\mathrm{ad}(x)$ is surjective for all $x \notin \mathfrak{z}$.*
(*iii*) *$J_z : \mathfrak{n}/\mathfrak{z} \longrightarrow \mathfrak{n}/\mathfrak{z}$ is an isomorphism for all $z \in \mathfrak{z} - \{0\}$.*

Proof. (*i*) $\Leftrightarrow$ (*ii*) follows by definition.

(*ii*) $\Leftrightarrow$ (*iii*) $\mathrm{ad}(x)$ is not onto iff $\exists z \in \mathfrak{z} - \{0\}$ such that $(z, [x, y]) = 0$ for all $y \in \mathfrak{n}$ iff $\exists z \in \mathfrak{z} - \{0\}$ such that $(J_z x, y) = 0$ for all $y \in \mathfrak{n}$ iff $\exists z \in \mathfrak{z} - \{0\}$ such that $J_z x = 0$. □

Following [Ka1] we have the following:

Definition 1.3. Let $\mathfrak{n}$ be a two-step nilpotent Lie algebra. We say that $\mathfrak{n}$ is an *H-type algebra* if there exists a scalar product $(,)$ such that for every x in $\mathfrak{v} = \mathfrak{z}^\perp$ (the orthogonal complement of $\mathfrak{z}$) of length 1, $\mathrm{ad}(x)$ is an isometry from $\ker(\mathrm{ad}(x))^\perp$ onto $\mathfrak{z}$. Equivalently, $\mathfrak{n}$ is an H-type algebra if there exists a scalar product $(,)$ such that for all z in $\mathfrak{z}$,

$$J_z^2 = -(z, z)\mathrm{Id}, \tag{1.1}$$

where Id denotes the identity mapping.

It follows from Lemma 1.2 that any H-type algebra is regular.

Let M be an F-vector space and let $f : M \times M \longrightarrow F$ be a symmetric bilinear form. Then we will say that (M, f) is a *quadratic form*. We say that (M, f) is *non-degenerate* if $f(x, y) = 0$ for all $y \in M$ implies that $x = 0$.

Definition 1.4. The *Clifford algebra* of a quadratic form (M, f) is a pair $(\mathcal{C}(M, f), \theta)$, where $\mathcal{C}(M, f)$ is an F-algebra, $\theta : M \longrightarrow \mathcal{C}(M, f)$ is a linear function such that $\theta(x)^2 = f(x, x)1$ for each $x \in M$ and $(\mathcal{C}(M, f), \theta)$ satisfies the following universal property: if (A, μ) is a pair such that A is a F-algebra and $\mu : M \longrightarrow A$ is linear and satisfies $\mu(x)^2 = f(x, x)1$, then there exists an algebra morphism $\mu' : \mathcal{C}(M, f) \longrightarrow A$ such that $\mu'\theta = \mu$ and μ' is unique with respect to this property.

It is not hard to see that a Clifford algebra $(\mathcal{C}(M, f), \theta)$ exists for each quadratic form (M, f) and can be obtained as a quotient of the tensor algebra $T(M)$ by the ideal generated by $x \otimes x - f(x, x)1$. Moreover, $(\mathcal{C}(M, f), \theta)$ is unique modulo isomorphism of F-algebras and M is naturally embedded in $\mathcal{C}(M, f)$ (for more details see [Hu]).

Let $\mathcal{C}(n)$ denote $\mathcal{C}(\mathbb{R}^n, -(x, y))$, where $(\, , \,)$ is the standard scalar product on $\mathbb{R}^n$. For the construction of H-type algebras one needs the following:

Theorem 1.5. *(see* [BtD]*) Every module over the Clifford algebra $\mathcal{C}(n)$ is the direct sum of irreducible modules. Up to isomorphism there is precisely one irreducible module $\mathfrak{v}_n$ over $\mathcal{C}(n)$ for $n \not\equiv 3 \mod 4$, and there are precisely two $(\mathfrak{v}_n^+, \mathfrak{v}_n^-)$ for $n \equiv 3 \mod 4$.*

Each Clifford module has a scalar product which makes the generators of $\mathcal{C}(n)$ skew symmetric.

Let $\mathfrak{n}$ be an H-type algebra. Then $\{J_z : z \in \mathfrak{z}\}$ generates a Clifford algebra isomorphic to $\mathcal{C}(n)$ $(\dim(\mathfrak{z}) = n)$, and $\mathfrak{z}^\perp$ becomes a Clifford module. Conversely, given $\mathfrak{v}$ a Clifford module of $\mathcal{C}(n)$, then one can give $\mathbb{R}^n \oplus \mathfrak{v}$ the structure of a two-step nilpotent Lie algebra in the following way: $(Z, [X, Y]) = (Z.X, Y)$ $(Z \in \mathbb{R}^n,\ X, Y \in \mathfrak{v}_n^{\pm})$ and $\mathbb{R}^n$ is the center. This structure gives an H-type algebra. Using Theorem 1.5 and some extra work one has:

Up to isomorphism any H-type algebra with center $\mathfrak{z}$ of dimension n is of the form:

(a) $\mathfrak{n} = (\mathfrak{v}_n)^k \oplus \mathfrak{z}$ for $n \not\equiv 3 \mod 4$,

(b) $\mathfrak{n} = (\mathfrak{v}_n^+)^p \oplus (\mathfrak{v}_n^-)^q \oplus \mathfrak{z}$ for $n \equiv 3 \mod 4$.
Two pairs of exponents p, q and r, s give isomorphic algebras if and only if $\{p, q\} = \{r, s\}$.

§2. Characterization of H-type algebras

Let $\mathfrak{n}$ be a regular Lie algebra with $(\, , \,)$ and $< \, , \, >$ two scalar products in

$\mathfrak{n}$. Let S and T be endomorphisms of $\mathfrak{n}$ defined as follows:

$$(S_z x, x') = (z, [x, x']), \qquad \forall x, x' \in \mathfrak{n}, \ \forall z \in \mathfrak{z},$$
$$< T_z x, x' > = < z, [x, x'] >, \qquad \forall x, x' \in \mathfrak{n}, \ \forall z \in \mathfrak{z}.$$

Let A be the unique endomorphism of $\mathfrak{n}$ that satisfies

$$(x, y) = < Ax, y > \qquad \forall x, y \in \mathfrak{n}.$$

Then A is positive definite (for both $(\,,\,)$ and $<\,,\,>$).

Lemma 2.1. *Let $\pi_{\mathfrak{z}} : \mathfrak{n} \longrightarrow \mathfrak{z}$ be the canonical projection of $\mathfrak{n}$ to $\mathfrak{z}$ with respect to $<\,,\,>$ and let $C = \pi_{\mathfrak{z}} A_{|\mathfrak{z}}$. Then*

$$AS_z = T_{Cz}.$$

Proof. For $z \in \mathfrak{z}$ and $x, y \in \mathfrak{n}$ we have that

$$(S_z x, y) = (z, [x, y]) = < Az, [x, y] > = < \pi_{\mathfrak{z}}(Az), [x, y] > = < T_{\pi_{\mathfrak{z}}(Az)}(x), y > .$$

On the other hand $(S_z x, y) = < AS_z x, y >$, so $< AS_z x, y > = < T_{\pi_{\mathfrak{z}}(Az)}(x), y >$ and the result follows. □

We notice that if $C : \mathfrak{z} \longrightarrow \mathfrak{z}$ is as in the previous lemma, then C is invertible; because if $Cz = 0$, then $AS_z \equiv 0$, so $z = 0$.

Definition 2.2. A regular Lie algebra $\mathfrak{n}$ is said to be of *$\tilde{H}$-type* if the operators $\tilde{T}_z : \mathfrak{n}/\mathfrak{z} \to \mathfrak{n}/\mathfrak{z}$ obtained from T_z defined above satisfy the following condition:

$$f_{<\,,\,>}(z) = |\det(\tilde{T}_z^2)|^{1/m} \text{ is a quadratic form for } m = \dim \mathfrak{n} - \dim \mathfrak{z}.$$

Remark 2.3. Due to Lemma 2.1, this definition doesn't depend on the scalar product used to define T_z.

Remark 2.4. Let $<\,,\,>'$ be a scalar product in $\mathfrak{n}$ of $\tilde{H}$-type. Let C' be defined by the following equation:

$$< z, (C')^{-1} z' >' = \frac{1}{4}(f_{<\,,\,>'}(z + z') - f_{<\,,\,>'}(z - z')) \qquad \forall z, z' \in \mathfrak{z}.$$

Define $<\,,\,>$ in the following way:

$$(**) \qquad < x, y > := \begin{cases} < C'x, y >', & \forall x, y \in \mathfrak{z} \\ < x, y >', & \forall x, y \in \mathfrak{v}' \\ 0, & \forall x \in \mathfrak{z} \text{ and } y \in \mathfrak{v} \end{cases}$$

It is easy to check that the new scalar product satisfies:

$$(***) \qquad < x, y > = \frac{1}{4}(f_{<\,,\,>}(x + y) - f_{<\,,\,>}(x - y)) \qquad \forall x, y \in \mathfrak{z}.$$

We will call a product satisfying (***) *special.* Then it is clear that $\mathfrak{n}$ is of $\tilde{H}$-type if and only if there exists a special scalar product.

Now we have the following (see [LT]):

Theorem 2.5. *Let $\mathfrak{n}$ be a Lie algebra of H-type, then $\mathfrak{n}$ is $\tilde{H}$-type and the special scalar products are determined on $\mathfrak{z}$ up to constant multiple.*

In the following we will look for conditions on an algebra $\mathfrak{n}$ of $\tilde{H}$-type to be an H-type algebra.

Let $\mathfrak{n}$ be an H-type algebra. Let $(\, , \,)$ be a scalar product that satisfies (1.1), and let $< , >$ be another special scalar product. In this way we have $< , >_{|\mathfrak{z}\times\mathfrak{z}} = \lambda(\, , \,)_{|\mathfrak{z}\times\mathfrak{z}}$ and by Lemma 2.1, we have $\lambda A\tilde{J}_z = \tilde{T}_z$. Let $\{z_i\}$ be an orthonormal basis of $\mathfrak{z}$ with respect to $< , >$. Let $\tilde{J}_i = \tilde{J}_{z_i}$ and $\tilde{T}_i = \tilde{T}_{z_i}$. Thus $\tilde{J}_i^{-1}\tilde{J}_k = \tilde{T}_i^{-1}\tilde{T}_k$, and using $\tilde{J}_i^{-1} = -\mu\tilde{J}_i$ for some scalar μ, we obtain $\mu\tilde{J}_i\tilde{J}_k = -\tilde{T}_i^{-1}\tilde{T}_k$. We recall from §1 that the set of operators $\{J_i\}$ generate an algebra isomorphic to the Clifford algebra $C(\mathfrak{z}, -(\, , \,))$. Then $\{\tilde{J}_i\tilde{J}_k\}_{i<k}$ generate the subalgebra $C^0(\mathfrak{z}, -(\, , \,))$ of *even* degree operators. We point out that this subalgebra is also a Clifford algebra with "canonical" generators $\{\tilde{J}_1\tilde{J}_k\}_{k=2}^{n}$. So we have the following criterion:

> *Let $\mathfrak{n}$ be an $\tilde{H}$-type algebra and $< , >$ a special scalar product. Then a necessary condition for $\mathfrak{n}$ to be H-type is that $\{\tilde{T}_i^{-1}\tilde{T}_k\}_{i<k}$ generates a Clifford algebra isomorphic to $C^0(\mathfrak{z}, - < , >)$.*

Remark 2.6. In particular $(\tilde{T}_i^{-1}\tilde{T}_k)^2 = \alpha \, Id$ $(\forall i \neq k)$, for some negative α, but $\det(\tilde{T}_i^2) = \det(\tilde{T}_k^2)$, so $(\tilde{T}_i^{-1}\tilde{T}_k)^2 = -\mathrm{Id}$.

Finally we can show that the above condition is also sufficient.

Theorem 2.7. *Let $\mathfrak{n}$ be a Lie algebra of $\tilde{H}$-type, together with a special scalar product $< , >$. Let $z_1, \ldots, z_n$ be an orthonormal basis of $\mathfrak{z}$ and $T_i = \tilde{T}_{z_i}$ the corresponding operators on $\mathfrak{n}/\mathfrak{z}$. Then $\mathfrak{n}$ is H-type if and only if $((T_i)^{-1}T_j)^2 = -\mathrm{Id}$ $(1 \leq i, j \leq n)$.*

Proof. See [LT] □

§3. Real skew pencils

We recall some elements of the classification of real skew pencils that will be need to classify regular two-step nilpotent Lie algebras with center of dimension 2. Details can be found in [Ga] and [Th].

Definition 3.1. Let F be a field and μ, λ be algebraically independent variables over F. The matrix $\mu A + \lambda B$ in $\mathrm{M}_r(F)$ is called an $r \times r$-pencil. A pencil $\mu A + \lambda B$ is called *regular* if $\det(\mu A + \lambda B)$ is not the zero polynomial. Otherwise the pencil is called *singular*. We say that two pencils $\mu A + \lambda B$ and $\mu C + \lambda D$ are *strictly equivalent* if there are two non singular matrices S, T in $\mathrm{M}_r(F)$ such that $S(\mu A + \lambda B)T = \mu C + \lambda D$ or equivalently $SAT = C$ and $SBT = D$. A pencil $\mu A + \lambda B$ is called *skew* if both A and B are skew-symmetric matrices. Two skew pencils $\mu A + \lambda B$ and $\mu C + \lambda D$ are *strictly congruent* if there is a non-singular matrix S with elements in F satisfying

$$S^t(\mu A + \lambda B)S = \mu C + \lambda D.$$

Here we denote by S^t the transpose matrix of S.

To study the strict equivalence of regular pencils it is necessary to introduce elementary divisors. Let $\mu A + \lambda B$ be a regular pencil. We denote by $D_j(\mu, \lambda)$ the greatest common divisor of all the minors of order j in $\mu A + \lambda B$ $(j = 1, \dots, r)$. Then in the series

$$D_r(\mu, \lambda), D_{r-1}(\mu, \lambda), \dots, D_1(\mu, \lambda), D_0(\mu, \lambda)$$

each polynomial is divisible by the preceding one. The corresponding quotients will be denoted by

$$i_1(\mu, \lambda) = \frac{D_r(\mu, \lambda)}{D_{r-1}(\mu, \lambda)}, i_2(\mu, \lambda) = \frac{D_{r-1}(\mu, \lambda)}{D_{r-2}(\mu, \lambda)}, \dots, i_r(\mu, \lambda) = D_1(\mu, \lambda).$$

Definition 3.2. The polynomials $i_1(\mu, \lambda), i_2(\mu, \lambda), \dots, i_r(\mu, \lambda)$ are called the *invariant polynomials* of $\mu A + \lambda B$. Each polynomial $i_j(\mu, \lambda)$ can be decomposed into powers of irreducible factors over the given field F. These powers of irreducible factors are called *elementary divisors.* If an elementary divisor is in $F[\mu]$, it is called *infinite*, otherwise it is called *finite.*

More explicitly, if

$$(*) \qquad \begin{aligned} i_1(\mu, \lambda) &= \phi_1^{n_{11}}, \dots, \phi_t^{n_{1t}}, \\ i_2(\mu, \lambda) &= \phi_1^{n_{21}}, \dots, \phi_t^{n_{2t}}, \\ &\dots \\ i_r(\mu, \lambda) &= \phi_1^{n_{r1}}, \dots, \phi_t^{n_{rt}}, \end{aligned}$$

then all the powers among $\phi_1^{n_{11}}, \dots, \phi_t^{n_{rt}}$ in (*), as far as they are distinct from 1, are the elementary divisors, where $\phi_1, \dots, \phi_t$ are all the distinct irreducible factors over F that occur in $i_1(\mu, \lambda), i_2(\mu, \lambda), \dots, i_r(\mu, \lambda)$.

If an elementary divisor $\phi_i^{n_{ij}}$ appears m times in (*), we will say that $\phi_i^{n_{ij}}$ has *multiplicity* m. The elementary divisors (and their multiplicities) are invariant under equivalence of pencils, see [Ga, I p.141]. We remark that in fact elementary divisors are defined up to product by a constant; furthermore we have

Theorem 3.4. *Let $\mu A + \lambda B$ and $\mu C + \lambda D$ be two regular skew pencils (of the same dimensions). Then they are strictly congruent if and only if they have the same elementary divisors (counting multiplicities).*

The proof for strict equivalence is due to Weierstrass and Kronecker. One can find a proof in [Ga, II pp. 37–40]. For the theorem about strict congruence see [Th]. Now, following [Th] we will describe the canonical form (with respect to strict congruence) of a real skew pencil.

Let $\mu A + \lambda B$ be a real skew pencil with elementary divisors $\phi_1^{n_{11}}, \dots, \phi_t^{n_{rt}}$.

Proposition 3.5. *Let $\mu A + \lambda B$ be a real skew pencil. Then the elementary divisors occur in pairs, i.e. we can arrange the elementary divisors $\phi_1^{n_{11}}, \dots, \phi_t^{n_{rt}}$ as*

$$\mu^{e_1}, \mu^{e_1}, \dots, \mu^{e_s}, \mu^{e_s}, (\lambda - \mu\alpha_1)^{f_1}, (\lambda - \mu\alpha_1)^{f_1}, \dots, (\lambda - \mu\alpha_l)^{f_l}, (\lambda - \mu\alpha_l)^{f_l},$$
$$\xi(a_1, b_1)^{m_1}, \xi(a_1, b_1)^{m_1}, \dots, \xi(a_p, b_p)^{m_p}, \xi(a_p, b_p)^{m_p},$$

where $\alpha_1, \dots, \alpha_l, a_1, \dots, a_p$ are real numbers, $b_1, \dots, b_p$ are nonzero real numbers and $\xi(a,b) = (\lambda - \mu(a+ib))(\lambda - \mu(a-ib))$ $(a, b \in \mathbb{R})$.

For the proof of the proposition see [Th, pp 334–335].

In order to classify regular algebras we construct the blocks corresponding to elementary divisors with roots in $\mathbb{C} - \mathbb{R}$. For simplicity we will denote $(a + |b|i, m) := \xi(a,b)^m$. For each pair of elementary divisors of the form $(a + bi, m)$ $(a, b \in \mathbb{R},\ b > 0)$ we construct the following $4m \times 4m$ block:

$$C(a,b,m) = \begin{pmatrix} 0_{2m} & \begin{bmatrix} & & & R \\ & & R & S \\ & \cdots & & \\ R & S & & \end{bmatrix} \\ \begin{bmatrix} & & & -R \\ & & -R & -S \\ & \cdots & & \\ -R & -S & & \end{bmatrix} & 0_{2m} \end{pmatrix},$$

where $R = \begin{pmatrix} -\mu b & \lambda - \mu a \\ \lambda - \mu a & \mu b \end{pmatrix}$ and $S = \begin{pmatrix} 0 & \mu \\ \mu & 0 \end{pmatrix}$.

Now Theorem 3.4 can be complemented with the following:

Theorem 3.6. *Let $\mu A + \lambda B$ be a real skew pencil with elementary divisors $(a_1 + b_1 i, m_1), \dots, (a_p + b_p i, m_p)$. Then $\mu A + \lambda B$ is strict congruent to the pencil given by a matrix with the following blocks on the diagonal:*

$$C(a_1, b_1, m_1), \dots, C(a_p, b_p, m_p)$$

§4. Regular Two-step nilpotent Lie algebras with 2-dimensional center

The classification of complex two-step nilpotent Lie algebras with 2-dimensional center can be found in [Gau]. In this section we consider the real case, and for simplicity we present only the regular Lie algebras. Let $\mathfrak{n} = \mathfrak{v} \oplus \mathfrak{z}$ be a real vector space with $\dim \mathfrak{z} = 2$. Let $<,>: \mathfrak{n} \times \mathfrak{n} \to \mathbb{R}$ be a scalar product. Now consider two 2-step nilpotent Lie algebras $\mathfrak{n}_1$ and $\mathfrak{n}_2$ with center $\mathfrak{z}_1$ and $\mathfrak{z}_2$ respectively; such that $\dim(\mathfrak{n}_i) = \dim(\mathfrak{n})$ and $\dim(\mathfrak{z}_i) = \dim(\mathfrak{z}) = 2$. Clearly $\mathfrak{n}_i = \mathfrak{n}_i/\mathfrak{z}_i \oplus \mathfrak{z}_i$ as Lie algebras. Now take $\phi_i : \mathfrak{n}_i \to \mathfrak{n}$, an isomorphism of vector spaces such that $\phi_i : \mathfrak{n}_i/\mathfrak{z}_i \longrightarrow \mathfrak{v}$ and $\phi_i : \mathfrak{z}_i \longrightarrow \mathfrak{z}$ are isomorphisms. This induces two brackets $[\, ,\,]_i$ on $\mathfrak{n}$. Then $\mathfrak{n}_1$ and $\mathfrak{n}_2$ are isomorphic iff $(\mathfrak{n}, [\, ,\,]_1)$ and $(\mathfrak{n}, [\, ,\,]_2)$ are isomorphic with an isomorphism preserving $\mathfrak{z}$ and $\mathfrak{v}$. Choose $\mathcal{B} = \{e_1, e_2\}$ to be an orthogonal basis of $\mathfrak{z}$ and consider T_1, T_2, T_1', T_2' in $\mathrm{End}(\mathfrak{v})$ such that:

$$\begin{aligned} &[u,v]_1 =< u, T_1 v > e_1 + < u, T_2 v > e_2, \\ &[u,v]_2 =< u, T_1' v > e_1 + < u, T_2' v > e_2 \end{aligned} \qquad \text{for all} \quad u, v \in \mathfrak{v}. \tag{4.1}$$

We will call $P_1 = \mu T_1 + \lambda T_2$ *a pencil associated to* $\mathfrak{n}_1$ and we will say that the elementary divisors of P_1 is *a complete set of elementary divisors of* $\mathfrak{n}_1$.

Remark 4.1. $\mathfrak{n}_1$ and $\mathfrak{n}_2$ are isomorphic (as Lie algebras) if and only if there exists $X \in \mathrm{End}(\mathfrak{v})$ invertible, such that

$$X^t \,\mathrm{Span}_{\mathbb{R}}\{T_1, T_2\} X = \mathrm{Span}_{\mathbb{R}}\{T_1', T_2'\}.$$

The following well-known result is essential for classifying two-step nilpotent algebras with center of dimension 2.

Lemma 4.2. *Let $P_1 = \mu T_1 + \lambda T_2$ and $P_2 = \mu T_1' + \lambda T_2'$ be two $r \times r$-regular pencils, related by the following formula:*

$$\begin{aligned} T_1' &= a_{11} T_1 + a_{12} T_2, \\ T_2' &= a_{21} T_1 + a_{22} T_2, \end{aligned}$$

where the matrix $A = [a_{ij}]$ is non-singular. Then the elementary divisors of P_2 can be obtained by performing the following linear substitutions:

$$\begin{aligned} \mu' &= a_{11}\mu + a_{12}\lambda, \\ \lambda' &= a_{21}\mu + a_{22}\lambda; \end{aligned}$$

i.e. $p(\mu, \lambda)$ is an elementary divisor of P_1 if and only if $p(a_{11}\mu + a_{12}\lambda, a_{21}\mu + a_{22}\lambda)$ is an elementary divisor of P_2.

The proof of this lemma can be found in [TA, pp. 117–120].

Now let $p(\mu, \lambda) = \lambda - \mu\alpha$. Then an easy calculation shows that

$$p(a_{11}\mu + a_{12}\lambda, a_{21}\mu + a_{22}\lambda) = \begin{cases} \lambda - \mu \left[\dfrac{a_{11}\alpha - a_{21}}{a_{12}\alpha - a_{22}}\right] & \text{if} \quad a_{12}\alpha \neq a_{22} \\ \mu & \text{if} \quad a_{12} = a_{22} \end{cases},$$

thus:

Proposition 4.3. *Let $P_1 = \mu T_1 + \lambda T_2$ be a pencil with elementary divisors $(\alpha_1, m_1), \ldots, (\alpha_p, m_p)$ ($\mathrm{Im}(\alpha_j) > 0$). Let $P_2 = \mu T_1' + \lambda T_2'$ be a pencil related to P_1 as in Lemma 4.2. Then P_2 has elementary divisors $(T\alpha_1, m_1), \ldots, (T\alpha_p, m_p)$, where T is a real homography, i. e., a map given by*

$$T(\omega) = \frac{a_{11}\omega - a_{21}}{a_{12}\omega - a_{22}}, \qquad \omega \in \mathbb{C} \cup \{\infty\}, \, a_{ij} \in \mathbb{R}, a_{11}a_{22} - a_{12}a_{21} \neq 0.$$

□

Proposition 4.4.

(i) Let $\mathfrak{n}$ *be a two-step nilpotent Lie algebra with a 2-dimensional center. Then* $\mathfrak{n}$ *is regular if and only if* $(\alpha_1, m_1), \ldots, (\alpha_p, m_p)$ *is a complete set of elementary divisors of* $\mathfrak{n}$ *with* $Im(\alpha_i) > 0$ *for all* i. $\dim(\mathfrak{n}) = 2 + \sum 4m_i$.

(ii) Let $\mathfrak{n}_1$ *and* $\mathfrak{n}_2$ *be two regular two-step nilpotent Lie algebras with a 2-dimensional center, and let* $(\alpha_1, m_1), \ldots, (\alpha_p, m_p)$ *and* $(\alpha'_1, m'_1), \ldots, (\alpha'_{p'}, m'_{p'})$ *be complete sets of elementary divisors associated to* $\mathfrak{n}_1$ *and* $\mathfrak{n}_2$ *respectively. Then* $\mathfrak{n}_1$ *and* $\mathfrak{n}_2$ *are isomorphic if and only if there exists a real homography* $T : \mathbb{C} \cup \{\infty\} \to \mathbb{C} \cup \{\infty\}$*, such that*

$$\{(T\alpha_1, m_1), \ldots, (T\alpha_p, m_p)\} = \{(T\alpha'_1, m'_1), \ldots, (T\alpha'_{p'}, m'_{p'})\}.$$

Proof. (*i*) It follows from the definitions.

(*ii*) By Remark 4.1 $\mathfrak{n}_1$ is isomorphic to $\mathfrak{n}_2$, if and only if there exists X in $\mathrm{End}(\mathfrak{v})$ invertible, such that if $S_1 = X^t T_1 X$ and $S_2 = X^t T_2 X$. Then

$$T'_1 = a_{11} S_1 + a_{12} S_2,$$
$$T'_2 = a_{21} S_1 + a_{22} S_2,$$

with $a_{11}a_{22} - a_{12}a_{21} \neq 0$.

Then the result is obvious from Lemma 4.2 and Proposition 4.3. □

Now, which are the H-type algebras? The answer will be given in the following paragraphs.

Using 2.5, 2.7 and 4.3, we obtain:

Lemma 4.5. *Let* $\mathfrak{n}$ *be a regular two-step nilpotent Lie algebra with center of dimension 2 and with the following elementary divisors associated to* $\mathfrak{n}$*:*

(1) (β, m), *with* $Im(\beta) > 0$ *and* $m > 1$, *or*

(2) $(\beta_1, 1), (\beta_2, 1)$ *with* $Im(\beta_i) > 0$ *and* $\beta_1 \neq \beta_2$.

Then $\mathfrak{n}$ *is not* H*-type.*

Let $\mathfrak{n}$ be a two-step nilpotent Lie algebra with center $\mathfrak{z}$. We will say that $\mathfrak{n}$ is *reducible* if $\mathfrak{n} = \mathfrak{v}_1 \oplus \mathfrak{v}_2 \oplus \mathfrak{z}$, where $\mathfrak{v}_1, \mathfrak{v}_2$ are subspaces of $\mathfrak{n}$ such that $[\mathfrak{v}_1, \mathfrak{v}_2] = 0$. If $\mathfrak{n}$ is not reducible we will say that $\mathfrak{n}$ is *irreducible*. Notice that the pairs of elementary divisors (β, m), with $\mathrm{Im}(\beta) > 0$, correspond to irreducible algebras. We remark also that if $\mathfrak{n}$ is H-type (regular) and reducible with $\mathfrak{n} = \mathfrak{v}_1 \oplus \mathfrak{v}_2 \oplus \mathfrak{z}$, then $\mathfrak{v}_1 \oplus \mathfrak{z}$ and $\mathfrak{v}_2 \oplus \mathfrak{z}$ are H-type (regular respectively).

Proposition 4.6. *Let* $\mathfrak{n}$ *be a two-step nilpotent Lie algebra of dimension* $2+4p$ *with center of dimension 2. Then* $\mathfrak{n}$ *is* H*-type if and only if the set[B of elementary divisors associated to* $\mathfrak{n}$ *is of the form* $(\beta, 1)$ p*-times, with* $Im(\beta) > 0$.

Proof. ($\Leftarrow$) As $\mathfrak{n}$ has the elementary divisors $(\beta, 1) \ldots, (\beta, 1)$, we can get a pencil $P = \mu T_1 + \lambda T_2$ associated to $\mathfrak{n}$ such that P is formed by blocks in the diagonal of the form $C(a,b,1)$ (where $a+bi=\beta$; see §3). An easy calculation shows that $(C(a,b,1)^{-1}_{(1,0)}C(a,b,1)_{(0,1)})^2 = -\operatorname{Id}_4$ and from Theorem 2.7, we have that $\mathfrak{n}$ is H-type.

($\Rightarrow$) Let $(\beta_1, m_1), \dots, (\beta_p, m_p)$ be the elementary divisors associated to $\mathfrak{n}$. If some $m_i > 1$ or $\beta_i \neq \beta_j$ then by Lemma 4.5 we have a component that is not H-type, so $\mathfrak{n}$ is not H-type. □

Proposition 4.4 (i) gives us the sets of elementary divisors for regular two step nilpotent Lie algebras with 2-dimensional center $\mathfrak{n}$ and part (ii) gives the isomorphism classes. Proposition 4.6 gives the conditions on the set of elementary divisors of $\mathfrak{n}$ to be of H-type. Using this it is easy to obtain the following:

Corollary 4.7. *Let $\mathfrak{n}$ be a regular Lie algebra:*

(i) *If* $\dim(\mathfrak{n}) = 6$, *then $\mathfrak{n}$ is the unique álgebra of H-type;*

(ii) *If* $\dim(\mathfrak{n}) = 10$, *then $\mathfrak{n}$ is isomorphic to one and only one of the following:*

the algebra $(i,1),(i,1)$, the unique H-type algebra of dimension 10;

the algebra $(i,2)$, the unique $\tilde{H}$-type of dimension 10, not H-type;

the algebras $(i,1),(\alpha_t,1)$ with $t>0$ and $d_h(i,\alpha_t)=t$, where d_h is the hyperbolic distance in the upper half plane.

(iii) *If $\mathfrak{n}$ has elementary divisors (β, m), with* $\operatorname{Im}(\beta) > 0$; *then $\mathfrak{n}$ is $\tilde{H}$-type and has dimension $4m+2$ and it is not H-type for $m>1$.*

Item (i) and the existence of non H-type regular algebras of dimension 10 were known ([KaS]).

Finally we can give an example of regular two-step nilpotent algebras that are not $\tilde{H}$-algebras.

Let $\mathfrak{n}$ be an H-type algebra and $(\ ,\ )$ a scalar product satisfying Definition 1.3. Let $\mathfrak{z}$ the center of $\mathfrak{n}$ and $\mathfrak{v}_1$ the orthogonal complement. Take $z_1, \dots, z_n$ an orthogonal basis of $\mathfrak{z}$, and denote $J_i = J_{z_i}$ so we have that

$$[x,y] = (x, J_1 y)z_1 + \dots + (x, J_n y)z_n, \qquad \text{for all } x, y \text{ in } \mathfrak{v}_1.$$

Now, we construct our example $\mathfrak{m}$ in the following way: let $\mathfrak{m} = \mathfrak{z} \oplus \mathfrak{v}_1 \oplus \mathfrak{v}_2$, where $\mathfrak{v}_2$ is a copy of $\mathfrak{v}_1$. Define the product as

$$[x,y] := \begin{cases} (x, J_1 y)z_1 + (x, J_2 y)z_2 + \dots + (x, J_n y)z_n, & \text{if } x, y \text{ in } \mathfrak{v}_1 \\ (x, 2J_1 y)z_1 + (x, J_2 y)z_2 + \dots + (x, J_n y)z_n, & \text{if } x, y \text{ in } \mathfrak{v}_2 \\ 0 & \text{otherwise.} \end{cases}$$

Then it is not difficult to check that $\mathfrak{m}$ is not an $\tilde{H}$-type algebra.

References

[B] A. Besse, *Einstein Manifolds*, Springer-Verlag, New York-Berlin-Heidelberg-Tokyo, 1987.

[BtD] T. Bröcker and T. tom Dieck, *Representations of compact Lie groups*, Springer-Verlag, New York-Berlin-Heidelberg-Tokyo, 1985.

[C] M. Cowling, *Harmonic analysis on some nilpotent groups (with applications to the representation theory of some semisimple Lie groups)*, in "Topics in Modern Harmonic Analysis, Vol. 1", Istituto Nazionale di Alta Matematica, Roma, 81–123.

[Ga] F. Gantmacher, *The theory of matrices*, GITTL, Moscow, 1953.

[Gau] M. A. Gauger, *On the classification of metabelian Lie algebras*, Trans. Amer. Math. Soc. **179** (1973), 293–328.

[Hu] D. Husemoller, *Fibre bundles*, Springer-Verlag, New York-Heidelberg-Berlin, 1975.

[Ka1] A. Kaplan, *Fundamental solutions for a class of hypoelliptic PDE generated by composition of quadratic forms*, Trans. Amer. Math. Soc. **258** (1980), 147–153.

[Ka2] ———, *Riemannian nilmanifolds attached to Clifford modules*, Geom. Dedicata **11** (1981), 127–136.

[KaR] A. Kaplan and F. Ricci, *Harmonic analysis on groups of Heisenberg type*, in "Harmonic Analysis". Lecture Notes in Math. 992, Springer-Verlag, Berlin-Heidelberg-New York (1983), 416–435.

[KaS] A. Kaplan and L. Saal, personal communication.

[LT] F. Levstein and A. Tiraboschi, *Classes of 2-step nilpotent Lie algebras*, submitted to publication.

[MW] C. Moore and J. Wolf, *Square integrable representations of nilpotent groups*, Trans. Amer. Math. Soc. **185** (1973), 445–462.

[R] ———, *Commutative algebras of invariant functions on groups of Heisenberg type*, J. London Math. Soc. (2) **32** (1985), 265–271.

[TA] H. W. Turnbull and A. C. Aitken, *An introduction to the theory of canonical matrices*, Dover, New York, 1961.

[Th] R. Thompson, *Pencils of complex and real symmetric and skew matrices*, Linear Alg. and its Applications **147** (1991), 323–371.

[TV] F. Tricerri and L. Vanhecke, *Homogeneous structures on Riemannian manifolds*, London Math. Soc. Lecture Note Series 83, Cambridge Univ. Press, Cambridge, 1983.

[W] T. Wolter, *Einstein metrics on solvable groups*, Math. Zeit. **206** (1991), 457–471.

FAMAF
Ciudad Universitaria
5000 Córdoba,
Argentina
email: F. L.: levstein@mate.uncor.edu,
A. T.:tirabo@mate.uncor.edu

EQUIVARIANT DERIVED CATEGORIES, ZUCKERMAN FUNCTORS AND LOCALIZATION

DRAGAN MILIČIĆ AND PAVLE PANDŽIĆ

INTRODUCTION

In this paper we revisit some now classical constructions of modern representation theory: Zuckerman's cohomological construction and the localization theory of Bernstein and Beilinson. These constructions made an enormous impact on our understanding of representation theory during the last decades (see, for example, [19]). Our present approach and interest is slightly different than usual. We approach these constructions from the point of view of a student in homological algebra and not representation theory. Therefore, we drop certain assumptions natural from the point of view of representation theorists and stress some unifying principles.

Although both constructions have common heuristic origins in various attempts to generalize the classical Borel-Weil-Bott realization of irreducible finite-dimensional representations of compact Lie groups, they are remarkably different in technical details. Still, the duality theorem of Hecht, Miličić, Schmid and Wolf [11] indicated that there must exist a strong common thread between these constructions. This paper is an attempt to explain the unifying homological principles behind these constructions, which lead to the duality theorem as a formally trivial consequence.

In the first section we present an elementary and self-contained introduction to a generalization of the Zuckerman construction. Let $\mathfrak{g}$ be a complex semisimple Lie algebra and K a complex algebraic group that is a finite covering of a closed algebraic subgroup of the complex algebraic group $\mathrm{Int}(\mathfrak{g})$. Let $\mathcal{M}(\mathfrak{g}, K)$ be the category of Harish-Chandra modules for the pair $(\mathfrak{g}, K)$. Let H be a closed algebraic subgroup of K. Zuckerman observed that the forgetful functor $\mathcal{M}(\mathfrak{g}, K) \to \mathcal{M}(\mathfrak{g}, H)$ has a right adjoint $\Gamma_{K,H} : \mathcal{M}(\mathfrak{g}, H) \to \mathcal{M}(\mathfrak{g}, K)$. The functor $\Gamma_{K,H}$ is left exact, and its right-derived modules are the core of Zuckerman's approach.

As we mentioned before, Zuckerman's inspiration was in the Borel-Weil-Bott theorem and he wanted to construct a formal analogue of the sheaf cohomology functor. Therefore in his approach it was natural to assume that both groups K and H are reductive. In our exposition we drop this assumption. The main result of this section is a formula for derived Zuckerman functors which is a generalization of a result of Duflo and Vergne [9]. This formula

---

1991 *Mathematics Subject Classification.* Primary 22E46.

allows us to realize the derived category of $\mathcal{M}(\mathfrak{g}, K)$ as a full triangulated subcategory of the derived category of $\mathcal{M}(\mathfrak{g}, L)$ for a Levi factor L of K.

To relate Zuckerman's construction to the localization theory, we have to reinterpret the construction in terms of the equivariant derived categories. This is done in the second and third sections, where we review the construction of the equivariant derived category $D(\mathfrak{g}, K)$ of Harish-Chandra modules due to Beilinson and Ginzburg, construct the equivariant analogues of Zuckerman functors and relate them to the classical Zuckerman construction [17].

In the fourth section we discuss the localization of the previous constructions. The idea of Beilinson and Bernstein was to generalize the Borel-Weil-Bott theorem in the algebro-geometric setting to an equivalence of categories of modules over the enveloping algebra of $\mathfrak{g}$ with categories of sheaves of $\mathcal{D}$-modules on the flag variety X of $\mathfrak{g}$. This approach ties representation theory with the theory of $\mathcal{D}$-modules. To relate this construction with the constructions of the preceding sections, we define the equivariant derived category of Harish-Chandra sheaves on the flag variety X of $\mathfrak{g}$ and discuss the corresponding version of the localization theory. This allows us, on purely formal grounds, to construct a geometric version of the equivariant Zuckerman functor.

The final section contains a sketch of the proof of the duality theorem of [11]. In our approach, this is just a formula for the cohomology of standard Harish-Chandra sheaves on X in terms of derived Zuckerman functors. While the original proof required a tedious and not very illuminating calculation, the argument sketched here is just slightly more than a diagram chase. It is inspired by Bernstein's argument to prove a special case of the duality theorem.

In this paper we freely use the formalism of derived categories. We think that this should be a necessary part of the toolbox of any representation theorist. An interested reader lacking this background should find [10] and [18] invaluable references.

We would like to thank David Vogan for his remarks and questions which led to considerable improvement and clarification of the results in the first and second sections.

1. Zuckerman functors

Let $\mathfrak{g}$ be a complex semisimple Lie algebra and K an algebraic group acting on $\mathfrak{g}$ by a morphism $\phi : K \longrightarrow \operatorname{Int}(\mathfrak{g})$ such that its differential $\mathfrak{k} \longrightarrow \mathfrak{g}$ is an injection. In this situation we can identify $\mathfrak{k}$ with a Lie subalgebra of $\mathfrak{g}$. A Harish-Chandra module (V, π) for the pair $(\mathfrak{g}, K)$ is

(HC1) a $\mathcal{U}(\mathfrak{g})$-module;

(HC2) an algebraic K-module, i.e., V is a union of finite-dimensional K-invariant subspaces V_i on which K acts algebraically, that is, via algebraic group morphisms $K \longrightarrow GL(V_i)$;

(HC3) the actions of $\mathfrak{g}$ and K are compatible; i.e.,

(a) the differential of the K-action agrees with the action of $\mathfrak{k}$ as a subalgebra of $\mathfrak{g}$;

(b)
$$\pi(k)\pi(\xi)\pi(k^{-1})v = \pi(\phi(k)\xi)v$$
for all $k \in K$, $\xi \in \mathfrak{g}$ and $v \in V$.

A morphism of Harish-Chandra modules is a linear map that intertwines the actions of $\mathfrak{g}$ and K. If V and W are two Harish-Chandra modules for $(\mathfrak{g}, K)$, $\mathrm{Hom}_{(\mathfrak{g},K)}(V, W)$ denotes the space of all morphisms between V and W. Let $\mathcal{M}(\mathfrak{g}, K)$ be the category of Harish-Chandra modules for the pair $(\mathfrak{g}, K)$. This is clearly an abelian $\mathbb{C}$-category.

Let T be a closed algebraic subgroup of K. Then we have a natural forgetful functor $\mathcal{M}(\mathfrak{g}, K) \longrightarrow \mathcal{M}(\mathfrak{g}, T)$. The Zuckerman functor $\Gamma_{K,T} : \mathcal{M}(\mathfrak{g}, T) \longrightarrow \mathcal{M}(\mathfrak{g}, K)$ is by definition the right adjoint functor to this forgetful functor.

First we describe a construction of this functor. Let $R(K)$ be the ring of regular functions on K. Then for any vector space V, we can view $R(K) \otimes V$ as the vector space of all regular maps from K into V and denote it by $R(K, V)$. We define an algebraic representation ρ of K on $R(K, V)$ as the tensor product of the right regular representation of K on $R(K)$ and trivial action on V.

Now let V be an algebraic K-module. Then we have the natural *matrix coefficient* map $c : V \longrightarrow R(K, V)$ which maps a vector $v \in V$ into the function $k \longmapsto \pi(k)v$. Clearly, c is an injective morphism of K-modules.

If we define the representation λ of K on $R(K, V)$ as the tensor product of the left regular representation of K on $R(K)$ with the natural action on V, it commutes with the action ρ. The image of c is in the space of all λ-invariant functions in $R(K, V)$. Moreover, c is an isomorphism of V onto the space of λ-invariants, and the inverse morphism is the evaluation at $1 \in K$.

If V is a Harish-Chandra module in $\mathcal{M}(\mathfrak{g}, K)$, we define a representation ν of $\mathfrak{g}$ on $R(K, V)$ by
$$(\nu(\xi)F)(k) = \pi(\phi(k)\xi)F(k), \quad k \in K,$$
for $\xi \in \mathfrak{g}$ and $v \in V$. By a direct calculation, we see that $c : V \longrightarrow R(K, V)$ intertwines $\mathfrak{g}$-actions. The representation ν also commutes with the λ-action.

Therefore, the Harish-Chandra module V can be reconstructed from the image of the matrix coefficient map. We use this observation to construct the Zuckerman functor.

Let W be a Harish-Chandra module in $\mathcal{M}(\mathfrak{g}, T)$. Then we can define the structure of a $\mathcal{U}(\mathfrak{g})$-module on $R(K, W)$ by the ν-action, and the structure of an algebraic K-module by the ρ-action as above. The action ν is K-equivariant, i.e.,
$$\rho(k)\nu(\xi)\rho(k^{-1}) = \nu(\phi(k)\xi)$$
for $\xi \in \mathcal{U}(\mathfrak{g})$ and $k \in K$. Let λ be the tensor product of the left regular representation of $\mathfrak{k}$ and T on $R(K)$ with the natural action on W. This defines a structure of Harish-Chandra module for $(\mathfrak{k}, T)$ on $R(K, W)$. One can check that these actions of $\mathfrak{k}$ and T commute with the representations ν and ρ.

Therefore, the subspace of $(\mathfrak{k}, T)$-invariants
$$\Gamma_{K,T}(W) = R(K, W)^{(\mathfrak{k},T)}$$

in $R(K,W)$ (with respect to λ) is a $\mathfrak{g}$- and K-submodule.

1.1. Lemma. *Let W be a Harish-Chandra module for $(\mathfrak{g},T)$. Then $\Gamma_{K,T}(W)$ is a Harish-Chandra module for $(\mathfrak{g},K)$.*

Proof. We already mentioned that ν is K-equivariant. Also, for $\xi \in \mathfrak{k}$ and $F \in \Gamma_{K,T}(W)$ we have

$$\begin{aligned}(\rho(\xi)F)(k) = \tfrac{d}{dt}F(k\exp(t\xi))\big|_{t=0} &= \tfrac{d}{dt}F(\exp(t(\phi(k)\xi))k)\big|_{t=0}\\ &= \pi(\phi(k)\xi)F(k) = (\nu(\xi)F)(k), \quad k \in K,\end{aligned}$$

since F is λ-invariant. Therefore, the differential of ρ agrees with the restriction of ν to $\mathfrak{k}$ on $\Gamma_{K,T}(W)$, i.e., the actions ν and ρ define a structure of Harish-Chandra module on $\Gamma_{K,T}(W)$. □

Let V and W be two Harish-Chandra modules for $(\mathfrak{g},T)$ and $\alpha \in \operatorname{Hom}_{(\mathfrak{g},T)}(V,W)$. Then α induces a linear map $1 \otimes \alpha : R(K,V) \longrightarrow R(K,W)$. Clearly, $1 \otimes \alpha$ intertwines the actions ν, ρ and λ on these modules. Hence, it induces a morphism $\Gamma_{K,T}(\alpha) : \Gamma_{K,T}(V) \longrightarrow \Gamma_{K,T}(W)$. It follows that $\Gamma_{K,T}$ is an additive functor from $\mathcal{M}(\mathfrak{g},T)$ into $\mathcal{M}(\mathfrak{g},K)$.

Let V be a Harish-Chandra module in $\mathcal{M}(\mathfrak{g},K)$. Then, as we saw above, the matrix coefficient map c_V of V is a $(\mathfrak{g},K)$-morphism of V into $\Gamma_{K,T}(V)$. It is easy to check that the maps c_V actually define a natural transformation of the identity functor on $\mathcal{M}(\mathfrak{g},K)$ into the composition of $\Gamma_{K,T}$ with the forgetful functor.

On the other hand, let W be a Harish-Chandra module for $(\mathfrak{g},T)$ and $e_W : \Gamma_{K,T}(W) \longrightarrow W$ the linear map given by $e_W(F) = F(1)$. Then e_W is a $(\mathfrak{g},T)$-morphism from $\Gamma_{K,T}(W)$ into W. Clearly, the maps e_W define a natural transformation of the composition of the forgetful functor with the functor $\Gamma_{K,T}$ into the identity functor on $\mathcal{M}(\mathfrak{g},T)$.

Using these natural transformations, we get the following result.

1.2. Proposition. *The functor $\Gamma_{K,T} : \mathcal{M}(\mathfrak{g},T) \longrightarrow \mathcal{M}(\mathfrak{g},K)$ is right adjoint to the forgetful functor from $\mathcal{M}(\mathfrak{g},K)$ into $\mathcal{M}(\mathfrak{g},T)$.*

Proof. Let V be a Harish-Chandra module in $\mathcal{M}(\mathfrak{g},K)$ and W a Harish-Chandra module in $\mathcal{M}(\mathfrak{g},T)$. For $\alpha \in \operatorname{Hom}_{(\mathfrak{g},T)}(V,W)$, the composition $\bar{\alpha} = \Gamma_{K,T}(\alpha) \circ c_V : V \longrightarrow \Gamma_{K,T}(W)$ is in $\operatorname{Hom}_{(\mathfrak{g},K)}(V,\Gamma_{K,T}(W))$. Thus we have a linear map $\alpha \longmapsto \bar{\alpha}$ of $\operatorname{Hom}_{(\mathfrak{g},T)}(V,W)$ into $\operatorname{Hom}_{(\mathfrak{g},K)}(V,\Gamma_{K,T}(W))$.

Also, if $\beta \in \operatorname{Hom}_{(\mathfrak{g},K)}(V,\Gamma_{K,T}(W))$, $\tilde{\beta} = e_W \circ \beta \in \operatorname{Hom}_{(\mathfrak{g},T)}(V,W)$. Thus we have a linear map $\beta \longmapsto \tilde{\beta}$ of $\operatorname{Hom}_{(\mathfrak{g},K)}(V,\Gamma_{K,T}(W))$ into $\operatorname{Hom}_{(\mathfrak{g},T)}(V,W)$.

By a direct calculation, we see that these maps are inverse to each other. □

The functor $\Gamma_{K,T}$ is called the *Zuckerman functor.*

Let $\Gamma_K = \Gamma_{K,\{1\}}$. Since Γ_K is right adjoint to the forgetful functor from $\mathcal{M}(\mathfrak{g},K)$ into $\mathcal{M}(\mathfrak{g})$, it maps injectives into injectives. This has the following consequence.

1.3. Lemma. *The category $\mathcal{M}(\mathfrak{g}, K)$ has enough injectives.*

Proof. Let V be an object in $\mathcal{M}(\mathfrak{g}, K)$. Then there exists an injective object I in $\mathcal{M}(\mathfrak{g})$ and a $\mathfrak{g}$-monomorphism $i : V \longrightarrow I$. Since Γ_K is left exact, $\Gamma_K(i) : \Gamma_K(V) \longrightarrow \Gamma_K(I)$ is also a monomorphism. By the arguments in the proof of 1.2, the adjunction morphism $V \longrightarrow \Gamma_K(V)$ is also a monomorphism. Therefore, the composition of these two morphisms defines a monomorphism $V \longrightarrow \Gamma_K(I)$ of V into an injective object in $\mathcal{M}(\mathfrak{g}, K)$. □

Let U be an algebraic representation of K. Then $P(U) = \mathcal{U}(\mathfrak{g}) \otimes_{\mathcal{U}(\mathfrak{k})} U$, with the K-action given by the tensor product of the action ϕ on $\mathcal{U}(\mathfrak{g})$ and the natural action on U, is an algebraic representation of K. Moreover, it also has a natural structure of a $\mathcal{U}(\mathfrak{g})$-module, given by left multiplication in the first factor. It is straightforward to check that $P(U)$ is a Harish-Chandra module. Therefore, P is an exact functor from the category of algebraic representations of K into $\mathcal{M}(\mathfrak{g}, K)$. In addition,

$$\mathrm{Hom}_{(\mathfrak{g},K)}(P(U), V) = \mathrm{Hom}_K(U, V)$$

for any Harish-Chandra module V, i.e., P is left adjoint to the forgetful functor from $\mathcal{M}(\mathfrak{g}, K)$ into the category of algebraic representations of K.

Assume now that K is reductive. Then the category of algebraic representations of K is semisimple, and every object in it is projective. Therefore, Harish-Chandra modules $P(U)$ are projective in $\mathcal{M}(\mathfrak{g}, K)$ for arbitrary algebraic representation U of K.

This leads to the following result.

1.4. Lemma. *If K is a reductive algebraic group, the category $\mathcal{M}(\mathfrak{g}, K)$ has enough projectives.*

In addition, every finitely generated object in $\mathcal{M}(\mathfrak{g}, K)$ is a quotient of a finitely generated projective object.

Proof. This is analogous to the proof of 1.3, using the fact that for any Harish-Chandra module V, the adjointness morphism $P(V) \longrightarrow V$ is surjective. The last remark is obvious from the previous discussion. □

Let $D^+(\mathcal{M}(\mathfrak{g}, K))$ and $D^+(\mathcal{M}(\mathfrak{g}, T))$ be the derived categories of complexes bounded from below corresponding to $\mathcal{M}(\mathfrak{g}, K)$ and $\mathcal{M}(\mathfrak{g}, T)$. Since the category $\mathcal{M}(\mathfrak{g}, T)$ has enough injectives, there exists the derived functor

$$R\Gamma_{K,T} : D^+(\mathcal{M}(\mathfrak{g}, T)) \longrightarrow D^+(\mathcal{M}(\mathfrak{g}, K))$$

of $\Gamma_{K,T}$. Also, $R\Gamma_{K,T}$ is the right adjoint of the natural "forgetful" functor from $D^+(\mathcal{M}(\mathfrak{g}, K))$ into $D^+(\mathcal{M}(\mathfrak{g}, T))$.

This immediately implies the following remark. Let H be a closed algebraic subgroup of K such that $T \subset H \subset K$. Then the functors $R\Gamma_{K,H} \circ R\Gamma_{H,T}$ and $R\Gamma_{K,T}$ are right adjoint to the natural functor $D^+(\mathcal{M}(\mathfrak{g}, K)) \longrightarrow D^+(\mathcal{M}(\mathfrak{g}, T))$, hence they are isomorphic, i.e., we have the following theorem.

1.5. Theorem. $R\Gamma_{K,T} = R\Gamma_{K,H} \circ R\Gamma_{H,T}.$

The next theorem is our version of the main result of [9]. In the following we assume that T is in addition reductive. For any Harish-Chandra module W in $\mathcal{M}(\mathfrak{k}, T)$ we denote by

$$H^p(\mathfrak{k}, T; W) = \operatorname{Ext}^p_{(\mathfrak{k},T)}(\mathbb{C}, W)$$

the pth relative Lie algebra cohomology group of W.

1.6. Theorem. *Assume that T is reductive. Let V be a Harish-Chandra module in $\mathcal{M}(\mathfrak{g}, T)$. Then*

$$R^p\Gamma_{K,T}(V) = H^p(\mathfrak{k}, T; R(K, V))$$

for $p \in \mathbb{Z}_+$, where the relative Lie algebra cohomology is calculated with respect to the λ-action.

To prove this result we need some preparation.

As we mentioned before, the category $\mathcal{M}(\mathfrak{g}, T)$ has enough injectives. Let V be a Harish-Chandra module for $(\mathfrak{g}, T)$ and $V \longrightarrow I^{\cdot}$ a right resolution of V by injective modules in $\mathcal{M}(\mathfrak{g}, T)$. Then

$$R^p\Gamma_{K,T}(V) = H^p(\operatorname{Hom}_{(\mathfrak{k},T)}(\mathbb{C}, R(K, I^{\cdot}))).$$

Here $R(K, I^{\cdot})$ are viewed as $(\mathfrak{k}, T)$-modules with respect to the actions λ of $\mathfrak{k}$ and T. To prove the theorem it is enough to prove that for any injective object I in $\mathcal{M}(\mathfrak{g}, T)$, the module $R(K, I)$ viewed as a Harish-Chandra module for $(\mathfrak{k}, T)$ with respect to the action λ, is acyclic for the functor $\operatorname{Hom}_{(\mathfrak{k},T)}(\mathbb{C}, -)$. This is proved in the next lemma.

1.7. Lemma. *Let I be an injective object in $\mathcal{M}(\mathfrak{g}, T)$. Then $R(K, I)$, viewed as a Harish-Chandra module for $(\mathfrak{k}, T)$ with respect to the action λ, satisfies*

$$\operatorname{Ext}^p_{(\mathfrak{k},T)}(\mathbb{C}, R(K, I)) = 0$$

for $p > 0$.

Proof. First, let S be a module in $\mathcal{M}(\mathfrak{k}, T)$. Then, we can define the action of $\mathfrak{g}$ on $\mathcal{U}(\mathfrak{g}) \otimes_{\mathcal{U}(\mathfrak{k})} S$ as left multiplication in the first variable and the action of T as the tensor product of the adjoint action on the first factor with the natural action on the second factor. It is easy to check that for any Harish-Chandra module W in $\mathcal{M}(\mathfrak{g}, T)$ we have

$$\operatorname{Hom}_{(\mathfrak{g},T)}(\mathcal{U}(\mathfrak{g}) \otimes_{\mathcal{U}(\mathfrak{k})} S, W) = \operatorname{Hom}_{(\mathfrak{k},T)}(S, W).$$

Therefore, the forgetful functor from $\mathcal{M}(\mathfrak{g}, T)$ into $\mathcal{M}(\mathfrak{k}, T)$ is the right adjoint of the exact functor $S \longrightarrow \mathcal{U}(\mathfrak{g}) \otimes_{\mathcal{U}(\mathfrak{k})} S$. Hence it preserves injectives; in particular, I is an injective object in $\mathcal{M}(\mathfrak{k}, T)$.

Second, if F is a finite-dimensional algebraic representation of K and F^* its contragredient, we have

$$\mathrm{Hom}_{(\mathfrak{k},T)}(R, F \otimes W) = \mathrm{Hom}_{(\mathfrak{k},T)}(R \otimes F^*, W)$$

for any two Harish-Chandra modules R and W in $\mathcal{M}(\mathfrak{k}, T)$. Therefore, $W \longmapsto F \otimes W$ is the right adjoint of the exact functor $R \longmapsto R \otimes F^*$. It follows that $F \otimes I$ is an injective object in $\mathcal{M}(\mathfrak{k}, T)$.

Let U be an algebraic representation of K considered as a Harish-Chandra module for $(\mathfrak{k}, T)$. Then on $U \otimes I$ we can define the tensor product structure of Harish-Chandra module for $(\mathfrak{k}, T)$. Since U is an algebraic representation, U is a union of finite-dimensional K-invariant subspaces U_j, $j \in J$; i.e., $U = \varinjlim_{j \in J} U_j$. Therefore, for any R in $\mathcal{M}(\mathfrak{k}, T)$, finitely generated over $\mathcal{U}(\mathfrak{k})$,

$$\mathrm{Hom}_{(\mathfrak{k},T)}(R, U \otimes I) = \varinjlim_{j \in J} \mathrm{Hom}_{(\mathfrak{k},T)}(R, U_j \otimes I).$$

By 1.4, since T is reductive, $\mathbb{C}$ has a left resolution $P^{\cdot}$ by projective finitely generated Harish-Chandra modules in $\mathcal{M}(\mathfrak{k}, T)$. Therefore, since the direct limit functor is exact,

$$\begin{aligned}\mathrm{Ext}^p_{(\mathfrak{k},T)}(\mathbb{C}, U \otimes I) &= H^p(\mathrm{Hom}_{(\mathfrak{k},T)}(P^{\cdot}, U \otimes I)) \\ &= H^p(\varinjlim_{j \in J} \mathrm{Hom}_{(\mathfrak{k},T)}(P^{\cdot}, U_j \otimes I)) = \varinjlim_{j \in J} H^p(\mathrm{Hom}_{(\mathfrak{k},T)}(P^{\cdot}, U_j \otimes I)) \\ &= \varinjlim_{j \in J} \mathrm{Ext}^p_{(\mathfrak{k},T)}(\mathbb{C}, U_j \otimes I).\end{aligned}$$

But the last expression is zero, since $U_j \otimes I$ is an injective $(\mathfrak{k}, T)$-module by the above discussion. In particular,

$$\mathrm{Ext}^p_{(\mathfrak{k},T)}(\mathbb{C}, U \otimes I) = 0$$

for $p > 0$. Applying this for $U = R(K)$, we get 1.7, and thus also 1.6. □

Now we want to study one of the adjointness morphisms attached to the adjoint pair consisting of the forgetful functor and the derived Zuckerman functor $R\Gamma_{K,T}$.

Let V be a Harish-Chandra module in $\mathcal{M}(\mathfrak{g}, K)$, with action π. We can view it as an object in $\mathcal{M}(\mathfrak{g}, T)$. We want to calculate the derived Zuckerman functors $R^p\Gamma_{K,T}(V)$, $p \in \mathbb{Z}_+$. To do this, we have to calculate the relative Lie algebra cohomology modules from 1.6. The calculation is based on the following observations. First, the matrix coefficient map $V \longrightarrow R(K, V)$ defines a linear map γ of $R(K, V)$ into itself, given by

$$\gamma(F)(k) = c(F(k))(k) = \pi(k)F(k), \quad k \in K,$$

for $F \in R(K,V)$. This map is clearly an isomorphism of linear spaces and its inverse is given by

$$\delta(F)(k) = \pi(k^{-1})F(k), \quad k \in K,$$

for $F \in R(K,V)$. We can define the following actions on $R(K,V)$: the representation $1 \otimes \pi$ of $\mathfrak{g}$ which is the tensor product of the trivial representation on $R(K)$ and the natural representation on V, the representation τ of K which is the tensor product of the right regular representation of K on $R(K)$ with the natural representation on V, and the representation μ of K which is the tensor product of the left regular representation of K on $R(K)$ with the trivial representation on V.

By a direct calculation we check that the following result holds.

1.8. Lemma. *For any V in $\mathcal{M}(\mathfrak{g},K)$, the linear map $\gamma : R(K,V) \to R(K,V)$ is a linear space automorphism. Also,*

(i) *γ intertwines the representation $1 \otimes \pi$ of $\mathfrak{g}$ with ν;*
(ii) *γ intertwines the representation τ of K with ρ;*
(iii) *γ intertwines the representation μ of K with λ.*

Therefore, to calculate $H^p(\mathfrak{k},T;R(K,V))$, we can assume that the actions of $\mathfrak{k}$ and T are given by μ. In this case, we have

$$R^p\Gamma_{K,T}(V) = H^p(\mathfrak{k},T;R(K,V)) = H^p(\mathfrak{k},T;R(K)) \otimes V.$$

Here the relative Lie algebra cohomology of $R(K)$ is calculated with respect to the left regular action. The action of $\mathfrak{g}$ on the last module is given as the tensor product of the trivial action on the first factor and the natural action on V, while the action of K is given as the tensor product of the action induced by the right regular representation on $R(K)$ with the natural action on V.

Let $K^\dagger$ be the subgroup of K generated by the identity component K_0 of K and T.

Assume first that K is reductive. Then the left regular representation on $R(K)$ is a direct sum of irreducible finite-dimensional representations of $K^\dagger$. It is well known that, for any nontrivial finite-dimensional irreducible representation F of $K^\dagger$, the relative Lie algebra cohomology modules $H^p(\mathfrak{k},T;F)$ are zero for any $p \in \mathbb{Z}_+$. Let $\mathrm{Ind}_{K^\dagger}^K(1)$ be the space of functions on K that are constant on right $K^\dagger$-cosets. It follows that the inclusion of $\mathrm{Ind}_{K^\dagger}^K(1)$ into $R(K)$ induces isomorphisms

$$H^p(\mathfrak{k},T;\mathrm{Ind}_{K^\dagger}^K(1)) = H^p(\mathfrak{k},T;R(K)), \quad p \in \mathbb{Z}_+.$$

Therefore, in this case we have

$$R^p\Gamma_{K,T}(V) = H^p(\mathfrak{k},T;\mathrm{Ind}_{K^\dagger}^K(1)) \otimes V = H^p(\mathfrak{k},T;\mathbb{C}) \otimes \mathrm{Ind}_{K^\dagger}^K(1) \otimes V, \quad p \in \mathbb{Z}_+,$$

and the action of K is the tensor product of the trivial action on the first factor, the right regular action on the second factor and the natural action on the third factor. The action of $\mathfrak{g}$ is given by the tensor product of trivial actions on the first two factors with the natural action on the third factor.

Now we drop the assumption that K is reductive. Denote by U the unipotent radical in K. Since T is reductive, $T \cap U = \{1\}$. Moreover, there exists a Levi factor L of K, such that $T \subset L$. Denote by $L^\dagger$ the subgroup of L generated by the identity component L_0 of L and T. By 1.5, we have the spectral sequence

$$R^p\Gamma_{K,L}(R^q\Gamma_{L,T}(V)) \Rightarrow R^{p+q}\Gamma_{K,T}(V).$$

Since L is reductive, by the above discussion we have

$$R^q\Gamma_{L,T}(V) = H^q(\mathfrak{l}, T; \mathbb{C}) \otimes \operatorname{Ind}_{L^\dagger}^L(1) \otimes V, \quad q \in \mathbb{Z}_+.$$

Since the restriction to L induces an isomorphism of $\operatorname{Ind}_{K^\dagger}^K(1)$ with $\operatorname{Ind}_{L^\dagger}^L(1)$, we also have

$$R^q\Gamma_{L,T}(V) = H^q(\mathfrak{l}, T; \mathbb{C}) \otimes \operatorname{Ind}_{K^\dagger}^K(1) \otimes V, \quad q \in \mathbb{Z}_+.$$

Hence, it remains to study $R^p\Gamma_{K,L}(V)$, $p \in \mathbb{Z}_+$, for V in $\mathcal{M}(\mathfrak{g}, K)$. Let $\mathfrak{u}$ and $\mathfrak{l}$ be the Lie algebras of U and L respectively. Let $N^\cdot(\mathfrak{u})$ be the standard complex of $\mathfrak{u}$. Then $N^p(\mathfrak{u}) = \mathcal{U}(\mathfrak{u}) \otimes_{\mathbb{C}} \bigwedge^{-p}\mathfrak{u}$, $p \in \mathbb{Z}$, are algebraic K-modules for the tensor products of the adjoint actions on $\mathcal{U}(\mathfrak{u})$ and $\bigwedge^\cdot \mathfrak{u}$. Therefore, if we write $N^p(\mathfrak{u})$ as $\mathcal{U}(\mathfrak{k}) \otimes_{\mathcal{U}(\mathfrak{l})} \bigwedge^{-p}\mathfrak{u}$, $p \in \mathbb{Z}$, we see that they are algebraic L-modules and also $\mathcal{U}(\mathfrak{k})$-modules for the left multiplication in the first factor. In this way we see that

$$\mathcal{U}(\mathfrak{k}) \otimes_{\mathcal{U}(\mathfrak{l})} \textstyle\bigwedge^{-\cdot}\mathfrak{u}$$

is a projective resolution of $\mathbb{C}$ in $\mathcal{M}(\mathfrak{k}, L)$. It is usually called the relative standard complex for the pair $(\mathfrak{k}, \mathfrak{l})$. It follows that for any object W in $\mathcal{M}(\mathfrak{k}, L)$ we have

$$\begin{aligned} H^p(\mathfrak{k}, L; W) = \operatorname{Ext}^p_{(\mathfrak{k},L)}(\mathbb{C}, W) &= H^p(\operatorname{Hom}_{(\mathfrak{k},L)}(\mathcal{U}(\mathfrak{k}) \otimes_{\mathcal{U}(\mathfrak{l})} \textstyle\bigwedge^\cdot \mathfrak{u}, W)) \\ &= H^p(\operatorname{Hom}_L(\textstyle\bigwedge^\cdot \mathfrak{u}, W)) = H^p(\operatorname{Hom}_{\mathbb{C}}(\textstyle\bigwedge^\cdot \mathfrak{u}, W)^L) \end{aligned}$$

for $p \in \mathbb{Z}_+$. Since L is reductive, we have

$$H^p(\mathfrak{k}, L; W) = H^p(\mathfrak{u}, W)^L, \quad p \in \mathbb{Z}_+.$$

In particular, we have

$$H^p(\mathfrak{k}, L; R(K)) = H^p(\mathfrak{u}, R(K))^L, \quad p \in \mathbb{Z}_+.$$

The quotient map $K \longrightarrow K/U$ induces a natural inclusion of $R(K/U)$ into $R(K)$ as algebraic K-modules for the left regular action.

1.9. Lemma. *We have*

$$H^p(\mathfrak{u}, R(K)) = \begin{cases} R(K/U) \text{ if } p = 0; \\ 0 \text{ if } p > 0. \end{cases}$$

Proof. We prove a slightly more general statement which allows induction. Let N be a normal unipotent subgroup of K and $\mathfrak{n}$ the Lie algebra of N. We claim that

$$H^p(\mathfrak{n}, R(K)) = \begin{cases} R(K/N) \text{ if } p = 0; \\ 0 \text{ if } p > 0. \end{cases}$$

Since $H^0(\mathfrak{n}, R(K))$ consists of functions constant on N-cosets, the statement for $p = 0$ is evident.

Hence, we just have to establish the vanishing for $p > 0$. The map $(u, l) \longmapsto u \cdot l$ from $U \times L$ into K is an isomorphism of varieties. Moreover, if we assume that N acts by left multiplication on the first factor in $U \times L$ and by left multiplication on K, it is an N-equivariant isomorphism. Therefore, the corresponding algebraic representations of N on $R(K)$ and $R(U \times L) = R(U) \otimes R(L)$ are isomorphic. This implies that

$$H^p(\mathfrak{n}, R(K)) = H^p(\mathfrak{n}, R(U) \otimes R(L)) = H^p(\mathfrak{n}, R(U)) \otimes R(L), \quad p \in \mathbb{Z}_+,$$

as linear spaces. Assume first that N is abelian. Since U is an affine space, $R(U)$ is a polynomial algebra. Moreover, $H^p(\mathfrak{n}, R(U))$, $p \in \mathbb{Z}_+$, is just the cohomology of the Koszul complex with coefficients in this algebra. By the polynomial version of the Poincaré lemma we see that $H^p(\mathfrak{n}, R(U)) = 0$ for $p > 0$.

Now we proceed by induction on $\dim N$. If $\dim N > 0$, the commutator subgroup $N' = (N, N)$ is a unipotent group and $\dim N' < \dim N$. Moreover, N' is a normal subgroup of K. Therefore, its Lie algebra $\mathfrak{n}' = [\mathfrak{n}, \mathfrak{n}]$ is an ideal in $\mathfrak{k}$. By the Hochschild-Serre spectral sequence of Lie algebra cohomology we have

$$H^p(\mathfrak{n}/\mathfrak{n}', H^q(\mathfrak{n}', R(K))) \Rightarrow H^{p+q}(\mathfrak{n}, R(K)).$$

Also, by the induction assumption this spectral sequence collapses, i.e., we have

$$H^p(\mathfrak{n}/\mathfrak{n}', R(K/N')) = H^p(\mathfrak{n}, R(K))$$

for $p \in \mathbb{Z}_+$. Finally, by the first part of the proof, the left side is 0 if $p > 0$ since N/N' is abelian. □

As an immediate consequence, we have

$$H^p(\mathfrak{k}, L; R(K)) = H^p(\mathfrak{u}, R(K))^L = \begin{cases} R(K/U)^L = R(L)^L = \mathbb{C} \text{ for } p = 0; \\ 0 \text{ for } p > 0. \end{cases}$$

It follows that for V in $\mathcal{M}(\mathfrak{g}, K)$, we have

$$\Gamma_{K,L}(V) = V$$

and

$$R^p\Gamma_{K,L}(V) = 0 \text{ for } p > 0.$$

Therefore, the spectral sequence we considered earlier collapses, and we immediately get the following consequence.

1.10. Proposition. *Let T be a reductive subgroup of K and L a Levi factor of K containing T. Let V be a module in $\mathcal{M}(\mathfrak{g}, K)$. Then we have*

$$R^p\Gamma_{K,T}(V) = H^p(\mathfrak{l}, T; \mathbb{C}) \otimes \operatorname{Ind}_{K^\dagger}^{K}(1) \otimes V, \quad p \in \mathbb{Z}_+.$$

Now we prove a result about derived categories of Harish-Chandra modules which reduces the case of general pairs $(\mathfrak{g}, K)$ to the study of categories with reductive group K.

Assume that K is arbitrary. Let U be the unipotent radical and L a Levi factor of K. The category $\mathcal{M}(\mathfrak{g}, K)$ is a subcategory of $\mathcal{M}(\mathfrak{g}, L)$. Moreover, since U is connected, any $(\mathfrak{g}, L)$-morphism between two objects in $\mathcal{M}(\mathfrak{g}, K)$ is automatically a $(\mathfrak{g}, K)$-morphism. Therefore, $\mathcal{M}(\mathfrak{g}, K)$ is a full subcategory of $\mathcal{M}(\mathfrak{g}, L)$. Moreover, $(\mathfrak{g}, L)$-subobjects and $(\mathfrak{g}, L)$-quotients of any object in $\mathcal{M}(\mathfrak{g}, K)$ are objects in $\mathcal{M}(\mathfrak{g}, K)$.

1.11. Lemma. *The category $\mathcal{M}(\mathfrak{g}, K)$ is a thick subcategory of $\mathcal{M}(\mathfrak{g}, L)$.*

Proof. It remains to show that the subcategory $\mathcal{M}(\mathfrak{g}, K)$ of $\mathcal{M}(\mathfrak{g}, L)$ is closed under extensions. Consider a short exact sequence

$$0 \longrightarrow V \longrightarrow V' \longrightarrow V'' \longrightarrow 0$$

in $\mathcal{M}(\mathfrak{g}, L)$. Then the adjointness morphism of the zero-th Zuckerman functor of a module into the module implies the commutativity of the following diagram:

$$\begin{array}{ccccccccc}
0 & \longrightarrow & \Gamma_{K,L}(V) & \longrightarrow & \Gamma_{K,L}(V') & \longrightarrow & \Gamma_{K,L}(V'') & \longrightarrow & R^1\Gamma_{K,L}(V) \\
 & & \downarrow & & \downarrow & & \downarrow & & \\
0 & \longrightarrow & V & \longrightarrow & V' & \longrightarrow & V'' & \longrightarrow & 0
\end{array}$$

Assume that V and V'' are objects in $\mathcal{M}(\mathfrak{g}, K)$. Then, by 1.10, we have the commutative diagram

$$\begin{array}{ccccccccc}
0 & \longrightarrow & V & \longrightarrow & \Gamma_{K,L}(V') & \longrightarrow & V'' & \longrightarrow & 0 \\
 & & \Vert & & \downarrow & & \Vert & & \\
0 & \longrightarrow & V & \longrightarrow & V' & \longrightarrow & V'' & \longrightarrow & 0
\end{array},$$

and the middle vertical arrow is also an isomorphism, i.e., $V' \cong \Gamma_{K,L}(V')$. □

Let $D^*(\mathcal{M}(\mathfrak{g}, K))$, where $*$ is either b, $+$, $-$ or nothing, be the derived category of $\mathcal{M}(\mathfrak{g}, K)$ consisting of bounded, bounded from below, bounded from above or arbitrary complexes, respectively.

We can consider the full subcategory $D^*_{\mathcal{M}(\mathfrak{g},K)}(\mathcal{M}(\mathfrak{g}, L))$ of all complexes in $D^*(\mathcal{M}(\mathfrak{g}, L))$ with cohomology in $\mathcal{M}(\mathfrak{g}, K)$. By a standard argument using 1.11 and the long exact sequence of cohomology modules attached to a distinguished triangle, we can conclude that for any distinguished triangle in $D^*(\mathcal{M}(\mathfrak{g},L))$, if two vertices are in $D^*_{\mathcal{M}(\mathfrak{g},K)}(\mathcal{M}(\mathfrak{g},L))$, the third one is there too, i.e., $D^*_{\mathcal{M}(\mathfrak{g},K)}(\mathcal{M}(\mathfrak{g},L))$ is a triangulated subcategory of $D^*(\mathcal{M}(\mathfrak{g},L))$.

In addition, we have the natural forgetful functor For from $D^*(\mathcal{M}(\mathfrak{g},K))$ to $D^*(\mathcal{M}(\mathfrak{g},L))$ and its image is inside $D^*_{\mathcal{M}(\mathfrak{g},K)}(\mathcal{M}(\mathfrak{g},L))$. Denote by α the induced functor from $D^*(\mathcal{M}(\mathfrak{g},K))$ into $D^*_{\mathcal{M}(\mathfrak{g},K)}(\mathcal{M}(\mathfrak{g},L))$.

1.12. Theorem. *The functor* $\alpha : D^*(\mathcal{M}(\mathfrak{g},K)) \to D^*_{\mathcal{M}(\mathfrak{g},K)}(\mathcal{M}(\mathfrak{g},L))$ *is an equivalence of categories.*

Proof. Clearly, it is sufficient to prove this statement for the derived categories of unbounded complexes. Since the functor $R\Gamma_{K,L}$ has finite right cohomological dimension by 1.6, the adjointness of the forgetful functor For : $\mathcal{M}(\mathfrak{g},K) \to \mathcal{M}(\mathfrak{g},L)$ and $\Gamma_{K,L} : \mathcal{M}(\mathfrak{g},L) \to \mathcal{M}(\mathfrak{g},K)$ implies that $R\Gamma_{K,L}$ is also the right adjoint of the forgetful functor from $D(\mathcal{M}(\mathfrak{g},K)) \to D(\mathcal{M}(\mathfrak{g},L))$. Since, by 1.10, the objects in $\mathcal{M}(\mathfrak{g},K)$ are $\Gamma_{K,L}$-acyclic, we have

$$R\Gamma_{K,L}(\mathrm{For}(V^\cdot)) = \Gamma_{K,L}(\mathrm{For}(V^\cdot)) = V^\cdot$$

for any complex $V^\cdot$ in $D(\mathcal{M}(\mathfrak{g},K))$. In addition, we have the adjointness morphism $\mathrm{For}(R\Gamma_{K,L}(U^\cdot)) \to U^\cdot$ for any $U^\cdot$ in $D(\mathcal{M}(\mathfrak{g},L))$. If $U^\cdot$ is a complex in $D_{\mathcal{M}(\mathfrak{g},K)}(\mathcal{M}(\mathfrak{g},L))$, its cohomology modules are $\Gamma_{K,L}$-acyclic by 1.10. Therefore, by a standard argument, $H^p(R\Gamma_{K,L}(U^\cdot)) = \Gamma_{K,L}(H^p(U^\cdot)) = H^p(U^\cdot)$, for $p \in \mathbb{Z}$, and the adjointness morphism is a quasiisomorphism. Hence, α is an equivalence of categories. $\square$

It follows that we can view $D^*(\mathcal{M}(\mathfrak{g},K))$ as a triangulated subcategory in $D^*(\mathcal{M}(\mathfrak{g},L))$.

Now we can discuss the consequences of 1.12. with respect to Zuckerman functors. Let H be a subgroup of K and T its Levi factor. Then we have the following commutative diagram

$$\begin{array}{ccc} D^+(\mathcal{M}(\mathfrak{g},T)) & \xrightarrow{R\Gamma_{H,T}} & D^+(\mathcal{M}(\mathfrak{g},H)) \\ {\scriptstyle R\Gamma_{K,T}}\downarrow & & \downarrow{\scriptstyle R\Gamma_{K,H}} \\ D^+(\mathcal{M}(\mathfrak{g},K)) & = & D^+(\mathcal{M}(\mathfrak{g},K)) \end{array} .$$

Finally, by replacing the top left corner with $D^+_{\mathcal{M}(\mathfrak{g},H)}(\mathcal{M}(\mathfrak{g},T))$ and inverting the top horizontal arrow, we get the commutative diagram

$$\begin{array}{ccc} D^+_{\mathcal{M}(\mathfrak{g},H)}(\mathcal{M}(\mathfrak{g},T)) & \xleftarrow{\alpha} & D^+(\mathcal{M}(\mathfrak{g},H)) \\ {\scriptstyle R\Gamma_{K,T}}\downarrow & & \downarrow{\scriptstyle R\Gamma_{K,H}} \\ D^+(\mathcal{M}(\mathfrak{g},K)) & = & D^+(\mathcal{M}(\mathfrak{g},K)) \end{array} ,$$

i.e., $R\Gamma_{K,H}$ is the restriction of $R\Gamma_{K,T}$ to $D^+(\mathcal{M}(\mathfrak{g},H))$. Since the right cohomological dimension of $R\Gamma_{K,T}$ is $\le \dim(K/T)$ by 1.6, the right cohomological dimension of $R\Gamma_{K,H}$ is also $\le \dim(K/T)$. Therefore, both functors extend to the categories of unbounded complexes, and we have the following result.

1.13. Theorem. *The Zuckerman functor $R\Gamma_{K,H}$ is the restriction of $R\Gamma_{K,T}$ to the subcategory $D(\mathcal{M}(\mathfrak{g},H))$ of $D(\mathcal{M}(\mathfrak{g},T))$.*

2. Equivariant derived categories

As we already remarked in the introduction, in certain instances the construction of Zuckerman functors is not sufficiently flexible for applications. The problem lies in the construction of the derived category $D(\mathcal{M}(\mathfrak{g},K))$. In this section we discuss a more appropriate construction due to Beilinson and Ginzburg [3].

The first, and critical, step is a "two-step" definition of Harish-Chandra modules.

A triple (V,π,ν) is called a *weak Harish-Chandra module* for the pair $(\mathfrak{g},K)$ if:

(W1) V is a $\mathcal{U}(\mathfrak{g})$-module with an action π;
(W2) V is an algebraic K-module with an action ν;
(W3) for any $\xi \in \mathfrak{g}$ and $k \in K$ we have

$$\pi(\phi(k)\xi) = \nu(k)\pi(\xi)\nu(k)^{-1};$$

i.e., the $\mathfrak{g}$-action map $\mathfrak{g} \otimes V \longrightarrow V$ is K-equivariant.

The action ν of K differentiates to an action of $\mathfrak{k}$ which we denote also by ν. We put $\omega(\xi) = \nu(\xi) - \pi(\xi)$ for $\xi \in \mathfrak{k}$. The following simple observation is critical.

2.1. Lemma. *Let V be a weak Harish-Chandra module. Then*

(i) *ω is a representation of $\mathfrak{k}$ on V;*
(ii) *ω is K-equivariant, i.e.,*

$$\omega(\mathrm{Ad}(k)\xi) = \nu(k)\omega(\xi)\nu(k)^{-1}$$

for $\xi \in \mathfrak{k}$ and $k \in K$;
(iii)

$$[\omega(\xi),\pi(\eta)] = 0,$$

for $\eta \in \mathfrak{g}$ and $\xi \in \mathfrak{k}$.

Proof. By (W3), the representation π of $\mathfrak{k}$ is K-equivariant. Since the representation ν of $\mathfrak{k}$ is obviously K-equivariant, (ii) follows immediately. By differentiating (W3) we also get

$$[\pi(\xi),\pi(\eta)] = \pi([\xi,\eta]) = [\nu(\xi),\pi(\eta)]$$

for $\xi \in \mathfrak{k}$ and $\eta \in \mathfrak{g}$. This implies that

$$[\omega(\xi),\pi(\eta)] = 0$$

for $\xi \in \mathfrak{k}$ and $\eta \in \mathfrak{g}$; i.e., (iii) holds.

$$[\omega(\xi), \omega(\eta)] = [\omega(\xi), \nu(\eta)] - [\omega(\xi), \pi(\eta)] = [\omega(\xi), \nu(\eta)] =$$
$$[\nu(\xi), \nu(\eta)] - [\pi(\xi), \nu(\eta)] = \nu([\xi, \eta]) - [\pi(\xi), \pi(\eta)] = \nu([\xi, \eta]) - \pi([\xi, \eta]) = \omega([\xi, \eta]),$$

for any $\xi, \eta \in \mathfrak{k}$. □

We see that a weak Harish-Chandra module V is a Harish-Chandra module if and only if $\omega = 0$. A morphism $\alpha : V \longrightarrow W$ of two weak Harish-Chandra modules is a linear map that is a morphism for both $\mathcal{U}(\mathfrak{g})$- and K-module structures. We denote by $\mathcal{M}(\mathfrak{g}, K)_w$ the category of all weak Harish-Chandra modules for the pair $(\mathfrak{g}, K)$. Clearly, the category $\mathcal{M}(\mathfrak{g}, K)$ of Harish-Chandra modules is a full subcategory of $\mathcal{M}(\mathfrak{g}, K)_w$. Also, $\mathcal{M}(\mathfrak{g}, K)_w$ is an abelian category.

Now we define a functor from $\mathcal{M}(\mathfrak{g})$ into $\mathcal{M}(\mathfrak{g}, K)_w$. Let V be a $\mathfrak{g}$-module. We consider the linear space $R(K, V)$ with the following actions of $\mathfrak{g}$ and K,

(i)
$$(\pi(\xi)F)(k) = \pi_V(\phi(k)\xi)F(k), \qquad k \in K,$$
for $\xi \in \mathfrak{g}$ and $F \in R(K, V)$;

(ii)
$$(\nu(k)F)(h) = F(hk), \quad h \in K,$$
for $k \in K$ and $F \in R(K, V)$.

As remarked in §1, the action π is K-equivariant. Hence, $\mathrm{Ind}_w(V) = R(K, V)$ is a weak Harish-Chandra module. If for a $\mathfrak{g}$-morphism $\alpha : V \longrightarrow W$ we define $\mathrm{Ind}_w(\alpha) = 1 \otimes \alpha$, Ind_w becomes an exact functor from $\mathcal{M}(\mathfrak{g})$ into $\mathcal{M}(\mathfrak{g}, K)_w$.

Let V be a weak Harish-Chandra module. As in §1, we see that the matrix coefficient map c_V from V into $\mathrm{Ind}_w(V)$, defined by $c_V(v)(k) = \nu_V(k)v$ for $v \in V$ and $k \in K$, is a $(\mathfrak{g}, K)$-morphism. Furthermore, the maps c_V define a natural transformation of the identity functor on $\mathcal{M}(\mathfrak{g}, K)_w$ into the composition of Ind_w and the forgetful functor from $\mathcal{M}(\mathfrak{g}, K)_w$ into $\mathcal{M}(\mathfrak{g})$.

On the other hand, for any W in $\mathcal{M}(\mathfrak{g})$, we define a linear map $e_W : \mathrm{Ind}_w(W) \longrightarrow W$ by $e_W(F) = F(1)$ for $F \in \mathrm{Ind}_w(W)$. As in §1, we see that e_W is a morphism of $\mathfrak{g}$-modules, and that the maps e_W define a natural transformation of the composition of the forgetful functor from $\mathcal{M}(\mathfrak{g}, K)_w$ into $\mathcal{M}(\mathfrak{g})$ with Ind_w into the identity functor on $\mathcal{M}(\mathfrak{g})$.

Proceeding as in the proof of 1.2, we get

2.2. Lemma. *The functor* $\mathrm{Ind}_w : \mathcal{M}(\mathfrak{g}) \longrightarrow \mathcal{M}(\mathfrak{g}, K)_w$ *is right adjoint to the forgetful functor from* $\mathcal{M}(\mathfrak{g}, K)_w$ *into* $\mathcal{M}(\mathfrak{g})$.

This immediately leads to the following result analogous to 1.3.

2.3. Lemma. *The category* $\mathcal{M}(\mathfrak{g}, K)_w$ *has enough injectives.*

Let U be in $\mathcal{M}(\mathfrak{g}, K)_w$. Denote by

$$U^{\mathfrak{k}} = \{u \in U \mid \omega(\xi)u = 0, \ \xi \in \mathfrak{k}\}.$$

Then, by 2.1, $U^{\mathfrak{k}}$ is the largest Harish-Chandra submodule of U. Clearly, for any Harish-Chandra module V, we have

$$\mathrm{Hom}_{(\mathfrak{g},K)}(V,U) = \mathrm{Hom}_{(\mathfrak{g},K)}(V,U^{\mathfrak{k}}).$$

Therefore, $U \longmapsto U^{\mathfrak{k}}$ is the right adjoint of the forgetful functor $\mathcal{M}(\mathfrak{g},K) \to \mathcal{M}(\mathfrak{g},K)_w$.

Therefore, the composition $V \longmapsto \mathrm{Ind}_w(V)^{\mathfrak{k}}$ is the right adjoint of the forgetful functor from the category $\mathcal{M}(\mathfrak{g},K)$ into $\mathcal{M}(\mathfrak{g})$. This is the Zuckerman functor Γ_K.

2.4. Lemma. *For any V in $\mathcal{M}(\mathfrak{g})$, we have*

$$\Gamma_K(V) = \mathrm{Ind}_w(V)^{\mathfrak{k}}.$$

To make this more explicit we calculate the ω-action on $\mathrm{Ind}_w(V)$. We have

$$\begin{aligned}(\lambda(\mathrm{Ad}(k)\xi)F)(k) &= (\gamma(\mathrm{Ad}(k)\xi)F)(k) + \pi_V(\mathrm{Ad}(k)\xi)F(k)\\ &= (\pi(\xi)F)(k) - (\nu(\xi)F)(k) = -(\omega(\xi)F)(k), \quad k \in K,\end{aligned}$$

for $\xi \in \mathfrak{k}$, where we denoted by γ the left regular representation of $\mathfrak{k}$ on $R(K)$ tensored by the trivial representation on V. Hence, we established the following formula:

$$(\omega(\xi)F)(k) = -(\lambda(\mathrm{Ad}(k)\xi)F)(k), \qquad k \in K.$$

This implies that the largest Harish-Chandra submodule $\Gamma_K(V) = \mathrm{Ind}_w(V)^{\mathfrak{k}}$ of $\mathrm{Ind}_w(V)$ can be characterized as the submodule of all λ-invariants in $\mathrm{Ind}_w(V)$.

This agrees with the construction in §1.

Denote by $D(\mathfrak{g},K)_w = D(\mathcal{M}(\mathfrak{g},K)_w)$ the derived category of the category $\mathcal{M}(\mathfrak{g},K)_w$ of weak Harish-Chandra modules. Then we can consider the forgetful functors $D(\mathcal{M}(\mathfrak{g},K)) \to D(\mathfrak{g},K)_w$ and $D(\mathfrak{g},K)_w \to D(\mathcal{M}(\mathfrak{g}))$. Their composition is the forgetful functor from $D(\mathcal{M}(\mathfrak{g},K))$ into $D(\mathcal{M}(\mathfrak{g}))$. Therefore, the right adjoint functor $R\Gamma_K : D(\mathcal{M}(\mathfrak{g})) \to D(\mathcal{M}(\mathfrak{g},K))$ is the composition of the right adjoint functor $V^{\cdot} \longmapsto \mathrm{Ind}_w(V^{\cdot})$ from $D(\mathcal{M}(\mathfrak{g}))$ into $D(\mathfrak{g},K)_w$ with the right derived functor of the functor $U \longmapsto U^{\mathfrak{k}}$. This leads us back to the setup of §1.

Instead of proceeding like in the last step, Beilinson and Ginzburg interpret the condition $\omega = 0$, which makes a weak Harish-Chandra module an ordinary Harish-Chandra module, as a homotopic condition.

An *equivariant Harish-Chandra complex* $V^{\cdot}$ is a complex of weak Harish-Chandra modules equipped with a linear map i from $\mathfrak{k}$ into graded linear maps from $V^{\cdot}$ to $V^{\cdot}$ of degree -1. This map satisfies the following conditions:

(E1) i_ξ, $\xi \in \mathfrak{k}$, are $\mathfrak{g}$-morphisms, i.e.,

$$\pi(\eta)i_\xi = i_\xi\pi(\eta), \text{ for } \eta \in \mathfrak{g};$$

(E2) i_ξ, $\xi \in \mathfrak{k}$, are K-equivariant, i.e.

$$i_{\mathrm{Ad}(k)\xi} = \nu(k) i_\xi \nu(k^{-1}) \text{ for } k \in K;$$

(E3)

$$i_\xi i_\eta + i_\eta i_\xi = 0$$

for $\xi, \eta \in \mathfrak{k}$;

(E4)

$$d i_\xi + i_\xi d = \omega(\xi)$$

for $\xi \in \mathfrak{k}$.

Clearly, (E4) implies that cohomology modules of equivariant Harish-Chandra complexes are Harish-Chandra modules.

A morphism ϕ of equivariant Harish-Chandra complexes is a morphism of complexes of weak Harish-Chandra modules that also satisfies

$$\phi \circ i_\xi = i_\xi \circ \phi$$

for all $\xi \in \mathfrak{k}$. Let $C^*(\mathfrak{g}, K)$ be the abelian category of equivariant Harish-Chandra complexes with the appropriate boundedness condition. Two morphisms $\phi, \psi : V^\cdot \to W^\cdot$ in this category are *homotopic* if there exists a homotopy Σ of the corresponding complexes of weak Harish-Chandra modules that in addition satisfies

$$\Sigma \circ i_\xi = -i_\xi \circ \Sigma$$

for any $\xi \in \mathfrak{k}$. We denote by $K^*(\mathfrak{g}, K)$ the corresponding homotopic category of equivariant complexes. This category has a natural structure of a triangulated category [18]. Quasiisomorphisms form a localizing class of morphisms in $K^*(\mathfrak{g}, K)$. The localization of $K^*(\mathfrak{g}, K)$ with respect to quasiisomorphisms is the *equivariant derived category* $D^*(\mathfrak{g}, K)$ of Harish-Chandra modules. Clearly, $D^*(\mathfrak{g}, K)$ inherits the structure of a triangulated category from $K^*(\mathfrak{g}, K)$, but a priori $D^*(\mathfrak{g}, K)$ doesn't have to be a derived category of an abelian category (still, in this particular case, the reader should consult 2.14 at this point).

We have a natural functor $\iota : D^*(\mathcal{M}(\mathfrak{g}, K)) \to D^*(\mathfrak{g}, K)$ that maps a complex of Harish-Chandra modules $V^\cdot$ into the equivariant Harish-Chandra complex $V^\cdot$ with $i_\xi = 0$ for all $\xi \in \mathfrak{k}$.

In particular, for a Harish-Chandra module V we denote by $D(V)$ the complex

$$\dots \to 0 \to V \to 0 \to \dots$$

where V is in degree zero, and the corresponding equivariant Harish-Chandra complex. A straightforward modification of the standard argument proves that $D : \mathcal{M}(\mathfrak{g}, K) \to D(\mathfrak{g}, K)$ is fully faithful, i.e,

$$\mathrm{Hom}_{(\mathfrak{g},K)}(V, W) = \mathrm{Hom}_{D(\mathfrak{g},K)}(D(V), D(W))$$

for any two Harish-Chandra modules V and W. Hence, if we equip $D^*(\mathfrak{g}, K)$ with the standard truncation functors, its core is isomorphic to $\mathcal{M}(\mathfrak{g}, K)$.

Therefore, we have a natural sequence of functors

$$D^*(\mathcal{M}(\mathfrak{g}, K)) \xrightarrow{\iota} D^*(\mathfrak{g}, K) \to D^*(\mathfrak{g}, K)_w \to D^*(\mathcal{M}(\mathfrak{g}))$$

where the last two are just the corresponding forgetful functors.

In various applications, like the ones we discuss later in the paper, it is necessary to consider simple variants of the above construction. Let $\mathcal{Z}(\mathfrak{g})$ be the center of the enveloping algebra $\mathcal{U}(\mathfrak{g})$. Let $\mathfrak{h}$ be an (abstract) Cartan algebra of $\mathfrak{g}$ ([14], §2). Denote by W the Weyl group of the root system Σ of $\mathfrak{g}$ in $\mathfrak{h}^*$. By a classical result of Harish-Chandra, the space of maximal ideals $\mathrm{Max}(\mathcal{Z}(\mathfrak{g}))$ is isomorphic to the space $\mathfrak{h}^*/W$ of W-orbits in $\mathfrak{h}^*$. Let θ be a W-orbit in $\mathfrak{h}^*$, and denote by J_θ the corresponding maximal ideal of $\mathcal{Z}(\mathfrak{g})$. Let $\lambda \in \theta$. We denote by χ_λ the unique homomorphism of $\mathcal{Z}(\mathfrak{g})$ into $\mathbb{C}$ with its kernel equal to J_θ. We denote by $\mathcal{U}_\theta$ the quotient of the enveloping algebra $\mathcal{U}(\mathfrak{g})$ by the ideal generated by J_θ. Then we can view the category $\mathcal{M}(\mathcal{U}_\theta)$ of $\mathcal{U}_\theta$-modules as a full subcategory of the category $\mathcal{M}(\mathfrak{g})$. Following the classical terminology, the objects of $\mathcal{M}(\mathcal{U}_\theta)$ are just $\mathcal{U}(\mathfrak{g})$-modules with *infinitesimal character* χ_λ.

Since the image of ϕ is in $\mathrm{Int}(\mathfrak{g})$, the group K acts trivially on $\mathcal{Z}(\mathfrak{g})$, hence also on $\mathrm{Max}(\mathcal{Z}(\mathfrak{g}))$. Therefore, we can define the category $\mathcal{M}(\mathcal{U}_\theta, K)$ as the full subcategory of $\mathcal{M}(\mathfrak{g}, K)$ of Harish-Chandra modules with infinitesimal character χ_λ. Clearly, for any Harish-Chandra module V in $\mathcal{M}(\mathfrak{g}, K)$, the module

$$\mathcal{U}_\theta \otimes_{\mathcal{U}(\mathfrak{g})} V = V/J_\theta V$$

is in $\mathcal{M}(\mathcal{U}_\theta, K)$. Therefore, we have the right exact functor $P_\theta : \mathcal{M}(\mathfrak{g}, K) \to \mathcal{M}(\mathcal{U}_\theta, K)$. It is straightforward to check the following result.

2.5. Lemma. *The functor $P_\theta : \mathcal{M}(\mathfrak{g}, K) \to \mathcal{M}(\mathcal{U}_\theta, K)$ is the left adjoint of the forgetful functor from $\mathcal{M}(\mathcal{U}_\theta, K)$ into $\mathcal{M}(\mathfrak{g}, K)$.*

Analogously, we can define the category $\mathcal{M}(\mathcal{U}_\theta, K)_w$ of weak Harish-Chandra modules with infinitesimal character χ_λ, and the corresponding derived categories $D^*(\mathcal{U}_\theta, K)_w$ and $D^*(\mathcal{U}_\theta, K)$. In addition, the definition of P_θ obviously extends to the corresponding categories of weak Harish-Chandra modules and we have an obvious analogue of 2.5. Let U be an algebraic K-module. Then $\mathcal{U}(\mathfrak{g}) \otimes U$, equipped with the $\mathcal{U}(\mathfrak{g})$-action by left multiplication on the first factor and the tensor product of natural algebraic actions of K, is a weak Harish-Chandra module. Since it is a flat $\mathcal{U}(\mathfrak{g})$-module, it is also acyclic for P_θ. Since any weak Harish-Chandra module V is a quotient of $\mathcal{U}(\mathfrak{g}) \otimes V$, P_θ has its left derived functor $LP_\theta : D^-(\mathfrak{g}, K)_w \to D^-(\mathcal{U}_\theta, K)_w$. Moreover, the homological dimension of $\mathcal{U}(\mathfrak{g})$ is finite, so LP_θ extends to the derived categories of unbounded complexes. Also, this functor is the left adjoint of the forgetful functor from $D(\mathcal{U}_\theta, K)_w$ into $D(\mathfrak{g}, K)_w$. Finally, it also induces the left derived functor[1]

$$LP_\theta : D(\mathfrak{g}, K) \to D(\mathcal{U}_\theta, K).$$

This is again the left adjoint of the forgetful functor:

2.6. Proposition. *The functor $LP_\theta : D(\mathfrak{g}, K) \to D(\mathcal{U}_\theta, K)$ is the left adjoint of the forgetful functor from $D(\mathcal{U}_\theta, K)$ into $D(\mathfrak{g}, K)$.*

Clearly, if H is a closed subgroup of K, we have the following commutative diagram:

$$\begin{array}{ccc} \mathcal{M}(\mathfrak{g}, K) & \xrightarrow{P_\theta} & \mathcal{M}(\mathcal{U}_\theta, K) \\ \downarrow & & \downarrow \\ \mathcal{M}(\mathfrak{g}, H) & \xrightarrow{P_\theta} & \mathcal{M}(\mathcal{U}_\theta, H) \end{array}.$$

This leads to the commutative diagram

$$\begin{array}{ccc} D(\mathfrak{g}, K) & \xrightarrow{LP_\theta} & D(\mathcal{U}_\theta, K) \\ \downarrow & & \downarrow \\ D(\mathfrak{g}, H) & \xrightarrow[LP_\theta]{} & D(\mathcal{U}_\theta, H) \end{array}$$

where the vertical arrows represent forgetful functors.

The following result is proved in [17].

2.7. Theorem. *The forgetful functor from $D^+(\mathfrak{g}, K)$ into $D^+(\mathfrak{g}, H)$ (resp. from $D^+(\mathcal{U}_\theta, K)$ into $D^+(\mathcal{U}_\theta, H)$) has the right adjoint $R\Gamma^{equi}_{K,H} : D^+(\mathfrak{g}, H) \to D^+(\mathfrak{g}, K)$ (resp. $R\Gamma^{equi}_{K,H} : D^+(\mathcal{U}_\theta, H) \to D^+(\mathcal{U}_\theta, K)$).*

If H is reductive, the amplitude of $R\Gamma^{equi}_{K,H}$ is finite. In this situation, the above claims hold also for unbounded equivariant derived categories.

We are going to discuss the construction of this *equivariant Zuckerman functor* $R\Gamma^{equi}_{K,H}$ in the next section.

As in the case of ordinary Zuckerman functors, the following result holds.

2.8. Proposition. *Let $T \subset H$ be algebraic subgroups of K. Then we have the isomorphism of functors*

$$R\Gamma^{equi}_{K,T} = R\Gamma^{equi}_{K,H} \circ R\Gamma^{equi}_{H,T}.$$

If H is reductive, the functor $R\Gamma^{equi}_{K,H}$ extends to the equivariant derived categories of unbounded complexes, and preserves its adjointness property. Therefore, by taking the adjoints of the above diagram, we get the following result (which also follows from the explicit formula for $R\Gamma^{equi}_{K,H}$ we are going to discuss in the next section).

[1] Since equivariant derived categories are not derived categories in the standard sense, this requires some additional care (see [8], [17]).

2.9. Lemma. *If H is a reductive subgroup of K, the following diagram commutes*

$$\begin{array}{ccc} D(\mathcal{U}_\theta, H) & \longrightarrow & D(\mathfrak{g}, H) \\ {\scriptstyle R\Gamma^{equi}_{K,H}}\downarrow & & \downarrow{\scriptstyle R\Gamma^{equi}_{K,H}} \\ D(\mathcal{U}_\theta, K) & \longrightarrow & D(\mathfrak{g}, K) \end{array} .$$

This explains the ambiguity in our notation.

Another simple consequence of the construction of the equivariant Zuckerman functor and 1.6 is the following result ([17], 6.2.7).

2.10. Lemma. *Assume that H is a reductive subgroup of K. Let V be a Harish-Chandra module in $\mathcal{M}(\mathfrak{g}, H)$. Then*

$$H^p(R\Gamma^{equi}_{K,H}(D(V))) = R^p\Gamma_{K,H}(V)$$

for any $p \in \mathbb{Z}_+$.

Let K be an arbitrary algebraic group and L a Levi subgroup of K. As we remarked in 1.11, $\mathcal{M}(\mathfrak{g}, K)$ is a thick subcategory of $\mathcal{M}(\mathfrak{g}, L)$. Clearly, the same applies to the subcategory $\mathcal{M}(\mathcal{U}_\theta, K)$ of $\mathcal{M}(\mathcal{U}_\theta, L)$. Therefore, we can define full triangulated subcategories $D^*_{\mathcal{M}(\mathfrak{g},K)}(\mathfrak{g}, L)$ and $D^*_{\mathcal{M}(\mathcal{U}_\theta,K)}(\mathcal{U}_\theta, L)$ of $D^*(\mathfrak{g}, L)$, resp. $D^*(\mathcal{U}_\theta, L)$, consisting of equivariant Harish-Chandra complexes with cohomology in $\mathcal{M}(\mathfrak{g}, K)$, resp. $\mathcal{M}(\mathcal{U}_\theta, K)$.

Clearly, we have natural forgetful functors $D^*(\mathfrak{g}, K) \to D^*_{\mathcal{M}(\mathfrak{g},K)}(\mathfrak{g}, L)$ and $D^*(\mathcal{U}_\theta, K) \to D^*_{\mathcal{M}(\mathcal{U}_\theta,K)}(\mathcal{U}_\theta, L)$. The following result is an equivariant analogue of 1.12.

2.11. Theorem. *The natural forgetful functors*

$$D^*(\mathfrak{g}, K) \to D^*_{\mathcal{M}(\mathfrak{g},K)}(\mathfrak{g}, L) \text{ and } D^*(\mathcal{U}_\theta, K) \to D^*_{\mathcal{M}(\mathcal{U}_\theta,K)}(\mathcal{U}_\theta, L)$$

are equivalences of categories.

Proof. The proofs of these equivalences are identical. Therefore, we discuss the case of modules over the enveloping algebra.

Also, it is sufficient to prove this statement for the derived categories of unbounded complexes. The functor $R\Gamma^{equi}_{K,L}$ is the right adjoint of the forgetful functor For : $D(\mathfrak{g}, K) \to D(\mathfrak{g}, L)$. Therefore, we have the natural transformation of the identity functor on $D(\mathfrak{g}, K)$ into $R\Gamma^{equi}_{K,L} \circ$ For. Assume that $V^\cdot$ is an equivariant complex in $D(\mathfrak{g}, K)$. Then we have a natural morphism $V^\cdot \to R\Gamma^{equi}_{K,L}(V^\cdot)$. Since $H^p(V^\cdot)$, $p \in \mathbb{Z}$, are Harish-Chandra modules in $\mathcal{M}(\mathfrak{g}, K)$, they are $\Gamma_{K,L}$-acyclic by 1.10. By a standard argument using 2.10, this implies that

$$H^p(R\Gamma^{equi}_{K,L}(V^\cdot)) = \Gamma_{K,L}(H^p(V^\cdot)) = H^p(V^\cdot)$$

and the natural morphism $V^\cdot \to R\Gamma^{equi}_{K,L}(V^\cdot)$ is a quasiisomorphism.

On the other hand, we have the adjointness morphism $\mathrm{For}(R\Gamma^{equi}_{K,L}(U^\cdot)) \to U^\cdot$ for any $U^\cdot$ in $D(\mathfrak{g},L)$. If $U^\cdot$ is a complex in $D_{\mathcal{M}(\mathfrak{g},K)}(\mathfrak{g},L)$, its cohomology modules are $\Gamma_{K,L}$-acyclic by 1.10. Therefore, as before, $H^p(R\Gamma^{equi}_{K,L}(U^\cdot)) = H^p(U^\cdot)$, for $p \in \mathbb{Z}$, and the adjointness morphism is a quasiisomorphism.

Hence, $D^*(\mathfrak{g},K) \to D^*_{\mathcal{M}(\mathfrak{g},K)}(\mathfrak{g},L)$ is an equivalence of categories. □

It follows that we can view $D^*(\mathfrak{g},K)$ and $D^*(\mathcal{U}_\theta,K)$ as triangulated subcategories in $D^*(\mathfrak{g},L)$ and $D^*(\mathcal{U}_\theta,L)$ respectively.

Now we can discuss the consequences of 2.11 with respect to equivariant Zuckerman functors. Let H be a subgroup of K and T a Levi factor of H. Then we have the following commutative diagram

$$\begin{array}{ccc} D^+(\mathfrak{g},T) & \xrightarrow{R\Gamma^{equi}_{H,T}} & D^+(\mathfrak{g},H) \\ {\scriptstyle R\Gamma^{equi}_{K,T}}\downarrow & & \downarrow{\scriptstyle R\Gamma^{equi}_{K,H}} \\ D^+(\mathfrak{g},K) & = & D^+(\mathfrak{g},K) \end{array} .$$

Finally, by replacing the top left corner with $D^+_{\mathcal{M}(\mathfrak{g},H)}(\mathfrak{g},T)$ and inverting the top horizontal arrow, we get the commutative diagram:

$$\begin{array}{ccc} D^+_{\mathcal{M}(\mathfrak{g},H)}(\mathfrak{g},T) & \longleftarrow & D^+(\mathfrak{g},H) \\ {\scriptstyle R\Gamma^{equi}_{K,T}}\downarrow & & \downarrow{\scriptstyle R\Gamma^{equi}_{K,H}} \\ D^+(\mathfrak{g},K) & = & D^+(\mathfrak{g},K) \end{array} ,$$

i.e., $R\Gamma^{equi}_{K,H}$ is the restriction of $R\Gamma^{equi}_{K,T}$ to $D^+(\mathfrak{g},H)$. Since the amplitude of $R\Gamma^{equi}_{K,T}$ is finite by 2.7, the amplitude of $R\Gamma^{equi}_{K,H}$ is also finite. Both functors extend to the categories of unbounded complexes, and we have the following result.

2.12. Theorem. *The equivariant Zuckerman functor $R\Gamma^{equi}_{K,H}$ is the restriction of $R\Gamma^{equi}_{K,T}$ to the subcategory $D(\mathfrak{g},H)$ (resp. $D(\mathcal{U}_\theta,H)$) of $D(\mathfrak{g},T)$ (resp. $D(\mathcal{U}_\theta,T)$).*

2.13. Remark. We can now eliminate the assumption of H being reductive from 2.7, 2.9 and 2.10. We already noted that above for 2.7. For 2.9 it follows from 2.12, and for 2.10 from 2.12 and 1.13.

Finally, we quote a result of Bernstein and Lunts [5] which completely explains 2.10.[2]

2.14. Theorem. *The natural functor $\iota : D(\mathcal{M}(\mathfrak{g},K)) \to D(\mathfrak{g},K)$ is an equivalence of categories.*

Proof. For reductive K, the claim is proved in [17], Section 5.8. For arbitrary K, let L be a Levi factor of K. The equivalence $\iota : D(\mathcal{M}(\mathfrak{g},L)) \to D(\mathfrak{g},L)$ induces an equivalence of subcategories $D_{\mathcal{M}(\mathfrak{g},K)}(\mathcal{M}(\mathfrak{g},L))$ and $D_{\mathcal{M}(\mathfrak{g},K)}(\mathfrak{g},L)$.

[2]Bernstein and Lunts treat the case of derived categories bounded from below. The general case follows from [17].

The first of these is equivalent to $D(\mathcal{M}(\mathfrak{g}, K))$ by 1.12, and the second is equivalent to $D(\mathfrak{g}, K)$ by 2.11. The induced functor is clearly $\iota : D(\mathcal{M}(\mathfrak{g}, K)) \to D(\mathfrak{g}, K)$, and it is an equivalence. □

3. Equivariant Zuckerman functors

In this section we sketch the construction of equivariant Zuckerman functors $R\Gamma^{equi}_{K,H}$. The details can be found in [17]. For simplicity, we will describe the definitions and the arguments for $\mathcal{U}(\mathfrak{g})$-modules. However, analogous statements hold also for the variant with $\mathcal{U}_\theta$-modules, with identical proofs.

Let $N^\cdot(\mathfrak{k}) = \mathcal{U}(\mathfrak{k}) \otimes \bigwedge^{-\cdot} \mathfrak{k}$ be the standard complex of $\mathfrak{k}$. It can be viewed as an equivariant $(\mathfrak{k}, H)$-complex in the following way: $\mathfrak{k}$ acts by left multiplication on the first factor, H acts by the tensor product of the adjoint actions on both factors, and the map i is given by

$$i_\xi(u \otimes \lambda) = -u \otimes \lambda \wedge \xi,$$

for $\xi \in \mathfrak{h}$ and $u \otimes \lambda \in N^\cdot(\mathfrak{k})$.

The standard complex has a natural structure of an algebra. It is generated by the subalgebra $\mathcal{U}(\mathfrak{k}) \otimes 1$ isomorphic to the enveloping algebra $\mathcal{U}(\mathfrak{k})$ and the subalgebra $1 \otimes \bigwedge \mathfrak{k}$ isomorphic to the exterior algebra $\bigwedge \mathfrak{k}$. The multiplication is defined by the relations

$$(u \otimes 1) \cdot (1 \otimes \lambda) = u \otimes \lambda,$$

for $u \in \mathcal{U}(\mathfrak{k})$ and $\lambda \in \bigwedge \mathfrak{k}$, and

$$(1 \otimes \xi) \cdot (\eta \otimes 1) - (\eta \otimes 1) \cdot (1 \otimes \xi) = 1 \otimes [\xi, \eta]$$

for $\xi, \eta \in \mathfrak{k}$. With this multiplication and with its natural grading and differential, $N^\cdot(\mathfrak{k})$ becomes a *differential graded algebra*, namely a graded algebra which is also a complex of vector spaces, such that for any two homogeneous elements x and y,

$$d(x \cdot y) = dx \cdot y + (-1)^{\deg x} x \cdot dy.$$

Furthermore, the principal antiautomorphism of $\mathcal{U}(\mathfrak{k})$ and the linear isomorphism of $\mathfrak{k} \subset \bigwedge \mathfrak{k}$ defined by $\xi \longmapsto -\xi$ extend to a *principal antiautomorphism* ι of $N^\cdot(\mathfrak{k})$; it satisfies

$${}^\iota(x \cdot y) = (-1)^{\deg x \deg y}\, {}^\iota y \cdot {}^\iota x$$

for homogeneous $x, y \in N^\cdot(\mathfrak{k})$.

Any equivariant $(\mathfrak{g}, K)$-complex can be viewed as a graded module over $N^\cdot(\mathfrak{k})$, if we let $\mathfrak{k} \subset \mathcal{U}(\mathfrak{k})$ act via the ω-action, and $\mathfrak{k} \subset \bigwedge \mathfrak{k}$ via the map i. We will denote this action of $N^\cdot(\mathfrak{k})$ by ω again.

Let $V^\cdot$ be an equivariant $(\mathfrak{g}, H)$-complex. If we forget the H-action and the equivariant structure, $R(K, V^\cdot) = R(K) \otimes V^\cdot = \mathrm{Ind}_w(V^\cdot)$ is a complex of weak Harish-Chandra modules for $(\mathfrak{g}, K)$. On the other hand, we can consider $R(K, V^\cdot)$ as an equivariant $(\mathfrak{k}, H)$-complex with respect to the λ-action of §1. That is, $\mathfrak{k}$ and H act by the left regular representation tensored by the natural action on V, and the map i on $R(K, V^\cdot)$ is given by

$$i_{R(K,V^\cdot),\xi}(f \otimes v) = f \otimes i_{V^\cdot,\xi}(v)$$

for $\xi \in \mathfrak{h}$, $f \in R(K)$ and $v \in V^\cdot$ (here $i_{V^\cdot}$ is the i-map of $V^\cdot$ as an equivariant $(\mathfrak{g}, H)$-complex).

We can now consider the complex of vector spaces

$$\Gamma_{K,H}^{equi}(V^\cdot) = \mathrm{Hom}^\cdot_{(\mathfrak{k},H)}(N^\cdot(\mathfrak{k}), R(K, V^\cdot)).$$

Here $\mathrm{Hom}_{(\mathfrak{k},H)}$ consists of graded linear maps of equivariant $(\mathfrak{k}, H)$-complexes $N^\cdot(\mathfrak{k})$ and $R(K, V^\cdot)$ described above. These linear maps f intertwine the actions of $\mathfrak{k}$ and H, and,

$$f \circ i_{N^\cdot(\mathfrak{k}),\xi} = (-1)^{\deg f} i_{R(K,V^\cdot),\xi} \circ f$$

for $\xi \in \mathfrak{h}$. The differential of this complex is given by

$$df = d_{R(K,V^\cdot)} \circ f - (-1)^{\deg f} f \circ d_{N^\cdot(\mathfrak{k})},$$

for a homogeneous $f \in \Gamma_{K,H}^{equi}(V^\cdot)$.

The $(\mathfrak{g}, K)$-action on $R(K, V^\cdot)$ defines, by composition, a $(\mathfrak{g}, K)$-action on $\Gamma_{K,H}^{equi}(V^\cdot)$. In this way, $\Gamma_{K,H}^{equi}(V^\cdot)$ becomes a complex of weak $(\mathfrak{g}, K)$-modules. Moreover, if we define the map i by

$$(i_\xi f)(u \otimes \lambda)(k) = -(-1)^{\deg f} f((1 \otimes \mathrm{Ad}(k)\xi) \cdot (u \otimes \lambda))(k),$$

for $\xi \in \mathfrak{k}$, a homogeneous $f \in \Gamma_{K,H}^{equi}(V^\cdot)$, $u \otimes \lambda \in N^\cdot(\mathfrak{k})$ and $k \in K$, the complex $\Gamma_{K,H}^{equi}(V^\cdot)$ becomes an equivariant $(\mathfrak{g}, K)$-complex. One can check that $\Gamma_{K,H}^{equi}$ is a functor from the category of equivariant $(\mathfrak{g}, H)$-complexes $C(\mathfrak{g}, H)$ into $C(\mathfrak{g}, K)$. It also induces a functor between the corresponding homotopic categories $K(\mathfrak{g}, H)$ and $K(\mathfrak{g}, K)$.

Moreover, we have

3.1. Theorem. *The functor $\Gamma_{K,H}^{equi}$ is right adjoint to the forgetful functor from $C(\mathfrak{g}, K)$ into $C(\mathfrak{g}, H)$ (resp. $K(\mathfrak{g}, K)$ into $K(\mathfrak{g}, H)$).*

Proof. We just define the adjointness morphisms, and leave tedious checking to the reader. For an equivariant $(\mathfrak{g}, K)$-complex $V^\cdot$, we define $\Phi_{V^\cdot} : V^\cdot \to \Gamma_{K,H}^{equi}(V^\cdot)$ by

$$\Phi_{V^\cdot}(v)(u \otimes \lambda)(k) = (-1)^{\deg v \deg \lambda} \omega_{V^\cdot}(\iota(u \otimes \lambda)) \nu_{V^\cdot}(k) v,$$

for homogeneous $v \in V^{\cdot}$, $\lambda \in \bigwedge \mathfrak{k}$, $u \in \mathcal{U}(\mathfrak{k})$ and $k \in K$. The other adjointness morphism is much simpler: for an equivariant $(\mathfrak{g}, H)$-complex $W^{\cdot}$, we define $\Psi_{W^{\cdot}} : \Gamma^{equi}_{K,H}(W^{\cdot}) \to W^{\cdot}$ by

$$\Psi_{W^{\cdot}}(f) = f(1 \otimes 1)(1)$$

for $f \in \Gamma^{equi}_{K,H}(W^{\cdot})$. □

By results of Bernstein and Lunts, $\Gamma^{equi}_{K,H}$ always has a right derived functor

$$R\Gamma^{equi}_{K,H} : D^{+}(\mathfrak{g}, H) \to D^{+}(\mathfrak{g}, K).$$

The proof of this uses the existence of *K-injective* resolutions of equivariant complexes. Their existence is established in [5] (see [17] for a more detailed account). However, these resolutions are very complicated and it is unclear when they are bounded above. The following theorem gives an explicit formula for $R\Gamma^{equi}_{K,H}$ in case H is reductive.

3.2. Lemma. *Assume that H is reductive. Then for any acyclic equivariant $(\mathfrak{g}, H)$-complex $V^{\cdot}$, the complex $\Gamma^{equi}_{K,H}(V^{\cdot})$ is acyclic.*

Proof. It is obvious that $R(K, V^{\cdot})$ is an acyclic equivariant $(\mathfrak{k}, H)$-complex. Therefore, it is enough to prove that the functor

$$\mathrm{Hom}^{\cdot}_{(\mathfrak{k},H)}(N^{\cdot}(\mathfrak{k}), -),$$

from equivariant $(\mathfrak{k}, H)$-complexes to complexes of vector spaces, preserves acyclicity.

This follows from the fact that the $(\mathfrak{k}, H)$-complex $N^{\cdot}(\mathfrak{k})$, is *K-projective* (see [4], [18][3] or [17]). It is proved by induction, using the *Hochschild-Serre filtration* $\mathrm{F}.N^{\cdot}(\mathfrak{k})$, associated to the subalgebra $\mathfrak{h}$, of the standard complex $N^{\cdot}(\mathfrak{k})$ ([17], 6.1). Namely, one can see that the graded pieces corresponding to this filtration are K-projective. Then one shows that the short exact sequences

$$0 \to \mathrm{F}_{p-1} N^{\cdot}(\mathfrak{k}) \to \mathrm{F}_p N^{\cdot}(\mathfrak{k}) \to \mathrm{Gr}_p N^{\cdot}(\mathfrak{k}) \to 0$$

define distinguished triangles in the homotopic category of equivariant $(\mathfrak{k}, H)$-complexes. If two vertices of a distinguished triangle are K-projective, then so is the third vertex. Since the filtration is finite, the theorem follows. □

Hence, the functor $\Gamma^{equi}_{K,H}$ preserves acyclic complexes. It follows that it also preserves quasiisomorphisms. Therefore, it is well defined on morphisms in the derived category. Hence, $\Gamma^{equi}_{K,H}$ defines a functor on the level of derived categories which is equal to $\Gamma^{equi}_{K,H}$ on objects, i.e., we have the following result.

3.3. Theorem. *Assume that H is reductive. Then for any equivariant $(\mathfrak{g}, H)$-complex $V^{\cdot}$, we have*

$$R\Gamma^{equi}_{K,H}(V^{\cdot}) = \Gamma^{equi}_{K,H}(V^{\cdot}) = \mathrm{Hom}^{\cdot}_{(\mathfrak{k},H)}(N^{\cdot}(\mathfrak{k}), R(K, V^{\cdot})).$$

In particular, $R\Gamma^{equi}_{K,H}$ has finite amplitude.

[3] Verdier calls such objects "free on the left."

4. Localization of Zuckerman functors

In this section we sketch the localization of equivariant Zuckerman functors. The details will appear in [15].

First we recall the basic constructions and results of the localization theory of Beilinson and Bernstein (cf. [1], [14]). Let X be the flag variety of $\mathfrak{g}$. For any λ in the dual $\mathfrak{h}^*$ of the abstract Cartan algebra $\mathfrak{h}$, Beilinson and Bernstein construct a *twisted sheaf of differential operators* $\mathcal{D}_\lambda$ on X and an algebra homomorphism $\mathcal{U}(\mathfrak{g}) \to \Gamma(X, \mathcal{D}_\lambda)$. They prove that

$$\Gamma(X, \mathcal{D}_\lambda) = \mathcal{U}_\theta \quad \text{and} \quad H^p(X, \mathcal{D}_\lambda) = 0 \text{ for } p > 0,$$

where $\theta = W \cdot \lambda$. Let $\mathcal{M}_{qc}(\mathcal{D}_\lambda)$ be the category of quasicoherent $\mathcal{D}_\lambda$-modules on X. Then, for any object $\mathcal{V}$ in $\mathcal{M}_{qc}(\mathcal{D}_\lambda)$ its global sections are an object in $\mathcal{M}(\mathcal{U}_\theta)$, i.e., the functor of global sections Γ is a left exact functor from $\mathcal{M}_{qc}(\mathcal{D}_\lambda)$ into $\mathcal{M}(\mathcal{U}_\theta)$. It has a left adjoint $\Delta_\lambda : \mathcal{M}(\mathcal{U}_\theta) \to \mathcal{M}_{qc}(\mathcal{D}_\lambda)$ defined by

$$\Delta_\lambda(V) = \mathcal{D}_\lambda \otimes_{\mathcal{U}_\theta} V$$

for any $\mathcal{U}_\theta$-module V. This functor is the *localization functor.*

The functor Γ has finite right cohomological dimension. Therefore, it defines a functor $R\Gamma : D(\mathcal{D}_\lambda) \to D(\mathcal{U}_\theta)$ from the derived category of $\mathcal{M}_{qc}(\mathcal{D}_\lambda)$ into the derived category of $\mathcal{M}(\mathcal{U}_\theta)$. On the other hand, the functor Δ_λ has finite left cohomological dimension if and only if the orbit θ is regular [12]. Therefore, in general, we have the left derived functor $L\Delta_\lambda : D^-(\mathcal{U}_\theta) \to D^-(\mathcal{D}_\lambda)$ which is the left adjoint of $R\Gamma : D^-(\mathcal{D}_\lambda) \to D^-(\mathcal{U}_\theta)$. If θ is regular, $L\Delta_\lambda$ extends to $D(\mathcal{U}_\theta)$ and is the left adjoint of $R\Gamma : D(\mathcal{D}_\lambda) \to D(\mathcal{U}_\theta)$. Moreover, for regular θ, we have the following result [2].

4.1. Theorem. *If $\theta \in \mathfrak{h}^*$ is regular, the functor $R\Gamma : D(\mathcal{D}_\lambda) \to D(\mathcal{U}_\theta)$ is an equivalence of categories. Its quasi-inverse is $L\Delta_\lambda : D(\mathcal{U}_\theta) \to D(\mathcal{D}_\lambda)$.*

Clearly, the group $\mathrm{Int}(\mathfrak{g})$ acts algebraically on the sheaf of algebras $\mathcal{D}_\lambda$. Let K be an algebraic group satisfying the conditions from §1. Then K acts algebraically on X and $\mathcal{D}_\lambda$. $\mathcal{V}$ is a *weak Harish-Chandra sheaf* on X if

(i) $\mathcal{V}$ is a quasicoherent $\mathcal{D}_\lambda$-module;
(ii) $\mathcal{V}$ is a K-equivariant $\mathcal{O}_X$-module (cf. [16]);
(iii) the action morphism $\mathcal{D}_\lambda \otimes_{\mathcal{O}_X} \mathcal{V} \to \mathcal{V}$ is a morphism of K-equivariant $\mathcal{O}_X$-modules.

A weak Harish-Chandra sheaf $\mathcal{V}$ is a *Harish-Chandra sheaf*[4] if the differential of the K-action on $\mathcal{V}$ agrees with the action of $\mathfrak{k}$ given by the map $\mathfrak{k} \to \mathcal{U}(\mathfrak{g}) \to \mathcal{D}_\lambda$. A morphism of weak Harish-Chandra sheaves is a $\mathcal{D}_\lambda$-module morphism which is also a morphism of K-equivariant $\mathcal{O}_X$-modules. We denote by $\mathcal{M}_{qc}(\mathcal{D}_\lambda, K)_w$ the abelian category of weak Harish-Chandra sheaves, and by $\mathcal{M}_{qc}(\mathcal{D}_\lambda, K)$ its full subcategory of Harish-Chandra sheaves.

[4]In [14] we assumed that a Harish-Chandra sheaf is a coherent $\mathcal{D}_\lambda$-module. No such restriction is convenient in our present setting.

For any $\mathcal{V}$ in $\mathcal{M}_{qc}(\mathcal{D}_\lambda, K)_w$, the module $\Gamma(X, \mathcal{V})$ of global sections of $\mathcal{V}$ is in $\mathcal{M}(\mathcal{U}_\theta, K)_w$. Conversely, the localization $\Delta_\lambda(V)$ of a weak Harish-Chandra module from $\mathcal{M}(\mathcal{U}_\theta, K)_w$ is in $\mathcal{M}_{qc}(\mathcal{D}_\lambda, K)_w$. Also,

$$\operatorname{Hom}_{(\mathcal{D}_\lambda, K)}(\Delta_\lambda(U), \mathcal{V}) = \operatorname{Hom}_{(\mathcal{U}_\theta, K)}(U, \Gamma(X, \mathcal{V})),$$

for any U in $\mathcal{M}(\mathcal{U}_\theta, K)_w$ and $\mathcal{V}$ in $\mathcal{M}_{qc}(\mathcal{D}_\lambda, K)_w$, i.e., the functors Δ_λ and Γ are again an adjoint pair. Moreover, if $\mathcal{V}$ is a Harish-Chandra sheaf, $\Gamma(X, \mathcal{V})$ is a Harish-Chandra module. Also, if V is a Harish-Chandra module, $\Delta_\lambda(V)$ is a Harish-Chandra sheaf.

Let Σ be the root system in $\mathfrak{h}^*$ attached to $\mathfrak{g}$, and let $Q(\Sigma)$ be the corresponding root lattice. For any $\nu \in Q(\Sigma)$, let $\mathcal{O}(\nu)$ be the corresponding $\operatorname{Int}(\mathfrak{g})$-homogeneous invertible $\mathcal{O}_X$-module on X. Then we have the natural *twist* functor $\mathcal{V} \longmapsto \mathcal{V}(\nu) = \mathcal{V} \otimes_{\mathcal{O}_X} \mathcal{O}(\nu)$ from $\mathcal{M}_{qc}(\mathcal{D}_\lambda)$ into $\mathcal{M}_{qc}(\mathcal{D}_{\lambda+\nu})$ ([14], §4). This functor is clearly an equivalence of categories and its quasi-inverse is $\mathcal{W} \longmapsto \mathcal{W}(-\nu)$. The twist functor preserves (weak) Harish-Chandra sheaves, hence it induces equivalences of $\mathcal{M}_{qc}(\mathcal{D}_\lambda, K)_w$ with $\mathcal{M}_{qc}(\mathcal{D}_{\lambda+\nu}, K)_w$ (resp. $\mathcal{M}_{qc}(\mathcal{D}_\lambda, K)$ with $\mathcal{M}_{qc}(\mathcal{D}_{\lambda+\nu}, K)$).

For a root $\alpha \in \Sigma$, let $\alpha\check{}$ be its dual root. By the Borel-Weil theorem, there exists a unique set of positive roots Σ^+ in Σ such that $H^p(X, \mathcal{O}(\nu))$ vanish for $p > 0$ for all $\nu \in Q(\Sigma)$ satisfying $\alpha\check{}(\nu) \leq 0$ for $\alpha \in \Sigma^+$. We say that $\lambda \in \mathfrak{h}^*$ is *antidominant* if $\alpha\check{}(\lambda)$ is not a positive integer for $\alpha \in \Sigma^+$. If λ is antidominant and regular, the functor $\Gamma : \mathcal{M}_{qc}(\mathcal{D}_\lambda) \longrightarrow \mathcal{M}(\mathcal{U}_\theta)$ is an equivalence of categories (cf. [14], 3.7). Hence, it induces an equivalence of $\mathcal{M}_{qc}(\mathcal{D}_\lambda, K)_w$ with $\mathcal{M}(\mathcal{U}_\theta, K)_w$. By the analogue of 2.2, we know that the forgetful functor $\mathcal{M}(\mathcal{U}_\theta, K)_w \longrightarrow \mathcal{M}(\mathcal{U}_\theta)$ has a right adjoint Ind_w and the adjointness morphism $V \longrightarrow \operatorname{Ind}_w(V)$ is a monomorphism. Therefore, for any regular antidominant λ, the forgetful functor $\mathcal{M}_{qc}(\mathcal{D}_\lambda, K)_w \longrightarrow \mathcal{M}_{qc}(\mathcal{D}_\lambda)$ has a right adjoint $Ind_w : \mathcal{M}_{qc}(\mathcal{D}_\lambda) \longrightarrow \mathcal{M}_{qc}(\mathcal{D}_\lambda, K)_w$ and the adjointness morphism $\mathcal{V} \longrightarrow Ind_w(\mathcal{V})$ is a monomorphism. By applying the twist functor we deduce that this statement holds for arbitrary λ in $\mathfrak{h}^*$.

The functor Ind_w can be described in geometric terms. Let $p : K \times X \longrightarrow X$ be the projection to the second variable. Let $\mu : K \times X \longrightarrow X$ be the action morphism, i.e., $\mu(k, x) = k \cdot x$ for $k \in K$ and $x \in X$. For a morphism f of algebraic varieties, we denote by f^* and f_* the inverse image and the direct image functors between the corresponding categories of $\mathcal{O}$-modules.

4.2. Lemma. *For any $\mathcal{V}$ in $\mathcal{M}_{qc}(\mathcal{D}_\lambda)$, we have $Ind_w(\mathcal{V}) = \mu_*(p^*(\mathcal{V}))$ as $\mathcal{O}_X$-modules.*

Since μ is an affine morphism,

$$\begin{aligned} H^p(X, Ind_w(\mathcal{V})) &= H^p(X, \mu_*(p^*(\mathcal{V}))) = H^p(K \times X, p^*(\mathcal{V})) \\ &= R(K) \otimes H^p(X, \mathcal{V}) = \operatorname{Ind}_w(H^p(X, \mathcal{V})), \end{aligned}$$

for $\mathcal{V}$ in $\mathcal{M}_{qc}(\mathcal{D}_\lambda)$ and $p \in \mathbb{Z}_+$. In particular, if $\mathcal{I}$ is an injective quasicoherent $\mathcal{D}_\lambda$-module, $\mathcal{I}$ is Γ-acyclic and $H^p(X, Ind_w(\mathcal{I})) = 0$ for $p > 0$, i.e., $Ind_w(\mathcal{I})$ is also Γ-acyclic.

4.3. Lemma.

(i) *The category $\mathcal{M}_{qc}(\mathcal{D}_\lambda, K)_w$ has enough injectives.*
(ii) *Injective weak Harish-Chandra sheaves are acyclic for the functor of global sections $\Gamma(X, -)$.*

Proof. (i) Let $\mathcal{V}$ be a weak Harish-Chandra sheaf. Since the category $\mathcal{M}_{qc}(\mathcal{D}_\lambda)$ has enough injectives, there exist an injective quasicoherent $\mathcal{D}_\lambda$-module $\mathcal{I}$ and a monomorphism $\mathcal{V} \longrightarrow \mathcal{I}$. Since Ind_w is exact, $Ind_w(\mathcal{V}) \longrightarrow Ind_w(\mathcal{I})$ is a monomorphism of weak Harish-Chandra sheaves. Therefore, $\mathcal{V} \longrightarrow Ind_w(\mathcal{I})$ is a monomorphism of weak Harish-Chandra sheaves. On the other hand, since Ind_w is the right adjoint of an exact functor, it preserves injectives. This implies that $Ind_w(\mathcal{I})$ is an injective weak Harish-Chandra sheaf.

(ii) Let $\mathcal{J}$ be an injective weak Harish-Chandra sheaf. Then, by the above argument, $\mathcal{J}$ is a submodule of $Ind_w(\mathcal{I})$ for some injective quasicoherent $\mathcal{D}_\lambda$-module $\mathcal{I}$. Therefore, $Ind_w(\mathcal{I}) = \mathcal{J} \oplus \mathcal{W}$ for some weak Harish-Chandra sheaf $\mathcal{W}$. But this implies that $H^p(X, \mathcal{J})$ is a direct summand of $H^p(X, Ind_w(\mathcal{I}))$ for $p \in \mathbb{Z}_+$. Hence, $H^p(X, \mathcal{J}) = 0$ for $p > 0$. □

Let $D^*(\mathcal{D}_\lambda, K)_w = D^*(\mathcal{M}_{qc}(\mathcal{D}_\lambda, K)_w)$ be the derived category of the abelian category of weak Harish-Chandra sheaves. Since $\mathcal{M}_{qc}(\mathcal{D}_\lambda, K)_w$ has enough injectives, the right derived functor $R\Gamma : D^+(\mathcal{D}_\lambda, K)_w \longrightarrow D^+(\mathcal{U}_\theta, K)_w$ of Γ exists. Moreover, by 4.3(ii), the following diagram is commutative:

$$\begin{array}{ccc} D^+(\mathcal{D}_\lambda, K)_w & \xrightarrow{R\Gamma} & D^+(\mathcal{U}_\theta, K)_w \\ \downarrow & & \downarrow \\ D^+(\mathcal{D}_\lambda) & \xrightarrow[R\Gamma]{} & D^+(\mathcal{U}_\theta) \end{array},$$

where the vertical arrows are the forgetful functors, and the lower horizontal arrow is the standard cohomology functor. This explains the ambiguity in our notation. Moreover, since the right cohomological dimension is finite, the standard truncation argument extends this statement to derived categories of unbounded complexes.

Now, let V be a weak Harish-Chandra module in $\mathcal{M}(\mathcal{U}_\theta, K)_w$. Let $P_w(V) = \mathcal{U}_\theta \otimes_{\mathbb{C}} V$ be the module on which K acts by the tensor product of the action ϕ on $\mathcal{U}_\theta$ with the natural action on V, and $\mathcal{U}_\theta$ by the multiplication in the first factor. Then $P_w(V)$ is a weak Harish-Chandra module in $\mathcal{M}(\mathcal{U}_\theta, K)_w$ since

$$k\cdot(ST\otimes v) = \phi(k)(ST)\otimes\nu(k)v = \phi(k)(S)\phi(k)(T)\otimes\nu(k)v = \phi(k)(S)(k\cdot(T\otimes v))$$

for $S, T \in \mathcal{U}_\theta$, $k \in K$ and $v \in V$. Moreover, the natural map $p : P_w(V) \longrightarrow V$ given by $p(T \otimes v) = \pi(T)v$, $T \in \mathcal{U}_\theta$, $v \in V$, satisfies

$$p(ST \otimes v) = \pi(ST)v = \pi(S)\pi(T)v = \pi(S)p(T \otimes v)$$

and

$$p(k\cdot(T\otimes v)) = p(\phi(k)T\otimes\nu(k)v) = \pi(\phi(k)T)\nu(k)v = \nu(k)\pi(T)v = \nu(k)p(T\otimes v)$$

for all $S, T \in \mathcal{U}_\theta$, $k \in K$ and $v \in V$; i.e., p is an epimorphism of weak Harish-Chandra modules. Clearly, $P_w(V)$ is a free $\mathcal{U}_\theta$-module and therefore Δ_λ-acyclic. This implies that $\Delta_\lambda : \mathcal{M}(\mathcal{U}_\theta, K)_w \to \mathcal{M}(\mathcal{D}_\lambda, K)_w$ has a left derived functor $L\Delta_\lambda : D^-(\mathcal{U}_\theta, K)_w \to D^-(\mathcal{D}_\lambda, K)_w$ and in addition the following diagram commutes:

$$\begin{array}{ccc} D^-(\mathcal{U}_\theta, K)_w & \xrightarrow{L\Delta_\lambda} & D^-(\mathcal{D}_\lambda, K)_w \\ \downarrow & & \downarrow \\ D^-(\mathcal{U}_\theta) & \xrightarrow[L\Delta_\lambda]{} & D^-(\mathcal{D}_\lambda) \end{array},$$

where the vertical arrows are forgetful functors and the lower horizontal arrow is the usual localization functor. In addition, we have

$$\mathrm{Hom}_{D^-(\mathcal{D}_\lambda,K)_w}(L\Delta_\lambda(U^\cdot), \mathcal{V}^\cdot) = \mathrm{Hom}_{D^-(\mathcal{U}_\theta,K)_w}(U^\cdot, R\Gamma(X, \mathcal{V}^\cdot)).$$

From the above discussion, it also follows that $R\Gamma \circ L\Delta_\lambda \cong 1$ on $D^-(\mathcal{U}_\theta, K)_w$, since this is obviously true on modules $P_w(V)$.

In addition, if θ is regular, the left cohomological dimension of Δ_λ is finite and $L\Delta_\lambda$ extends to $D(\mathcal{U}_\theta, K)_w$. Moreover, we have the following result, which is a variant of 4.1.

4.4. Theorem. *Let θ be regular. Then $R\Gamma : D(\mathcal{D}_\lambda, K)_w \to D(\mathcal{U}_\theta, K)_w$ is an equivalence of categories. Its quasiinverse is $L\Delta_\lambda : D(\mathcal{U}_\theta, K)_w \to D(\mathcal{D}_\lambda, K)_w$.*

Remark. The preceding results show that weak Harish-Chandra modules behave nicely with respect to the cohomology and localization functors, in sharp contrast to the case of Harish-Chandra modules. To see this, the reader should consider the case of $\mathcal{M}_{qc}(\mathcal{D}_\lambda, K)$, with $K = \mathrm{Int}(\mathfrak{g})$, which is clearly a semisimple abelian category.

In complete analogy with the constructions in §2, we can define equivariant complexes of Harish-Chandra sheaves, and corresponding categories $C^*(\mathcal{D}_\lambda, K)$ and $K^*(\mathcal{D}_\lambda, K)$. By localizing $K^*(\mathcal{D}_\lambda, K)$ with respect to quasi-isomorphisms we get the equivariant derived category $D^*(\mathcal{D}_\lambda, K)$ of Harish-Chandra sheaves.

The natural functors Γ from $K(\mathcal{D}_\lambda, K)$ into $K(\mathcal{U}_\theta, K)$ and Δ_λ from $K^-(\mathcal{U}_\theta, K)$ into $K^-(\mathcal{D}_\lambda, K)$ have right, resp. left, derived functors. More precisely, we have the following result.

4.5. Theorem. *The functors Γ and Δ_λ define the corresponding derived functors*

$$R\Gamma : D(\mathcal{D}_\lambda, K) \to D(\mathcal{U}_\theta, K)$$

and

$$L\Delta_\lambda : D^-(\mathcal{U}_\theta, K) \to D^-(\mathcal{D}_\lambda, K)$$

such that the following diagrams commute

$$\begin{array}{ccc} D(\mathcal{D}_\lambda, K) & \xrightarrow{R\Gamma} & D(\mathcal{U}_\theta, K) \\ \downarrow & & \downarrow \\ D(\mathcal{D}_\lambda, K)_w & \xrightarrow[R\Gamma]{} & D(\mathcal{U}_\theta, K)_w \end{array} ;$$

$$\begin{array}{ccc} D^-(\mathcal{U}_\theta, K) & \xrightarrow{L\Delta_\lambda} & D^-(\mathcal{D}_\lambda, K) \\ \downarrow & & \downarrow \\ D^-(\mathcal{U}_\theta, K)_w & \xrightarrow[L\Delta_\lambda]{} & D^-(\mathcal{D}_\lambda, K)_w \end{array} ;$$

where the vertical arrows represent forgetful functors.

Moreover, $L\Delta_\lambda$ is the left adjoint of $R\Gamma$, i.e.,

$$\mathrm{Hom}_{D^-(\mathcal{D}_\lambda,K)}(L\Delta_\lambda(U^\cdot), \mathcal{V}^\cdot) = \mathrm{Hom}_{D^-(\mathcal{U}_\theta,K)}(U^\cdot, R\Gamma(X, \mathcal{V}^\cdot))$$

for $U^\cdot$ in $D^-(\mathcal{U}_\theta, K)$ and $\mathcal{V}^\cdot$ in $D^-(\mathcal{D}_\lambda, K)$.

If θ is regular, Δ_λ has finite left cohomological dimension and $L\Delta_\lambda$ extends to $D(\mathcal{U}_\theta, K)$. This leads to the following equivariant version of 4.1.

4.6. Theorem. *Let θ be regular. Then $R\Gamma : D(\mathcal{D}_\lambda, K) \to D(\mathcal{U}_\theta, K)$ is an equivalence of categories. Its quasi-inverse is $L\Delta_\lambda : D(\mathcal{U}_\theta, K) \to D(\mathcal{D}_\lambda, K)$.*

For any $\nu \in Q(\Sigma)$, the twist functor $\mathcal{V} \longmapsto \mathcal{V}(\nu)$ induces equivalences of the corresponding derived categories, i.e., the equivalences $D(\mathcal{D}_\lambda, K)_w \to D(\mathcal{D}_{\lambda+\nu}, K)_w$ and $D(\mathcal{D}_\lambda, K) \to D(\mathcal{D}_{\lambda+\nu}, K)$.

Let H be a closed subgroup of K. Then we have the "forgetful" functor For : $D(\mathcal{D}_\lambda, K) \to D(\mathcal{D}_\lambda, H)$. Clearly, it commutes with twists, i.e., the following diagram is commutative:

$$\begin{array}{ccc} D(\mathcal{D}_\lambda, K) & \xrightarrow{\text{For}} & D(\mathcal{D}_\lambda, H) \\ {\scriptstyle -(\nu)}\downarrow & & \downarrow{\scriptstyle -(\nu)} \\ D(\mathcal{D}_{\lambda+\nu}, K) & \xrightarrow[\text{For}]{} & D(\mathcal{D}_{\lambda+\nu}, H) \end{array}$$

for any $\nu \in Q(\Sigma)$. Moreover, by 4.5 and 4.6, we have a commutative diagram

$$\begin{array}{ccc} D(\mathcal{D}_\lambda, K) & \xrightarrow{\text{For}} & D(\mathcal{D}_\lambda, H) \\ {\scriptstyle R\Gamma}\downarrow & & \downarrow{\scriptstyle R\Gamma} \\ D(\mathcal{U}_\theta, K) & \xrightarrow[\text{For}]{} & D(\mathcal{U}_\theta, H) \end{array} ,$$

where the vertical arrows are equivalences for λ regular. This, combined with 2.7 and 2.13, implies the following result.

4.7. Theorem.

(i) *The forgetful functor* For : $D(\mathcal{D}_\lambda, K) \to D(\mathcal{D}_\lambda, H)$ *has a right adjoint* $\Gamma^{geo}_{K,H} : D(\mathcal{D}_\lambda, H) \to D(\mathcal{D}_\lambda, K)$ *of finite amplitude.*

(ii) *The functor* $\Gamma^{geo}_{K,H}$ *commutes with twists, i.e., the following diagram is commutative:*

$$\begin{array}{ccc} D(\mathcal{D}_\lambda, H) & \xrightarrow{\Gamma^{geo}_{K,H}} & D(\mathcal{D}_\lambda, K) \\ {\scriptstyle -(\nu)}\downarrow & & \downarrow{\scriptstyle -(\nu)} \\ D(\mathcal{D}_{\lambda+\nu}, H) & \xrightarrow[\Gamma^{geo}_{K,H}]{} & D(\mathcal{D}_{\lambda+\nu}, K) \end{array}$$

for any $\nu \in Q(\Sigma)$.

(iii) *The following diagram is commutative:*

$$\begin{array}{ccc} D(\mathcal{D}_\lambda, H) & \xrightarrow{\Gamma^{geo}_{K,H}} & D(\mathcal{D}_\lambda, K) \\ {\scriptstyle R\Gamma}\downarrow & & \downarrow{\scriptstyle R\Gamma} \\ D(\mathcal{U}_\theta, H) & \xrightarrow[R\Gamma^{equi}_{K,H}]{} & D(\mathcal{U}_\theta, K) \end{array}.$$

We call the functor $\Gamma^{geo}_{K,H}$ the *geometric Zuckerman functor*. By 4.7 (iii), we can view it as the localization of the equivariant Zuckerman functor $R\Gamma^{equi}_{K,H}$. This functor can be described in $\mathcal{D}$-module theoretic terms using techniques analogous to [4] (the details will appear in [15]). It can be viewed as a generalization of Bernstein's functor of "integration along K-orbits".

5. Cohomology of standard Harish-Chandra sheaves

First we recall the construction of the standard Harish-Chandra sheaves (cf. [14], §6). Let $\lambda \in \mathfrak{h}^*$. Let Q be a K-orbit in X and τ an irreducible K-homogeneous connection on Q compatible with $\lambda + \rho$.[5] Then the direct image of τ with respect to the inclusion $Q \to X$ is the standard Harish-Chandra sheaf $\mathcal{I}(Q, \tau)$. Since τ is holonomic, $\mathcal{I}(Q, \tau)$ is also a holonomic $\mathcal{D}_\lambda$-module and therefore of finite length. This implies that its cohomologies $H^p(X, \mathcal{I}(Q, \tau))$, $p \in \mathbb{Z}_+$, are Harish-Chandra modules of finite length [13]. In this section, we calculate these cohomology modules in terms of "classical" Zuckerman functors.

Fix $x \in Q$. Denote by $\mathfrak{b}_x$ the Borel subalgebra of $\mathfrak{g}$ corresponding to x, and by S_x the stabilizer of x in K. Then the geometric fiber $T_x(\tau)$ of τ at x is an irreducible finite-dimensional representation ω of S_x. We can view it as an S_x-equivariant connection over the S_x-orbit $\{x\}$. Therefore, we can consider the standard Harish-Chandra sheaf $\mathcal{I}(\omega) = \mathcal{I}(\{x\}, \omega)$. It is an S_x-equivariant $\mathcal{D}_\lambda$-module. The following lemma is critical.

[5]Here ρ is the half-sum of positive roots in Σ^+.

5.1. Lemma.

$$\Gamma^{geo}_{K,S_x}(D(\mathcal{I}(\omega))) = D(\mathcal{I}(Q,\tau))[-\dim Q].$$

The proof of this lemma follows from the geometric description of the functor Γ^{geo}_{K,S_x} which we mentioned at the end of §4. This construction makes sense on any smooth algebraic variety with a K-action. By specialization, the linear form $\lambda+\rho$ determines a linear form on $\mathfrak{b}_x$. By restriction, it determines a linear form μ on the Lie algebra $\mathfrak{s}_x$ of S_x. In turn, μ determines a homogeneous twisted sheaf of differential operators $\mathcal{D}_{Q,\mu}$ on Q. Therefore, we can consider the equivariant derived categories $D^b(\mathcal{D}_{Q,\mu}, S_x)$ and $D^b(\mathcal{D}_{Q,\mu}, K)$ and the functor $\Gamma^{geo}_{K,S_x} : D^b(\mathcal{D}_{Q,\mu}, S_x) \longrightarrow D^b(\mathcal{D}_{Q,\mu}, K)$. Also, the following diagram commutes

$$\begin{array}{ccc} D^b(\mathcal{D}_{Q,\mu}, S_x) & \xrightarrow{i_{Q,+}} & D^b(\mathcal{D}_\lambda, S_x) \\ \Gamma^{geo}_{K,S_x}\big\downarrow & & \big\downarrow\Gamma^{geo}_{K,S_x} \\ D^b(\mathcal{D}_{Q,\mu}, K) & \xrightarrow{i_{Q,+}} & D^b(\mathcal{D}_\lambda, K) \end{array}.$$

Let $j_x : \{x\} \longrightarrow Q$ be the natural immersion. Then we have the $\mathcal{D}$-module direct image module $\mathcal{J}(\omega) = j_{x,+}(\omega)$. To establish 5.1, by the above diagram, it is enough to show that

$$\Gamma^{geo}_{K,S_x}(D(\mathcal{J}(\omega))) = D(\tau)[-\dim Q].$$

Clearly, $\mathcal{D}_{Q,0}$ is the sheaf of differential operators $\mathcal{D}_Q$ on Q. If ω is trivial, $\tau = \mathcal{O}_Q$, and we put $\mathcal{J} = \mathcal{J}(\omega) = j_{x,+}(\mathbb{C})$. Moreover, the general formula follows by tensoring with τ, from the special case

$$\Gamma^{geo}_{K,S_x}(D(\mathcal{J})) = D(\mathcal{O}_Q)[-\dim Q].$$

We define an action of $K \times K$ on K by

$$(k,l)\cdot h = khl^{-1}, \text{ for } k,h,l \in K.$$

Consider the orbit map $o_x : K \longrightarrow Q$ given by $o_x(k) = k \cdot x$ for $k \in K$. Let $\mathcal{D}_K$ be the sheaf of differential operators on K. Then the inverse image o_x^* is a functor from $D^b(\mathcal{D}_Q, K)$ into $D^b(\mathcal{D}_K, K \times S_x)$. It is an equivalence of categories (compare [4]). Also, o_x^* induces an equivalence of $D^b(\mathcal{D}_Q, S_x)$ with $D^b(\mathcal{D}_K, S_x \times S_x)$. Therefore we have the following commutative diagram:

$$\begin{array}{ccc} D^b(\mathcal{D}_Q, S_x) & \xrightarrow{o_x^*} & D^b(\mathcal{D}_K, S_x \times S_x) \\ \Gamma^{geo}_{K,S_x}\big\downarrow & & \big\downarrow\Gamma^{geo}_{K\times S_x, S_x\times S_x} \\ D^b(\mathcal{D}_Q, K) & \xrightarrow{o_x^*} & D^b(\mathcal{D}_K, K \times S_x) \end{array}.$$

The map $k \mapsto k^{-1}$ of K we induces the equivalences $D^b(\mathcal{D}_K, S_x \times S_x) \longrightarrow D^b(\mathcal{D}_K, S_x \times S_x)$ and $D^b(\mathcal{D}_K, K \times S_x) \longrightarrow D^b(\mathcal{D}_K, S_x \times K)$. Let $\pi : K \longrightarrow pt$ be the projection of K onto a point pt. Then, as before, the inverse image $\pi^* : D^b(\mathcal{D}_{pt}, S_x) \longrightarrow D^b(\mathcal{D}_K, S_x \times K)$ is an equivalence of categories. Let $q : Q \longrightarrow pt$. This leads to the following commutative diagram:

$$\begin{array}{ccc} D^b(\mathcal{D}_Q, S_x) & \xrightarrow{o_x^*} & D^b(\mathcal{D}_K, S_x \times S_x) \\ {\scriptstyle q^*}\big\uparrow & & \big\uparrow{\scriptstyle \mathrm{For}} \\ D^b(\mathcal{D}_{pt}, S_x) & \xrightarrow{\pi^*} & D^b(\mathcal{D}_K, S_x \times K) \end{array}$$

where the horizontal arrows are equivalences of categories. This diagram implies that q^* has a right adjoint Φ and that the following diagram commutes:

$$\begin{array}{ccc} D^b(\mathcal{D}_Q, S_x) & \xrightarrow{o_x^*} & D^b(\mathcal{D}_K, S_x \times S_x) \\ {\scriptstyle \Phi}\big\downarrow & & \big\downarrow{\scriptstyle \Gamma^{geo}_{S_x \times K, S_x \times S_x}} \\ D^b(\mathcal{D}_{pt}, S_x) & \xrightarrow{\pi^*} & D^b(\mathcal{D}_K, S_x \times K) \end{array} .$$

Since q is a smooth morphism, the shifted $\mathcal{D}$-module direct image functor $q_+[-\dim Q]$ is the right adjoint of the inverse image q^*. Therefore, we have the following commutative diagram:

$$\begin{array}{ccc} D^b(\mathcal{D}_Q, S_x) & \xrightarrow{\mathrm{For}} & D^b(\mathcal{D}_Q) \\ {\scriptstyle \Phi}\big\downarrow & & \big\downarrow{\scriptstyle q_+[-\dim Q]} \\ D^b(\mathcal{D}_{pt}, S_x) & \xrightarrow{\mathrm{For}} & D^b(\mathcal{D}_{pt}) \end{array} .$$

This allows us to calculate $\Gamma^{geo}_{K,S_x}(D(\mathcal{J}))$. Following the above equivalences and forgetting the equivariant structure, we see that it corresponds to

$$q_+(D(\mathcal{J}))[-\dim Q] = D((q_+ \circ j_{x,+})(\mathbb{C}))[-\dim Q] = D(\mathbb{C})[-\dim Q]$$

in $D^b(\mathcal{D}_{pt})$. By following the equivalences in the reverse order, we see that this object corresponds to $D(\mathcal{O}_Q)[-\dim Q]$. This completes the sketch of the proof of 5.1.[6]

The formula from 5.1 immediately leads to the following result:

$$R\Gamma(D(\mathcal{I}(Q,\tau))) = R\Gamma(\Gamma^{geo}_{K,S_x}(D(\mathcal{I}(\omega))))[\dim Q] = \Gamma^{equi}_{K,S_x}(R\Gamma(D(\mathcal{I}(\omega))))[\dim Q].$$

---

[6]The starting point of our investigation was Bernstein's argument to prove a special case of the duality theorem of [11]. Bernstein explained that argument in a seminar at the Institute for Advanced Study in the fall of 1985. It can be used to prove 5.1. If λ is antidominant and regular, the functor Γ and the localization functor Δ_λ are exact and the formula in 5.1 follows from the results explained in ([6], II.4). The general case of 5.1 follows immediately, since Γ^{geo}_{K,S_x} commutes with the twists by 4.7.

On the other hand, if δ is a finite-dimensional algebraic representation of S_x compatible with $\lambda+\rho$, we can view it as a $(\mathfrak{b}_x, S_x)$-module where $\mathfrak{b}_x$ acts by $\lambda+\rho$. Let Ω_X be the invertible $\mathcal{O}_X$-module of top degree differential forms on X. Then, its geometric fiber $T_x(\Omega_X)$ is one-dimensional and the Borel subgroup B_x of $\mathrm{Int}(\mathfrak{g})$ acts on it by a character. The differential of this action is equal to the specialization of 2ρ. Therefore, $\delta \otimes T_x(\Omega_X^{-1})$ can be viewed as a $(\mathfrak{b}_x, S_x)$-module where $\mathfrak{b}_x$ acts by the specialization of $\lambda-\rho$. Let

$$M(\delta) = \mathcal{U}(\mathfrak{g}) \otimes_{\mathcal{U}(\mathfrak{b}_x)} (\delta \otimes T_x(\Omega_X^{-1})),$$

where $\mathfrak{g}$ acts by left multiplication on the first factor and S_x by the tensor product of the action ϕ on $\mathcal{U}(\mathfrak{g})$ with the natural representation on $\delta \otimes T_x(\Omega_X^{-1})$. Then $M(\delta)$ is a Harish-Chandra module for the pair $(\mathfrak{g}, S_x)$. As a $\mathcal{U}(\mathfrak{g})$-module, $M(\delta)$ is a direct sum of $\dim\delta$ copies of the Verma module $M(\lambda) = \mathcal{U}(\mathfrak{g}) \otimes_{\mathcal{U}(\mathfrak{b}_x)} \mathbb{C}_{\lambda-\rho}$.

The following result is well known (see, for example, [13]).

5.2. Lemma. *We have*

$$H^p(X, \mathcal{I}(\omega)) = \begin{cases} M(\omega) \text{ for } p=0; \\ 0 \text{ for } p>0. \end{cases}$$

Therefore, the above relation implies that

$$R\Gamma(D(\mathcal{I}(Q,\tau))) = \Gamma^{equi}_{K,S_x}(D(\Gamma(\mathcal{I}(\omega))))[\dim Q] = \Gamma^{equi}_{K,S_x}(D(M(\omega)))[\dim Q].$$

Taking the cohomology of this complex and using the generalization of 2.10 from 2.13, we get

$$H^p(X, \mathcal{I}(Q,\tau)) = H^{p+\dim Q}(\Gamma^{equi}_{K,S_x}(D(M(\omega))) = R^{p+\dim Q}\Gamma_{K,S_x}(M(\omega))).$$

This proves the following result which computes the cohomology of standard Harish-Chandra sheaves.

5.3. Theorem. *Let $\lambda \in \mathfrak{h}^*$, Q a K-orbit in X and τ an irreducible K-homogeneous connection compatible with $\lambda+\rho$. Let $x \in Q$ and let S_x be the stabilizer of x in K. Let ω be the representation of S_x in the geometric fiber $T_x(\tau)$. Then we have*

$$H^p(X, \mathcal{I}(Q,\tau)) = R^{p+\dim Q}\Gamma_{K,S_x}(M(\omega))$$

for any $p \in \mathbb{Z}$.

Assume now that, in addition, K is reductive. Denote by T a Levi factor of S_x and by U_x the unipotent radical of S_x. Then we have

$$\dim(K/T) - \dim Q = \dim K - \dim T - (\dim K - \dim S_x) = \dim S_x - \dim T = \dim U_x.$$

By 1.13, we have

$$R^p\Gamma_{K,S_x}(M(\omega)) = R^p\Gamma_{K,T}(M(\omega)), \quad p \in \mathbb{Z}.$$

Denote by $V \longmapsto V\tilde{}$ the contragredient functor on the categories $\mathcal{M}(\mathfrak{g},T)$ and $\mathcal{M}(\mathfrak{g},K)$. Then, by the duality theorem for derived Zuckerman modules (see, for example, [9], [17]), we have

$$\begin{aligned} H^p(X,\mathcal{I}(Q,\tau))\tilde{} &= R^{p+\dim Q}\Gamma_{K,T}(M(\omega))\tilde{} \\ &= R^{\dim(K/T)-\dim Q-p}\Gamma_{K,T}(M(\omega)\tilde{}) = R^{\dim U_x - p}\Gamma_{K,T}(M(\omega)\tilde{}), \quad p \in \mathbb{Z}. \end{aligned}$$

This is exactly the statement of the duality theorem of Hecht, Miličić, Schmid and Wolf [11].

References

1. A. Beilinson, J. Bernstein, *Localisation de $\mathfrak{g}$-modules*, C. R. Acad. Sci. Paris, Ser. I **292** (1981), 15–18.
2. A. Beilinson, J. Bernstein, *A generalization of Casselman's submodule theorem*, Representation Theory of Reductive Groups (P.C. Trombi, ed.), Progress in Math., vol. 40, Birkhäuser, Boston, 1983, pp. 35–52.
3. A. Beilinson, J. Bernstein, *A proof of the Jantzen conjecture*, (preprint) (1989).
4. J. Bernstein, V. Lunts, *Equivariant sheaves and functors*, Lecture Notes in Math., vol. 1578, Springer-Verlag, Berlin-Heidelberg-Tokyo, 1994.
5. J. Bernstein, V. Lunts, *Localization for derived categories of $(\mathfrak{g},K)$-modules*, Journal of Amer. Math. Soc. **8** (1995), 819–856.
6. F. Bien, *$\mathcal{D}$-modules and spherical representations*, Mathematical Notes, vol. 39, Princeton University Press, Princeton, N.J., 1990.
7. A. Borel, N. Wallach, *Continuous cohomology, discrete subgroups, and representations of reductive groups*, Annals of Math. Studies, vol. 94, Princeton University Press, 1980.
8. P. Deligne, *Cohomologie à supports propres*, SGA 4, Lecture Notes in Math., vol. 305, Springer-Verlag, Berlin-Heidelberg-Tokyo, 1973.
9. M. Duflo, M. Vergne, *Sur le functeur de Zuckerman*, C. R. Acad. Sci. Paris **304** (1987), 467–469.
10. S. I. Gelfand, Yu.I. Manin, *Methods of homological algebra*, Springer-Verlag, Berlin, Heidelberg, New York, 1996.
11. H. Hecht, D. Miličić, W. Schmid, J. A. Wolf, *Localization and standard modules for real semisimple Lie groups I: The duality theorem*, Inventiones Math. **90** (1987), 297–332.
12. H. Hecht, D. Miličić, *On the cohomological dimension of the localization functor*, Proc. Amer. Math. Soc. **108** (1990), 249-254.
13. D. Miličić, *Localization and representation theory of reductive Lie groups*, (mimeographed notes), to appear.
14. D. Miličić, *Algebraic $\mathcal{D}$-modules and representation theory of semisimple Lie groups*, Analytic Cohomology and Penrose Transform (M. Eastwood, J.A. Wolf, R. Zierau, eds.), Contemporary Mathematics, vol. 154, Amer. Math. Soc., 1993, pp. 133–168.
15. D. Miličić, P. Pandžić, *Cohomology of standard Harish-Chandra sheaves*, (in preparation).
16. D. Mumford, *Geometric invariant theory*, Ergebnisse der Mathematik, vol. 34, Springer-Verlag, Berlin, Heidelberg, New York, 1965.
17. P. Pandžić, *Equivariant analogues of Zuckerman functors*, Ph.D. Thesis, University of Utah, Salt Lake City, 1995.
18. J.L. Verdier, *Catégories dérivées, état 0*, SGA $4\frac{1}{2}$, Lecture Notes in Math., vol. 569, Springer-Verlag, Berlin, Heidelberg, New York, 1977.

19. D.A. Vogan, *Representations of real reductive Lie groups*, Progress in Math., vol. 15, Birkhäuser, Boston, 1981.

(MILIČIĆ) DEPARTMENT OF MATHEMATICS, UNIVERSITY OF UTAH, SALT LAKE CITY, UTAH 84112

E-mail address: milicic@math.utah.edu

(PANDŽIĆ) DEPARTMENT OF MATHEMATICS, MASSACHUSETTS INSTITUTE OF TECHNOLOGY, CAMBRIDGE, MA 02139

E-mail address: pandzic@math.mit.edu

A COMPARISON OF GEOMETRIC THETA FUNCTIONS FOR FORMS OF ORTHOGONAL GROUPS

JÜRGEN ROHLFS *and BIRGIT SPEH †

ABSTRACT. Let F be a totally real extension of $\mathbb{Q}$, denote by V a finite dimensional F–vector space and by q a nondegenerate anisotropic form on V over F. We assume that $G = \mathrm{Res}_{F|\mathbb{Q}}\ \mathrm{Spin}(q)$ has an $\mathbb{R}$-fundamental torus of split rank 1 and fix a congruence subgroup Γ of $G(\mathbb{Q})$.

We consider a theta function $\Theta(s)$ for Γ which is related to the analytic torsion and the length spectrum of closed geodesics on the corresponding locally symmetric space. We compare these functions for inner forms of G which differ only at the infinite places using the stabilisation of the trace formula. As an application we obtain a relationship between multiplicities of certain representations of these groups.

1. INTRODUCTION

Let G be a real non compact semisimple Lie group and let $\Gamma \subset G$ be a co-compact congruence subgroup. Using heat trace methods, H. Moscovici and R. Stanton [M – S] defined a zeta function $\zeta_{tor}(s), s \in \mathbb{C}$, and expressed the analytic torsion of the corresponding locally symmetric space $M = \Gamma\backslash G/K$ essentially as a special value of this function. The zeta function is related to the length spectrum of closed geodesics on M and can be expressed as a sum over certain orbital integrals. It was first introduced by Selberg [Se] and the connection with analytic torsion was already established for hyperbolic manifolds by Fried [F]. A different approach to $\zeta_{tor}(s)$ is as follows:

Let $L^2(\Gamma\backslash G) = \widehat{\bigoplus}\pi^{m(\pi,\Gamma)}$ be the decomposition of the space of square integrable functions on $\Gamma\backslash G$ with respect to an invariant measure in unitary irreducible representations π with multiplicity $m(\pi, \Gamma)$ of π in $L^2(\Gamma\backslash G)$. In [Sp] the torsion $tor(\pi)$ of π was defined by

$$tor(\pi) = \sum_{i=0}^{\infty}(-1)^i\, i \dim\ \mathrm{Hom}_K(\wedge^i \mathbf{g}/\mathbf{k}, \pi)$$

---

*Supported by a grant of the Deutsche Forschungsgemeinschaft
†Supported by an NSF grant

where $\mathbf{g}$ resp. $\mathbf{k}$ are the Lie algebras of G resp. K. This notion allows one to express $\zeta_{tor}(s)$ on the harmonic side as

$$\zeta_{tor}(s) = \sum_{\pi(C)\neq 0} m(\pi,\Gamma)tor(\pi)\pi(C)^{-s}$$

for $Re(s) > 0$, where $\pi(C)$ is the value of the Casimir operator C on π. We prefer to work with a theta function

$$\Theta(s) := \sum_{\pi} m(\pi,\Gamma)tor(\pi)e^{-\pi(C)s},$$

where the representations with nontrivial cohomology also enter. We note that $\Theta(s)$ is identically zero if G does not contain a fundamental Cartan subgroup of $\mathbb{R}$–split rank at most one. It is known that $\Theta \neq 0$ only if G is isogenous to a product of simple factors where at most one factor is of the form $SO(p,q)(\mathbb{R})$, pq odd, or $SL_3(\mathbb{R})$ and where all other factors have a compact Cartan subgroup.

We want to compare $\Theta(s) = \Theta_G(s)$ for groups $G = G_1$ and $G = G_2$ that are inner forms of each other and differ only at the archimedean places. The subgroups $\Gamma_i \subset G_i$ are then defined by the same congruence conditions.

To do this, one at first constructs a Schwartz function tor_s on G such that

$$\pi(tor_s) = tor(\pi)e^{-\pi(C)s}$$

for all unitary representations π of G, see § 5. For this one uses the Paley–Wiener theorem. One has

$$tr\rho(tor_s) = \Theta(s)$$

where the left side denotes the trace of the convolution with tor_s on $L^2(G/\Gamma)$. Then one applies the Selberg trace formula

$$tr\rho(tor_s) = \sum_{\{\gamma\}} \mathrm{vol}(\Gamma_\gamma\backslash G_\gamma)O_G(\gamma, tor_s)$$

which computes $tr(\rho(tor_s))$ as a sum of orbital integral. The sum of orbital integrals is rewritten as a sum of stable orbital integrals.

To do this, we work with adelic methods and assume for simplicity that G is simply connected. The function tor_s then has to be replaced by a Schwartz function φ_s on the adelic points $G(\mathbb{A})$ of a group $G/\mathbb{Q}$ where the infinite component of φ_s is tor_s. In order to get only stable expressions we require that one finite factor of φ_s is an Euler–Poincaré function. We recall at the end of § 2 the properties of such functions. One arrives at a sum of stable orbital integrals

$$tr\rho_G(\varphi_s) = \sum_{\{\gamma\}_{st}\in E} SO_G(\gamma,\varphi_s)$$

where we sum over stable conjugacy classes of $G(\mathbb{A})$.

In Section 6 we show that two groups G_i which are inner forms of each other and differ only at the archimedean places have essentially the same fundamental Cartan subgroups. Moreover we describe in detail the orthogonal groups with which we shall be working.

The index set in the last sum is the same for such $G = G_1$ and $G = G_2$.

Finally in Section 5 we show that

$$|W(G_1,C)|SO_{G_1}(\gamma,\varphi_s^{(1)}) = |W(G_2,C)|SO_{G_2}(\gamma,\varphi_s^{(2)})$$

where $\varphi_s^{(i)} = \varphi_{G_i}$ are Schwartz functions on $G_i(\mathbb{A})$ as above which have $tor_s = tor_{G_i}(s)$ as the archimedean factor and coincide on all other non archimedean factors. Here $|W(G_i,C)|$ is the order of the Weyl group of a fundamental Cartan subgroup C of $G_i(\mathbb{R})$.

As a main result we obtain

$$|W(G_1,C)|\Theta_{G_1}(\varphi_s^{(1)}) = |W(G_2,C)|\Theta_{G_2}(\varphi_s^{(2)})\,.$$

As a consequence of this we get a formula which compares the torsion contributions of representation to $L^2(\Gamma_i\backslash G_i)$,; see § 2. Cor. for a precise statement.

2. Adelic notation and main results

In this section we describe our assumptions in detail and introduce the adelic language. Then we formulate our main results. The details of the proofs are given in the last four sections.

For general facts concerning algebraic groups over number fields or adeles, see [P – R]. Let F be a totally real extension of $\mathbb{Q}$. A typical place of F or $\mathbb{Q}$ is denoted by v. By S_∞ we denote the set of classes of archimedean places of F. The completion of F resp. $\mathbb{Q}$ with respect to v is denoted by F_v resp. $\mathbb{Q}_v$.

Let V be a finite dimensional F–vector space and let

$$b : V\otimes V \longrightarrow F$$

be a nondegenerate bilinear form with associated quadratic form $q(x) = b(x,x)$, $x\in V$. We denote by $O(q)$ and $SO(q)$ the orthogonal, respectively special orthogonal group of q and by $\mathrm{Spin}(q)$ the simply connected covering group of $SO(q)$. We put $n = \dim V$.

If v is a place of F we denote by $q = q_v$ the quadratic form induced by q on $V \otimes_F F_v$. The theorem of Hasse–Minkowski says that q is isotropic over F if and only if all q_v are isotropic; see for example [P – R].

We put $G =$ Spin(q) and assume that G is F–anisotropic and that $G(F \times_{\mathbb{Q}} \mathbb{R})$ contains a fundamental Cartan subgroup of $\mathbb{R}$–split rank 1.

If q_1, q_2 are two anisotropic forms over F we denote the corresponding groups by G_1, G_2. We assume that these groups are non isomorphic inner forms of each other which differ only at the infinite place. In §6 we show that then $n \geq 6$ is necessarily even and that

$$\mathrm{Spin}(q_i)(F \otimes_{\mathbb{Q}} \mathbb{R}) = \Pi_{v \in S_\infty} \mathrm{Spin}(q_i)(F_v)\,,$$

where for at least one $v_0 \in S_\infty$ the factor $\mathrm{Spin}(q_i)(F_{v_0})$ is compact and where there is a place $v_1 \in S_\infty$ such that q_i is of signature (r, s) with $r \cdot s$ odd over F_{v_1}. Moreover then for all $v \in S_\infty$, $v \neq v_0$, $v \neq v_1$, the group $\mathrm{Spin}(q_i)(F_v)$ contains a compact Cartan subgroup.

Let $\mathbb{A}$ be the ring of adeles over F and denote by $\mathbb{A}_f \subset \mathbb{A}$ the ring of finite adeles. By assumption we have an isomorphism $\Psi : G_1 \times_F \overline{F} \xrightarrow{\sim} G_2 \times_F \overline{F}$, $\overline{F}$ an algebraic closure of F, which induces an isomorphism $G_1(\mathbb{A}_f) \xrightarrow{\sim} G_2(\mathbb{A}_f)$. We identify $G_1(\mathbb{A}_f)$ and $G_2(\mathbb{A}_f)$. If $\prod_{v \notin S_\infty} K_v = K_f \subset G_1(\mathbb{A}_f)$ is an open and compact subgroup it defines arithmetic subgroups $\Gamma_i = G_i(F) \cap K_f$ of $G_i(F)$. We choose K_f such that the groups Γ_i are torsion free. If we consider only one group we drop the index i.

Let $K \subset G(F \otimes_{\mathbb{Q}} \mathbb{R})$ be a maximal compact subgroup. By strong approximation then

$$X(K_f) := \Gamma \backslash G(F \otimes_{\mathbb{Q}} \mathbb{R})/K \xrightarrow{\sim} G(F)\backslash G(\mathbb{A})/KK_f$$

where $\Gamma = G(F) \cap K_f$.

Let $\mathbf{g}, \mathbf{k}$ be the Lie algebras of $G(F \otimes_{\mathbb{Q}} \mathbb{R})$, respectively K, and let $\mathbf{g} = \mathbf{k} \oplus \mathbf{p}$ be the corresponding Cartan decomposition. For an admissible $(\mathbf{g}, K)$–module U we define

$$\mathrm{tor}(U) := \sum_i (-1)^i\, i \dim\ \mathrm{Hom}_K(\wedge^i \mathbf{p}, U)$$

and we denote for an irreducible $(\mathbf{g}, K)$–module U the eigenvalue of the Casimir–operator C on U by $U(C)$.

Let $L^2(G(F)\backslash G(\mathbb{A}))$ be the space of square integrable functions on $G(F)\backslash G(\mathbb{A})$ viewed as $G(\mathbb{A})$–module under right translation. Denote by $L^2(G(F)\backslash G(\mathbb{A}))^{K_f}$ its subspace of K_f–invariant functions. Then $G(F \otimes_{\mathbb{Q}} \mathbb{R})$ acts on this space and we denote the multiplicity of an irreducible unitary representation U of $G(F \otimes_{\mathbb{Q}} \mathbb{R})$ in this space by $m(U, K_f)$. Following [M – S] we define

$$\Theta(s, K_f) := \sum_U \operatorname{tor}(U) m(U, K_f) e^{-sU(C)}$$

where now we sum over all classes of irreducible unitary representations of $G(F \otimes_{\mathbb{Q}} \mathbb{R})$. This sum converges if the real part of s is positive, see [M – S]. We prove in § 5.2.

Theorem 1. If $s \in \mathbb{C}$, $\operatorname{Re}(s) > 0$, *then there exists a Schwartz function* tor_s *on* $G(F \otimes_{\mathbb{Q}} \mathbb{R})$ *such that*

$$tr(U(\operatorname{tor}_s)) = \operatorname{tor}(U) e^{-sU(C)}$$

for all irreducible unitary representations U of $G(F \otimes_{\mathbb{Q}} \mathbb{R})$.

Let $\varphi_s = \varphi_{\infty,s} \cdot \varphi_f = \varphi_{\infty,s} \cdot \prod_{v \notin S_\infty} \varphi_v$ be a continuous $\mathbb{C}$–valued function on $G(\mathbb{A})$ where $\varphi_{\infty,s} = \operatorname{tor}_s$ and where φ_v is a function with compact support on $G(F_v)$. Here for almost all v the function φ_v is the normalized characteristic function on K_v. We recall in § 3 that convolution with φ_s defines an action $\rho(\varphi_s)$ on the space $L^2(G(F)\backslash G(\mathbb{A}))$ of square integrable functions on $G(F)\backslash G(\mathbb{A})$ and that $\rho(\varphi_s)$ is of trace class. We define

$$\Theta(\varphi_s) = tr\rho(\varphi_s).$$

Let $U = U_\infty \otimes U_f$ be the product decomposition of an irreducible representation of $G(\mathbb{A}) = G(F \otimes_{\mathbb{Q}} \mathbb{R}) \times G(\mathbb{A}_f)$ occurring in $L^2(G(F)\backslash G(\mathbb{A}))$ into irreducible representations of $G(F \otimes_{\mathbb{Q}} \mathbb{R})$ and $G(\mathbb{A}_f)$. Then

$$tr\rho(\varphi_s) = \sum_U m(U) \operatorname{tor}(U_\infty) tr(U_f(\varphi_f)) e^{-sU_\infty(C)}$$

where $m(U)$ is the multiplicity of an irreducible representation U of $G(\mathbb{A})$ in $L^2(G(F)\backslash G(\mathbb{A}))$.

For the notion of an Euler–Poincaré function φ_{v_0} on $G(F_{v_0})$ and a proof of the following; see § 3.

Theorem 2. Let φ_s *be as above and assume that* φ_{v_0} *is an Euler–Poincaré function at one non archimedean place* v_0. *Then*

$$\Theta(\varphi_s) = \sum_{\{\gamma\}_{st} \in E} SO(\gamma, \varphi_s)$$

is a sum of stable orbital integrals indexed by the stable fundamental conjugacy classes of $G(F)$.

We assume now that the G_i , $i = 1, 2$ are as above and view $G_i(F)$ as subgroups of $G_1(\overline{F}) = G_2(\overline{F})$. In § 6 we show:

Theorem 3. *The sets of stable fundamental conjugacy classes of* $G_1(F)$ *and* $G_2(F)$ *coincide.*

To emphasize the dependence on the group G we now write $SO_G(\gamma, \varphi_s) = SO(\gamma, \varphi_s)$. We write $W(G_i, C)$ for the Weyl group of a fundamental Cartan $C \subset G_i(F \otimes_{\mathbb{Q}} \mathbb{R})$ and $\varphi_s^{(i)}$ for the function φ_s on $G_i(\mathbb{A})$. In § 5 we show:

Theorem 4. *Let* $\{\gamma\}_{st}$ *be a stable fundamental conjugacy class of* $G_1(F)$ *and of* $G_2(F)$, *see Thm. 3. Then*

$$|W(G_1, C)| SO_{G_1}(\gamma, \varphi_s^{(1)}) = |W(G_2, C)| SO_{G_2}(\gamma, \varphi_s^{(2)}) .$$

Now we can use Thm. 2 and arrive at the main result of this paper:

Theorem 5. *With the notation introduced above we have*

$$|W(G_1, C)| \Theta(\varphi_s^{(1)}) = |W(G_2, C)| \Theta(\varphi_s^{(2)}) .$$

Using the expression for $\Theta(\varphi_s^{(i)})$ on the spectral side and Thm. 5 we can compare the multiplicities of certain representations. To see this we recall some elementary properties of the generalized Dirichlet series and of the p–adic Euler–Poincaré function.

Let $\{\lambda_n\}_{n \in \mathbb{N}}$ be a monotone increasing sequence of positive real numbers with $\lim_{n \to \infty} \lambda_n = \infty$ and assume that $\{a_n\}_{n \in \mathbb{N}}$ is a sequence of complex numbers. We consider a Dirichlet series

$$f(s) := \sum_{n=1}^{\infty} a_n e^{-\lambda_n s}$$

for $s \in \mathbb{C}$. The following is well known, see for example [Z]. If the above series converges for $s = s_0 \in \mathbb{C}$, then it converges uniformly for all $s \in \mathbb{C}$ with $|arg(s - s_0)| \leq \pi/2 - \varepsilon; \varepsilon > 0$. Furthermore in this domain $\lim_{Re(s) \to \infty} f(s) = 0$. In the domain of convergence $s \mapsto f(s)$ is an analytic function which uniquely determines the sequences $\{a_n\}_{n \in \mathbb{N}}$ and $\{\lambda_n\}_{n \in \mathbb{N}}$. Of course this result extends to Dirichlet series of the form $a_0 + f(s)$, $a_0 \in \mathbb{C}$, and applies to both sides of the equality in Thm. 5.

Kottwitz introduced in [Ko 4: § 2] the notion of an Euler–Poincaré–function $f_{EP}(= \varphi_{v_0}$ in our notation) on a group over a local p-adic field F_{v_0} . We summarize some of its properties:

- $\varphi_0 \in C_c^\infty(G(F_{v_0}))$ provided G/F_{v_0} has anisotropic center.

- the orbital integrals of φ_0 are zero except for elliptic semisimple elements.

- if π is an irreducible and admissible representation of $G(F_{v_0})$ then $tr(\pi(\varphi_{v_0})) = \chi(\pi)$ is the Euler–Poincaré number of the continuous cohomology of π.

- if G/F_{v_0} is simple and if π is an unitary representation of $G(F_{v_0})$ then $tr(\pi(\varphi_{v_0})) = 0$ unless $\pi = \mathbb{C}$ is the trivial representation or $\pi = St_{v_0}$ is the Steinberg representation. One has $\chi(\mathbb{C}) = 1$ and $\chi(St_{v_0}) = (-1)^{l_{v_0}}$. where l_{v_0} is the F_{v_0}–rank of $G \times_F F_{v_0}$.

We assume now that the functions $\varphi_s^{(i)}$ are for all non archimedean factors different from φ_{v_0} normalized characteristic functions of K_v. The representations $\Pi^i = \Pi^i_\infty \otimes \Pi_f$ of $G_i(\mathbb{A})$ contribute to $\Theta(\varphi_s^{(i)})$ only if $\Pi_f = \otimes_v \Pi_v$, $\Pi_v^{K_v} \neq \{0\}$, $v \neq v_0$, $\Pi_{v_0} \cong St_{v_0}$ or $\Pi_{v_0} \cong \mathbb{C}$ and $\mathrm{tor}(\Pi_\infty^{(i)}) \neq 0$. Put $K_f = \Pi_{v \neq v_0} K_v \times I_{v_0}$, where I_{v_0} is an Iwahori subgroup of $G(F_{v_0})$. Then the coefficient $a_\lambda^{(i)}$ of the Dirichlet series $\Theta(\varphi_s^{(i)}) = \sum_\lambda a_\lambda^{(i)} e^{-\lambda s}$ is

$$\begin{aligned} a_\lambda^{(i)} := & \sum_{\substack{\Pi^i \\ \Pi^i_\infty(C)=\lambda \\ \Pi_{v_0}\cong\mathbb{C}}} \mathrm{tor}(\Pi^i_\infty) m(\Pi^i) \dim(\Pi_f^{K_f}) + \\ & +(-1)^{\ell_{v_0}} \sum_{\substack{\Pi^i \\ \Pi^i_\infty(C)=\lambda \\ \Pi_{v_0}\cong St_{v_0}}} \mathrm{tor}(\Pi^i_\infty) m(\Pi^i) \dim(\Pi_f^{K_f}) \end{aligned}$$

where $m(\Pi^i)$ is the multiplicity of Π^i in $L^2(G_i(F)\backslash G_i(\mathbb{A}))$. Here we use that the subspace of I_{v_0}–fixed vectors of St_{v_0} is 1–dimensional; see [B]. Using Thm. 5 we get:

Corollary. *We use the above notation. Then we have for all* $\lambda \in \mathbb{R}_+^*$ *the equation*

$$|W(G_1, C)| a_\lambda^{(1)} = |W(G_2, C)| a_\lambda^{(2)} .$$

Remark. For $\lambda = 0$ the corollary compares cohomological invariants of the discrete groups $\Gamma_i := G_i(F) \cap K_f$.

3. An elementary trace formula and its stabilisation

In this section we recall the elementary trace formula in the cocompact case where we consider traces of the action of Schwartz–functions instead of compactly supported functions. Moreover we describe the stabilisation process of Kottwitz for Schwartz–functions φ on the adelic points of the groups where φ has a p-adic local factor that is an Euler–Poincaré function. We assume thoughout that G is a semisimple, simply connected algebraic group defined over F which is F-anisotropic.

3.0. Let G be a above and assume that $\varphi \in C_c^\infty(G(\mathbb{A}))$ is a compactly supported smooth function $\varphi = \Pi_v \varphi_v$ where $\varphi_v \in C_c^\infty(G(F_v))$. Let $K_f = \Pi_{v \notin S_\infty} K_v$ be a compact open subgroup of $G(\mathbb{A}_f)$. Then φ_v is constant on K_v for almost all v. Now $G(\mathbb{A})$ acts by right translation on the space of square integrable functions $L^2(G(F)\backslash G(\mathbb{A}))$ by $\rho(a)f(g) = f(ga), g \in G(F)\backslash G(\mathbb{A}), a \in G(\mathbb{A}), f \in L^2(G(F)\backslash G(\mathbb{A}))$ and by definition $\rho(\varphi) = \int_{G(\mathbb{A})} \varphi(g)\rho(g)dg$. Here we use on $G(\mathbb{A})$ and $G(F)\backslash G(\mathbb{A})$ integration with respect to the Tamagawa measure. Then $\rho(\varphi) : L^2(G(F)\backslash G(\mathbb{A})) \longrightarrow L^2(G(F) \backslash G(\mathbb{A}))$ is known to be of trace class; see [G – G – P]. It is known that $L^2(G(F)\backslash G(\mathbb{A}))$ is a Hilbert direct sum of irreducible representations π of $G(\mathbb{A})$ which occur with finite multiplicity $m(\pi)$. Moreover for each π the trace $tr(\pi(\varphi))$ of the action $\pi(\varphi)$ of φ on π is defined and

$$tr(\varphi) = \sum_\pi m(\pi) tr \pi(\varphi) .$$

Here we use that $G(F)\backslash G(\mathbb{A})$ is compact which holds since G is F–anisotropic by assumption.

The elementary trace formula is an equality of absolutely convergent sums

$$\sum_\pi m(\pi) tr\pi(\varphi) = \sum_{\{\gamma\}} vol(G_\gamma(F)\backslash G_\gamma(\mathbb{A})) \int_{G_\gamma(\mathbb{A})\backslash G(\mathbb{A})} \varphi(x^{-1}\gamma x) d\bar{x} . \qquad (1)$$

Here $\{\gamma\}$ denotes the set of $G(F)$–conjugacy classes of elements of $G(F)$ and $G_\gamma \subset G$ is the centralizer of γ. The measure $d\bar{x}$ on $G_\gamma(\mathbb{A})\backslash G(\mathbb{A})$ is a quotient measure given by the Tamagawa measure of dg on $G(\mathbb{A})$ and the Tamagawa measure $d_\gamma(g)$ on $G_\gamma(\mathbb{A})$. The volume of $G_\gamma(F)\backslash G_\gamma(\mathbb{A})$ is computed with respect to the invariant measure induced by $d_\gamma(g)$. For all this see [G – G – P].

Finally we note that the trace formula (1) also holds verbatim if $\varphi \in C_c^\infty(G(\mathbb{A}))$ is replaced by a function $\varphi = \Pi_v \varphi_v \in C^\infty(G(\mathbb{A}))$ where for $v \notin S_\infty$ the functions φ_v are as above and where each φ_v, $v \in S_\infty$, is a Schwartz function. This can be deduced from the more general statement in [O – W, § 11] as follows. Since we work with simply connected groups we see that the claim holds for functions $\varphi = \varphi_\infty \otimes \chi_{K_f}$ where φ_∞ is a Schwartz function on $G(F \otimes_{\mathbb{Q}} \mathbb{R})$ and where χ_{K_f} is a normalized characteristic function on K_f. The discrete group occurring in the classical setting here is $\Gamma = G(F) \cap K_f$. Our general φ is a finite linear combination of translates of such functions. Hence (1) also holds for general φ.

3.1. The right side of the trace formula (1) has been stabilized by Kottwitz, [Ko 3]. We recall a very special case of his result and fix the necessary notation. Let F be a field and let $\overline{F}$ be an algebraic closure of F. Elements $x, y \in G(F)$ are called stably conjugate if x and y are conjugate in $G(\overline{F})$. We then write $x \sim_{st} y$ and denote by $\{x\}_{st}$ the stable conjugacy class of x.

We assume that G/F is a semisimple simply connected and anisotropic algebraic group. Let $\varphi \in C_c^\infty(G(\mathbb{A}))$, $\varphi = \Pi \varphi_v$, be as above and assume that φ_{v_0} is an Euler–Poincare function [Ko 4] for some place v_0. Let H be a connected reductive group over a local field K. Then Kottwitz has defined a sign $e(H) \in \pm\{1\}$. For H defined over F one puts $e_v(H) = e(H \times_F F_v)$. Then $\Pi_v e_v(H) = 1$, [Ko 2]. If $\gamma \in G(F_v)$ the centralizer G_γ of γ in G is connnected, see [St]. Hence $e(G_\gamma) =: e_v(\gamma)$ makes sense. Kottwitz puts

$$SO(\gamma_v, \varphi_v) := \sum_{\{\gamma\} \in \{\gamma_v\}_{st}} e_v(\gamma) O(\gamma, \varphi_v).$$

Here $\{\gamma\}$ runs in the finitely many $G(F_v)$–conjugacy classes within the stable conjugacy class of γ_v. The numbers $e_v(\gamma)$ and

$$O(\gamma, \varphi_v) = \int_{G_\gamma(F_v)\backslash G(F_v)} \varphi_v(x^{-1}\gamma x) d(\overline{x}_v)$$

depend only on the $G(F_v)$–conjugacy class of γ. The measure $d(\overline{x}_v)$ on $G_\gamma(F_v)\backslash G(F_v)$ is the quotient measure of some fixed invariant measure $d(g_v)$ on $G(F_v)$ by an invariant measure on $G_\gamma(F_v)$ and for the different $\gamma \sim_{st} \gamma_v$ one chooses compatible measures on $G_\gamma(F_v)$, [Ko 3]. For $\gamma \in G(F)$ one puts $SO(\gamma, \varphi) = \Pi_v SO(\gamma, \varphi_v)$ and calls this the stable orbital integral of γ. One chooses the occurring measures as factors of Tamagawa measures on G, resp. G_γ. Then the number $SO(\gamma, \varphi)$ is independent of all choices and depends only on the stable conjugacy class of γ. If $\gamma, \eta \in G(\mathbb{A})$ we call γ stably conjugate to η if they are conjugate in $G(\overline{\mathbb{A}})$, $\overline{\mathbb{A}} = \mathbb{A} \otimes_F \overline{F}$. This means that all components of γ and η are stably conjugate. Let $\{\gamma\}_{st}$ be the stable conjugacy class of $\gamma \in G(\mathbb{A})$. We write $\{\gamma\}_{st} \in E$ if there is an $\eta \in G(F) \subset G(\mathbb{A})$ which is stably conjugate to γ.

Following Kottwitz we now assume that one factor φ_{v_0} of φ is an Euler–Poincaré function. This assumption implies that all nonstable contributions to $tr(\rho(\varphi))$ vanish, see [Ko 4]. One arrives at

$$tr\rho(\varphi) = \sum_{\{\gamma\}_{st} \in E} SO(\gamma, \varphi) \tag{2}$$

Since G is anisotropic only elliptic conjugacy classes occur.

3.2. We need a formula like (2) where $\varphi_\infty = \Pi_{v \in S_\infty} \varphi_v$ is a Schwartz function which has non trivial real orbital integrals $O(\gamma, \varphi_\infty)$ only for fundamental conjugacy classes $\{\gamma\}$. For the notion of "fundamental," see §6.

Let $T|\mathbb{R}$ be a fundamental torus defined over $\mathbb{R}$ of $\mathrm{res}_{\mathbb{Q}} G \times_{\mathbb{Q}} \mathbb{R}$. Choose a basis $\alpha_1, ..., \alpha_r$ of the root system of the $T(\mathbb{C})$ –action on the complex Lie algebra g of $G(F \otimes_{\mathbb{Q}} \mathbb{C})$. For $1 < t \in \mathbb{R}$ put

$$T(t) = \{z \in T(\mathbb{C})/1 \leq |\alpha_i(z)| \leq t \text{ for } i = 1, ..., r\}.$$

Then $T(t)$ is a compact subset of $T(\mathbb{C})$ and $G_\infty(t) := G(F \otimes_{\mathbb{Q}} \mathbb{R}) \cap \bigcup_{w \in W_{\mathbb{C}}} wT(t)w^{-1}$ is a compact subset of $T(\mathbb{R})$. Here w also denotes an element of the normalizer of $T(\mathbb{C})$ in $G(F \otimes_{\mathbb{Q}} \mathbb{C})$ which represents the element w of the Weyl group $W_{\mathbb{C}}$ of $T(\mathbb{C})$.

3.4. Lemma. (i) *Every $G(F \otimes_{\mathbb{Q}} \mathbb{R})$ conjugacy class $\{\gamma\}$, $\gamma \in G(F \otimes_{\mathbb{Q}} \mathbb{R})$ fundamental, meets some $G_\infty(t)$.*
(ii) *If $\gamma \in G(F \otimes_{\mathbb{Q}} \mathbb{R})$ is fundamental and if some conjugacy class $\{\eta\}$, $\eta \in G(F \otimes_{\mathbb{Q}} \mathbb{R})$ contained in the stable conjugacy class $\{\gamma\}_{st}$ meets $G_\infty(t)$ then all conjugacy classes of $\{\gamma\}_{st}$ meet $G_\infty(t)$.*

Proof. Since all fundamental tori of $G(F \otimes_{\mathbb{Q}} \mathbb{R})$ are $G(F \otimes_{\mathbb{Q}} \mathbb{R})$ conjugate $\{\gamma\}$ meets $T(\mathbb{R}) \subset T(\mathbb{C})$. So we can assume that $\gamma \in T(\mathbb{R})$. But

$$T(\mathbb{C}) = \bigcup_{w \in W_{\mathbb{C}}} w\{z \in T(\mathbb{C}) | 1 \leq |\alpha_i(Z)| \text{ for } i = 1, 2, ..., r\}w^{-1}.$$

Choose a $w \in W_{\mathbb{C}}$ such that $1 \leq |\alpha_i(w\gamma w^{-1})|$ for all i and put $t = \max_i |\alpha_i(w\gamma w^{-1})|$. Then $\gamma \in G_\infty(t)$ and i) holds. To prove (ii) we can assume that $\{\gamma\}_{st}$ with $\gamma \in G_\infty(t)$ is given. Assume that $\eta \in T(\mathbb{R})$ is stably conjugate to γ. Since the semisimple conjugacy classes in $G(F \otimes_{\mathbb{Q}} \mathbb{C})$ are in one-to-one correspondence to $T(\mathbb{C})/W_{\mathbb{C}}$ we have $\gamma = w\eta w^{-1}$ for some w, i.e., $\eta \in G_\infty(t)$. qed.

3.5. To stabilize (1) for a function $\varphi = \Pi\varphi_v \in C^\infty(G(\mathbb{A}))$ where φ_∞ is a Schwartz function with non trivial orbital integrals only on fundamental conjugacy classes and where $\prod_{v \text{ finite}} \varphi_v$ has compact support $G_f \subset G(\mathbb{A}_f)$ and one factor of $\prod_v \varphi_v$ is an Euler–Poincaré function, we proceed as in [Ko 3, 4]. For $\gamma \in G(\mathbb{A})$ we write $\{\gamma\}_{st} \in E(t)$ if the stable conjugacy class meets $G_\infty(t)$ i.e. if there exists an $\eta \in G(F)$ such that $\{\eta\}_{st} = \{\gamma\}_{st}$, for a γ in a fundamental Cartan which satisfies $\{\gamma\} \cap C_\infty(t) \neq \emptyset$ (in $G(F \otimes_{\mathbb{Q}} \mathbb{R})$). Then 6.5 shows that all classes $\{\eta\} \in \{\gamma\}_{st}$ are fundamental and by 3.4 all these classes meet $G_\infty(t)$.

Since (1) is absolutely convergent we can regroup and get

$$tr\, \rho(\varphi) = \lim_{t \to \infty} \sum_{\{\gamma\} \in E(t)} \mathrm{vol}(G_\gamma(F) \backslash G_\gamma(\mathbb{A})) \int_{G_\gamma(\mathbb{A}) \backslash G(\mathbb{A})} \varphi(x^{-1}\gamma x) dx.$$

We put $G(t) = G_\infty(t) \times G_f \subset G(\mathbb{A})$. Then $G(t)$ is a compact subset of $G(\mathbb{A})$ and using [Ko. 3, Prop. 8.5] there are only finitely many $G(\mathbb{A})$–conjugacy classes that meet $G(t)$ and meet some fundamental conjugacy class of $G(F)$. This means that the formal part of the stabilisation process in [Ko. 3, Thm. 9.6] can be applied to

$$\sum_{\{\gamma\}\in E(t)} \mathrm{vol}(G_\gamma(F)\backslash G_\gamma(\mathbb{A}))O(\gamma,\varphi)$$

and one arrives at a formula like loc. cit. 9.6.5. The considerations in [Ko 4: § 9] concerning the local orbital integrals of the Euler–Poincaré–functions apply verbatim and one gets

$$\sum_{\{\gamma\}\in E(t)} \mathrm{vol}(G_\gamma(F)\backslash G_\gamma(\mathbb{A}))O(\gamma,\varphi) = \sum_{\{\gamma\}_{st}\in E(t)} SO(\gamma,\varphi).$$

Note that on the right side there are only finitely many local factors $SO(\gamma,\varphi_v)$ which are not equal to 1 and the non trivial factors are finite sums of not necessarily positive terms. So the sum $\sum_{\{\gamma\}_{st}\in E} SO(\gamma,\varphi)$ is not known to be absolutely convergent but because of the absolute convergence of (1) it makes sense to define

$$\sum_{\{\gamma\}_{st}\in E} SO(\gamma,\varphi) := \lim_{t\to\infty} \sum_{\{\gamma\}_{st}\in E(t)} SO(\gamma,\varphi).$$

With this notation we now have shown:

3.5. Proposition. *If G/F is semisimple simply connected and anisotropic over F and if φ is as above then*

$$tr\rho(\varphi) = \sum_\pi m(\pi)tr(\pi(\varphi)) = \sum_{\{\gamma\}_{st}\in E} SO(\gamma,\varphi).$$

4. Torsion on L-packets of representations

We prove that the torsion of a representation is constant on L-packets and compare the torsion of L–packets for inner forms of spin groups.

4.1. The torsion of principal series representations.

Let G be a connected real semisimple Lie group with a maximal compact subgroup K. We will assume that the complexification $G_{\mathbb{C}}$ is simply connected. The Lie algebra of G will be denoted by $\mathbf{g}$ and the Cartan decomposition by $\mathbf{g} = \mathbf{k} \oplus \mathbf{p}$. We discuss the torsion of irreducible unitary representations of G and of principal series representations $I(Q,\Pi,\nu)$ induced from a cuspidal parabolic subgroup $Q = MAN$, a discrete series or limit of discrete series representation Π of M and a character ν of A.

4.1.1. The representation of K on $\wedge^*\mathbf{p}$. Let $\mathbf{t}$ be a Cartan subalgebra of $\mathbf{k}$ and $\mathbf{c} = \mathbf{t} \oplus \mathbf{a}$ a fundamental Cartan subalgebra. We write $\sum(\mathbf{t},\mathbf{p})$ for the roots of $\mathbf{t}$ on $\mathbf{p}$. The multiplicity of the 0-weight space is equal the dimension of $\mathbf{a}$. If $\alpha \in \sum(\mathbf{t},\mathbf{p})$ then so is $-\alpha$. The character ρ_E of the virtual representation of K on $\sum_i(-1)^i \wedge^i \mathbf{p}$ equals $\Pi_{\alpha\in\sum(\mathbf{t},\mathbf{p})}(1-e^\alpha)$.

4.1.2. Lemma. *Define the function* $\rho_E(t) = \Pi_{\alpha\in\sum(t,p)}(1-t\cdot e^\alpha)$. *Then* $\frac{d}{dt}\rho_E(t)_{|t=1}$ *is the character* ρ_{dE} of K *on the virtual representation* $dE = \sum_i(-1)^i i \wedge^i \mathbf{p}$.

Proof. The coefficient of t^i in $\rho_E(t)$ is the character of K on $\wedge^i\mathbf{p}$ multiplied with $(-1)^i$. q.e.d.

4.1.3. Reduction to simple Lie algebras. Now suppose that $\mathbf{g}$ is a direct sum of two Lie algebras $\mathbf{g}_1$ and $\mathbf{g}_2$. Then $\mathbf{p} = \mathbf{p}_1 \oplus \mathbf{p}_2$ and $\mathbf{k} = \mathbf{k}_1 \oplus \mathbf{k}_2$. Put $\rho_E^{(j)}(t) = \Pi_{\alpha\in\sum(\mathbf{t},\mathbf{p}_j)}(1-t\cdot e^\alpha)$ and write $\rho_E^{(j)}$, $\rho_{dE}^{(j)}$ for the corresponding characters of $\mathbf{k}_j$.

4.1.4. Lemma. *Under the above assumptions the character of* K *on* $\sum_i(-1)^i i \wedge^i \mathbf{p}$ *is*

$$\rho_E^{(1)}\rho_{dE}^{(2)} + \rho_E^{(2)}\rho_{dE}^{(1)}.$$

Proof. This is a consequence of the product rule. q.e.d.

4.1.5. Lemma. *The character* ρ_{dE} *is zero unless* $\mathbf{g}$ *has a fundamental Cartan algebra* $\mathbf{t}$ *of real rank at most 1.*

Proof. By 4.1.2 $\rho_{dE} = 0$ if $\dim \mathbf{a} \geq 2$ and the first claim holds. q.e.d.

Remark. If $\mathbf{g} = \mathbf{g}_1 \oplus \mathbf{g}_2$ where $\mathbf{g}_1$, has a compact Cartan subgroup and $\mathbf{g}_2$ has a fundamental Cartan subgroup $C_2 = T_2A_2$ of real rank 1 then

$$\rho_{dE} = \rho_E^{(1)} \cdot \rho_{dE}^{(2)}.$$

For an admissible $(\mathbf{g},K)$–module U we define the torsion by

$$tor(U) = \sum_i (-1)^i\, i \dim \operatorname{Hom}_K(\wedge^i\mathbf{p}, U).$$

4.1.6. Proposition. *Suppose that* $\mathbf{g}$ *does not have a Cartan subgroup of real rank at most 1. Then the torsion of its admissible* $(\mathbf{g}, K)$*-modules is* 0. *Suppose* $\mathbf{g} = \mathbf{g}_1 \oplus \mathbf{g}_2$ *where* $\mathbf{g}_1$ *has a compact Cartan subalgebra and* $\mathbf{g}_2$ *has a fundamental Cartan subalgebra of real rank* 1. *Let* $\Pi = \Pi_1 \otimes \Pi_2$ *be a tensor product of admissible* $(\mathbf{g}_i, K_i)$*-modules. Then*

$$\mathrm{tor}(\Pi) = \chi(\Pi_1) \cdot \mathrm{tor}(\Pi_2)$$

where $\chi(\Pi_1)$ *denotes the Euler–Poincaré characteristic of the represesentation* Π_1.

Proof. Let Π be an admissible $(\mathbf{g}, K)$–module. The torsion of Π is the coefficient of the trivial representation in the Grothendieck group of $K = K_1 \times K_2$ of the tensor product

$$\sum_i (-1)^i i \wedge^i \mathbf{p}^* \otimes \Pi.$$

Hence the first claim follows from 4.1.5. We get for the second claim

$$\mathrm{tor}(\Pi) = \sum_i (-1)^i \dim\ \mathrm{Hom}_{K_1}(\wedge^i \mathbf{p}_1, \Pi_1) \cdot \mathrm{tor}(\Pi_2)\,.$$

By the definition of the $(\mathbf{g}_1, K_1)$–cohomology the first factor can be identified with the Euler–Poincaré characteristic of Π_1 (see also 4.2.3). q.e.d.

Remark. The classification of real simple Lie algebras shows that the only simple real Lie algebras which have a fundamental Cartan subalgebra $\mathbf{t} \oplus \mathbf{a}$ with $\dim \mathbf{a} = 1$ are isomorphic to $\mathbf{sl}(3, \mathbb{R})$ or $\mathbf{so}(p, q)$ with pq odd.

We will from now on assume that $\mathbf{g}$ *is isomorphic to* $\mathbf{so}(p, q), pq$ *odd, and that the real Lie group* $G =$ Spin $(p, q)(\mathbb{R})$ *is the corresponding connected spin group, see 4.1.*

4.1.7. Let $Q = MAN$ be a parabolic subgroup of G and let Π be a tempered representation of M. If ν is a character of AN we define using normalized induction the principal series representation $I(Q, \Pi, \nu)$. In [Sp] it was proved that $\mathrm{tor} I(Q, \Pi, \nu) = 0$ unless $Q = MAN$ is a cuspidal parabolic subgroup and has $\dim A = 1$ or $Q = G$.

4.1.8. Lemma. *Suppose that* $Q = MAN$ *is a fundamental cuspidal parabolic subgroup of* G *with* $\dim A = 1$. *As* $M \cap K$*-module* $\mathbf{p} = \mathbf{m} \cap \mathbf{p} \oplus \mathbf{a} \oplus \mathbf{p}_0$ *and* $\mathrm{tor}\ I(Q, \Pi, \nu)$ *is the coefficient of the trivial representation in*

$$(-1) \sum_i ((-1)^i \wedge^i (\mathbf{p} \cap \mathbf{m})^*) \otimes \sum_j (-1)^j (\wedge^j \mathbf{p}_0^*) \otimes \Pi\,.$$

Proof. Using Frobenius reciprocity we see that the torsion of a $(\mathbf{g}, K)$-module $I(Q, \Pi, \nu)$ is the coefficient of the trivial representation of $K \cap M$ in

$$\sum_i (-1)^i i \wedge^i \mathbf{p}^* \otimes \Pi .$$

Let $\sum_M = \sum(\mathbf{t}, \mathbf{p} \cap \mathbf{m})$ and $\sum_0 = \sum(\mathbf{t}, \mathbf{p}_0)$. In the Grothendieck group of $K \cap M$ the character of $\sum_i i(-1)^i \wedge^i \mathbf{p}$ is a product of characters of virtual representations $\Pi_{\beta \in \sum_M}(1 - e^\beta) = \sum_i (-1)^i \wedge^i (\mathbf{p} \cap \mathbf{m})$ and $\Pi_{\beta \in \sum_0}(1 - e^\beta) = \sum_i (-1)^i \wedge^i \mathbf{p}_0$ and $(1 - t)$. Thus the lemma follows. q.e.d.

This formula also appears in the analysis of the heat kernel in [M – S].

4.1.9. Proposition. *There are only finitely many equivalence classes of discrete series representations* Π *of* M *so that*

$$\operatorname{tor} I(Q, \Pi, \nu) \neq 0 .$$

Proof. There are only finitely many inequivalent discrete series representations which contain a given K-type. [H-Ch 1] q.e.d.

4.1.10. *Remark.* Suppose that $Q = MAN$ is a cuspidal parabolic subgroup of Spin(p,q) ($\mathbb{R}$) with pq odd and with dim A = 1. Then Q is obtained as follows: let $H = T_H A_H$ be a fundamental Cartan subalgebra invariant under the Cartan involution. Then MA is conjugate to the centralizer of A_H. The Lie algebra of M is isomorphic to $\mathbf{so}(p-1, q-1)$. Thus if $\mathbf{g} = \mathbf{k} \oplus \mathbf{p}$ is the Cartan decomposition of $\mathbf{g}$ then we have a decomposition $\mathbf{p} = \mathbf{m} \cap \mathbf{p} \oplus \mathbf{a} \oplus \mathbf{p}_0$ as $M \cap K$-module. There is a natural isomorphism of $K \cap M$-modules

$$\mathbf{p}_0 \xrightarrow{\sim} \mathbf{p}/\mathbf{m} \cap \mathbf{p} \oplus \mathbf{a} \underset{\sim}{\xrightarrow{exp}} K \backslash G / MA \xrightarrow{\sim} N \underset{\sim}{\xrightarrow{log}} \mathbf{n}$$

where $\mathbf{n}$ is the Lie algebra of N. Hence $\mathbf{p}_0$ is isomorphic as $K \cap M$ module to the restriction of the M–module $\mathbf{n}$. Thus we can view $\mathbf{p}_0$ as M-module where the M-action extends the obvious $M \cap K$-action. We call Q a *fundamental parabolic subgroup* of the *Spin group.*

4.2. The torsion of an L-packet. Let M^0 be the connected component of the identity of M of a fundamental parabolic $Q = MAN$ of $\mathrm{Spin}(p, q)(\mathbb{R}) = G$ with pq odd.

4.2.1. Lemma. *The group* M/M^0 *has order* 1 *if the quadratic form* which defines G *has signature* $(p, q) = (1, n-1)$. *Otherwise it is cyclic of order 2.*

Proof. The spin groups are connected algebraic groups and their real points are connected. Hence the fundamental Cartan subgroup TA is connected

[W 1.4.1.4] and thus contains the center of G. The Levi factor MA of the parabolic Q is a covering of the Levi factor of a parabolic subgroup of the connected component of the identity of $SO(p,q)(\mathbb{R})$. This parabolic subgroup is connected if $(p,q)=(1,n-1)$ and it has 2 connected components otherwise. Since the center of G is contained in M^0, the result follows. q.e.d.

4.2.2. Suppose that M has 2 connected components. Then we can write $M=M^0+M^0m_0$ where m_0 is in the normalizer of TA. In $SO(p,q)^0(\mathbb{R})$ then m_0 can be represented by a diagonal matrix whose entries are all 1 except at place (p,p) and $(p+1,p+1)$ where they are -1.

Conjugation of thc compact torus T of M^0 by m_0 corresponds to a reflection in a noncompact root and defines an automorphism of the Weyl group of $(M^0\cap K,T)$. The Weyl group of (M,T) is generated by the Weyl group of $(M^0\cap K,T)$ and by $Ad(m_0)$.

Let Π_0 be a representation of M^0. The induced representation $ind_{M^0}^{M}\Pi_0$ is irreducible iff Π_0 and $\Pi_0^{m_0}$ are not equivalent. This follows from Frobenius reciprocity. If Π_0 is a discrete series representation to M^0 then $\Pi_0^{m_0}$ is inequivalent to Π_0. Thus every discrete series representation Π of M is induced from a discrete series representation Π_0 of the connected component M^0.

4.2.3. We use the following observation in 4.2.4 and 4.3.5: Let Π_0 be a discrete series representation of M^0 and let F be an irreducible finite dimensional representation of M^0. We write $\mathbf{m}\cap\mathbf{k}\oplus\mathbf{m}\cap\mathbf{p}$ for the Cartan decompostion of $\mathbf{m}$. Then

$$\sum_i(-1)^i \dim \operatorname{Hom}_{K\cap M^0}(\wedge^i(\mathbf{p}\cap\mathbf{m}),\ \Pi_0\otimes F)=$$
$$=\sum_i(-1)^i \dim H^i(\mathbf{m},\ K\cap M^0,\ \Pi_0\otimes F)\,.$$

The left hand side has an interpretation in terms of characters. The character of Θ_λ of Π_0 is of the form

$$\Theta_\lambda=\frac{(-1)^q\sum_{w\in W_{M\cap K}}(\det w)e^{w\lambda}}{\prod_{\alpha\in\sigma^+}(e^{\alpha/2}-e^{-\alpha/2})}$$

where $q=1/2\dim\mathbf{p}\cap\mathbf{m}$ and λ is a regular character on the compact Cartan subgroup H [H – Ch 1]. The character of $K\cap M^0$ on $(-1)^i\wedge^i\mathbf{p}\cap\sigma$ is $\prod(1-e^\beta)$ where we take the product over all roots β of H on $\mathbf{p}\cap\sigma$, i.e. over all noncompact roots σ_n. Let σ^+ be the system of positive roots determined by λ. We write σ_c^+ for the compact roots in σ^+ and

$$\chi_\mu=\frac{\sum_{w\in W_M}(\det w)e^{w(\mu+\rho)}}{\prod_{\alpha\in\sigma^+}(e^{\alpha/2}-e^{-\alpha/2})}$$

for the character of the finite dimensional representation F. Since

$$\prod_{\beta\in\sigma_n}(1-e^{\beta}) = (-1)^q \prod_{\beta\in\sigma_n^+}(e^{\beta/2}-e^{-\beta/2})^2$$

we have $(*)$:

$$\Theta_\lambda\cdot\chi_\mu\cdot\prod_{\alpha\in\sigma_n}(1-e^{\alpha}) = (-1)^q\frac{(-1)^q\sum_{w\in W_{M\cap K}}(\det w)e^{w\lambda}}{\prod_{\alpha\in\sigma_c^+}(e^{\alpha/2}-e^{-\alpha/2})}\frac{\sum_{w\in W_M}(\det w)e^{w(\mu+\rho)}}{\prod_{\alpha\in\sigma_c^+}(e^{\alpha/2}-e^{-\alpha/2})}$$

We can write

$$\frac{\sum_{w\in W_M}(\det w)e^{w(\mu+\rho)}}{\prod_{\alpha\in\sigma_c^+}(e^{\alpha/2}-e^{-\alpha/2})}$$

as a sum of $|W_M/W_{M\cap K}|$ inequivalent representations of $\mathbf{m}\cap\mathbf{k}$. The coefficient of the trivial representation of $\mathbf{k}\cap\mathbf{m}$ in $\Theta_\lambda\cdot\chi_\nu\cdot\prod_{\alpha\in\Sigma_n}(1-e^{\alpha})$ is $(-1)^q$ if $\mu+\rho=\lambda$ and zero otherwise since the Euler Poincaré characteristic of Π is equal to $(-1)^q$ (see [B-W], II, 5.4).

4.2.4. Proposition. *Let G be Spin(p,q) with pq odd and $Q = MAN$ the fundamental parabolic subgroup. Suppose that Π_1,Π_2 are discrete series representations of M. If Π_1,Π_2 are in the same L-packet of M, then*

$$\operatorname{tor} I(Q,\Pi_1,\nu) = \operatorname{tor} I(Q,\Pi_2,\nu)$$

Proof. We write $\Pi_1 = \operatorname{ind}_{M^0}^M\Pi_0$, see 4.2.2.

By Frobenius reciprocity

$$\Big(\sum_i(-1)^i i\wedge^i\mathbf{p}^*\otimes ind_{M^0}^M\Pi_0\Big)^{K\cap M} = \Big(\sum_i(-1)^i i\wedge^i\mathbf{p}^*\otimes\Pi_0\Big)^{K\cap M^0}.$$

Next we use 4.1.8. Thus it suffices to show that the coefficient of

$$\Big(\sum(-1)^i\wedge^i(\mathbf{p}\cap\mathbf{m})^*\otimes\Pi_0\otimes\sum_j(-1)^j\wedge^j\mathbf{p}_0^*\Big)^{K\cap M^0}$$

depends only on the L-packet of Π_0. Since $\mathbf{p}_0^*$ is isomorphic to the restriction to $M^0\cap K$ of a finite dimensional representation of M^0 (see 4.1.10) this is the Euler–Poincaré characteristic of the representation $\Pi_0\otimes\sum_j(-1)^j\wedge^j\mathbf{p}_0^*$ (see 4.2.3) which is constant on the L-packet of Π_0 in M^0 and therefore on the L-group of Π. (see [B-W] II, 5.4) q.e.d.

The torsion of a principal series representation is independent of the continuous parameter ν. Therefore we denote the torsion of the induced representation $I(Q,\Pi,\nu)$ by $\operatorname{tor}(Q,\Pi)$.

4.3.1. The torsion of representations $I(Q,\Pi,\nu)$ with singular data. Let Π be a limit of discrete series representation of M^0 and $\lambda \in \hat{T}$ its Harish Chandra parameter. Then λ is singular with respect to a noncompact root of $\sigma_M = \sigma(\mathbf{m}\otimes\mathbf{C}, \mathbf{h}\otimes\mathbf{C})$. Let $\mathcal{C}$ be a dominant Weyl chamber determined by λ amd write σ_M^+ for the system of positive roots of σ_M corresponding to $\mathcal{C}$. Write σ_c^+ resp. σ_n^+ for the compact respectively noncompact roots in σ^+. We write W_M for the Weyl group of σ_M and $W_{K\cap M}$ for the real Weyl group of (M^0, T).

Since Π_0 is a limit of discrete series representation its character is of the form

$$\Theta_\lambda = \frac{(-1)^q \sum_{w\in W_{M\cap K}} (\det w) e^{w\lambda}}{\prod_{\alpha\in\sigma^+}(e^{\alpha/2} - e^{-\alpha/2})}$$

where $q = 1/2 \dim \mathbf{p}\cap\mathbf{m}$ [H – Ch 1].

4.3.2. Proposition. *Suppose that* Π *is a limit of a discrete series representations. Then*

$$\operatorname{tor} I(Q,\Pi,\nu) = 0.$$

Proof. We have to compute the coefficient of the trivial representation of $K\cap M^0$ in the tensor product

$$\Pi_0 \otimes (\sum_i ((-1)^i \wedge^i \mathbf{p}^* \cap \mathbf{m}^*) \otimes (\sum_i ((-1)^i \wedge^i \mathbf{p}_0^*) .$$

The character of the above representation is

$$\frac{(-1)^q \sum_{w\in W_{M\cap K}} (\det w) e^{w\lambda}}{\prod_{\alpha\in\sigma^+}(e^{\alpha/2} - e^{-\alpha/2})} \cdot \prod_{\beta\in\sigma_n} (1 - e^\beta) \cdot \chi$$

where λ is singular with respect to σ_n^+ and χ is the character of the virtual representation of $M^0\cap K$ on $\sum_i ((-1)^i \wedge^i \mathbf{p}_0^*$. Since β and $-\beta$ are in σ_n we have

$$\prod_{\beta\in\sigma_n} (1 - e^\beta) = (-1)^q \prod_{\beta\in\sigma_n^+} (e^{\beta/2} - e^{-\beta/2})^2 .$$

But

$$\Theta_\lambda \cdot \prod_{\beta\in\sigma_n^+} (e^{\beta/2} - e^{-\beta/2}) = \frac{\sum_{w\in W_{M\cap K}} (\det w) e^{w\lambda}}{\prod_{\alpha\in\sigma_c^+}(e^{\alpha/2} - e^{-\alpha/2})} .$$

This is the character of a finite dimensional representation of $M^0\cap K$ with highest weight $\lambda - 1/2\sum_{\alpha\in\sigma_c^+} \alpha = \lambda - \rho_c$.

The representation of $M^0 \cap K$ on $\mathbf{p}_0$ is isomorphic to the restriction of a representation of M^0 (see 4.1.10). Hence χ is the restriction to $M^0 \cap K$ of the character χ_M of a finite dimensional representation of M^0. Let μ be the highest weight of an irreducible summand χ_F of χ_M and put $\rho_M = 1/2 \sum_{\alpha \in \sigma^+} \alpha$. Then

$$\chi_F \cdot \prod_{\beta \in \sigma_M \cap \sigma^+} (e^{\beta/2} - e^{-\beta/2}) = \frac{\sum_{w \in W_M} (\det w) e^{w(\mu+\rho_M)}}{\prod_{\alpha \in \sigma_c^+} (e^{\alpha/2} - e^{-\alpha/2})}.$$

Thus it is a sum of characters of irreducible representations of $M^0 \cap K$ whose highest weights are of the form $w(\mu + \rho_M) - \rho_c$ for $w \in W_M / W_{M \cap K}$, each appearing with coefficient 1 or -1. So we have to compute the coefficient of the trivial representation in

$$\frac{\sum_{w \in W_{M \cap K}} (\det w) e^{w\lambda}}{\prod_{\alpha \in \sigma_c^+} (e^{\alpha/2} - e^{-\alpha/2})} \cdot \frac{\sum_{w \in W_M} (\det w) e^{w(\mu+\rho_M)}}{\prod_{\alpha \in \sigma_c^+} (e^{\alpha/2} - e^{-\alpha/2})}.$$

This coefficient is non zero iff

$$\lambda - \rho_c = w(\mu + \rho_M) - \rho_c$$

for some $w \in W_M$. Since λ and μ are both in the closure of the same dominant Weyl chamber of σ^+ this holds iff $\lambda = \mu + \rho_M$. Since λ is singular and $\mu + \rho_M$ is regular this is impossible. q.e.d.

4.4.1. Comparison of the torsion of L-packets for inner form of the Spin groups. Let q_1 and q_2 be quadratic forms with signatures (r_1, s_1) and (r_2, s_2). We assume that $r_1 + s_1 = r_2 + s_2 = n$ and that the integers r_1, s_1, r_2, s_2 are odd. Let G_1 and G_2 be the corresponding spin groups which are inner forms of each other. We view both groups as subgroups of the complex Lie group $G_1(\mathbb{C}) = G_2(\mathbb{C})$ and assume that the G_i share a common fundamental Cartan subgroup $T_1 A_1 = T_2 A_2$. The subgroups M_1 and M_2 of the Levi factor of the fundamental parabolics have the same complexification and are inner forms of each other. In particular we can identify the complex Weyl groups W_i of M_i.

Suppose that Π_1^0 is a discrete series representation of M_1^0. Let Π_2^0 be a discrete series representation of M_2^0 with the same infinitesimal character (or a finite dimensional representation if M_2^0 is compact). We write Π_1, Π_2 for the induced representations to M_1, M_2. Let Q_1, Q_2 be the maximal parabolic subgroups with Levi factor $M_1 A, M_2 A$. Let $W_{i,c}$ be the compact Weyl groups of M_i and denote by $\mathbf{m}_i$ the Lie algebra of M_i.

4.4.2. Proposition. *Under these assumptions*

$$\text{tor } I(Q_1, \Pi_1, \nu) = \text{tor } I(Q_2, \Pi_2, \nu).$$

Proof. We use 4.1.8 and have to compare the coefficients of the trivial representation of $K_i \cap M_i^0$ in the tensor product

$$\Pi_i^0 \otimes \sum_j (-1)^j \wedge^j (\mathbf{p} \cap \mathbf{m}_i)^* \otimes \sum_j (-1)^j \wedge^j \mathbf{p}_0^*$$

for $i = 1,2$. Let $d_i, i = 1,2$ denote the dimensions of the symmetric spaces $M_i/K \cap M_i, i = 1,2$. The groups M_1^0 and M_2^0 are inner forms of each other and their representations on $\mathbf{p}_0 \otimes \mathbb{C}$ have the same highest weights, see 4.1.10.

The coefficient of the trivial representation of $K \cap M_i^0$ in

$$\Pi_i^0 \otimes \sum_j ((-1)^j \wedge^j (\mathbf{p} \cap m_i)^* \otimes \wedge^r \mathbf{p}_0^*$$

is equal to

$$\sum_j (-1)^j \dim \text{ Hom }_{K \cap M_i}(\wedge^j (\mathbf{p} \cap m_i)^*, \Pi_i^0 \otimes \wedge^r \mathbf{p}_0^*) =$$

$$= \sum_j (-1)^j \dim H^j(\mathbf{m}_i, K \cap M_i, \Pi_i^0 \otimes \wedge^r \mathbf{p}_0^*) .$$

Since discrete series representations have non trivial cohomology only in the middle degree [B – W: II Thm. 5.4], we have

$$\sum_j (-1)^j \dim H^j(\mathbf{m}_1, K \cap M_1, \Pi_1^0 \otimes \wedge^r \mathbf{p}_0^*) =$$

$$(-1)^{(d_1-d_2)/2} \sum_j (-1)^j \dim H^j(\mathbf{m}_2, K \cap M_2, \Pi_2^0 \otimes \wedge^r \mathbf{p}_0^*)$$

The claim follows from $d_1 - d_2 = (r_1 - 1)(s_1 - 1) - (r_2 - 1)(s_2 - 1) \equiv 0 \bmod 4$. q.e.d.

Remark. It is possible to prove 4.2.3, 4.3.2 and 4.4.2 directly using characters without referring to the Euler Poincaré characteristic. We prefer these proofs because they show the relationship between the torsion of the induced representation $I(Q, \Pi, \nu)$ and the Euler–Poincaré characteristic of Π.

5. THE FUNCTION TOR_s AND ITS ORBITAL INTEGRALS

In this entire section we will assume unless otherwise stated that G is a product of real spin groups. We use a Paley–Wiener theorem to introduce a Schwartz function tor_s. We discuss the orbital integrals tor_s for semisimple

γ. We compare the orbital integrals of tor_s and semisimple γ for different inner forms of G.

5.1. Existence of tor_s : Suppose that G is a Spin group of a quadratic form with signature $(p,q), pq$ odd. We fix on G the measure we obtain from the Killing form and the choice of an Cartan involution We denote by $\mathcal{P}$ the set of equivalence classes of parabolic subgroups $P \subset G$. For $P \in \mathcal{P}$ we write $\mathcal{D}_{\mathcal{P}}$ for the equivalence classes of generalized principal series representations induced from P. We write $C_{P,\pi,\nu}$ for the eigenvalue of the Casimir C on the principal series representation $I(P,\Pi,\nu)$.

We fix $s \in \mathbb{C}$ with a positive real part. In particular we showed in the last section that the function

$$PW : \bigcup_{P \in \mathcal{P}} \mathcal{D}_{\mathcal{P}} \longrightarrow C$$

defined by

$$PW(I(P,\Pi,\nu)) = \text{tor}(I(P,\Pi,\nu))e^{-sC_{P,\Pi,\nu}}$$

satisfies the assumptions of the Paley–Wiener theorem, see [D: Thm. 1.9]. Thus for every $s \in \mathbb{C}$ with $Re(s) > 0$ there exists a Schwartz function tor_s on $\text{Spin}(p,q)(\mathbb{R}) = G$ so that

$$trI(P,\Pi,\nu)(\text{tor}_s) = \text{tor}(I(P,\Pi,\nu))e^{-sC_{P,\Pi,\nu}}$$

5.2. Proposition. *For $s \in \mathbb{C}$ with a positive real part there exists a Schwartz function tor_s on G so that for every irreducible unitary representation U*

$$trU(\text{tor}_s) = \text{tor}(U)e^{-sU(C)}.$$

Proof. Since the torsion depends only on the K–types it is an additive function on the Grothendieck group of admissible $(\mathbf{g},K)$–modules. It suffices to prove the theorem for generators of the Grothendieck group, i.e. the generalized principal series principal series representations $I(P,\Pi,\nu)$ where $C_{P,\pi,\nu} = \lambda$ is fixed. Hence the claim follows from 5.1. q.e.d.

Remark. Let k_{-s} be the heat kernel constructed in [M – S]. The measure dx used there is proportional to our measure dg with a constant c_G. By loc. cit. Prop. 2.1

$$trU(\text{tor}_s) = trU(c_G^{-1}k_{-s})$$

for all principal series representations U, hence for all irreducible unitary representations.

5.3. Let $G = \Pi_{i \in S} G_i$ be a product of real spin groups, so that all factors but G_1 have a compact Cartan subgroup. Let $\chi_{G_i} \in C_c^\infty(G_i)$ be an Euler–Poincaré function so that $tr\Pi(\chi_{G_i}) = (-1)^{\frac{1}{2}\dim X_i}$ if Π is a discrete series representation and $\Pi(\chi_{G_i}) = 0$ if Π is any other tempered representation. Here X_i is the symmetric space corresponding to G_i. For the existence of χ_{G_i}, see [C – D]. We define the torsion function tor_s by

$$\mathrm{tor}_s = \mathrm{tor}_s^1 \times \prod_{i \in S, i \neq 1} \chi_{G_i}$$

where tor_s^1 is the function constructed in 5.3 for G_1. We note that $\Pi(C) = 0$ if Π is a discrete series representation with $\mathrm{tor}(\Pi) \neq 0$. Hence:

5.4. Corollary. *Let $G = \prod_{i \in S} G_i$ be a product of spin groups, so that all factors but one have a compact Cartan subgroup. If $Re(s) > 0$ then*

$$trU(\mathrm{tor}_s) = \mathrm{tor}(U)e^{-sU(C)}$$

for every unitary representation U.

Let q_1 and q_2 be quadratic forms with signatures (r_1, s_1) and (r_2, s_2). We assume that $r_1 + s_1 = r_2 + s_2 = n$ and that the integers r_1, s_1, r_2, s_2 are odd. Let G_1 and G_2 be the corresponding spin groups and W_1, W_2 the complex Weyl groups of the Levi factor of a maximal parabolic subgroup P^i. For i in a finite set S let G_i be a n-dimensional spin group with a compact Cartan subgroup. In $G^1 = G_1 \times \Pi_{i \in S} G_i$, $G^2 = G_2 \times \Pi_{i \in S} G_i$ we fix parabolic subgroups P^1, P^2 corresponding to each other (see 4.3.1 for details). Let $I(P^1, \Pi_1, \nu)^L$, $I(P^2, \Pi, \nu)^L$ be L–packets of corresponding principal series representations. The L–packets have $|W_i|/|W_{i,c}|$ elements. From 4.4.2 we conclude:

5.5. Corollary. *We keep the above assumptions and denote by $\mathrm{tor}_s^{(1)}$, respectively $\mathrm{tor}_s^{(2)}$ the torsion functions of G^1, respecitvely of G^2. Then*

$$|W_{1,c}| I(P^1, \Pi_1, \nu)^L(\mathrm{tor}_s^{(1)}) = |W_{2,c}| I(P^2, \Pi_2, \nu)^L(\mathrm{tor}_s^{(2)}).$$

Here for short $I(P^i, \Pi_i, \nu)^L(\mathrm{tor}_s^{(i)})$ denotes the sum of the traces $tr(\Psi(\mathrm{tor}_s^i))$ for Ψ in the L-Packet $I(P^i, \Pi_i, \nu)^L$.

Next, we consider the orbital integral

$$O_G(\gamma, \mathrm{tor}_s) = \int_{G/G_\gamma} \mathrm{tor}_s(g\gamma g^{-1}) d(\overline{g}) .$$

Here $d(\overline{g})$ is an invariant measure, see 3.1.

5.6. Proposition. *The orbital integral* $O_G(\gamma, \mathrm{tor}_s)$ *is zero if* γ *is not conjugate to an element in the fundamental Cartan subgroup.*

Proof. The claim follows from the Fourier expansion of orbital integrals as explained in [H – Ch 2] and [He]. The orbital integral is a product of orbital integrals of the factors of G. For the factor G_1 one can also use the remark following 5.2 and [M – S]. q.e.d.

5.7. Theorem. *Let* h *be a regular element in the fundamental Cartan subgroup* $C = TA$. *Then*

$$O_G(h, \mathrm{tor}_s)$$

depends only on the inner class of G.

Proof. The orbital integral of tor_s is a product of orbital integrals of the factors of G. One proceeds as in the proof of 5.6. We sketch the argument for $G = G_1$. We choose a unit vector $y \in \mathbf{a}$ as in [M – S]. Let $h = t \exp xY$ with $t \in T$, $a = \exp xY \in A, x \in \mathbb{R}$. Let ℓ be the Lie algebra of MA and define $\Delta(h) = |\det(1 - Ada^{-1}|g/\ell)|$ and for a positive set σ^+ of roots of $(\mathbf{m}_{\mathbb{C}}, \mathbf{t}_{\mathbb{C}})$ define $\Delta_M^+ = \Pi_{a \in \sigma^+}(1 - e^{-\alpha})$. By the results of Harish–Chandra and Herb mentioned in 5.6 we get

$$O_G(h, \mathrm{tor}_s) =$$

$$= \frac{1}{\Delta(h) \cdot \Delta_M^+(h)} \frac{e^{-x^2/4s}}{(4\pi s)^{1/2}} \sum_{\Pi_\wedge \in \hat{M}^0_{dis}} e^{-sC_{\Pi_\wedge}} \mathrm{tor}(I(Q, \Pi_\wedge) \sum_{w \in W(M,C)} \det(w) e^{-w(\wedge + \rho_M)(h)}$$

$$= \frac{1}{\Delta(h) \cdot \Delta_M^+(h)} \frac{e^{-x^2/4s}}{(4\pi s)^{1/2}} \sum_{\wedge \in P^+(M^0)} e^{-sC_{\Pi_\wedge}} \mathrm{tor}(I(Q, \Pi_\wedge) \sum_{w \in W_{\mathbb{C}}(M,C)} \det(w) e^{-w(\wedge + \rho_M)(h)}$$

which is independent of the inner class of G. Here $C_{\Pi_\wedge}$ denotes the eigenvalue of the Casimir operator for M^0 on $\Pi_\wedge$ and $\hat{M}^0_{dis}$ denotes the set of equivalence classes of discrete series representations of M^0 where $\Pi_\wedge$ has parameter $\wedge$ in the set $P^+(M^0)$ of dominant integral weights of M^0. We use the normalisation of measures given in the above references and note that the occurring measures are compatible on inner forms of G. q.e.d.

5.8. Theorem. *Let* $G_i, i = 1, 2,$ *and* $C \subset G_i$ *be as above. For* $\gamma \in C$ *we have*

$$|W(G_1, C)| SO_{G_1}(\gamma, tor_s^{(1)}) = |W(G_2, C)| SO_{G_2}(\gamma, tor_s^{(2)}) .$$

Proof. At first we assume that $\gamma \in C$ is regular, i.e. its centralizer is C. Then the set of G_1–conjugacy classes in the stable conjugacy class $\{\gamma\}_{st}$ corresponds to

$$\ker(H^1(\mathbb{C}|\mathbb{R}, C(\mathbb{C})) \longrightarrow H^1(\mathbb{C}|\mathbb{R}, G_1(\mathbb{C})))$$

which is known to be isomorphic to $W_{\mathbb{C}}/W(G_1, C)$, see [Ko 3, 4]. Here $W_{\mathbb{C}} = W_{\mathbb{C}}(G_1, C)$ is the complex Weyl group of G_1. For a regular $\eta \in C$ the Kottwitz sign $e(\eta)$ is 1 by definition. Hence

$$SO_{G_1}(\gamma, tor_s) = \sum_{s \in W_{\mathbb{C}}/W(G_1,C)} O_{G_1}(w\gamma w^{-1}, tor_s)$$
$$= \frac{1}{|W(G_1,C)|} \sum_{w \in W_{\mathbb{C}}} O_{G_1}(w\gamma w^{-1}, tor_s) \,.$$

Since $O_{G_1}(w\gamma w^{-1}, tor_s) = O_{G_2}(w\gamma w^{-1}, tor_s)$ for all $w \in W_{\mathbb{C}}$ by 5.7 the claim holds for regular γ.
If $\gamma \in C$ is singular we use the technique developed by Harish-Chandra to the deduce the result from the result for regular γ, see [Sh] and [H – Ch 1]. qed.

6. Transfer of stable fundamental conjugacy classes and quadratic forms

If G_1, G_2 are algebraic groups over F which are inner forms of each other then there is a corresponding isomorphism $\psi : G_1 \times_F \overline{F} \longrightarrow G_2 \times_F \overline{F}$ which maps a $G_1(\overline{F})$–conjugacy class $\{\gamma_1\}_{G_1(\overline{F})}$ to the $\mathrm{gal}(\overline{F}|F)$–stable $G_2(\overline{F})$ conjugacy class $\{\psi(\gamma_1)\}_{G_2(\overline{F})}$. If there is a $\gamma_2 \in G_2(F)$ such that $\{\psi(\gamma_1)\}_{G_2(\overline{F})} = \{\gamma_2\}_{G_2(\overline{F})}$ one says that $\{\gamma_1\}_{G_1(\overline{F})}$ or γ_1 *transfers* to G_2. In this section we give conditions on G_1, G_2 and γ_1 such that γ_1 transfers to G_2 and determine for which special orthogonal groups $G_i = SO(q_i)$ these conditions are fulfilled.

6.1. Let G be a connected reductive group defined over a number field F. Let T be a maximal torus of G which is defined over F. If v is a place of F we call T fundamental at v if $T \times_F F_v$ is a torus with minimal F_v–rank in $G \times_F F_v$. It is well known that fundamental tori in $G \times_F F_v$ are $G(F_v)$–conjugate. Moreover the natural map $H^1(F_v, T) \longrightarrow H^1(F_v, G)$ is surjective if T is fundamental at v. For all this see for example [Ko 3: 10.2]. Let T be a maximal over F defined torus in G. We call T *$\mathbb{R}$–fundamental* if $T \times_F F_v$ is fundamental for all $v \in S_\infty(F)$.

6.2. An element $\gamma \in G(F)$ is called *fundamental* if there exists an *$\mathbb{R}$–fundamental* torus T of G such that $\gamma \in T(F)$. Obviously then all elements of the $G(F)$–conjugacy class $\{\gamma\}$ are fundamental and $\{\gamma\}$ is called a *fundamental conjugacy class.*

6.3. We recall that semisimple elements $\gamma, \eta \in G(F)$ are called *stably conjugate* if there is a $g \in G(\overline{F})$ such that $g\gamma g^{-1} = \eta$ and such that for all

$\varphi \in \text{gal}(\overline{F}|F)$ one has $g^{-1}\,{}^{\varphi}g \in G^0_\gamma(\overline{F})$, where G^0_γ is the connected component of the centralizer G_γ of γ in G. The $G(F)$–conjugacy classes within the stable conjugacy class $\{\gamma\}_{st}$ of γ are in bijection to

$$im(\ker(H^1(F,G^0_\gamma) \longrightarrow H^1(F,G)) \longrightarrow \ker(H^1(F,G_\gamma) \longrightarrow H^1(F,G)))\,.$$

For all this see [Ko 1: § 3].

The following result can be found between the lines in work of Steinberg and Kneser. A version for groups with simply connected derived group whose real points contain a compact Cartan was proved by Harder (unpublished letter). We need in 6.5 and 6.7 the following result for all reductive centralizers of fundamental elements.

6.4. Proposition. *Let G be a connected reductive group defined over F. Let $\xi \in H^1(F,G)$ and denote by S the finite set of places of F such that $S_\infty(F) \subset S$ and such that $v \notin S$ if the image of ξ in $H^1(F_v,G)$ is trivial. Then*

i) *there exist a maximal torus T of G, defined over F, such that $T \times F_v$ is fundamental for all $v \in S$.*

ii) *if T is as in* i) *the class ξ is in the image of $H^1(F,T)$ in $H^1(F,G)$.*

Proof. The existence of T is proved in [Ha 1, Lemma 5.3.3]. So it suffices to prove ii). According to [Ko 1] there exists a z-extension $G' \xrightarrow{\alpha} G$ and a $\xi' \in H^1(F,G')$ such that $\alpha(\xi') = \xi$. This means that here G' is a connected reductive group with simply connected derived group and that α is a surjective F–rational homomorphism whose kernel is central and isomorphic to an induced torus. If S is as above and $v \notin S$ then $\xi' \in H^1(F,G')$ maps to zero in $H^1(F_v,G')$. This follows from $H^1(F, \ker(\alpha)) = H^1(F_v, \ker(\alpha)) = \{1\}$ since $\ker(\alpha)$ is an induced torus. We use i) and find a maximal torus T' in G' such that $T' \times_F F_v$ is fundamental for $v \in S$. Hence it suffices to prove ii) for T' and G' and ξ'.

We assume now that G has a simply connected derived group G_{Der} choose T according to i) and consider the diagram

$$\begin{array}{ccccccc} H^1(F,G_{\text{Der}}) & \longrightarrow & H^1(F,G) & \longrightarrow & H^1(F,D) & & \\ \uparrow & & \uparrow & & \uparrow & & \\ H^1(F,T_{\text{Der}}) & \longrightarrow & H^1(F,T) & \longrightarrow & H^1(F,D) & \longrightarrow & H^2(F,T_{\text{Der}}) \end{array}$$

where $G/G_{\text{Der}} = D$ is a torus and where $T_{\text{Der}} = (T \cap G_{\text{Der}})^0$. Then $T_{\text{Der}} \times_F F_v$ is fundamental in $G_{\text{Der}} \times_F F_v$ for $v \in S$. Moreover it is known that $H^2(F_v, T_{\text{Der}}) =$

1 for all $v \in S$. See [Ha 1], [Ha 2] or [Ko 3]. It follows from this and by definition of S that the image of ξ in $\oplus_v H^2(F_v, T_{\mathrm{Der}})$ is trivial.

Using $H^2(F_v, T_{\mathrm{Der}}) = \{1\}, v \in S_\infty(F)$, and Tate–Nakayama duality we get that the natural map

$$H^2(F, T_{\mathrm{Der}}) \longrightarrow \oplus_v H^2(F_v, T_{\mathrm{Der}})$$

is injective, see [Ha 1, A.]. Hence there is a $\eta \in H^1(F,T)$ whose image in $H^1(F,D)$ coincides with the image of ξ. We choose cocycles ξ and η representing ξ and η. Let ${}^\eta G$ denote the η-twisted F-form of G. Then $\xi\eta^{-1}$ represents a class which maps to the trivial class in $H^1(F,D)$ and hence is the image of a class in $H^1(F, {}^\eta G_{\mathrm{Der}})$. By the Hasse-principle $H^1(F, {}^\eta G_{\mathrm{Der}}) \xrightarrow{\sim} \oplus_{v \in S_\infty(T)} H^1(F_v, {}^\eta G_{\mathrm{Der}})$, see [P – R]. Since ${}^\eta T_{\mathrm{Der}} = T_{\mathrm{Der}}$ is $\mathbb{R}$-fundamental we have a surjection $H^1(F_v, T_{\mathrm{Der}}) \longrightarrow H^1(F_v, {}^\eta G_{\mathrm{Der}})$ for all $v \in S_\infty(T)$. But we also have a surjection $H^1(F, T_{\mathrm{Der}}) \longrightarrow \oplus_{v \in S_\infty(F)} H^1(F_v, T_{\mathrm{Der}})$, see [Ha 1: A 1.2]. Hence $\xi\eta^{-1}$ is represented by a class $v \in H^1(F, T_{\mathrm{Der}})$ and then $v\eta \in H^1(F,T)$ maps to ξ. q.e.d.

6.5. Proposition. *If $\gamma \in G(F)$ is fundamental, then all $G(F)$-conjugacy classes of $\{\gamma\}_{st}$ are fundamental.*

Proof. Let $\xi \in G(F)$ represent a $G(F)$-conjugacy class in $\{\gamma\}_{st}$. Then $\xi = g\gamma g^{-1}$ for some $g \in G(\overline{F})$ such that $g^{-1}\,{}^\varphi g = a_\varphi \in G^0_\gamma(\overline{F})$ for all $\varphi \in \mathrm{gal}(\overline{F}|F)$. By Prop. 6.4 there exists a maximal torus $T|F$ which in particular is $\mathbb{R}$-fundamental in G^0_γ such that the class of a_φ is in the image of $H^1(F,T)$. Since γ is $\mathbb{R}$-fundamental in G then G^0_γ contains an $\mathbb{R}$-fundamental torus of G. Hence T is also an $\mathbb{R}$-fundamental torus of G. Since γ lies in the connected part of the center $Z(G^0_\gamma)^0$ which is contained in all maximal tori of G^0_γ we have $\gamma \in T(F)$. Moreover $a_\varphi = h^{-1} t_\varphi {}^\varphi h$ for a cocycle $\{t_\varphi\}$ for $H^1(F,T)$ and some $h \in G^0_\gamma(\overline{F})$. Now $t_\varphi = (gh^{-1})^{-1}\, {}^\varphi(gh^{-1})$; $\xi = (gh^{-1})\gamma(gh^{-1})^{-1}$ and $(gh^{-1})T(gh^{-1})^{-1}$ is a maximal torus of G which is defined over F. It contains ξ and is $\mathbb{R}$-fundamental since T and $gh^{-1}T(gh^{-1})^{-1}$ are isomorphic over F. qed.

6.6. Let G_1, G_2 be connected reductive groups which are inner forms of each other, i.e. there is an isomorphism $\psi : G_1 \times_F \overline{F} \longrightarrow G_2 \times_F \overline{F}$ such that $\psi^{-1}\,{}^\varphi\psi = int(a_\varphi), \varphi \in \mathrm{gal}(\overline{F}|F), a_\varphi \in G_1(\overline{F})$. Then $\{a_\varphi\}, \varphi \in \mathrm{gal}(\overline{F}|F)$, defines a cocycle representing a class $a \in H^1(F, G_{1_{ad}})$, where $G_{1_{ad}}$ is the adjoint quotient of G_1.

Let $\mathbb{A}_f \subset \mathbb{A}$ be the ring of finite adeles over F and let us put $\bar{\mathbb{A}}_f = \mathbb{A}_f \otimes_F \overline{F}$. We denote the kernel of the natural map $H^1(\overline{F}|F, G_{1_{ad}}(\overline{F})) \longrightarrow H^1\left(\overline{F}|F,\, G_{1_{ad}}(\bar{\mathbb{A}}_f)\right)$ by $H^1(F, G_{1_{ad}})^\infty$. If ψ gives rise to a class in $H^1(F, G_{1_{ad}})^\infty$ then for all $v \notin S_\infty(T)$ we have isomorphisms $\psi_v : G_1 \times_F F_v \xrightarrow{\sim} F_v$ such that $\prod_{v \notin S_\infty(F)} \psi_v = \psi_f : G_1(\mathbb{A}_f) \xrightarrow{\sim} G_2(\mathbb{A}_f)$ is an isomorphism. For short we then say that G_1 and G_2 are *inner forms of each other which only differ at the archimedean places.*

6.7. Proposition. *Let G_1, G_2 be connected reductive groups defined over F which are inner forms of each other and differ only at the archimedean places of F. Then all stable fundamental conjugacy classes of G_1 transfer to G_2 and vice versa.*

Proof. We write $G = G_1$. Let $\xi \in H^1(F, G_{ad})^\infty$ be the class determined by G_2 as in 4.6. Write $\xi = \{\xi_\varphi)\}, \varphi \in \text{gal}(\overline{F}|F)$, for a cocycle representing ξ. We identify G_2 with ${}^\xi G$. Then $G_2(F)$ is identified with a subgroup of $G(\overline{F})$. *Let*$\{\gamma\}_{st}$ be a stable fundamental conjugacy class. We have to show that $\{\gamma\}_{st}$ contains an element of ${}^\xi G(F)$. Let $G_{ad,\gamma}$ be the centralizer of γ in G_{ad}. Since γ is fundamental then $G_{ad,\gamma}$ contains an $\mathbb{R}$–fundamental torus T which is defined over F. By 6.4. we have $\xi \in \text{ im}\,(H^1(F,T) \longrightarrow H^1(F, G_{ad}))$. Hence there is a cocycle $\eta_\varphi \in G^0_{ad,\gamma}(\overline{F})$ such that $\eta_\varphi = g^{-1}\xi_\varphi\, {}^\varphi g$ for some $g \in G(\overline{F})$. But then $\eta_\varphi\, {}^\varphi\gamma\eta_\varphi^{-1} = \eta_\varphi\gamma\,\eta_\varphi^{-1} = \gamma$. We get

$$\xi_\gamma\, {}^\varphi\left(g\gamma g^{-1}\right)\xi_\varphi^{-1} = g\gamma g^{-1}$$

for all $\varphi \in \text{gal}(\overline{F}|F)$. Hence $g\gamma g^{-1} \in {}^\xi G(F)$. q.e.d.

6.8. *Remark.* We use the notation of 6.7 and replace the cocycle $\xi = \{\xi_\varphi\}$ by $\eta = \{\eta_\varphi\} = \{g^{-1}\xi_\varphi\, {}^\varphi g\}$. Then ${}^\xi G$ is replaced by ${}^\eta G$ and $\gamma \in {}^\eta G(F)$. We have an $\mathbb{R}$–fundamental torus $T \subset G^0_\gamma$ such that $\eta_\varphi \in T(\overline{F})$. Then T is also a torus defined over F of ${}^\eta G$ and $\gamma \in^\eta T(F) = T(F)$. We have $\{\gamma\}_{st} = \{h\gamma h^{-1} | h \in G(F),\ h^{-1}\, {}^\varphi h \in G^0_\gamma(\overline{F})$ for all $\varphi \in \text{gal}(\overline{F}|F)\}$. If we write $\varphi(\eta)$ for the $\{\eta_\varphi\}$– twisted φ–action of $\text{gal}(\overline{F}|F)$ on ${}^\eta G$ then $\{\gamma\}_{st} = \{h\gamma h^{-1} | h \in G(\overline{F}), h^{-1\varphi(\eta)}h \in G^0_\gamma(\overline{F})$ for all $\varphi \in \text{gal}(\overline{F}|F)\}$. However, in general it is not true that there is a bijection of the set of $G(F)$–conjugacy classes in $\{\gamma\}_{st}$ to the set of ${}^\eta G(F)$–conjugacy classes in $\{\gamma\}_{st}$.

6.9. For the following well known facts, see for example [Kn 3] or [Ha 3]. We make $H^1(F,G)^\infty = \ker(H^1(F,G) \longrightarrow \oplus_{v \notin S_\infty(F)} H^1(F_v, G))$ more explicit if G is a special orthogonal group.

i) We assume that F is a totally real number field and we assume that $q^0 : F^n \longrightarrow F$ is a non degenerate n–dimensional quadratic form, where $n \geq 3$. For $x \in F^n$ we write $q^0(x) = b_{q^0}(x,x)$ where b_{q^0} is the symmetric $n \times n$-matrix representing q. Then $\det(b^0_q) \bmod F^{*^2}$ is called the determinant $d(q^0)$ of q^0. Let ${}^t a$ be the transposed matrix of an $n \times n$–Matrix a. Since a non degenerate n–dimensional quadratic form q becomes equivalent to q^0 over $\overline{F}$ there is an $a \in GL_n(\overline{F})$ such that $a^{-1}\, {}^\varphi a \in O(q^0)(\overline{F})$ for all $\varphi \in \text{gal}(\overline{F}|F)$ and $b_q = a\, b_{q^0}\, {}^t a$ for the symmetric matrix given by q. The cocycle $a^{-1}\, {}^\varphi a$ represents a class in $H^1(F, O(q^0))$ and we obtain a bijection between the equivalence classes of nondegenerate quadratic forms over F with $H^1(F, O(q^0))$. Over $\overline{F}$ we have an exact sequence

$$1 \longrightarrow SO(q^0) \longrightarrow O(q^0) \longrightarrow \mu_2 \longrightarrow 1$$

where $A \in O(q^0)(\overline{F})$ is mapped to $\det(A) \in \{\pm 1\} = \mu_2(\overline{F})$. As usual one identifies $H^1(F, \mu_2) \xrightarrow{\sim} F^*/F^{*^2}$ and there results an exact sequence of pointed sets

$$1 \longrightarrow H^1(F, SO(q^0)) \longrightarrow H^1(F, O(q^0)) \longrightarrow H^1(F, \mu_2) \longrightarrow 1 .$$

If we twist the $\mathrm{gal}(\overline{F}|F)$-action on $O(q^0)(\overline{F})$ by a cocycle $\{\xi_\varphi\}$ for $H^1(F, SO(q^0))$ we get a corresponding exact sequence. From this it follows that the map $H^1(F, SO(q^0)) \longrightarrow H^1(F, O(q^0))$ is injective. Hence $H^1(F, SO(q^0))$ is in bijection to the classes of n–dimensional quadratic forms q where $\det(q^0) \equiv \det(q) \bmod F^{*^2}$.

ii) If q_v is a quadratic form over F_v the Hasse–Witt invariant $\varepsilon_v(q_v) \in \{\pm 1\}$ is defined. If $F_v \cong \mathbb{R}$ and if $(r_v, s_v) = \mathrm{sig}(q_v)$ is the signature of q_v then $\varepsilon_v(q_v) = (-1)^{s_v(s_v-1)/2}$. If q is a quadratic form over F we can view q also as quadratic form over F_v and $\varepsilon_v(q)$ is defined. We put $c_v(q_v) = \varepsilon_v(q_v)\varepsilon_v(q^0)$. The Hasse-principle for quadratic forms and the existence theorem for quadratic forms with given local data then can be reformulated as: There is an exact sequence of pointed sets

$$1 \longrightarrow H^1(F, SO(q^0)) \longrightarrow \bigoplus_v H^1(F_v, SO(q^0)) \xrightarrow{c} \{\pm 1\} \longrightarrow 1 .$$

Here $c(\{q_v\}) = \prod_v c_v(q_v)$. By the Hasse-principle the first arrow actually is injective. There results the following exact sequence

$$1 \longrightarrow H^1(F, SO(q^0))^\infty \longrightarrow \bigoplus_{v \in S_\infty(F)} H^1(F_v, SO(q^0)) \xrightarrow{c} \{\pm 1\} \longrightarrow 1 .$$

where now $c = \prod_{v \in S_\infty(F)} c_v$

iii) We assume that $\prod_{v \in S_\infty(F)} SO(q^0)(F_v)$ contains a fundamental Cartan subgroup of $\mathbb{R}$– split rank 1. Then there is exactly one place $u \in S_\infty(F)$ such that $SO(q^0)(F_u)$ contains such a Cartan and all $SO(q^0)(F_v)$, $v \neq u$, contain a compact Cartan. If $(r_v^0, s_v^0) = sig_v(q^0)$ is the signature of q^0 over F_v then $SO(q^0)(F_v)$ contains no compact Cartan iff $r_v^0 s_v^0 \equiv 1 \bmod 2$ and if $r_u^0 s_u^0 \equiv 1 \bmod 2$ then $SO(q^0)(F_u)$ contains a fundamental Cartan of $\mathbb{R}$–split rank 1. Hence $n = \dim q^0 = r_u^0 + s_u^0$ is even and for $v \neq u$ we have $r_v^0 \equiv s_v^0 \equiv 0 \bmod 2$.

iv) Let q represent an element of $H^1(F, SO(q^0))^\infty$ such that $\prod_{v \in S_\infty(F)} SO(q)(F_v)$ contains a fundamental Cartan of $\mathbb{R}$–split rank 1. Since $SO(q)$ is in particular an inner form of $SO(q^0)$ the group $SO(q)(F_v)$ contains a compact Cartan for all $v \neq u$, u as in iii). Let (r_u, s_u) be the signature of q over F_u. Then $r_u \cdot s_u \equiv 1 \bmod 2$.

v) Let $H^1(F, SO(q^0))_1^\infty$ be the set of classes of $H^1(F, SO(q^0))^\infty$ which represent a quadratic form q such that $\prod_{v \in S_\infty(F)} SO(q)(F_v)$ contains a fundamental Cartan of $\mathbb{R}$–split rank 1 and such that at least one factor $SO(q)(F_w)$ is compact. We assume that q^0 has these properties. Then q is anisotropic. By the strong Hasse–principle every anisotropic form q of $\dim q \geq 5$ has to have such a compact factor.

6.10. Proposition. *Let $r, s \in \mathbb{N}$ with $r + s = n$, $r \cdot s$ odd. Then there is a quadratic form representing a class of $H^1(F, SO(q^0))_1^\infty$ with $sig_u(q) = (r, s)$ if and only if $[F : \mathbb{Q}] \geq 3$ or if $[F : \mathbb{Q}] = 2$ and $s \equiv s_u^0 \bmod 4$.*

Proof. We put $s = s_u^0 + 2j$. Then $s(s-1) \equiv s_u^0(s_u^0 - 1) + 2j \bmod 4$. We let s_v be even for $v \neq u$. Then $s_v(s_v - 1) \equiv s_v \bmod 4$. Hence we get from ii) an equation for the $s_v, v \neq u$ as follows:

$$\sum_{\substack{v \in S_\infty(F) \\ v \neq u}} s_v + 2j \equiv \sum_{\substack{v \in S_\infty(F) \\ v \neq u}} s_v^0 \bmod 4$$

where $w \neq u$, $w \in S_\infty(F)$ and $s_w = 0$ or $s_w = n$. Then this equation is solvable if $[F : \mathbb{Q}] \geq 3$ or if j is even. If $[F : \mathbb{Q}] = 2$ and if j is odd the equation is

$$s_w + 2 \equiv s_w^0 \bmod 4$$

where $s_w, s_w^0 \in \{0, n\}$ and this equation is not solvable. q.e.d.

6.11. Corollary. *If $[F : \mathbb{Q}] \geq 3$ or if $[F : \mathbb{Q}] = 2$ and $n \equiv 0 \bmod 4$ or if $[F : \mathbb{Q}] = 2$ and $s_u \equiv 1 \bmod 4$ then there exists an anisotropic quadratic form q such that $SO(q)$ and $SO(q^0)$ are inner forms of each other which differ only at the infinite primes such that $sig_u(q) = (n-1, 1)$.*

Proof. If $[F : \mathbb{Q}] \geq 3$ this follows from 6.10. If $[F : \mathbb{Q}] = 2$ and $sig_u(q^0) = (r_u^0, s_u^0)$ and $s_u^0 \equiv 1 \bmod 4$ again 6.10 applies. If $[F : \mathbb{Q}] = 2$ and $n \equiv 0 \bmod 4$. Then $sig_u(-q^0) = (s_u^0, r_u^0)$ and if $s_u^0 \equiv 3 \bmod 4$ we have $r_u^0 \equiv 1 \bmod 4$. Since $SO(-q) = SO(q)$ and $SO(-q^0) = SO(q^0)$ again 6.10 applies. q.e.d.

References

[B] Borel, A., Admissible representations of a semisimple group over a local field with vectors fixed under an Iwahori subgroup, *Invent. Math.* **35** (1976), 233 – 259.

[B – W] Borel, A., Wallach, N., *Continuous cohomology, discrete subgroups, and representations of reductive groups,* Princeton University Press, 1980.

[C – D] Clozel, L., Delorme, P., Le théorème de Paley–Wiener invariant pour les groupes de Lie reductifs, *Invent. Math.* **77** (1984), 427–453.

[D] Delorme, P., Formules limites et formules asyptotiques pour les multiplicités dans $L^2(G/\Gamma)$, *Duke Math. J.* **53** (1986), 691–731.

[F] Fried, D., Analytic torsion and closed geodesics on hyperbolic manifolds, *Invent. Math.* **84** (1986), 523–540.

[G – G – P] Gelfand, I.M., Graev, M.I., Piatetski–Shapiro, I., *Representation theory and automorphic functions,* W.B. Saunders Co. Philadelphia, 1969.

[Ha 1] Harder, G., Über die Galoiskohomologie halbeinfacher Matrizengruppen I, *Math. Z.* **90** (1965), 404–428.

[Ha 2] Harder, G., Über die Galoiskohomologie halbeinfacher Matrizengruppen II, *Math. Z.* **92** (1966), 396 – 415.

[Ha 3] Harder, G., Bericht über neuere Resultate der Galoiskohomologie. *Jahresber. Dtsch. Math.-Ver.* **70** (1968), 182–216.

[H – Ch 1] Harish-Chandra, Discrete series for semisimple Lie groups II. Explicit determination of the characters, *Acta Math.* **116** (1966), 1–111.

[H – Ch 2] Harish-Chandra, Supertempered distributions on reductive groups, *Advances in Math. Suppl. Studies* **8** (1983), 139–152.

[He] Herb, R., Fourier inversion of invariant integrals on semisimple real Lie groups, *Trans. Am. Math. Soc.* **249** (1979), 281–302.

[Kn 1] Kneser, M, Galoiskohomologie halbeinfacher algebraischer Gruppen über *p*-adischen Körpern I., *Math. Z.* **88** (1965), 40–47.

[Kn 2] Kneser, M., Galoiskohomologie halbeinfacher algebraischer Gruppen II, *Math. Z.* **89** (1965), 250–272.

[Kn 3] Kneser, M., Lectures on Galois cohomology of classical groups. Tata Inst. of Fund. Research, Bombay (1969).

[Ko 1] Kottwitz, R.: Rational conjugacy classes in reductive groups, *Duke Math. J.* **49** (1982), 785–806.

[Ko 2] Kottwitz, R., Sign changes in harmonic analysis on reductive groups, *Trans. Am. Math. Soc.* **278** (1983), 289–297.

[**Ko 3**] Kottwitz, R., Stable trace formula: elliptic singular terms, *Math. Ann.* **275** (1986), 265–399.

[**Ko 4**] Kottwitz, R., Tamagawa numbers, *Ann. Math.* **127** (1988), 629–646.

[**L**] Labesse, J.–P., Pseudo–coefficients très cuspidaux et K–theory, *Math. Ann.* **291** (1991), 607–616.

[**M - S**] Moscovici, H., Stanton, R., R–torsion and zeta functions for locally symmetric manifolds, *Invent. Math.* **105** (1991), 185–216.

[**O – W**] Osborne, M., S., Warner, G., The Selberg trace formula. I: Γ–rank one lattices, *Advances in Math. Suppl. Studies* **6** (1978), 1– 113.

[**P - R**] Platonov, V., Rapinchuk, A., *Algebraic groups and number theory,* Academic Press, Boston MA, 1994.

[**Se**] Selberg, A., Harmonic analysis and discontinuous groups in weakly symmetric Riemannian spaces with applications to Dirichlet series, *J. Indian. Math. Soc.* **20** (1956), 47–87.

[**Sh**] Shelstad, D., Orbital integrals, endoscopic groups and L–indistinguishability, Journées Automorphes, Pub Math. Univ. Paris VII, Vol. 15, Paris (1983).

[**Sp**] Speh, B., Analytic torsion and automorphic forms. in: *Noncompact Lie groups and some of their applications*, ed. E. Tanner, R. Wilson, Kluwer Academic Publishers, 157–168.

[**St**] Steinberg, R., *Endomorphisms of algebraic groups.* Mem. Amer. Math. Soc. 80, 1968.

[**V**] Vogan, D., *Representations of Real Reductive Lie Groups*, Birkhäuser, Boston, Basel, Stuttgart 1981.

[**W**] Warner, G., *Harmonic analysis on semisimple Lie groups* I. Springer Verlag, Berlin, Heidelberg, New York, 1972.

[**Z**] Zagier, D.B., *Zetafunktionen und quadratische Körper.* Hochschultext. Springer Verlag. Berlin – Heidelberg – New York, 1981.

Jürgen Rohlfs
Katholische Universität Eichstätt
Mathematisch–Geographische Fakultät
Ostenstr. 26-28
85072 Eichstätt
Germany

Birgit Speh
Cornell University
Department of Mathematics
Ithaca, New York 14853
U S A

FLAG MANIFOLDS AND REPRESENTATION THEORY

JOSEPH A. WOLF

CONTENTS

This article is an expanded version of lectures at the "Fifth Workshop on Representation Theory of Lie Groups and Its Applications," Córdoba, Argentina, August 1995. The topics were complex flag manifolds, real group orbits, and linear cycle spaces, with applications to the geometric construction of representations of semisimple Lie groups. These topics come up in many aspects of complex differential geometry and harmonic analysis.

Research partially supported by N.S.F. Grant DMS 93 21285.

PART 1. COMPLEX FLAG MANIFOLDS.

In this part we indicate the basic facts for real group orbits on complex flag manifolds.

§1. PARABOLIC SUBALGEBRAS AND COMPLEX FLAGS.

Fix a complex semisimple Lie algebra $\mathfrak{g}$ and a Cartan subalgebra $\mathfrak{h} \subset \mathfrak{g}$. Let $\Sigma = \Sigma(\mathfrak{g}, \mathfrak{h})$ denote the corresponding root system, and fix a positive subsystem $\Sigma^+ = \Sigma^+(\mathfrak{g}, \mathfrak{h})$. The corresponding **Borel subalgebra**

$$\mathfrak{b} = \mathfrak{h} + \sum_{\alpha \in \Sigma^+} \mathfrak{g}_{-\alpha} \subset \mathfrak{g} \tag{1.1}$$

has its nilradical[1] $\mathfrak{b}^{-n} = \sum \mathfrak{g}_{-\alpha}$ and a Levi complement $\mathfrak{h}$.

In general a subalgebra $\mathfrak{s} \subset \mathfrak{g}$ is called a **Borel subalgebra** if it is Int $(\mathfrak{g})$–conjugate to a subalgebra of the form (1.1), in other words if there exist choices of $\mathfrak{h}$ and $\Sigma^+(\mathfrak{g}, \mathfrak{h})$ such that $\mathfrak{s}$ is given by (1.1).

Let G denote the (unique) connected simply connected Lie group with Lie algebra $\mathfrak{g}$. The Cartan subgroup of G corresponding to $\mathfrak{h}$ is $H = Z_G(\mathfrak{h})$. It has Lie algebra $\mathfrak{h}$, and it is connected because G is connected, complex and semisimple. The **Borel subgroup** $B \subset G$ corresponding to a Borel subalgebra $\mathfrak{b} \subset \mathfrak{g}$ is defined to be the G–normalizer of $\mathfrak{b}$, that is,

$$B = \{g \in G \mid \mathrm{Ad}\,(g)\mathfrak{b} = \mathfrak{b}\}\ . \tag{1.2}$$

Here are the basic facts on these Borel subgroups.

1.3. Lemma. *B has Lie algebra $\mathfrak{b}$, B is a closed connected subgroup of G, and B is its own normalizer in G.*

Proof. B is closed in G by definition (1.2). It follows that the normalizer $E = N_G(B)$ is closed in G, so E is a Lie subgroup. Let $\mathfrak{e}$ denote the Lie algebra of E. Then $\mathfrak{b} \subset \mathfrak{e}$ and $[\mathfrak{e}, \mathfrak{b}] \subset \mathfrak{b}$. Any subalgebra of $\mathfrak{g}$ that properly contains $\mathfrak{b}$ must be of the form $\mathfrak{b} + \sum_{\sigma \in S} \mathfrak{g}_\alpha$ with $S \subset \Sigma^+$, because $\mathfrak{h} \subset \mathfrak{b}$. Thus it would contain a 3–dimensional simple subalgebra and could not normalize $\mathfrak{b}$. Now $\mathfrak{e} = \mathfrak{b}$, in particular E normalizes $\mathfrak{b}$, so $E = B$. This shows both that B is its own normalizer and that B has Lie algebra $\mathfrak{b}$. Finally, B is connected because the Weyl group $W(\mathfrak{g}, \mathfrak{h})$ is **simply** transitive on the set of all positive subsystems of $\Sigma(\mathfrak{g}, \mathfrak{h})$. □

The other basic facts are not quite as obvious.

[1] Here we describe the nilradical as a sum of negative root spaces, rather than positive, so that, in applications, positive functionals on $\mathfrak{h}$ will correspond to positive bundles (instead of negative bundles), and holomorphic discrete series representations will be highest weight (instead of lowest weight) representations.

1.4. Lemma. *Let $G_u \subset G$ be a compact real form. Then G_u is transitive on $X = G/B$, and X has a G_u–invariant Kaehler metric. In particular X has the structure of compact Kaehler manifold.*

Proof. It suffices to consider a G_u constructed by means of a Weyl basis of $\mathfrak{g}$ using $\mathfrak{h}$ and Σ^+. This yields a real form $\mathfrak{g}_u \subset \mathfrak{g}$ on which the Killing form is negative definite. Then the G–normalizer of $\mathfrak{g}_u$ coincides with the real analytic subgroup of G for $\mathfrak{g}_u$, that is, G_u. By construction $\mathfrak{h}_u = \mathfrak{g}_u \cap \mathfrak{h}$ is the real form of $\mathfrak{h}$ on which the roots take pure imaginary values, and $\mathfrak{g} \cap \mathfrak{b} = \mathfrak{h}_u$. Now a dimension count shows that the G_u–orbit of the identity coset $x_0 = 1B \in G/B = X$ is open in X. It is also closed in X because G_u is compact. This proves the transitivity, and thus proves that X is compact.

Let $\lambda \in \mathfrak{h}^*$ such that $\langle \lambda, \alpha \rangle > 0$ for every $\alpha \in \Sigma^+$. Extend λ to a linear functional on $\mathfrak{g}$ by $\lambda(\mathfrak{g}_\gamma) = 0$ for every $\gamma \in \Sigma$, and view it as a 1–cochain for Lie algebra cohomology of $(\mathfrak{g}, \mathfrak{h})$. Then $d\lambda$ is a 2–cocycle on $G_u/H_u = X$, and as a 2–form it combines with the complex structure to define a Kaehler metric. Thus, for every λ in the positive Weyl chamber of $(\mathfrak{g}, \mathfrak{h}, \Sigma^+)$, we have a G_u–invariant Kaehler metric on X. □

1.5. Lemma. *There is a finite dimensional irreducible representation π of G with the following property: Let $[v]$ be the image of a lowest weight vector in the projective space $\mathbb{P}(V_\pi)$ corresponding to the representation space of π. Then the action of G on V_π induces a holomorphic action of G on $\mathbb{P}(V_\pi)$, and B is the G–stabilizer of $[v]$. In particular $X = G/B$ is a complete projective variety.*

Proof. For example, let $\rho = \frac{1}{2} \sum_{\alpha \in \Sigma^+} \alpha$ as usual and let π be the irreducible representation of highest weight ρ. The lowest weight is $-\rho$ and the assertions are immediate. □

1.6. Lemma. *B is a maximal solvable subgroup of G.*

Proof. The argument of Lemma 1.3 shows that $\mathfrak{b}$ is a maximal solvable subalgebra of $\mathfrak{g}$. If $E \subset G$ is a solvable subgroup, and $B \subsetneq E$ then the closure of E in G has those same properties, so we may assume E closed in G. But then E has a Lie algebra that is not solvable, so E is not solvable. We conclude that B is maximal solvable. □

A theorem of Borel says that any solvable subgroup of G has a fixed point on the complete projective variety X; this is conjugate to a subgroup of B. This gives another proof of Lemma 1.6, in fact it shows that the Borel subgroups are exactly the maximal solvable subgroups of G. That's how Borel originally defined them. The Borel subalgebras and subgroups given by (1.1) and (1.2) are the **standard** Borels.

A subalgebra $\mathfrak{p} \subset \mathfrak{g}$ is called **parabolic** if it contains a Borel subalgebra. For example, let Ψ be the simple root system corresponding to Σ^+ and let Φ be an arbitrary subset of Ψ. Every root $\alpha \in \Sigma$ has a unique expression

$$\alpha = \sum_{\psi \in \Psi} n_\psi(\alpha)\psi \tag{1.7}$$

where the $n_\psi(\alpha)$ are integers, all $\geqq 0$ if $\alpha \in \Sigma^+$ and all $\leqq 0$ if $\alpha \in \Sigma^- = -\Sigma^+$. Set

$$\Phi^r = \{\alpha \in \Sigma \mid n_\psi(\alpha) = 0 \text{ whenever } \psi \notin \Phi\} \tag{1.8}$$

and

$$\Phi^n = \{\alpha \in \Sigma^+ \mid \alpha \notin \Phi^r\} = \{\alpha \in \Sigma \mid n_\psi(\alpha) > 0 \text{ for some } \psi \notin \Phi\}. \tag{1.9}$$

Now define

$$\mathfrak{p}_\Phi = \mathfrak{p}_\Phi^r + \mathfrak{p}_\Phi^{-n} \text{ with } \mathfrak{p}_\Phi^r = \mathfrak{h} + \sum_{\alpha\in\Phi^r} g_\alpha \text{ and } \mathfrak{p}_\Phi^{-n} = \sum_{\alpha\in\Phi^n} \mathfrak{g}_{-\alpha}\,. \tag{1.10}$$

Then $\mathfrak{p}_\Phi$ is a subalgebra of $\mathfrak{g}$ that contains the Borel subalgebra (1.1), so it is a parabolic subalgebra of $\mathfrak{g}$.

1.11. Proposition. *Let $\mathfrak{p} \subset \mathfrak{g}$ be a subalgebra that contains the Borel subalgebra $\mathfrak{b} = \mathfrak{h} + \sum_{\alpha\in\Sigma^+} \mathfrak{g}_{-\alpha}$ of $\mathfrak{g}$. Then there is a set Φ of simple roots such that $\mathfrak{p} = \mathfrak{p}_\Phi$.*

Proof. Define $\Phi = \{\psi \in \Psi \mid \mathfrak{g}_\psi \subset \mathfrak{p}\}$. Then $\mathfrak{p}_\Phi \subset \mathfrak{p}$, and we must prove $\mathfrak{p} \subset \mathfrak{p}_\Phi$. Both contain $\mathfrak{b}$, so this comes down to showing that $\alpha \in \Sigma^+, \mathfrak{g}_\alpha \subset \mathfrak{p}$ implies $n_\psi(\alpha) = 0$ whenever $\psi \in \Psi \setminus \Phi$. We will prove this by induction on the level $\ell(\alpha) = \sum n_\psi(\alpha)$.

If $\ell(\alpha) = 1$ then α is simple, so $\mathfrak{g}_\alpha \subset \mathfrak{p}$ implies $\alpha \in \Phi$. Then $\psi \notin \Phi$ implies $\psi \neq \alpha$ so $n_\psi(\alpha) = 0$.

Now let $\ell(\alpha) = \ell_0 > 1$ and suppose that $n_{\psi'}(\gamma) = 0$ for all $\psi' \in \Psi \setminus \Phi$, whenever $\gamma \in \Sigma^+$ and $\mathfrak{g}_\gamma \subset \mathfrak{p}$ with $\ell(\gamma) < \ell_0$. Suppose first that we can (and do) choose $\psi \in \Phi$ such that $\gamma = \alpha - \psi$ is a root. Then

$$\mathfrak{g}_\gamma = [\mathfrak{g}_\alpha, \mathfrak{g}_{-\psi}] \subset [\mathfrak{p}, \mathfrak{b}^{-n}] \subset [\mathfrak{p}, \mathfrak{p}] \subset \mathfrak{p}.$$

If $\psi' \in \Psi \setminus \Phi$, then $n_{\psi'}(\alpha) = n_{\psi'}(\gamma)$, which is zero by the induction hypothesis. Suppose second that we cannot (and do not) choose ψ from among the elements of Φ. Then

$$\mathfrak{g}_\psi = [\mathfrak{g}_\alpha, \mathfrak{g}_{-\gamma}] \subset [\mathfrak{p}, \mathfrak{b}^{-n}] \subset [\mathfrak{p}, \mathfrak{p}] \subset \mathfrak{p},$$

so $\psi \in \Phi$, a contradiction. We have proved $n_{\psi'}(\gamma) = 0$ for all $\psi' \in \Psi \setminus \Phi$. Proposition 1.11 is proved. □

The **parabolic subgroup** $P \subset G$ corresponding to a parabolic subalgebra $\mathfrak{p} \subset \mathfrak{g}$ is defined to be the G–normalizer of $\mathfrak{p}$, that is,

$$P = \{g \in G \mid \mathrm{Ad}\,(g)\mathfrak{p} = \mathfrak{p}\}\,. \tag{1.12}$$

The basic facts on parabolic subgroups are most easily derived from the corresponding results for Borel subgroups. However, the two notions were developed separately, and from different viewpoints, in the 1950s.

1.13. Lemma. *The parabolic subgroup $P \subset G$ defined by* (1.12) *has Lie algebra $\mathfrak{p}$. That group P is a closed connected subgroup of G, and P is its own normalizer in G. In particular, a Lie subgroup of G is parabolic if and only if it contains a Borel subgroup.*

Proof. The argument of Lemma 1.3 shows that P has Lie algebra $\mathfrak{p}$, is closed and connected, and is equal to its own G–normalizer. Let $S \subset G$ be a Lie subgroup that contains a Borel subgroup B. Then its Lie algebra $\mathfrak{s}$ contains $\mathfrak{b}$, hence is parabolic. Because S is pinched between the analytic subgroup of G for $\mathfrak{s}$ and the G–normalizer of $\mathfrak{s}$, which coincide because parabolic subgroups are closed and connected, S is the parabolic subgroup of G for $\mathfrak{s}$. □

Let $B \subset P \subset G$ consist of a Borel subgroup contained in a parabolic subgroup. Then we have complex homogeneous quotient spaces $X = G/B$ and $Z = G/P$ and a G–equivariant holomorphic projection $X \to Z$ given by $gB \mapsto gP$. In particular, transitivity of G_u on X gives transitivity of G_u on Z in

1.14. Lemma. *Let $G_u \subset G$ be a compact real form. Then G_u is transitive on $Z = G/P$, and Z has a G_u–invariant Kaehler metric. In particular Z has the structure of compact Kaehler manifold.*

The argument of Lemma 1.4 is easily modified to prove the Kaehler statement in Lemma 1.14. Just take λ in the dual space of the center of $\mathfrak{p}^r$ such that $\langle \lambda, \alpha \rangle > 0$ for all $\alpha \in \Phi^n$.

1.15. Lemma. *Fix a standard parabolic subgroup $P = P_\Phi$ in G. Then there is a finite dimensional irreducible representation π of G with the following property: Let $[v]$ be the image of a lowest weight vector in the projective space $\mathbb{P}(V_\pi)$ corresponding to the representation space of π. Then the action of G on V_π induces a holomorphic action of G on $\mathbb{P}(V_\pi)$, and P is the G–stabilizer of $[v]$. In particular $Z = G/P$ is a complete projective variety.*

Proof. We use the argument of Lemma 1.5, with a different choice of highest weight. Recall $\rho = \frac{1}{2}\sum_{\Sigma^+} \alpha$ and set $\rho_\Phi = \frac{1}{2}\sum_{\Phi^r \cap \Sigma^+} \alpha$. If $\psi \in \Psi$ now $\frac{2\langle \rho_\Phi, \psi \rangle}{\langle \psi, \psi \rangle}$ is 1 if $\psi \in \Phi$, is 0 if $\psi \notin \Phi$. Now let π be the irreducible representation of G with lowest weight $-(\rho - \rho_\Phi)$, in other words highest weight $w(\rho - \rho_\Phi)$ where w is the element of the Weyl group that sends Σ^+ to its negative. Then the assertions are immediate. □

At this point we summarize, as follows.

1.16. Proposition. *Let P be a complex Lie subgroup of G. Then the following conditions are equivalent.* (1) *G/P is a compact complex manifold.* (2) *G/P is a complete projective variety.* (3) *If G_u denotes a compact real form of G then G/P is a G_u–homogeneous compact Kaehler manifold.* (4) *G/P is the projective space orbit of an extremal weight vector in an irreducible finite dimensional representation of G.* (5) *G/P is a G–equivariant quotient manifold of G/B, for some Borel subgroup $B \subset G$.* (6) *P is a parabolic subgroup of G.*

We will simply refer to these spaces $Z = G/P$ as **complex flag manifolds.**

References for §1.

- A. Borel, "Groupes linéaires algébriques", Ann. of Math **64** (1956), 20–82.
- A. Borel, "Linear algebraic groups", Benjamin, 1961.
- A. Borel, "Linear algebraic groups", Second Enlarged Edition, GTM 126, Springer–Verlag, 1991.
- A. Borel & J. Tits, "Groupes réductifs", Publ. Math. I.H.E.S. **27** (1965), 55–150.
- J. Tits, "Sur certains classes d'espaces homogènes de groupes de Lie", Memoir, Belgian Academy of Sciences, 1955.
- J. Tits, "Espaces homogènes complexes compacts", Comment. Math. Helv. **37** (1962), 111–120.

§2. Intersections of Parabolics.

In order to examine the orbit structure of a complex flag manifold $Z = G/P$ under the action of a real form G_0 of G, we need to know that the intersection of any two parabolic subgroups of G contains a Cartan subgroup.

The Bruhat Lemma for the complex flag manifold $X = G/B$ is as follows. We may assume B given by (1.1) and (1.2). Consider the Weyl group $W = W(\mathfrak{g}, \mathfrak{h}) = N_G(H)/H$. Given $w \in W$ choose a representative $s_w \in N_G(H)$. Let $x_0 = 1B \in G/B = X$. The crudest form of the Bruhat decomposition is sufficient for our needs. Here is the statement; I won't give a proof.

2.1. Lemma. *X is the disjoint union of the B–orbits $B(s_w x_0), w \in W$.*

In fact this decomposes X as a union of cells. To see that, one first notes that the isotropy subgroup of B at $s_w x_0$ is the analytic subgroup B_w of G with Lie algebra $\mathfrak{b}_w = \mathfrak{h} + \sum_{\beta \in w(\Sigma^+)} \mathfrak{g}_{-\beta}$. One then checks that this decomposes $B = N_w(B \cap B_w)$ where N_w is the unipotent analytic subgroup of G with Lie algebra

$$\mathfrak{n}_w = \sum_{\alpha \in \Sigma^+ \cap w(\Sigma^-)} \mathfrak{g}_{-\alpha} \, .$$

Thus the map $\xi \mapsto \exp(\xi) s_w x_0$ gives a diffeomorphism of the real vector space $\mathfrak{n}_w$ onto the orbit $B(s_w x_0)$.

2.2. Lemma. *If P_1 and P_2 are parabolic subgroups of G then $P_1 \cap P_2$ contains a Cartan subgroup of G.*

Proof. Let $\mathfrak{b}$ and $\mathfrak{b}'$ be Borel subalgebras of $\mathfrak{g}$. We will show that $\mathfrak{b} \cap \mathfrak{b}'$ contains a Cartan subalgebra of $\mathfrak{g}$. For this, we may assume that $\mathfrak{b}$ is our standard Borel $\mathfrak{h} + \sum_{\alpha \in \Sigma^+} \mathfrak{g}_{-\alpha}$. Let B and B' be the corresponding Borel subgroups of G. Then B' is the G–stabilizer of a point $x' \in X = G/B$. Following the Bruhat Lemma 2.1 we may take $x' = b s_w x_0$ for some $b \in B$ and $w \in W$. Without loss of generality we conjugate by b^{-1}. Now we may assume $x' = s_w x_0$. Then $B' = \mathrm{Ad}\,(s_w) B$ so $\mathfrak{b}' = \mathrm{ad}\,(s_w)\mathfrak{b}$, which contains $\mathfrak{h}$.

If $h \in H$ then h normalizes both $\mathfrak{b}$ and $\mathfrak{b}'$, so $h \in B \cap B'$. Thus the intersection of two Borel subgroups contains a Cartan subgroup. The lemma follows. □

2.3. Corollary. *Let τ denote complex conjugation of $\mathfrak{g}$ over a real form $\mathfrak{g}_0$. Let $\mathfrak{p}$ be a parabolic subalgebra of $\mathfrak{g}$. Then $\mathfrak{p} \cap \tau\mathfrak{p}$ contains a τ–stable Cartan subalgebra of $\mathfrak{g}$.*

Proof. Set $\mathfrak{q} = \mathfrak{p} \cap \tau\mathfrak{p}$. It is a τ–stable complex subalgebra of $\mathfrak{g}$, so $\mathfrak{q}_0 = \mathfrak{g}_0 \cap \mathfrak{q}$ is a real form of $\mathfrak{q}$ and τ induces the complex conjugation of $\mathfrak{q}$ over $\mathfrak{q}_0$. Choose a Cartan subalgebra $\mathfrak{j}_0$ of $\mathfrak{q}_0$. Its complexification $\mathfrak{j}$ is a Cartan subalgebra of $\mathfrak{q}$. Lemma 2.2 says that $\mathfrak{q}$ contains Cartan subalgebras of $\mathfrak{g}$. Thus $\mathfrak{j}$ is a τ–stable Cartan subalgebra of $\mathfrak{g}$. □

References for §2.

- F. Bruhat, "Sur les représentations induites des groupes de Lie", Bull. Soc. Math. France **84** (1956), 97–205.
- Harish–Chandra, "On a lemma of F. Bruhat", J. **35** (1956), 203–210.
- J. A. Wolf, "The action of a real semisimple Lie group on a complex flag manifold, I: Orbit structure and holomorphic arc components", Bull. Amer. Math. Soc. **75** (1969), 1121–1237.

§3. Real Group Actions.

Let G_0 be a real form of G. In other words, G_0 is a Lie subgroup of G whose Lie algebra $\mathfrak{g}_0$ is a real form of $\mathfrak{g}$. Although G is connected, G_0 does not have to be connected. We write τ both for the complex conjugation of $\mathfrak{g}$ over $\mathfrak{g}_0$ and for the corresponding conjugation of G over G_0.

Fix a parabolic subgroup $P \subset G$ and let Z denote the corresponding complex flag manifold. Since P is its own normalizer in G, we may view Z as the space of all G–conjugates of $\mathfrak{p}$, by the correspondence $gP \leftrightarrow \mathrm{Ad}(g)\mathfrak{p}$. We will write $\mathfrak{p}_z$ for the parabolic subalgebra of $\mathfrak{g}$ corresponding to $z \in Z$, and will write P_z for the corresponding parabolic subgroup of G.

Here is the principal trick for dealing with G_0–orbits on Z. We will use it constantly. Consider the orbit $G_0(z)$. The isotropy subgroup of G_0 at z is $G_0 \cap P_z$. That isotropy subgroup has Lie algebra $\mathfrak{g}_0 \cap \mathfrak{p}_z$, which is a real form of $\mathfrak{p}_z \cap \tau\mathfrak{p}_z$. Lemma 2.3 says that $\mathfrak{p}_z \cap \tau\mathfrak{p}_z$ contains a τ–stable Cartan subalgebra $\mathfrak{h}$ of $\mathfrak{g}$. Now $\mathfrak{p}_z$ contains a Borel subalgebra of $\mathfrak{g}$ that contains $\mathfrak{h}$. Express that Borel as $\mathfrak{b} = \mathfrak{h} + \sum_{\alpha \in \Sigma^+} \mathfrak{g}_\alpha$ for an appropriate choice of positive root system $\Sigma^+ = \Sigma^+(\mathfrak{g}, \mathfrak{h})$. We have proved

3.1. Theorem. *Let G_0 be a real form of the complex semisimple Lie group G, let τ denote complex conjugation of $\mathfrak{g}$ over $\mathfrak{g}_0$, and consider an orbit $G_0(z)$ on a complex flag manifold $Z = G/P$. Then there exist a τ–stable Cartan subalgebra $\mathfrak{h} \subset \mathfrak{p}_z$ of $\mathfrak{g}$, a positive root system $\Sigma^+ = \Sigma^+(\mathfrak{g}, \mathfrak{h})$, and a set Φ of simple roots, such that $\mathfrak{p}_z = \mathfrak{p}_\Phi$ and $P_z = P_\Phi$.*

3.2. Corollary. *In the notation of* Theorem 3.1, $\mathfrak{p}_z \cap \tau\mathfrak{p}_z$ *is the semidirect sum of its nilpotent radical*

$$(\mathfrak{p}_\Phi^{-n} \cap \tau\mathfrak{p}_\Phi^{-n}) + (\mathfrak{p}_\Phi^{r} \cap \tau\mathfrak{p}_\Phi^{-n}) + (\mathfrak{p}_\Phi^{-n} \cap \tau\mathfrak{p}_\Phi^{r})$$

with the Levi complement

$$\mathfrak{p}_\Phi^r \cap \tau\mathfrak{p}_\Phi^r = \mathfrak{h} + \sum_{\Phi^r \cap \tau\Phi^r} \mathfrak{g}_\alpha .$$

In particular, $\dim_{\mathbb{R}} \mathfrak{g}_0 \cap \mathfrak{p}_z = \dim_{\mathbb{C}} \mathfrak{p}_\Phi^r + |\Phi^n \cap \tau\Phi^n|$.

Proof. The subspace $(\mathfrak{p}_\Phi^{-n} \cap \tau\mathfrak{p}_\Phi^{-n}) + (\mathfrak{p}_\Phi^{r} \cap \tau\mathfrak{p}_\Phi^{-n}) + (\mathfrak{p}_\Phi^{-n} \cap \tau\mathfrak{p}_\Phi^{r})$ of $\mathfrak{p}_\Phi \cap \tau\mathfrak{p}_\Phi$ is the sum of all root spaces $\mathfrak{g}_\alpha \subset \mathfrak{p}_\Phi \cap \tau\mathfrak{p}_\Phi$ such that $\mathfrak{g}_{-\alpha} \not\subset \mathfrak{p}_\Phi \cap \tau\mathfrak{p}_\Phi$. So it is the nilradical of $\mathfrak{p}_\Phi \cap \tau\mathfrak{p}_\Phi$. The subspace $\mathfrak{p}_\Phi^r \cap \tau\mathfrak{p}_\Phi^r = \mathfrak{h} + \sum_{\Phi^r \cap \tau\Phi^r} \mathfrak{g}_\alpha$ is a reductive subalgebra that is a vector space complement, so it is a Levi complement. Now compute

$$\begin{aligned}
&\dim_{\mathbb{R}} \mathfrak{g}_0 \cap \mathfrak{p}_z \\
&\qquad = \dim_{\mathbb{C}} \mathfrak{p}_\Phi \cap \tau\mathfrak{p}_\Phi = \dim_{\mathbb{C}} \mathfrak{h} + |(\Phi^r \cup \Phi^n) \cap \tau(\Phi^r \cup \Phi^n)| \\
&\qquad = \left(\dim_{\mathbb{C}} \mathfrak{h} + |\Phi^r \cap \tau\Phi^r| + |\Phi^n \cap \tau\Phi^r| + |\Phi^r \cap \tau\Phi^n|\right) + |\Phi^n \cap \tau\Phi^n| \\
&\qquad = \dim_{\mathbb{C}} \mathfrak{p}_\Phi^r + |\Phi^n \cap \tau\Phi^n|
\end{aligned}$$

as asserted. □

3.3. Corollary. *In the notation of* Theorem 3.1, $\operatorname{codim}_{\mathbb{R}}(G_0(z) \subset Z) = |\Phi^n \cap \tau\Phi^n|$. *In particular,* $G_0(z)$ *is open in* Z *if and only if* $\Phi^n \cap \tau\Phi^n$ *is empty.*

Proof. In view of Corollary 3.2, the codimension in question is given by

$$\begin{aligned}
&\operatorname{codim}_{\mathbb{R}}(G_0(z) \subset Z) \\
&\qquad = \dim_{\mathbb{R}} Z - \dim_{\mathbb{R}} G_0(z) \\
&\qquad = 2|\Phi^n| - [\dim_{\mathbb{R}} G_0 - \dim_{\mathbb{R}}(G_0 \cap P_z)] \\
&\qquad = 2|\Phi^n| - [(\dim_{\mathbb{R}} \mathfrak{h} + |\Phi^r| + 2|\Phi^n|) - (\dim_{\mathbb{R}} \mathfrak{h} + |\Phi^r| + |\Phi^n \cap \tau\Phi^n|)] \\
&\qquad = |\Phi^n \cap \tau\Phi^n|
\end{aligned}$$

as asserted. □

3.4. Corollary. *The number of* G_0*-orbits on* Z *is finite. The maximal-dimensional orbits are open and the minimal-dimensional orbits are closed.*

Proof. The number of G_0–conjugacy classes of Cartan subalgebras $\mathfrak{h}_0 \subset \mathfrak{g}_0$ is finite. So the number of G_0–conjugacy classes of τ–stable Cartan subalgebras $\mathfrak{h} \subset \mathfrak{g}$ is finite. Given such an $\mathfrak{h}$, the number of positive root systems Σ^+ is finite. Given $(\mathfrak{h}, \Sigma^+)$, the number of sets Φ of simple roots is finite. Thus the number of possibilities for P_Φ is finite up to G_0–conjugacy. This proves that the number of G_0–orbits on Z is finite. It also gives a (very) rough upper bound on the number. The other statements follow because the closure of an orbit is a union of orbits. □

References for §3.

- J. A. Wolf, "The action of a real semisimple Lie group on a complex flag manifold, I: Orbit structure and holomorphic arc components", Bull. Amer. Math. Soc. **75** (1969), 1121–1237.
- J. A. Wolf, "Fine structure of hermitian symmetric spaces", in "Symmetric Spaces", ed. W. M. Boothby & G. L. Weiss, Marcel Dekker, 1972, pp. 271–357

§4. Open Orbits.

Fix a Cartan involution θ of $\mathfrak{g}_0$ and G_0. In other words θ is an automorphism of square 1 and, using $G_0 \subset G$ so that $\mathfrak{g}_0$ is semisimple and G_0 has finite center, the fixed point set $K_0 = G_0^\theta$ is a maximal compact subgroup of G_0. Thus $\mathfrak{g}_0 = \mathfrak{k}_0 + \mathfrak{s}_0$ where $\mathfrak{k}_0$ is the Lie algebra of K_0 and is the $(+1)$–eigenspace of θ on $\mathfrak{g}_0$, and $\mathfrak{s}_0$ is the (-1)–eigenspace. The Killing form of $\mathfrak{g}_0$ is negative definite on $\mathfrak{k}_0$ and is positive definite on $\mathfrak{s}_0$, and $\mathfrak{k}_0 \perp \mathfrak{s}_0$ under the Killing form.

Every Cartan subalgebra of $\mathfrak{g}_0$ is $\mathrm{Ad}(G_0)$–conjugate to a θ–stable Cartan subalgebra. A θ–stable Cartan subalgebra $\mathfrak{h}_0 \subset \mathfrak{g}_0$ is called **fundamental** if it maximizes $\dim(\mathfrak{h}_0 \cap \mathfrak{k}_0)$, **compact** if it is contained in $\mathfrak{k}_0$, which is a more stringent condition. More generally, a Cartan subalgebra of $\mathfrak{g}_0$ is called **fundamental** if it is conjugate to a θ–stable fundamental Cartan subalgebra.

4.1. Lemma. *The following conditions are equivalent for a θ–stable Cartan subalgebra $\mathfrak{h}_0 \subset \mathfrak{g}_0$.*

(i) *$\mathfrak{h}_0$ is a fundamental Cartan subalgebra of $\mathfrak{g}_0$,*
(ii) *$\mathfrak{h}_0 \cap \mathfrak{k}_0$ contains a regular element of $\mathfrak{g}_0$, and*
(iii) *there is a positive root system $\Sigma^+ = \Sigma^+(\mathfrak{g}, \mathfrak{h}), \mathfrak{h} = \mathfrak{h}_0 \otimes \mathbb{C}$, such that $\tau\Sigma^+ = \Sigma^-$.*

A θ–stable Cartan subalgebra $\mathfrak{h}_0 \subset \mathfrak{g}_0$ is compact if and only if $\tau\Sigma^+ = \Sigma^-$ for every positive root system $\Sigma^+(\mathfrak{g}, \mathfrak{h})$.

4.2. Theorem. *Let $Z = G/P$ be a complex flag manifold, G semisimple and simply connected, and let G_0 be a real form of G. The orbit $G_0(z)$ is open in Z if and only if $\mathfrak{p}_z = \mathfrak{p}_\Phi$ where*

(i) *$\mathfrak{p}_z \cap \mathfrak{g}_0$ contains a fundamental Cartan subalgebra $\mathfrak{h}_0 \subset \mathfrak{g}_0$ and*
(ii) *Φ is a set of simple roots for a positive root system $\Sigma^+(\mathfrak{g}, \mathfrak{h}), \mathfrak{h} = \mathfrak{h}_0 \otimes \mathbb{C}$, such that $\tau\Sigma^+ = \Sigma^-$.*

Fix $\mathfrak{h}_0 = \theta\mathfrak{h}_0$, $\Sigma^+(\mathfrak{g}, \mathfrak{h})$ and Φ as above. Let $W(\mathfrak{g}, \mathfrak{h})^{\mathfrak{h}_0}$ and $W(\mathfrak{p}_\Phi^r, \mathfrak{h})^{\mathfrak{h}_0}$ denote the respective subgroups of Weyl groups that stabilize $\mathfrak{h}_0$. Then the open G_0–orbits on Z are parameterized by $W(\mathfrak{k}, \mathfrak{h} \cap \mathfrak{k}) \backslash W(\mathfrak{g}, \mathfrak{h})^{\mathfrak{h}_0} / W(\mathfrak{p}_z^r, \mathfrak{h})^{\mathfrak{h}_0}$.

4.3. Corollary. *Suppose that G_0 has a compact Cartan subgroup, i.e. that $\mathfrak{k}_0$ contains a Cartan subalgebra of $\mathfrak{g}_0$. Then an orbit $G_0(z)$ is open in Z if and only if $\mathfrak{g}_0 \cap \mathfrak{p}_z$ contains a compact Cartan subalgebra $\mathfrak{h}_0$ of $\mathfrak{g}_0$, and then,* *in the notation of* Theorem 4.2, *the open* G_0*-orbits on* Z *are parameterized by* $W(\mathfrak{k},\mathfrak{h})\backslash W(\mathfrak{g},\mathfrak{h})/W(\mathfrak{p}_z^\tau,\mathfrak{h})$.

A careful examination of the way $\mathfrak{k}_0$ sits in both $\mathfrak{k}$ and $\mathfrak{g}_0$ gives us

4.4. Theorem. *Let* $Z = G/P$ *be a complex flag manifold,* G *semisimple and simply connected, and let* G_0 *be a real form of* G. *Let* $z \in Z$ *such that* $G_0(z)$ *is open in* Z, *and let* $\mathfrak{h}_0 \subset \mathfrak{g}_0 \cap \mathfrak{p}_z$ *be a* θ*-stable fundamental Cartan subalgebra of* $\mathfrak{g}_0$. *Then* $K_0(z)$ *is a compact complex submanifold of* $G_0(z)$. *Let* K *be the complexification of* K_0, *analytic subgroup of* G *with Lie algebra* $\mathfrak{k} = \mathfrak{k}_0 \otimes \mathbb{C}$. *Then* $K_0(z) = K(z) \cong K/(K \cap P_z)$, *complex flag manifold of* K.

The compact subvariety $K_0(z)$ controls the topology of an open orbit $G_0(z) \subset Z$, as follows. As we saw before, or by Corollary 4.3, the compact real form $G_u \subset G$ is transitive on Z. This gives us a realization $Z = G_u/V_u$ where $V_u \subset G_u$ is the centralizer of a torus subgroup. In particular, V_u is connected. Since $G_0 \subset G$, Z is compact and simply connected. In view of Theorem 4.4, one can apply this argument to the compact subvariety $K_0(z) \subset G_0(z)$, so it is simply connected. Now a deformation argument shows that the open orbit $G_0(z) \subset Z$ has $K_0(z)$ as a deformation retract, so $G_0(z)$ is simply connected. Thus one obtains

4.5. Proposition. *Let* $Z = G/P$ *be a complex flag manifold,* G *semisimple and simply connected, and let* G_0 *be a real form of* G. *Let* $z \in Z$ *such that* $G_0(z)$ *is open in* Z. *Then* $G_0(z)$ *is simply connected and* G_0 *has connected isotropy subgroup* $(P_z \cap \tau P_z)_0$ *at* z.

The compact subvariety $Y = K_0(z)$ also has a strong influence on the function theory for an open orbit $D = G_0(z) \subset Z$. The idea is that a holomorphic function on D must be constant on gY whenever $g \in G$ and $gY \subset D$, so if there are "too many" translates of Y inside D then that holomorphic function must be constant on D. But this has to be formulated carefully.

Let $Z = G/P$ be a complex flag manifold, G semisimple and simply connected, and let G_0 be a real form of G. Let $z \in Z$ such that $G_0(z)$ is open in Z. Then there are decompositions $G = G_1 \times \cdots \times G_m$ and $P = P_1 \times \cdots \times P_m$ with $P_i = P \cap G_i$ and each G_i simple. Consider the corresponding decompositions $Z = Z_1 \times \cdots \times Z_m$ with $Z_i = G_i/P_i$ and $z = (z_1, \ldots, z_m)$, $G_0 = G_{1,0} \times \cdots \times G_{m,0}$, $G_0(z) = G_{1,0}(z_1) \times \cdots \times G_{m,0}(z_m)$ and $K_0(z) = K_{1,0}(z_1) \times \cdots \times K_{m,0}(z_m)$. If

(i) $G_{i,0} \cap (P_i)_{z_i} = ((P_i)_{z_i} \cap \tau(P_i)_{z_i})_0$ is compact, thus contained in $K_{i,0}$,
(ii) $G_{i,0}/K_{i,0}$ is an hermitian symmetric coset space, and
(iii) $G_{i,0}(z_i) \to G_{i,0}/K_{i,0}$ is holomorphic for one of the two invariant complex structures on $G_{i,0}/K_{i,0}$,

then we set $L_i = K_i$ so $L_{i,0} = K_{i,0}$. Otherwise we set $L_i = G_i$ so $L_{i,0} = G_{i,0}$. Note that each $G_{i,0}/L_{i,0}$ is a bounded symmetric domain, irreducible or reduced to a point. Set $L = L_0 \times \cdots \times L_m$ so $L_0 = L_{1,0} \times \cdots \times L_{m,0}$. Then we say that

$$D(G_0, z) = G_0/L_0 = (G_{1,0}/L_{1,0}) \times \cdots \times (G_{m,0}/L_{m,0}) \tag{4.6}$$

is the **bounded symmetric domain subordinate to** $G_0(z)$. Now we can state a precise result for holomorphic functions on $G_0(z)$.

4.7. Theorem. *Let $Z = G/P$ be a complex flag manifold, G semisimple and simply connected, and let G_0 be a real form of G. Let $z \in Z$ with $G_0(z)$ be open in Z. Let $D(G_0, z)$ be the bounded symmetric domain subordinate to $G_0(z)$. Then $\pi : g(z) \mapsto gL_0$ is a holomorphic map of $G_0(z)$ onto $D(G_0, z)$, and the holomorphic functions on $G_0(z)$ are just the $\widetilde{f} = f \cdot \pi$ where $f : D(G_0, z) \to \mathbb{C}$ is holomorphic.*

Thus, in most cases there are no nonconstant holomorphic functions on $G_0(z)$, but in fact this depends on some delicate structure.

References for §4.

- A. Borel and F. Hirzebruch, "Characteristic classes and homogeneous spaces, I", Amer. J. Math. **30** (1958), 458–538.
- A. Borel and F. Hirzebruch, "Characteristic classes and homogeneous spaces, II", Amer. J. math. **31** (1959), 315–382.
- A. Frölicher, "Zur Differentialgeometrie der komplexen Strukturen", Math. Ann. **129** (1955) 50–95.
- J. Tits, "Espaces homogènes complexes compacts", Comment. Math. Helv. **37** (1962), 111–120.
- J. A. Wolf and A. Gray, "Homogeneous spaces defined by Lie group automorphisms, I", J. Diff. Geometry **2** (1968), 77–114.
- J. A. Wolf and A. Gray, "Homogeneous spaces defined by Lie group automorphisms, II", J. Diff. Geometry **2** (1968), 115–159.
- J. A. Wolf, "The action of a real semisimple Lie group on a complex flag manifold, I: Orbit structure and holomorphic arc components", Bull. Amer. Math. Soc. **75** (1969), 1121–1237.

§5. Example: Hermitian Symmetric Spaces.

In this section, $Z = G_u/K_0$ is an irreducible hermitian symmetric space of compact type. Thus $Z = G/P$ where G is a connected simply connected complex simple Lie group with a real form $G_0 \subset G$ of hermitian type, as follows. Fix a Cartan involution θ of G_0 and the corresponding eigenspace decomposition $\mathfrak{g}_0 = \mathfrak{k}_0 + \mathfrak{s}_0$ where $\mathfrak{k}_0$ is the Lie algebra of the fixed point set $K_0 = G_0^\theta$. Then $G_u \subset G$ is the compact real form of G that is the analytic subgroup for the compact real form $\mathfrak{g}_u = \mathfrak{k}_0 + \mathfrak{s}_u$ of $\mathfrak{g}$ where $\mathfrak{s}_u = \sqrt{-1}\,\mathfrak{s}_0$ of $\mathfrak{g}$.

There is a compact Cartan subalgebra $\mathfrak{t}_0 \subset \mathfrak{k}_0$ of $\mathfrak{g}_0$. If $\alpha \in \Sigma(\mathfrak{g}, \mathfrak{t})$ then either $\mathfrak{g}_\alpha \subset \mathfrak{k}$ and we say that the root α is **compact**, or $\mathfrak{g}_\alpha \subset \mathfrak{s}$ and we say that α is **noncompact**. There is a simple root system $\Psi = \{\psi_0, \ldots, \psi_m\}$ such that ψ_0 is noncompact and the other ψ_i are compact. Furthermore, ψ_0 is a long root, and every noncompact positive root is of the form $\psi_0 + \sum_{1 \leqq i \leqq m} n_i \psi_i$ with each integer $n_i \geqq 0$. Thus $\mathfrak{g} = \mathfrak{k} + \mathfrak{s}^+ + \mathfrak{s}^-$ where

$$\mathfrak{k} = \mathfrak{t} + \sum_{n_0=0} \mathfrak{g}_\alpha\,, \mathfrak{s}^+ = \sum_{n_0=1} \mathfrak{g}_\alpha\,, \text{ and } \mathfrak{s}^- = \sum_{n_0=-1} \mathfrak{g}_\alpha\,. \tag{5.1}$$

Here $\mathfrak{p} = \mathfrak{p}_{\{\psi_1,\dots,\psi_m\}}$, in other words

(5.2) $$\mathfrak{p}^r = \mathfrak{k}\ ,\ \mathfrak{p}^n = \mathfrak{s}^+\ ,\ \text{and}\ \mathfrak{p}^{-n} = \mathfrak{s}^-\ ;\ \text{so}\ \mathfrak{p} = \mathfrak{k} + \mathfrak{s}^-\ .$$

The Cartan subalgebras of $\mathfrak{g}_0$ all are $\mathrm{Ad}(G_0)$–conjugate to one of the $\mathfrak{h}_{\Gamma,0}$ given as follows. Let $\Gamma = \{\gamma_1,\dots,\gamma_r\}$ be a set of noncompact positive roots that is **strongly orthogonal** in the sense that

(5.3) $$\text{if } 1 \leqq i < j \leqq r \text{ then none of } \pm\gamma_i \pm \gamma_j \text{ is a root.}$$

Then each $\mathfrak{g}[\gamma_i] = [\mathfrak{g}_{\gamma_i}, \mathfrak{g}_{-\gamma_i}] + \mathfrak{g}_{\gamma_i} + \mathfrak{g}_{-\gamma_i} \cong \mathfrak{sl}(2,\mathbb{C})$, say with

$$h_{\gamma_i} \leftrightarrow \begin{pmatrix} 1 & 0 \\ 0 & -1 \end{pmatrix},\quad e_{\gamma_i} \leftrightarrow \begin{pmatrix} 0 & 1 \\ 0 & 0 \end{pmatrix},\quad f_{\gamma_i} \leftrightarrow \begin{pmatrix} 0 & 0 \\ 1 & 0 \end{pmatrix},$$

where $h_{\gamma_i} \in [\mathfrak{g}_{\gamma_i}, \mathfrak{g}_{-\gamma_i}]$, $e_{\gamma_i} \in \mathfrak{g}_{\gamma_i}$ and $f_{\gamma_i} \in \mathfrak{g}_{-\gamma_i}$ as usual, and such that $\mathfrak{g}_0[\gamma_i] = \mathfrak{g}_0 \cap \mathfrak{g}_{\gamma_i} \cong \mathfrak{su}(1,1)$ is spanned by $\sqrt{-1}\,h_{\gamma_i}$, $e_{\gamma_i} + f_{\gamma_i}$ and $\sqrt{-1}\,(e_{\gamma_i} - f_{\gamma_i})$. Thus $\sqrt{-1}\,h_{\gamma_i}$ spans the compact Cartan subalgebra $\mathfrak{t}_{\gamma_i} = \mathfrak{g}_0[\gamma_i] \cap \mathfrak{t}$ of $\mathfrak{g}_0[\gamma_i]$ and $e_{\gamma_i} + f_{\gamma_i}$ spans the noncompact Cartan subalgebra $\mathfrak{a}_{\gamma_i} = \mathfrak{g}_0[\gamma_i] \cap \mathfrak{s}$ of $\mathfrak{g}_0[\gamma_i]$. Strong orthogonality (5.3) says $[\mathfrak{g}_{\gamma_i}, \mathfrak{g}_{\gamma_j}] = 0$ for $1 \leqq i < j \leqq r$. Define

(5.4) $$\mathfrak{t}_\Gamma = \sum_{1\leqq i \leqq r} \mathfrak{t}_{\gamma_i} \text{ and } \mathfrak{a}_\Gamma = \sum_{1 \leqq i \leqq r} \mathfrak{a}_{\gamma_i}\ .$$

Then $\mathfrak{g}$ has Cartan subalgebras

(5.5) $$\mathfrak{t} = \mathfrak{t}_\Gamma + (\mathfrak{t} \cap \mathfrak{t}_\Gamma^\perp) \quad \text{and} \quad \mathfrak{h}_\Gamma = \mathfrak{a}_\Gamma + (\mathfrak{t} \cap \mathfrak{t}_\Gamma^\perp)$$

They are $\mathrm{Int}(\mathfrak{g})$–conjugate, for the **partial Cayley transform**

(5.5) $$c_\Gamma = \prod_{1 \leqq i \leqq r} \exp\left(\frac{\pi}{4}\sqrt{-1}\,(e_{\gamma_i} - f_{\gamma_i})\right) \text{ satisfies } \mathrm{Ad}(c_\Gamma)\mathfrak{t}_\Gamma = \mathfrak{a}_\Gamma\ .$$

However, their real forms

(5.6) $$\mathfrak{t}_0 = \mathfrak{g}_0 \cap \mathfrak{t} \text{ and } \mathfrak{h}_{\Gamma,0} = \mathfrak{g}_0 \cap \mathfrak{h}_\Gamma$$

are not $\mathrm{Ad}(G_0)$–conjugate except in the trivial case where Γ is empty, for the Killing form has rank $m = \dim \mathfrak{t}_0$ and signature $2|\Gamma| - m$ on $\mathfrak{h}_{\Gamma,0}$. More precisely,

5.7. Proposition. *Every Cartan subalgebra of $\mathfrak{g}_0$ is $Ad(G_0)$–conjugate to one of the $\mathfrak{h}_{\Gamma,0}$, and Cartan subalgebras $\mathfrak{h}_{\Gamma,0}$ and $\mathfrak{h}_{\Gamma',0}$ are $Ad(G_0)$–conjugate if and only if the cardinalities $|\Gamma| = |\Gamma'|$.*

We recall Kostant's "cascade construction" of a maximal set of strongly orthogonal noncompact positive roots in $\Sigma(\mathfrak{g}, \mathfrak{t})$. This set has cardinality $\ell = \mathrm{rank}_{\mathbb{R}}\mathfrak{g}_0$ and is given by

$\Xi = \{\xi_1, \dots, \xi_\ell\}$, where

(5.8) ξ_1 is the maximal (necessarily noncompact positive) root and

ξ_{m+1} is a maximal noncompact positive root $\perp \{\xi_1, \dots, \xi_m\}$.

The roots ξ_i are long, and any set of strongly orthogonal noncompact positive long roots in $\Sigma(\mathfrak{g}, \mathfrak{t})$ is $W(G_0, T_0)$–conjugate to a subset of Ξ. Further, the Weyl group $W(G_0, T_0)$ induces every permutation of Ξ.

Let $z_0 = 1 \cdot P \in G/P = Z$, the base point of our flag manifold Z when Z is viewed as a homogeneous space. The Cartan subalgebra $\mathfrak{h}_{\Gamma,0} \subset \mathfrak{g}_0$ leads to the orbits $G_0(c_\Gamma c_\Delta^2 z_0) \subset Z$ where $\Gamma \cup \Delta$ is a set of strongly orthogonal noncompact positive roots in $\Sigma(\mathfrak{g}, \mathfrak{t})$ with Γ and Δ disjoint. In view of the Weyl group equivalence just discussed, we may take $\Gamma = \{\xi_1, \dots, \xi_r\}$ and $\Delta = \{\xi_{r+1}, \dots, \xi_{r+s}\}$, both inside Ξ. Using $G_0 = K_0 \exp(\mathfrak{a}_{\Xi,0}) K_0$ one arrives at

5.9. Theorem. *The G_0–orbits on Z are just the orbits $D_{\Gamma,\Delta} = G_0(c_\Gamma c_\Delta^2 z_0)$ where Γ and Δ are disjoint subsets of Ξ. Two such orbits $D_{\Gamma,\Delta} = D_{\Gamma',\Delta'}$ if and only if cardinalities $|\Gamma| = |\Gamma'|$ and $|\Delta| = |\Delta'|$. An orbit $D_{\Gamma,\Delta}$ is open if and only if Γ is empty, closed if and only if $(\Gamma, \Delta) = (\Xi, \emptyset)$. An orbit $D_{\Gamma',\Delta'}$ is in the closure of $D_{\Gamma,\Delta}$ if and only if $|\Delta'| \leqq |\Delta|$ and $|\Gamma \cup \Delta| \leqq |\Gamma' \cup \Delta'|$.*

References for §5.

- A. Borel and J. de Siebenthal, "Les sous–groupes fermés de rang maximum des groupes de Lie clos", Comment. Math. Helv. **23** (1949), 200–221.
- A. Korányi and J. A. Wolf, "Realization of hermitian symmetric spaces as generalized half–planes", Annals of Math. **81** (1965), 265–288.
- C. C. Moore, "Compactifications of symmetric spaces II (The Cartan domains)", Amer. J. Math. **86** (1964), 358–378.
- J. A. Wolf, "The action of a real semisimple Lie group on a complex flag manifold, I: Orbit structure and holomorphic arc components", Bull. Amer. Math. Soc. **75** (1969), 1121–1237.
- J. A. Wolf, "Fine structure of hermitian symmetric spaces", in "Symmetric Spaces", ed. W. M. Boothby & G. L. Weiss, Marcel Dekker, 1972, pp. 271–357.
- J. A. Wolf and A. Korányi, "Generalized Cayley transformations of bounded symmetric domains", Amer. J. Math. **87** (1965), 899–939.

§6. The Closed Orbit.

There must be at least one closed G_0–orbit on Z, by Corollary 3.4. In the examples of §5 it is unique. We will see that it is unique in general and that it has some interesting structure.

First look at the case where $G = SL(2;\mathbb{C}), G_0 = SU(1,1)$, and X is the Riemann sphere. G acts as usual by linear fractional transformations. Then

$$G_0 = \left\{ \begin{pmatrix} a & b \\ \bar{b} & \bar{a} \end{pmatrix} \,\middle|\, |a|^2 - |b|^2 = 1 \right\}. \tag{6.1}$$

and are three G_0–orbits, as follows.

The interior of the unit disk $G_0(0)$:

$$P_0 = \left\{ \begin{pmatrix} a & 0 \\ c & a^{-1} \end{pmatrix} \right\} \text{ and } H_0 = \left\{ \begin{pmatrix} e^{i\theta} & 0 \\ 0 & e^{-i\theta} \end{pmatrix} \middle| \, \theta \text{ real} \right\}.$$

The exterior of the unit disk $G_0(\infty)$:

$$P_\infty = \left\{ \begin{pmatrix} a & b \\ 0 & a^{-1} \end{pmatrix} \right\} \text{ and } H_0 = \left\{ \begin{pmatrix} e^{i\theta} & 0 \\ 0 & e^{-i\theta} \end{pmatrix} \middle| \, \theta \text{ real} \right\}.$$

(6.2) The unit circle $G_0(1)$:

$$P_1 = \left\{ \begin{pmatrix} a & b \\ c & d \end{pmatrix} \middle| \, ad - bc = 1, a + b = c + d \right\}$$

$$\text{so } \mathfrak{p}_1^r = \left\{ \begin{pmatrix} a & 0 \\ 0 & -a \end{pmatrix} \right\}, \mathfrak{p}_1^{-n} = \left\{ \begin{pmatrix} -b & b \\ -b & b \end{pmatrix} \right\}$$

$$\text{and } H_0 = \left\{ \pm \begin{pmatrix} \cosh(t) & \sinh(t) \\ \sinh(t) & \cosh(t) \end{pmatrix} \middle| \, t \text{ real} \right\}.$$

The first two give the open orbits, with H_0 compact, and the third gives the closed orbit, where H_0 is the T_0A_0 of an Iwasawa decomposition of G_0. This mirrors the general case for closed orbits:

6.3. Theorem. *Let $X = G/P$ be a complex flag manifold and let G_0 be a real form of G. Then there is a unique closed orbit $G_0(z) \subset Z$. Further, there is an Iwasawa decomposition $G_0 = K_0A_0N_0$ such that $G_0 \cap P_z$ contains H_0N_0 whenever H_0 is a Cartan subgroup of G_0 that contains A_0. (In other words, whenever $H_0 = T_0A_0$ where T_0 is a Cartan subgroup of the K_0–centralizer M_0 of A_0.)*

Proof. We first consider the case where $P = B$, Borel subgroup of G. Fix a closed orbit $G_0(x) \subset X$. Then $G_0(x)$ is compact. I claim that $G_0 \cap B_x$ contains the A_0N_0 of an Iwasawa decomposition $G_0 = K_0A_0N_0$. Let $H_0' \subset G_0 \cap B_x$ be a Cartan subgroup. Suppose that it is not conjugate to the T_0A_0 of a fixed minimal parabolic subalgebra $\mathfrak{q}_0 = \mathfrak{m}_0 + \mathfrak{a}_0 + \mathfrak{n}_0 \subset \mathfrak{g}_0$. Replacing $\mathfrak{q}_0$ by a G_0–conjugate we then have $H_0' = T_0'A_0'$ with $T_0 \subsetneqq T_0' \subset K_0$ and $A_0' \subsetneqq A_0$. Then we have a root $\alpha \in \Sigma = \Sigma(\mathfrak{g}, \mathfrak{h})$ that vanishes on $\mathfrak{t}$, and such that the intersection of

$$\mathfrak{g}[\alpha] = \mathfrak{g}_\alpha + \mathfrak{g}_{-\alpha} + [\mathfrak{g}_\alpha, \mathfrak{g}_{-\alpha}] \tag{6.4}$$

with $\mathfrak{h}$ is contained in $\mathfrak{a}$ while the intersection with $\mathfrak{h}'$ is contained in $\mathfrak{t}'$. This is exactly the example of (6.1) and (6.2). Now $G_0[\alpha](x)$ is an open hemisphere in $G[\alpha](x)$. But $G_0[\alpha]$ is closed in G_0 and has compact isotropy at x, so $G_0[\alpha](x)$ is closed in $G_0(x)$. With $G_0(x)$ closed in X now $G_0[\alpha](x)$ is closed in X, thus closed in $G[\alpha](x)$, where in fact it is an open hemisphere. This contradicts our hypothesis that $G_0(x)$ is closed in X. We have proved[2] that $G_0 \cap B_x$ contains the T_0A_0 of an Iwasawa decomposition of G_0.

Denote complexification by dropping the subscript 0. Since $T_0A_0 \subset B_x$ now $B_x = B_{M,x}AN$ where $B_{M,x}$ is a Borel subgroup of $M = Z_K(A)$. It follows that $G_0 \cap B_x = T_0A_0N_0$. Now let $G_0(x')$ be another closed orbit on X. Then $B_{x'} = B_{M',x'}A'N'$ for another Iwasawa decomposition $G_0 = K_0'A_0'N_0'$ and a choice of Borel subgroup $B_{M',x'} \subset M'$. But any two Iwasawa decompositions of G_0 are conjugate by an element of G_0, and using compactness of M_0 we have that any two Borel subalgebras of M are conjugate by an element of M_0. Thus $x' \in G_0(x)$ and $G_0(x) = G_0(x')$.

We have proved uniqueness of the closed orbit when P is a Borel subgroup of G. For the general case, choose a Borel subgroup $B \subset P$ and note that the G–equivariant holomorphic fibration $\pi : X = G/B \to G/P = Z$ has compact fibres. Now the closed G–orbits in Z are just the $\pi(G_0(x))$ where $G_0(x)$ is a closed G–orbit in X. The latter is unique. This completes the proof. □

Another interesting fact about the structure and geometry of closed orbits is

6.5. Theorem. *Let $Z = G/P$ be a complex flag manifold and let G_0 be a real form of G. Let $G_0(z)$ be the unique closed G_0-orbit on Z. Then* $\dim_{\mathbb{R}} G_0(z) \geqq \dim_{\mathbb{C}} Z$, *and the following conditions are equivalent.*

1. $\dim_{\mathbb{R}} G_0(z) = \dim_{\mathbb{C}} Z$.

2. $\tau\Phi^n = \Phi^n$.

3. View G as the group of complex points, and G_0 as an open subgroup in the group of real points, of a linear algebraic group defined over $\mathbb{R}$. Then P_z is the group of complex points in an algebraic subgroup defined over $\mathbb{R}$.

4. Z is the set of complex points in a projective variety defined over $\mathbb{R}$, and $G_0(z)$ is the set of real points.

References for §6.

- J. A. Wolf, "The action of a real semisimple Lie group on a complex flag manifold, I: Orbit structure and holomorphic arc components", Bull. Amer. Math. Soc. **75** (1969), 1121–1237.

Part 2. The Bott–Borel–Weil Theorem and the Principal Series.

In this part we combine the Bott–Borel–Weil Theorem with unitary induction, realizing the unitary principal series on the closed orbit, in order to indicate the pattern used later for geometric realization of the standard tempered representations.

---

[2]Here is a shorter, but less elementary, proof. A_0N_0 is a solvable group acting birationally on the complete variety $G_0(x)$, so it has a fixed point by a theorem of Borel. If $g(x)$ is that fixed point then $\mathrm{Ad}(g^{-1})(A_0N_0)$ fixes x.

§7. Principal Series and the Closed Orbit.

In order to introduce the connection between unitary representations of G_0 and G_0–orbits on the complex flag manifold $Z = G/P$, we look at the **principal series** of G_0.

A subalgebra $\mathfrak{q}_0 \subset \mathfrak{g}_0$ is a **parabolic subalgebra** of $\mathfrak{g}_0$ if it is a real form of a parabolic subalgebra $\mathfrak{q} \subset \mathfrak{g}$. A subgroup $Q_0 \subset G_0$ is a **parabolic subgroup** of G_0 if it is a real form of a parabolic subalgebra $Q \subset G$, that is, if $Q_0 = G_0 \cap Q$ and its Lie algebra $\mathfrak{q}_0$ is a parabolic subalgebra of $\mathfrak{g}_0$. For example, fix an Iwasawa decomposition $G_0 = K_0 A_0 N_0$, and let $M_0 = Z_{K_0}(A_0)$, as usual. Then $Q_0 = M_0 A_0 N_0$ is minimal among the parabolic subgroups of G_0 and is called a **minimal parabolic subgroup**. From the construction, any two minimal parabolic subgroups of G_0 are conjugate. Now fix a minimal parabolic subgroup $Q_0 = M_0 A_0 N_0$.

Whenever E is a topological group we write $\widehat{E}$ for its unitary dual. Thus $\widehat{E}$ consists of the unitary equivalence classes of (strongly continuous) topologically irreducible unitary representations of E. Now $[\eta] \in \widehat{M_0}$ and $\sigma \in \mathfrak{a}_0^*$ determine $[\alpha_{\eta,\sigma}] \in \widehat{Q_0}$ by

$$\alpha_{\eta,\sigma}(man) = \eta(m) e^{i\sigma(\log a)}. \tag{7.1}$$

The corresponding **principal series representation of** G_0 is

$$\pi_{\eta,\sigma} = \operatorname{Ind}_{Q_0}^{G_0}(\alpha_{\eta,\sigma}), \quad \text{unitarily induced representation.} \tag{7.2}$$

The **principal series** of G_0 consists of the unitary equivalence classes of these representations. A famous result of Bruhat says that if σ satisfies a certain nonsingularity condition then $\pi_{\eta,\sigma}$ is irreducible.

In order to realize the principal series of G_0 on closed orbits, we need the Bott–Borel–Weil Theorem for M_0. We have to be careful here because the compact group M_0 need not be connected. We will first decompose M_0 as the product $Z_{M_0}(M_0^0)M_0^0$ where M_0^0 is its identity component, then indicate the analog of the Cartan highest weight description for $\widehat{M_0}$. That done, the standard Bott–Borel–Weil Theorem for M_0^0 will carry over to M_0.

Choose a Cartan subgroup $T_0 \subset M_0$. It specifies a Cartan subgroup $H_0 = T_0 A_0 \cong T_0 \times A_0$ in G_0. Our choice of Q_0 specifies a choice of positive restricted root system $\Sigma^+(\mathfrak{g}_0, \mathfrak{a}_0)$: The Lie algebra of N_0 is given by $\mathfrak{n}_0 = \sum_{\alpha \in \Sigma^+(\mathfrak{g}_0,\mathfrak{a}_0)} (\mathfrak{g}_0)_{-\alpha}$. Now any positive root system $\Sigma^+(\mathfrak{m}, \mathfrak{t})$ specifies a positive system $\Sigma^+(\mathfrak{g}, \mathfrak{h})$ by

$$\begin{aligned} &\alpha \in \Sigma^+(\mathfrak{g}, \mathfrak{h}) \text{ if and only if either } \alpha|_{\mathfrak{a}_0} = 0 \text{ and } \alpha|_{\mathfrak{t}} \in \Sigma^+(\mathfrak{m}, \mathfrak{t}) \\ &\qquad\qquad\qquad \text{or } \alpha|_{\mathfrak{a}_0} \neq 0 \text{ and } \alpha|_{\mathfrak{a}_0} \in \Sigma^+(\mathfrak{g}_0, \mathfrak{a}_0). \end{aligned} \tag{7.3}$$

7.4. Lemma. $M_0 = Z_{M_0}(M_0^0)M_0^0$. *Given a unitary representation class* $[\eta] \in \widehat{M_0}$, *there exist unique classes* $[\chi] \in \widehat{Z_{M_0}(M_0^0)}$ *and* $[\eta^0] \in \widehat{M_0^0}$ *such* *that* $[\eta] = [\chi \otimes \eta^0]$, *and* $[\chi]$ *and* $[\eta^0]$ *restrict to multiples of the same unitary character on the center of* M_0^0.

Remark. The argument will show that T_0 *meets every topological component of* M_0.

Proof. The first assertion is equivalent to the statement that if $m \in M_0$ then the coset mM_0^0 meets $Z_{M_0}(M_0^0)$. Replacing m by some mm' with $m' \in M_0^0$ we may assume that $\mathrm{Ad}(m)$ preserves both T_0 and a positive root system $\Sigma^+(\mathfrak{m}, \mathfrak{t})$. By the definition of M_0, $\mathrm{Ad}(m)$ acts trivially on A_0, so it preserves the positive restricted root system $\Sigma^+(\mathfrak{g}_0, \mathfrak{a}_0)$. Now $\mathrm{Ad}(m)$ preserves the positive root system $\Sigma^+(\mathfrak{g}, \mathfrak{h})$ defined in (7.3). Thus m centralizes $\mathfrak{h}$, that is, $m \in H$. Now $m \in M_0 \cap H = T_0$ In particular $\mathrm{Ad}(m)$ induces an inner automorphism on M_0^0. Thus mM_0^0 meets $Z_{M_0}(M_0^0)$, as claimed.

The second assertion follows from the first. □

Let $\Psi_{\mathfrak{m}}$ denote the set of simple roots in $\Sigma^+(\mathfrak{m}, \mathfrak{t})$. Every subset $\Phi \subset \Psi_{\mathfrak{m}}$ defines

$$\begin{aligned}
&\mathfrak{z}_\Phi = \{\xi \in \mathfrak{t} \mid \phi(\xi) = 0 \text{ for all } \phi \in \Phi\} \\
&\qquad \text{and } \mathfrak{z}_{\Phi,0} = \mathfrak{m}_0 \cap \mathfrak{z}_\Phi, \text{ real form of } \mathfrak{z}_\Phi, \\
&U_\Phi = Z_M(\mathfrak{z}_\Phi), U_{\Phi,0} = M_0 \cap U_\Phi, \text{ and their Lie algebras } \mathfrak{u}_\Phi \text{ and } \mathfrak{u}_{\Phi,0}, \\
(7.5)\quad &\mathfrak{r}_\Phi = \mathfrak{u}_\Phi + \sum_{\gamma \in \Sigma^+(\mathfrak{m},\mathfrak{t})} \mathfrak{m}_{-\gamma}, \text{ parabolic subalgebra of } \mathfrak{m}, \\
&R_\Phi = N_M(\mathfrak{r}_\Phi), \text{ corresponding parabolic subgroup of } M, \text{ and} \\
&S_\Phi = M/R_\Phi, \text{ associated complex flag manifold.}
\end{aligned}$$

Lemma 7.4 holds for $U_{\Phi,0}$ By Lemma 1.14, M_0 acts transitively on S_Φ, so $M_0 \cap R_\Phi = U_{\Phi,0}$ implies

7.6. Lemma. S_Φ *is a compact homogeneous Kaehler manifold under the action of* M_0, *and* $S_\Phi = M_0/U_{\Phi,0}$ *as coset space. Furthermore* $U_{\Phi,0} = Z_{M_0}(M_0^0)U_{\Phi,0}^0$, *so* $\widehat{U_{\Phi,0}}$ *decomposes as does* $\widehat{M_0}$ *in* Lemma 7.4.

An irreducible unitary representation μ of $U_{\Phi,0}$, say with representation space V_μ, gives us

$$\begin{aligned}
&\mathbb{V}_\mu \to S_\Phi : U_{\Phi,0}\text{–homogeneous hermitian holomorphic vector bundle,} \\
(7.7)\quad &A^{p,q}(S_\Phi; \mathbb{V}_\mu) : \text{ space of } C^\infty\ \mathbb{V}_\mu\text{–valued } (p,q)\text{–forms on } S_\Phi, \\
&\mathcal{O}(\mathbb{V}_\mu) : \text{ sheaf of germs of holomorphic sections of } \mathbb{V}_\mu \to S_\Phi.
\end{aligned}$$

If $\mathbb{T} \to S_\Phi$ is the holomorphic tangent bundle then $A^{p,q}(S_\Phi; \mathbb{V}_\mu)$ is the space of C^∞ sections of

$$\mathbb{V}_\mu^{p,q} = \mathbb{V}_\mu \otimes \Lambda^p(\mathbb{T}^*) \otimes \Lambda^q(\overline{\mathbb{T}}^*) \to S_\Phi. \tag{7.8}$$

As M_0 is compact, $\mathbb{V}_\mu^{p,q}$ has an M_0–invariant hermitian metric, so we also have the Hodge–Kodaira orthocomplementation operators

$$(7.9)\qquad \begin{aligned}&\sharp : A^{p,q}(S_\Phi;\mathbb{V}_\mu) \to A^{n-p,n-q}(S_\Phi;\mathbb{V}_\mu^*)\\ &\qquad\text{and } \tilde{\sharp} : A^{n-p,n-q}(S_\Phi;\mathbb{V}_\mu^*) \to A^{p,q}(S_\Phi;\mathbb{V}_\mu)\end{aligned}$$

where $n = \dim_{\mathbb{C}} S_\Phi$. The global M_0–invariant hermitian inner product on $A^{p,q}(S_\Phi;\mathbb{V}_\mu)$ is given by taking the inner product in each fibre of $\mathbb{V}_\mu^{p,q}$ and integrating over S_Φ. It can also be expressed in terms of the $\sharp$ operator,

$$(7.10)\qquad \langle F_1, F_2\rangle_{S_\Phi} = \int_{M_0} \langle F_1, F_2\rangle_{mU_{\Phi,0}} d(mU_{\Phi,0}) = \int_{S_\Phi} F_1 \bar{\wedge} \sharp F_2$$

where $\bar{\wedge}$ means exterior product followed by contraction of V_μ against V_μ^*. The last equality of (7.10) is essentially the definition of $\sharp$. Now the Cauchy–Riemann operator

$$\overline{\partial} : A^{p,q}(S_\Phi;\mathbb{V}_\mu) \to A^{p,q+1}(S_\Phi;\mathbb{V}_\mu)$$

has formal adjoint

$$(7.11)\qquad \overline{\partial}^* : A^{p,q+1}(S_\Phi;\mathbb{V}_\mu) \to A^{p,q}(S_\Phi;\mathbb{V}_\mu) \text{ given by } \overline{\partial}^* = -\tilde{\sharp}\overline{\partial}\sharp\ .$$

This in turn defines an operator that is elliptic S_Φ, the Kodaira–Hodge–Laplace operator

$$(7.12)\qquad \square = \overline{\partial}\,\overline{\partial}^* + \overline{\partial}^*\overline{\partial} : A^{p,q}(S_\Phi;\mathbb{V}_\mu) \to A^{p,q}(S_\Phi;\mathbb{V}_\mu).$$

We have the space of square integrable $\mathbb{V}_\mu$–valued (p,q)–forms on S_Φ,

$$(7.13)\qquad \begin{aligned}&L_2^{p,q}(S_\Phi;\mathbb{V}_\mu) :\\ &\qquad L_2 \text{ completion of } A^{p,q}(S_\Phi;\mathbb{V}_\mu) \text{ for the inner product (7.10)}\ .\end{aligned}$$

Weyl's Lemma says that the closure of $\tilde{\square}$ of $\square$, as a densely defined operator on $L_2^{p,q}(S_\Phi;\mathbb{V}_\mu)$ from the domain $A^{p,q}(S_\Phi;\mathbb{V}_\mu)$, is essentially self–adjoint. Its kernel

$$(7.14)\qquad \mathcal{H}_2^{p,q}(S_\Phi;\mathbb{V}_\mu) = \{\omega \in \text{ Domain}(\tilde{\square}) \mid \tilde{\square}\omega = 0\}$$

is the space of **square integrable harmonic** (p,q)–forms on S_Φ with values in $\mathbb{V}_\mu$. Harmonic forms are smooth by elliptic regularity, i.e., $\mathcal{H}_2^{p,q}(S_\Phi;\mathbb{V}_\mu) \subset A^{p,q}(S_\Phi;\mathbb{V}_\mu)$. Everything is invariant under the action of M_0, and the natural action of the group M_0 on $\mathcal{H}_2^{p,q}(S_\Phi;\mathbb{V}_\mu)$ is a unitary representation.

We write $\mathcal{H}_2^q(S_\Phi;\mathbb{V}_\mu)$ for $\mathcal{H}_2^{0,q}(S_\Phi;\mathbb{V}_\mu)$, because those are the only harmonic spaces that we will use, and because $\mathcal{H}_2^q(S_\Phi;\mathbb{V}_\mu)$ is naturally isomorphic to the sheaf cohomology $H^q(S_\Phi, \mathcal{O}(V_\mu))$.

Just to avoid confusion, we state some conventions explicitly. We will use (unless we state otherwise) χ for representations of $Z_{M_0}(M_0^0)$. We will use μ for representations of $U_{\Phi,0}$ and μ^0 for representations of its identity component $U_{\Phi,0}^0$, $\rho_{\mathfrak{u}_\Phi}$ for half the sum of the roots in $\Sigma^+(\mathfrak{u}_\Phi, \mathfrak{t})$, and μ_β^0 for the irreducible representation of $U_{\Phi,0}^0$ of highest weight $\beta - \rho_{\mathfrak{u}_\Phi}$ (corresponding to infinitesimal character β). Similarly, we will use η for representations of M_0 and η^0 for representations of its identity component M_0^0, $\rho_{\mathfrak{m}}$ for half the sum of the roots in $\Sigma^+(\mathfrak{m}, \mathfrak{t})$, and η_ν^0 for the irreducible representation of M_0^0 of highest weight $\nu - \rho_{\mathfrak{m}}$ (corresponding to infinitesimal character ν). With these conventions, the Bott–Borel–Weil Theorem for M_0 is

7.15. Theorem. *Let $[\mu] = [\chi \otimes \mu_\beta^0] \in \widehat{U_{\Phi,0}}$ and fix an integer $q \geqq 0$.*
1. *If $\langle \beta - \rho_{\mathfrak{u}_\Phi} + \rho_{\mathfrak{m}}, \alpha \rangle = 0$ for some $\alpha \in \Sigma(\mathfrak{m}, \mathfrak{t})$ then $\mathcal{H}_2^q(S_\Phi; \mathbb{V}_\mu) = 0$.*
2. *If $\langle \beta - \rho_{\mathfrak{u}_\Phi} + \rho_{\mathfrak{m}}, \alpha \rangle \neq 0$ for all $\alpha \in \Sigma(\mathfrak{m}, \mathfrak{t})$, let w be the unique element in $W(\mathfrak{m}, \mathfrak{t})$ such that $\nu = w(\beta - \rho_{\mathfrak{u}_\Phi} + \rho_{\mathfrak{m}})$ is in the positive Weyl chamber, i.e. satisfies $\langle \nu, \alpha \rangle > 0$ for all $\alpha \in \Sigma^+(\mathfrak{m}, \mathfrak{t})$. So $q_0 = \text{length}(w) = |\{\alpha \in \Sigma^+(\mathfrak{m}, \mathfrak{t}) \mid \langle \beta - \rho_{\mathfrak{u}_\Phi} + \rho_{\mathfrak{m}}, \alpha \rangle < 0\}|$. Then $\mathcal{H}_2^q(S_\Phi; \mathbb{V}_\mu) = 0$ for $q \neq q_0$, and M_0 acts irreducibly on $\mathcal{H}_2^{q_0}(S_\Phi; \mathbb{V}_\mu)$ by $[\chi \otimes \eta_\nu^0]$.*

Fix $[\mu] = [\chi \otimes \mu_\beta^0] \in \widehat{U_{\Phi,0}}$ as before. Given $\sigma \in \mathfrak{a}_0^*$ we will use the Bott–Borel–Weil Theorem to find the principal series representation $\pi_{\chi \otimes \eta_\nu^0, \sigma}$ on a cohomology space related to the closed orbit in the complex flag manifold $Z_\Phi = G/P_\Phi$. Here the simple root system $\Psi_{\mathfrak{m}} \subset \Psi$ by the coherence in our choice of $\Sigma^+(\mathfrak{g}, \mathfrak{h})$, so $\Phi \subset \Psi$ and Φ defines a parabolic subgroup $P_\Phi \subset G$.

Let $z_\Phi = 1P_\Phi \in G/P_\Phi = Z_\Phi$. As $A_0 N_0 \subset G_0 \cap P_\Phi$ we have $G_0 \cap P_\Phi = U_{\Phi,0} A_0 N_0$. Thus $Y_\Phi = G_0(z_\Phi)$ is the closed G_0–orbit on Z_Φ, and S_Φ sits in Y_Φ as the orbit $M_0(z_\Phi)$. Here note that $Q_0 = M_0 A_0 N_0 = \{g \in G_0 \mid gS_\Phi = S_\Phi\}$.

7.16. Lemma. *The map $Y_\Phi \to G_0/Q_0$, given by $g(z_\Phi) \mapsto gQ_0$, defines a G_0-equivariant fibre bundle with structure group M_0 and whose fibres gS_Φ are the maximal complex analytic submanifolds of Y_Φ.*

The data (μ, σ) defines a representation $\gamma_{\mu,\sigma}$ of $U_{\Phi,0} A_0 N_0$ by

(7.17a) $$\gamma_{\mu,\sigma}(uan) = e^{(\rho_{\mathfrak{a}} + i\sigma)(\log a)} \mu(u) \quad \text{where } \rho_{\mathfrak{a}} = \tfrac{1}{2} \textstyle\sum_{\alpha \in \Sigma^+} \mathfrak{g}_\alpha \, .$$

That defines a G_0–homogeneous complex vector bundle

(7.17b) $$\mathbb{V}_{\mu,\sigma} \to G_0/U_{\Phi,0} A_0 N_0 = Y_\Phi \text{ such that } \mathbb{V}_{\mu,\sigma}|_{S_\Phi} = \mathbb{V}_\mu \, .$$

Each $\mathbb{V}_{\mu,\sigma}|_{gS_\Phi}$ is an $\text{Ad}(g)Q_0$–homogeneous holomorphic vector bundle.

Since $[\mu]$ is unitary and K_0 acts transitively on G_0/Q_0 we have a K_0–invariant hermitian metric on $\mathbb{V}_{\mu,\sigma}$. We will use it without explicit reference.

Consider the subbundle $\mathbb{T} \to Y_\Phi$ of the complexified tangent bundle of Y_Φ, defined by

(7.18a) $$\mathbb{T}|_{gS_\Phi} \to gS_\Phi \text{ is the holomorphic tangent bundle of } gS_\Phi \, .$$

It defines

$$\begin{aligned}
&\mathbb{V}^{p,q}_{\mu,\sigma} = \mathbb{V}_{\mu,\sigma} \otimes \Lambda^p(\mathbb{T}^*) \otimes \Lambda^q(\overline{\mathbb{T}}^*) \to Y_\Phi \ , \\
&A^{p,q}(Y_\Phi; \mathbb{V}_{\mu,\sigma}) : C^\infty \text{ sections of } \mathbb{V}^{p,q}_{\mu,\sigma} \to Y_\Phi \ , \text{ and} \\
&\mathcal{O}(\mathbb{V}_{\mu,\sigma}) : \text{ sheaf of germs of } C^\infty \text{ sections of } \mathbb{V}_{\mu,\sigma} \to Y_\Phi \\
&\qquad\qquad\qquad\qquad\qquad \text{that are holomorphic over every } gS_\Phi \ .
\end{aligned} \tag{7.18b}$$

$A^{p,q}(Y_\Phi; \mathbb{V}_{\mu,\sigma})$ is the space of $\mathbb{V}_{\mu,\sigma}$–valued partially (p,q)–forms on Y_Φ.

The fibre V_μ of $\mathbb{V}_\mu \to S_\Phi$ has a positive definite $U_{\Phi,0}$–invariant hermitian inner product because μ is unitary; we translate this around by K_0 to obtain a K_0–invariant hermitian structure on the vector bundle $\mathbb{V}^{p,q}_{\mu,\sigma} \to Y_\Phi$. Similarly $\mathbb{T} \to Y_\Phi$ carries a K_0–invariant hermitian metric. Using these hermitian metrics we have K_0–invariant Hodge–Kodaira orthocomplementation operators

$$\begin{aligned}
&\sharp : A^{p,q}(Y_\Phi; \mathbb{V}_{\mu,\sigma}) \to A^{n-p,n-q}(Y_\Phi; \mathbb{V}^*_{\mu,\sigma}) \\
&\tilde{\sharp} : A^{n-p,n-q}(Y_\Phi; \mathbb{V}^*_{\mu,\sigma}) \to A^{p,q}(Y_\Phi; \mathbb{V}_{\mu,\sigma})
\end{aligned} \tag{7.19}$$

where $n = \dim_{\mathbb{C}} S_\Phi$. The global G_0–invariant hermitian inner product on $A^{p,q}(Y_\Phi; \mathbb{V}_{\mu,\sigma})$ is given by taking the M_0–invariant inner product along each fibre of $Y_\Phi \to G_0/Q_0$ and integrating over G_0/Q_0,

$$\langle F_1, F_2 \rangle_{Y_\Phi} = \int_{K_0/M_0} \left(\int_{kS_\Phi} F_1 \bar{\wedge} \sharp F_2 \right) d(kM_0). \tag{7.20}$$

where $\bar{\wedge}$ means exterior product followed by contraction of V_μ against V_μ^*.

The $\overline{\partial}$ operator of Z_Φ induces the $\overline{\partial}$ operators on each of the gS_Φ, so they fit together to give us an operator

$$\overline{\partial} : A^{p,q}(Y_\Phi; \mathbb{V}_{\mu,\sigma}) \to A^{p,q+1}(Y_\Phi; \mathbb{V}_{\mu,\sigma}) \tag{7.21a}$$

that has formal adjoint

$$\overline{\partial}^* : A^{p,q+1}(Y_\Phi; \mathbb{V}_{\mu,\sigma}) \to A^{p,q}(Y_\Phi; \mathbb{V}_{\mu,\sigma}) \text{ given by } \overline{\partial}^* = -\tilde{\sharp}\overline{\partial}\sharp \ . \tag{7.21b}$$

This in turn defines an elliptic operator, the "partial Kodaira–Hodge–Laplace operator"

$$\square = \overline{\partial}\,\overline{\partial}^* + \overline{\partial}^*\overline{\partial} : A^{p,q}(Y_\Phi; \mathbb{V}_{\mu,\sigma}) \to A^{p,q}(Y_\Phi; \mathbb{V}_{\mu,\sigma}). \tag{7.21c}$$

$A^{p,q}(Y_\Phi; \mathbb{V}_{\mu,\sigma})$ is a pre Hilbert space with the global inner product (7.20). Denote

$$L_2^{p,q}(Y_\Phi; \mathbb{V}_{\mu,\sigma}) : \text{Hilbert space completion of } A^{p,q}(Y_\Phi; \mathbb{V}_{\mu,\sigma}). \tag{7.22}$$

Apply Weyl's Lemma along each gS_Φ to see that the closure $\widetilde{\Box}$ of $\Box$, as a densely defined operator on $L_2^{p,q}(Y_\Phi; \mathbb{V}_{\mu,\sigma})$ from the domain $A^{p,q}(Y_\Phi; \mathbb{V}_{\mu,\sigma})$, is essentially self-adjoint. Its kernel

$$\mathcal{H}_2^{p,q}(Y_\Phi; \mathbb{V}_{\mu,\sigma}) = \{\omega \in \text{Domain}(\widetilde{\Box}) \mid \widetilde{\Box}\omega = 0\} \tag{7.23}$$

is the space of **square integrable partially harmonic** (p,q)–forms on Y_Φ with values in $\mathbb{V}_{\mu,\sigma}$.

The factor $e^{\rho_\mathfrak{s}}$ in the representation $\gamma_{\mu,\sigma}$ that defines $\mathbb{V}_{\mu,\sigma}$ insures that the global inner product on $A^{p,q}(Y_\Phi; \mathbb{V}_{\mu,\sigma})$ is invariant under the action of G_0. The other ingredients in the construction of $\mathcal{H}_2^{p,q}(Y_\Phi; \mathbb{V}_{\mu,\sigma})$ are invariant as well, so G_0 acts naturally on $\mathcal{H}_2^{p,q}(Y_\Phi; \mathbb{V}_{\mu,\sigma})$ by isometries. This action is a unitary representation of G_0.

Essentially as before, we write $\mathcal{H}_2^{q}(Y_\Phi; \mathbb{V}_{\mu,\sigma})$ for $\mathcal{H}_2^{0,q}(Y_\Phi; \mathbb{V}_{\mu,\sigma})$, because those are the only harmonic spaces that we will use, and because $\mathcal{H}_2^{q}(Y_\Phi; \mathbb{V}_{\mu,\sigma})$ is closely related to the sheaf cohomology $H^q(Y_\Phi, \mathcal{O}(V_{\mu,\sigma}))$. The relation, which we will see later, is that they have the same underlying Harish–Chandra module.

We can now combine the Bott–Borel–Weil Theorem 7.15 with the definition ((7.1) and (7.2)) of the principal series, obtaining

7.24. Theorem. *Let* $[\mu] = [\chi \otimes \mu_\beta^0] \in \widehat{U_{\Phi,0}}$ *and* $\sigma \in \mathfrak{a}_0^*$, *and fix an integer* $q \geqq 0$.
1. *If* $\langle \beta - \rho_{\mathfrak{u}_\Phi} + \rho_\mathfrak{m}, \alpha\rangle = 0$ *for some* $\alpha \in \Sigma(\mathfrak{m}, \mathfrak{t})$ *then* $\mathcal{H}_2^{q}(Y_\Phi; \mathbb{V}_{\mu,\sigma}) = 0$.
2. *If* $\langle \beta - \rho_{\mathfrak{u}_\Phi} + \rho_\mathfrak{m}, \alpha\rangle \neq 0$ *for all* $\alpha \in \Sigma(\mathfrak{m}, \mathfrak{t})$, *let* w *be the unique element in* $W(\mathfrak{m}, \mathfrak{t})$ *such that* $\nu = w(\beta - \rho_{\mathfrak{u}_\Phi} + \rho_\mathfrak{m})$ *is in the positive Weyl chamber, i.e. satisfies* $\langle \nu, \alpha\rangle > 0$ *for all* $\alpha \in \Sigma(\mathfrak{m}, \mathfrak{t})$. *So* $q_0 = \text{length}(w) = |\{\alpha \in \Sigma^+(\mathfrak{m}, \mathfrak{t}) \mid \langle \beta - \rho_{\mathfrak{u}_\Phi} + \rho_\mathfrak{m}, \alpha\rangle < 0\}|$. *Then* $\mathcal{H}_2^{q}(Y_\Phi; \mathbb{V}_{\mu,\sigma}) = 0$ *for* $q \neq q_0$, *and the natural action of* G_0 *on* $\mathcal{H}_2^{q_0}(Y_\Phi; \mathbb{V}_{\mu,\sigma})$ *is the principal series representation* $\pi_{\chi \otimes \eta_\nu^0, \sigma}$.

References for §7.

- R. Bott, "Homogeneous vector bundles", Annals of Math. **66** (1957), 203–248.
- F. Bruhat, "Sur les représentations induites des groupes de Lie", Bull. Soc. Math. France **84** (1956), 97–205.
- Harish–Chandra, "Harmonic analysis on real reductive Lie groups, III: The Maass-Selberg relations and the Plancherel formula", Annals of Math. **104** (1976), 117–201.
- B. Kostant, "Lie algebra cohomology and the generalized Borel–Weil theorem", Annals of Math. **74** (1961), 329–387.
- N. Wallach, "Real reductive groups", I, Academic Press, 1988.
- N. Wallach, "Real reductive groups", II, Academic Press, 1992.
- J. A. Wolf, "The action of a real semisimple Lie group on a complex flag manifold, I: Orbit structure and holomorphic arc components", Bull. Amer. Math. Soc. **75** (1969), 1121–1237.

• J. A. Wolf, "The action of a real semisimple Lie group on a complex flag manifold, II: Unitary representations on partially holomorphic cohomology spaces", Memoirs. Amer. Math. Soc., No. 138, 1974.

Part 3. Tempered Series and the Plancherel Formula.

In this part we indicate the basic facts on tempered representations and see just how the tempered series suffice for harmonic analysis on the real group.

§8. The Discrete Series.

We recall the definition and Harish–Chandra parametrization of the discrete series for reductive Lie groups. This can be viewed as a noncompact group version of Cartan's theory of the highest weight for representations of compact Lie groups.

The **discrete series** of a unimodular locally compact group G_0 is the subset $\widehat{G_{0,d}} \subset \widehat{G_0}$ consisting of all classes $[\pi]$ for which π is equivalent to a subrepresentation of the left regular representation of G_0. The following are equivalent: (i) π is a discrete series representation of G_0, (ii) every coefficient $f_{u,v}(x) = \langle u, \pi(x)v\rangle$ belongs to $L^2(G_0)$, (iii) for some nonzero u, v in the representation space H_π, the coefficient $f_{u,v} \in L^2(G_0)$. Then one has orthogonality relations much as in the case of finite groups: There is a real number $\deg(\pi) > 0$ such that the $L^2(G_0)$-inner product of coefficients of π is given by

$$(8.1a) \qquad \langle f_{u,v}, f_{s,t}\rangle = \frac{1}{\deg(\pi)}\langle u,s\rangle\overline{\langle v,t\rangle} \text{ for } s,t,u,v \in H_\pi.$$

Furthermore, if π' is another discrete series representation of G_0, and is not equivalent to π, then

$$(8.1b) \qquad \langle f_{u,v}, f_{u',v'}\rangle = 0 \text{ for } u,v \in H_\pi \text{ and } u',v' \in H_{\pi'}.$$

In fact these orthogonality relations come out of convolution formulae. With the usual

$$f * h(x) = [L(f)h](x) = \int_G f(y)h(y^{-1}x)\, dy$$

we have

$$(8.2a) \qquad f_{u,v} * f_{s,t} = \frac{1}{\deg(\pi)}\langle u,t\rangle f_{s,v} \quad \text{for} \quad s,t,u,v \in H_\pi$$

and

$$(8.2b) \qquad f_{u,v} * f_{u',v'} = 0 \text{ for } u,v \in H_\pi \text{ and } u',v' \in H_{\pi'}$$

whenever π and π' are inequivalent discrete series representations of G_0.

If G_0 is compact, then every class in $\widehat{G_0}$ belongs to the discrete series, and if Haar measure is normalized as usual to total volume 1 then $\deg(\pi)$ has the usual meaning, the dimension of H_π. The orthogonality relations for irreducible unitary representations of compact groups are more or less equivalent to the Peter–Weyl Theorem.

More generally, if G_0 is a unimodular locally compact group then $L^2(G_0) = {}^0L^2(G_0) \oplus {}'L^2(G_0)$, orthogonal direct sum, where ${}^0L^2(G_0) = \sum_{[\pi]\in\widehat{G_{0,d}}} H_\pi \otimes H_\pi^*$, the "discrete" part, and ${}'L^2(G_0) = {}^0L^2(G_0)^\perp$, the "continuous" part. If, further, G_0 is a group of type I then ${}'L^2(G_0)$ is a continuous direct sum (direct integral) over $\widehat{G_0} \setminus \widehat{G_{0,d}}$ of the Hilbert spaces $H_\pi \otimes H_\pi^*$.

We will need the discrete series, not only for G_0 but for certain reductive subgroups as well. (A Lie group is called **reductive** if its Lie algebra is the direct sum of a semisimple Lie algebra and a commutative Lie algebra.) These reductive subgroups generally will not be semisimple, and even if G_0 is connected they will generally not be connected. So we want to work with a class of groups that is hereditary in the sense that it includes all the connected semisimple Lie groups of finite center, and also includes the above-mentioned subgroups of groups in the class. This is the **Harish–Chandra class**, or **class $\mathcal{H}$**.

While I'll state results for Harish–Chandra class, I'll set things up so that the statements remain valid without essential change for the larger hereditary class that contains all connected semisimple groups, whether of finite or of infinite center.

Let G_0 be a reductive Lie group, G_0^0 its identity component, $\mathfrak{g}_0$ its Lie algebra, and $\mathfrak{g} = \mathfrak{g}_0 \otimes_{\mathbb{R}} \mathbb{C}$. Suppose that $[G_0^0, G_0^0]$ has finite center, that G_0/G_0^0 is finite, and that if $x \in G_0$ then $\mathrm{Ad}(x)$ is an inner automorphism of $\mathfrak{g}$. Then we say that G_0 belongs to class $\mathcal{H}$. From now on we will assume that G_0 belongs to class $\mathcal{H}$.

If π is a unitary representation of G_0, and if $f \in L^1(G_0)$, we have the bounded operator $\pi(f) = \int_G f(x)\pi(x)dx$ on H_π. Now suppose that π has finite composition series, i.e., is a finite sum of irreducible representations. If $f \in C_c^\infty(G_0)$ then $\pi(f)$ is of trace class. Furthermore, the map

$$\Theta_\pi : C_c^\infty(G_0) \to \mathbb{C} \text{ defined by } \Theta_\pi(f) = \text{ trace } \pi(f) \tag{8.3}$$

is a distribution on G_0. Θ_π is called the **character**, the **distribution character** or the **global character** of π.

Let $\mathcal{Z}(\mathfrak{g})$ denote the center of the universal enveloping algebra $\mathcal{U}(\mathfrak{g})$. If we interpret $\mathcal{U}(\mathfrak{g})$ as the algebra of all left-invariant differential operators on G_0 then $\mathcal{Z}(\mathfrak{g})$ is the subalgebra of those that are also invariant under right translations. If π is irreducible then $d\pi|_{\mathcal{Z}(\mathfrak{g})}$ is an associative algebra homomorphism $\chi_\pi : \mathcal{Z}(\mathfrak{g}) \to \mathbb{C}$ called the **infinitesimal character** of π. We say that π is **quasi-simple** if it has an infinitesimal character, i.e. if it is a direct sum of irreducible representations that have the same infinitesimal character.

Let π be quasi-simple. Then the distribution character Θ_π satisfies a system of differential equations

$$z \cdot \Theta_\pi = \chi_\pi(z)\Theta_\pi \text{ for all } z \in \mathcal{Z}(\mathfrak{g}). \tag{8.4}$$

The regular set

$$G_0' = \{x \in G_0 : \mathfrak{g}^{\mathrm{Ad}(x)} \text{ is a Cartan subalgebra of } \mathfrak{g}\}$$

is a dense open subset whose complement has codimension $\geqq 2$. Every $x \in G_0'$ has a neighborhood on which at least one of the operators $z \in \mathcal{Z}(\mathfrak{g})$ is elliptic. It follows that $\Theta_\pi|_{G_0'}$ is integration against a real analytic function T_π on G_0'. A much deeper result of Harish–Chandra says that Θ_π has only finite jump singularities across the singular set $G_0 \setminus G_0'$, so T_π is locally L^1 and Θ_π is integration against it,

$$\Theta_\pi(f) = \int_{G_0} f(x)T_\pi(x)dx \text{ for all } f \in C_c^\infty(G_0). \tag{8.5}$$

So we may (and do) identify Θ_π with the function T_π. This key element of Harish–Chandra's theory allows the possibility of *a priori* estimates on characters and coefficients as well as explicit character formulae.

Fix a Cartan involution θ of G_0. In other words, θ is an automorphism of G_0, θ^2 is the identity, and the fixed point set $K_0 = G_0^\theta$ is a maximal compact subgroup of G_0. The choice is essentially unique, because the Cartan involutions of G_0 are just the $\mathrm{Ad}(x) \cdot \theta \cdot \mathrm{Ad}(x)^{-1}$, $x \in G_0^0$. If $G_0 = U(p,q)$ then $\theta(x) = {}^t x^{-1}$ and $K_0 = U(p) \times U(q)$.

Every Cartan subgroup of G_0 is $\mathrm{Ad}(G_0^0)$-conjugate to a θ-stable Cartan subgroup. In particular, G_0 has compact Cartan subgroups if and only if K_0 contains a Cartan subgroup of G_0.

Harish–Chandra proved that G_0 has discrete series representations if and only if it has a compact Cartan subgroup. Suppose that this is the case and fix a compact Cartan subgroup $T_0 \subset K_0$ of G_0. Let $\Sigma = \Sigma(\mathfrak{g}, \mathfrak{t})$ be the root system, $\Sigma^+ = \Sigma^+(\mathfrak{g}, \mathfrak{t})$ a choice of positive root system, and let $\rho = \frac{1}{2}\sum_{\alpha \in \Sigma^+} \alpha$. If $\xi \in \mathfrak{t}$ then $\rho(\xi)$ is half the trace of $ad(\xi)$ on $\sum_{\alpha\in\Sigma^+} \mathfrak{g}_\alpha$.

If π is a discrete series representation of G_0 and Θ_π is its distribution character, then the equivalence class of π is determined by the restriction of Θ_π to $T_0 \cap G_0'$. Harish–Chandra parameterizes the discrete series of G_0 by parameterizing those restrictions.

Let $G_0^\dagger$ denote the finite index subgroup $T_0 G_0^0 = Z_{G_0}(G_0^0)G_0^0$ of G_0. In fact the argument of Lemma 7.4 is easily modified here to prove $T_0 .= Z_{G_0}(G_0^0)T_0^0$, so $T_0 = T_0^\dagger$. Lemma 7.4 says that the group M_0 of a minimal parabolic subgroup of G_0 satisfies $M_0 = M_0^\dagger$, and similarly, we have $U_{\Phi,0} = U_{\Phi,0}^\dagger$. In general, where M_0 may be noncompact, this need not hold.

The Weyl group $W^\dagger = W(G_0^\dagger, T_0)$ coincides with $W^0 = W(G_0^0, T_0^0)$ and is a normal subgroup of $W = W(G_0, T_0)$.

Every irreducible unitary representation of $T_0 = Z_{G_0}(G_0^0)T_0^0$ is of the form $\chi \otimes e^{i(\lambda-\rho)}$ where $\lambda \in i\mathfrak{t}_0^*$ and $\lambda - \rho$ satisfies an integrality condition, where $\chi \in \widehat{Z_{G_0}(G_0^0)}$, and where χ and $e^{i(\lambda-\rho)}$ restrict to (multiples of) the same unitary character on the center of G_0^0.

Let $\chi \otimes e^{i(\lambda-\rho)} \in \widehat{T_0}$ as above. Suppose that λ is regular, i.e., that $\langle\lambda, \alpha\rangle \neq 0$ for all $\alpha \in \Sigma$. Then there are unique discrete series representations π_λ^0 of G_0^0 and $\pi_{\chi,\lambda}^\dagger$ of $G_0^\dagger$ whose distribution characters satisfy

$$\Theta_{\pi_\lambda^0}(x) = (-1)^{q(\lambda)} \frac{\sum_{w\in W^0} \operatorname{sign}(w) e^{w(\lambda)}}{\prod_{\alpha\in\Sigma^+}(e^{\alpha/2} - e^{-\alpha/2})} \tag{8.6a}$$
$$\text{and } \Theta_{\pi_{\chi,\lambda}^\dagger}(zx) = \operatorname{trace} \chi(z)\Theta_{\pi_\lambda^0}(x)$$

for $z \in Z_{G_0}(G_0^0)$ and $x \in T_0^0 \cap G_0'$, where

$$q(\lambda) = |\{\alpha \in \Sigma^+(\mathfrak{k}, \mathfrak{t}) \mid \langle\alpha, \lambda\rangle < 0\}| + |\{\beta \in \Sigma^+(\mathfrak{g}, \mathfrak{t}) \setminus \Sigma^+(\mathfrak{k}, \mathfrak{t}) \mid \langle\beta, \lambda\rangle > 0\}|. \tag{8.6b}$$

Here note that $\pi_{\chi,\lambda}^\dagger = \chi \otimes \pi_\lambda^0$.

The same datum (χ, λ) specifies a discrete series representation $\pi_{\chi,\lambda}$ of G_0, by the formula $\pi_{\chi,\lambda} = \operatorname{Ind}_{G_0^\dagger}^{G_0}(\pi_{\chi,\lambda}^\dagger)$. This induced representation is irreducible because its conjugates by elements of $G_0/G_0^\dagger$ are mutually inequivalent, consequence of regularity of λ. $\pi_{\chi,\lambda}$ is characterized by the fact that its distribution character is supported in $G_0^\dagger$ and is given on $G_0^\dagger$ by

$$\Theta_{\pi_{\chi,\lambda}} = \sum_{1\leqq i\leqq r} \Theta_{\pi_{\chi,\lambda}^\dagger} \cdot \gamma_i^{-1} \tag{8.7}$$

with $\gamma_i = \operatorname{Ad}(g_i)|_{G_0^\dagger}$ where $\{g_1, \ldots, g_r\}$ is any system of coset representatives of G_0 modulo $G_0^\dagger$. To combine these into a single formula one chooses the g_i so that they normalize T_0, i.e. chooses the γ_i to be a system of coset representatives of W modulo $W^\dagger$.

Every discrete series representation of G_0 is equivalent to a representation $\pi_{\chi,\lambda}$ as just described. Discrete series representations $\pi_{\chi,\lambda}$ and $\pi_{\chi',\lambda'}$ are equivalent if and only if $\chi' \otimes e^{i\lambda'} = (\chi \otimes e^{i\lambda}) \cdot w^{-1}$ for some $w \in W$. And λ is both the infinitesimal character and the Harish–Chandra parameter for the discrete series representation $\pi_{\chi,\lambda}$.

References for §8.

- J. Dixmier, "Les C*-algèbres et leurs représentations", Gauthier–Villars, Paris, 1964.
- R. Godement, "Sur les rélations d'orthogonalité de V. Bargmann", C. R. Acad. Sci. Paris **225**, (1947), 521–523 and 657–659.
- Harish–Chandra, "Representations of a semisimple Lie group on a Banach space, I", Trans. Amer. Math. Soc. **75** (1953), 185–243.

- Harish–Chandra, "Representations of a semisimple Lie group on a Banach space, II", Trans. Amer. Math. Soc. **76** (1954), 26–65.
- Harish–Chandra, "Representations of a semisimple Lie group on a Banach space, III", Trans. Amer. Math. Soc. **76** (1954), 234–253.
- Harish–Chandra, "Representations of a semisimple Lie groups, IV", Amer. J. Math. **77** (1955), 743–777.
- Harish–Chandra, "Representations of a semisimple Lie groups, V", Amer. J. Math. **78** (1956), 1–41.
- Harish–Chandra, "Representations of semisimple Lie groups, VI", Amer. J. Math. **78** (1956), 564–628.
- Harish–Chandra, "Characters of semisimple Lie groups", Trans. Amer. Math. Soc. **83** (1956), 98–163.
- Harish–Chandra, "Invariant eigendistributions on a semisimple Lie group", Trans. Amer. Math. Soc. **119** (1965), 457–508.
- Harish–Chandra, "Discrete series for semisimple Lie groups, I", Acta Math. **113** (1965), 241–317.
- Harish–Chandra, "Discrete series for semisimple Lie groups, II", Acta Math. **116** (1966), 1–111.
- R. A. Herb, "Fourier inversion and the Plancherel formula for semisimple Lie groups", Amer. J. Math. **104** (1982), 9–58.
- R. A. Herb, "Discrete series characters and Fourier inversion on semisimple real Lie groups", Trans. Amer. Math. Soc. **277** (1983), 241–261.
- R. A. Herb and J. A. Wolf, "The Plancherel theorem for general semisimple Lie groups", Compositio Math. **57** (1986), 271–355.
- M. A. Rieffel, "Square integrable representations of Hilbert algebras", J. Funct. Anal. **3** (1969), 265–300.
- J. A. Wolf, "The action of a real semisimple Lie group on a complex manifold, II: Unitary representations on partially holomorphic cohomology spaces", Memoirs Amer. Math. Soc. **138**, 1974.

§9. The Tempered Series.

The representations of G_0 that enter into its Plancherel formula are the **tempered representations**. They are constructed from a class of real parabolic subgroups of G_0 called the **cuspidal parabolic subgroups**. One constructs a standard tempered representation by first constructing a relative discrete series representation for the reductive part of cuspidal parabolic subgroup, and then by unitary induction from the parabolic subgroup up to G_0. We start by recalling the definitions.

Let H_0 be a Cartan subgroup of G_0. Fix a Cartan involution θ of G_0 such that $\theta(H_0) = H_0$. We write K_0 for the fixed point set G_0^θ, which is a maximal compact subgroup of G_0. Decompose

$$\begin{aligned} &\mathfrak{h}_0 = \mathfrak{t}_0 \oplus \mathfrak{a}_0 \text{ and } H_0 = T_0 \times A_0 \\ &\quad \text{where} \quad T_0 = H_0 \cap K_0,\ \theta(\xi) = -\xi \text{ on } \mathfrak{a}_0, \text{ and } A_0 = \exp_G(\mathfrak{a}_0). \end{aligned} \tag{9.1}$$

Then the centralizer $Z_{G_0}(A_0)$ of A_0 in G_0 has form $M_0 \times A_0$ where $\theta(M_0) = M_0$. The group M_0 is a reductive Lie group of Harish–Chandra class. T_0 is a compact Cartan subgroup of M_0, so M_0 has discrete series representations.

Suppose that our positive root system $\Sigma^+ = \Sigma^+(\mathfrak{g}, \mathfrak{h})$ is defined by positive root systems $\Sigma^+(\mathfrak{m}, \mathfrak{t})$ and $\Sigma^+(\mathfrak{g}_0, \mathfrak{a}_0)$ as in (7.3).

A (real) parabolic subgroup $P_0 \subset G_0$ is called **cuspidal** if the commutator subgroup of the Levy component (reductive part) has a compact Cartan subgroup.

The Cartan subgroup $H_0 \subset G_0$ defines a cuspidal parabolic subgroup $P_0 = M_0A_0N_0$ of G_0 as follows. The Lie algebra of N_0 is $\mathfrak{n}_0 = \sum_{\alpha \in \Sigma^+(\mathfrak{g}_0, \mathfrak{a}_0)} (\mathfrak{g}_0)_{-\alpha}$, M_0 and A_0 are as above, and $M_0A_0 = M_0 \times A_0$ is the Levi component of P_0. One extreme is the case where $\dim \mathfrak{a}_0$ is maximal; then P_0 is a minimal parabolic subgroup of G_0. The other extreme is where $\dim \mathfrak{a}_0$ is minimal; if $\mathfrak{a}_0 = 0$ then $P_0 = G_0$.

Every cuspidal parabolic subgroup of G_0 is produced by the construction just described, as H_0 varies. Two cuspidal parabolic subgroups of G_0 are **associated** if they are constructed as above from G_0–conjugate Cartan subgroups; then we say that the G_0–conjugacy class of Cartan subgroups is **associated** to the G_0–association class of cuspidal parabolic subgroups.

As in (7.1),

$$(9.2) \qquad \begin{array}{l} [\eta] \in \widehat{M_0} \text{ and } \sigma \in \mathfrak{a}_0^* \\ \qquad\qquad \text{determine } [\alpha_{\eta,\sigma}] \in \widehat{P_0} \text{ by } \alpha_{\eta,\sigma}(man) = \eta(m)e^{i\sigma(\log a)}. \end{array}$$

Then we have

$$(9.3) \qquad \pi_{\eta,\sigma} = \operatorname{Ind}_{P_0}^{G_0}(\alpha_{\eta,\sigma}), \qquad \text{unitarily induced representation.}$$

The H_0**–series** or **principal** H_0**–series** of G_0 consists of the unitary equivalence classes of the representations (9.3) for which η is a discrete series representation of M_0. Harish–Chandra extended Bruhat's irreducibility results to all the H_0–series.

As the terminology indicates, $\pi_{\eta,\sigma} = \operatorname{Ind}_{P_0}^{G_0}(\alpha_{\eta,\sigma})$ is independent of the choice of $\Sigma^+(\mathfrak{g}_0, \mathfrak{a}_0)$. In fact this is the case even if η does not belong to the discrete series of M_0, and is a consequence of the character formula, which we now describe.

If J_0 is a Cartan subgroup of G_0 we write G'_{J_0} for the set of G_0–regular elements that are G_0–conjugate to an element of J_0. If further we fix a positive root system $\Sigma^+(\mathfrak{g}, \mathfrak{j})$ then we write $\Delta_{G_0,J_0} = \prod_{\gamma \in \Sigma^+(\mathfrak{g},\mathfrak{j})}(e^{\gamma/2} - e^{-\gamma/2})$. Passing to a 2–sheeted cover if necessary (it is not necessary if $G_0 \subset G$ with G complex and simply connected), e^ρ and Δ_{G_0,J_0} are well defined functions on J_0.

When dealing with both G_0 and M_0, we write M_0'' for the M_0–regular subset of M_0. If L_0 is a Cartan subgroup of M_0 we write M''_{L_0} for the set of elements of M_0'' that are M_0–conjugate to an element of L_0.

9.4. Theorem. *Fix a cuspidal parabolic subgroup* $P_0 = M_0A_0N_0$ *of* G_0, *let* $[\eta] \in \widehat{M_0}$, *and let* $\sigma \in \mathfrak{a}_0^*$. *Let* χ_ν, *with* $\nu \in \mathfrak{t}^*$, *be the infinitesimal character of* η *and let* Ψ_η *be the distribution character.*

1. $[\pi_{\eta,\sigma}]$ *has infinitesimal character* $\chi_{\nu+i\sigma}$ *relative to* $\mathfrak{h}$.

2. $[\pi_{\eta,\sigma}]$ *is a finite sum of classes from* $\widehat{G_0}$. *So it has well defined distribution character* $\Theta_{\pi_{\eta,\sigma}}$ *that is a locally summable function analytic on the regular set* G_0'.

3. $\Theta_{\pi_{\eta,\sigma}}$ *has support in the closure of* $\bigcup G'_{J_0}$ *where* J_0 *runs over a system of representatives of the* G_0*–conjugacy classes of Cartan subgroups of* M_0A_0.

4. *Fix a Cartan subgroup* $J_0 = J_{M,0} \times A_0$ *of* M_0A_0. *Let* $\{J_{i,0} = g_iJ_0g_i^{-1} \mid 1 \leqq i \leqq \ell(J_0)\}$ *be a system of representatives of the* M_0A_0*–conjugacy classes of Cartan subgroups of* M_0A_0 *that are* G_0*–conjugate to* J_0. *For each index* i *let* $N_{G_0}(J_{i,0})$ *and* $N_{M_0A_0}(J_{i,0})$ *denote normalizers in* G_0 *and* M_0A_0. *Let* $h \in J_0 \cap G_0'$ *and define* $h_i = g_ihg_i^{-1} \in J_{i,0}$ *Then the sets* $N_{G_0}(J_{i,0})(h_i)$ *and* $N_{M_0A_0}(J_{i,0})(h_i)$ *are finite, and* $\Theta_{\pi_{\eta,\sigma}}(h)$ *is given by*

$$\sum_{i=1}^{\ell(J_0)} \frac{1}{|\Delta_{G_0,J_{i,0}}(h_i)|} \times \tag{9.5}$$
$$\times \sum_{N_{G_0}(J_{i,0})(h_i)} \frac{|\Delta_{M_0A_0,J_{i,0}}(wh_i)|}{|N_{M_0A_0}(J_{i,0})(wh_i)|} \Psi_\eta((wh_i)_{M_0}) e^{i\sigma(\log(wh_i)_{A_0})} .$$

If $h \in J_0^0$, *so each* $h_i \in J_{i,0}^0$, *then the second sum runs over the Weyl group* $W(G_0, J_{i,0})$.

5. *If* $t \in T_0$ *and* $a \in A_0$ *with* $ta \in G_0'$ *then* (9.5) *reduces to*

(9.6)
$$\Theta_{\pi_{\eta,\sigma}}(ta) = \frac{|\Delta_{M_0,T_0}(t)}{|\Delta_{G_0,H_0}(ta)|} \sum_{N_{G_0}(H_0)(ta)} \frac{1}{|N_{M_0}(T_0)(wt)|} \Psi_\eta(wt) e^{i\sigma(\log(wa))} .$$

The formula (9.5) shows in particular that the distribution $\Theta_{\pi_{\eta,\sigma}}$ is independent of the choice of cuspidal parabolic subgroup P_0 associated to the G_0–conjugacy class of H_0. As $[\pi_{\eta,\sigma}]$ is a finite sum from $\widehat{G_0}$, now $[\pi_{\eta,\sigma}]$ also is independent of choice of P_0 for the given H_0. So Theorem 9.4 implies

9.7. Corollary. *The class* $[\pi_{\eta,\sigma}]$ *is independent of choice of cuspidal parabolic subgroup* $P_0 = M_0A_0N_0$ *for the given Cartan subgroup* $H_0 = T_0 \times A_0$.

The proof of Theorem 9.4 is a bit technical. It is based on the Harish–Chandra transform $\mathcal{F}_{P_0} : C_0^\infty(G_0) \to C_0^\infty(M_0A_0)$, given by

$$\mathcal{F}_{P_0}(b)(ma) = e^{-\rho(\log a)} \int_{K_0} \left\{ \int_{N_0} b(kmank^{-1})dn \right\} dk . \tag{9.8}$$

One first proves that $\pi_{\eta,\sigma}(b)$ is of trace class with

$$\operatorname{trace} \pi_{\eta,\sigma}(b) = \int_{M_0A_0} \mathcal{F}_{P_0}(b)(ma)\Psi_\eta(m)e^{i\sigma(\log a)}dmda . \tag{9.9}$$

Then one can calculate the infinitesimal character. From that, a look at K_0–types proves finiteness of the composition series. Then one has to extend the Weyl integration formula appropriately in order to compute the character formulae.

Theorem 9.4 specializes to the H_0–series as follows. Express

$$\eta = \eta_{\chi,\nu} = \mathrm{Ind}\,_{M_0^\dagger}^{M_0} (\chi \otimes \eta_\nu^0) \tag{9.10}$$

corresponding to $\chi \in Z_{M_0}(M_0^0)$ and $e^{\nu-\rho_m} \in \widehat{T_0^0}$ that restrict to multiples of the same unitary character on the center of M_0^0. Choose coset representatives $\{x_1, \ldots, x_\ell\}$ of M_0 modulo $M_0^\dagger$ that normalize $\mathfrak{t}_0$. They represent Weyl group elements $w_i \in W(M_0, T_0)$ that form a system of representatives of $W(M_0, T_0)$ modulo $W(M_0^0, T_0^0)$. Now, following (8.6) and (8.7), the distribution character of η is supported on $M_0^\dagger$, and it satisfies

$$\begin{aligned}\Psi_{\eta_{\chi,\nu}}(zt) = &\sum_{i=1}^{\ell} (-1)^{q_m(w_i\nu)} \mathrm{trace}\,\chi(x_i z x_i^{-1}) \times \\ &\times \frac{1}{\Delta_{M_0,T_0}(t)} \sum_{W(M_0^0,T_0^0)} \det(ww_i) e^{ww_i\nu}(t)\end{aligned} \tag{9.11}$$

for $z \in Z_{M_0}(M_0^0)$ and $t \in T_0^0 \cap G_0'$. The formula (9.11) characterizes $[\eta_{\chi,\nu}]$. With Theorem 9.4 it gives

9.12. Theorem. *Let $[\eta_{\chi,\nu}] \in \widehat{M_{0,d}}$ as in (9.10) and let $\sigma \in \mathfrak{a}_0^*$. Then $[\pi_{\chi,\nu,\sigma}] = [\mathrm{Ind}\,_{P_0}^{G_0}(\eta_{\chi,\nu} \otimes e^{i\sigma})]$ is the unique H_0–series representation class on G_0 whose distribution character satisfies*

$$\begin{aligned}&\Theta_{\pi_{\eta_{\chi,\nu,\sigma}}}(zta) = \\ &\frac{|\Delta_{M_0,T_0}(zt)|}{|\Delta_{G_0,H_0}(zta)|} \sum_{w(zta)} \frac{1}{|N_{M_0}(T_0)(w(zt))|} \Psi_{\eta_{\chi,\nu}}(w(zt)) e^{i\sigma(\log(wa))}\end{aligned} \tag{9.13}$$

where $w(zta)$ runs over $N_{G_0}(H_0)(zta)$, the $\Psi_{\eta_{\chi,\nu}}(w(zt))$ are given by (9.11), $z \in Z_{M_0}(M_0^0)$, $t \in T_0^0 \cap M_0''$ and $a \in A_0$.

Two H_0–series representations $[\pi_{\eta_{\chi,\nu,\sigma}}]$, $[\pi_{\eta_{\chi',\nu',\sigma'}}]$ of G_0 are equal if and only if $([\chi'], \nu', \sigma')$ is in the Weyl group orbit $W(G_0, H_0)([\chi], \nu, \sigma)$.

The H_0–series representations $[\pi_{\eta_{\chi,\nu,\sigma}}]$ has dual $[\pi^*_{\eta_{\chi,\nu,\sigma}}] = [\pi_{\eta_{\overline{\chi},-\nu,-\sigma}}]$ and has infinitesimal character $\chi_{\nu+i\sigma}$ relative to $\mathfrak{h}$, In particular it sends the Casimir element of $\mathcal{U}(\mathfrak{g})$ to $||\nu||^2 + ||\sigma||^2 - ||\rho||^2$.

Two complements to Theorem 9.12. First, one can check that if H_0 and $'H_0$ are non–conjugate Cartan subgroups of G_0 then every H_0–series representation is disjoint (no composition factors in common) from every $'H_0$–series representation. This is seen by examining the real and imaginary parts of the infinitesimal character. Second, the Harish–Chandra condition for irreducibility of $[\pi_{\eta_{\chi,\nu,\sigma}}]$ is that σ be regular for $(\mathfrak{g}_0, \mathfrak{a}_0)$.

References for §9.

- Harish–Chandra, "Harmonic analysis on real reductive groups, I", J. Funct. Anal. **19** (1975), 104–204.
- Harish–Chandra, "Harmonic analysis on real reductive groups, II", Inventiones Math. **36** (1976), 1–55.
- Harish–Chandra, "Harmonic analysis on real reductive groups, III", Annals of Math. **104** (1976), 117–201.
- R. A. Herb and J. A. Wolf, "The Plancherel theorem for general semisimple Lie groups", Compositio Math. **57** (1986), 271–355.
- R. A. Herb and J. A. Wolf, "Rapidly decreasing functions on general semisimple Lie groups", Compositio Math. **58** (1986), 73–110.
- N. Wallach, "Real reductive groups", I, Academic Press, 1988.
- N. Wallach, "Real reductive groups", II, Academic Press, 1992.
- J. A. Wolf, "The action of a real semisimple Lie group on a complex flag manifold, II: Unitary representations on partially holomorphic cohomology spaces", Memoirs. Amer. Math. Soc., No. 138, 1974.

§10. INDICATION OF THE PLANCHEREL FORMULA.

We start with Kostant's "cascade construction" for the conjugacy classes of Cartan subgroups of G_0. Suppose first that G_0 has a compact Cartan subgroup T_0. Fix a Cartan involution θ of G_0 such that $\theta(T_0) = T_0$ and the corresponding ± 1 eigenspace decomposition $\mathfrak{g}_0 = \mathfrak{k}_0 + \mathfrak{s}_0$ where $\mathfrak{k}_0$ is the Lie algebra of the maximal compact subgroup $K_0 = \{g \in G_0 \mid \theta(g) = g\}$. If $\alpha \in \Sigma(\mathfrak{g}, \mathfrak{t})$ then either $\mathfrak{g}_\alpha \subset \mathfrak{k}$ and we say that α is **compact**, or $\mathfrak{g}_\alpha \subset \mathfrak{s}$ and we say that α is **noncompact**.

Let $\alpha \in \Sigma(\mathfrak{g}, \mathfrak{t})$ be noncompact. Let $\mathfrak{g}[\alpha] = \mathfrak{g}_\alpha + \mathfrak{g}_{-\alpha} + [\mathfrak{g}_\alpha, \mathfrak{g}_{-\alpha}]$ as in (6.4), let $G[\alpha]$ denote the corresponding analytic subgroup of G, and consider the corresponding real forms $\mathfrak{g}_0[\alpha] = \mathfrak{g}_0 \cap \mathfrak{g}[\alpha]$ and $G_0[\alpha] = G_0 \cap G[\alpha]$. Then $G_0[\alpha] \cap T_0$ is a compact Cartan subgroup, and we can simply replace it by the noncompact Cartan subgroup of $G_0[\alpha]$. Let $\mathfrak{a}_0[\alpha]$ denote the Lie algebra of that noncompact Cartan subgroup. Then we have a new Cartan subgroup

$$\mathfrak{h}_0\{\alpha\} = \left(\mathfrak{t}_0 \cap (\mathfrak{g}_0[\alpha] \cap \mathfrak{t}_0)^\perp\right) + \mathfrak{a}_0[\alpha] \tag{10.1a}$$

and the corresponding Cartan subgroup

$$H_0\{\alpha\} = \{g \in G_0 \mid \mathrm{Ad}\,(g)\xi = \xi \text{ for all } \xi \in \mathfrak{h}_0\{\alpha\}. \tag{10.1b}$$

The point is that $H_0\{\alpha\}$ has one compact dimension less than that of T_0 and one noncompact dimensions more.

Let $\alpha, \beta \in \Sigma(\mathfrak{g}, \mathfrak{t})$ be noncompact. We can carry out the construction (10.1) for α and β independently, one after the other, in α and β are **strongly orthogonal** in the sense that α and β are linearly independent and neither of $\alpha \pm \beta$ are roots. We write this relation as $\alpha \perp\!\!\!\perp \beta$. If $\alpha \perp\!\!\!\perp \beta$ then we have the new Cartan subgroup $H_0\{\alpha, \beta\}$ given by

$$\mathfrak{h}_0\{\alpha, \beta\} = \left(\mathfrak{t}_0 \cap ((\mathfrak{g}_0[\alpha] \oplus \mathfrak{g}_0[\beta]) \cap \mathfrak{t}_0)^\perp\right) + \left(\mathfrak{a}_0[\alpha] \oplus \mathfrak{a}_0[\beta]\right) \tag{10.2a}$$

and

(10.2b) $$H_0\{\alpha,\beta\} = \{g \in G_0 \mid \mathrm{Ad}\,(g)\xi = \xi \text{ for all } \xi \in \mathfrak{h}_0\{\alpha,\beta\}.$$

Here $H_0\{\alpha,\beta\}$ has two compact dimensions less than that of T_0 and two noncompact dimension more.

We say that a set S of noncompact roots is **strongly orthogonal** if it is linearly independent and if any two of its elements are strongly orthogonal. Then as above we have a Cartan subgroup $H_0\{S\}$ given by

(10.3a) $$\mathfrak{h}_0\{S\} = (\mathfrak{t}_0 \cap ((\textstyle\sum_{\alpha\in S}\mathfrak{g}_0[\alpha]) \cap \mathfrak{t}_0)^{\perp}) + (\textstyle\sum_{\alpha\in S}\mathfrak{a}_0[\alpha])$$

and

(10.3b) $$H_0\{S\} = \{g \in G_0 \mid \mathrm{Ad}\,(g)\xi = \xi \text{ for all } \xi \in \mathfrak{h}_0\{S\}.$$

Here $H_0\{S\}$ has $|S|$ compact dimensions fewer than T_0 has, and $H_0\{S\}$ has $|S|$ noncompact dimensions more than T_0 has.

Cartan subgroups $H_0\{S_1\}$ and $H_0\{S_2\}$ are G_0–conjugate if and only if some $w \in W(G_0,T_0)$ sends S_1 to S_2. Kostant proved that every Cartan subgroup of G_0 is conjugate to $\mathfrak{h}_0\{S\}$ for some set S of strongly orthogonal noncompact roots.

This sets up a hierarchy among the conjugacy classes of Cartan subgroups of G_0 : $H_0\{S_1\} \leqq H_0\{S_2\}$ if and only if some Weyl group element $w \in W(G_0,T_0)$ sends S_2 to a subset of S_1. That in turn sets up a hierarchy among parts of the regular set G_0'. If H_0 is any Cartan subgroup of G_0 we denote $G'_{H_0} = G_0' \cap \mathrm{Ad}(G)H_0$, the set of all regular elements G_0' that are conjugate to an element of H_0. Now $G'_{H_0\{S_1\}} \leqq G'_{H_0\{S_2\}}$ if and only if some Weyl group element $w \in W(G_0,T_0)$ sends S_2 to a subset of S_1. Here G'_{T_0} sits at the top, the $G'_{H_0\{\alpha\}}$ sit just below, the $G'_{H_0\{\alpha,\beta\}}$ are on the next level down, and finally the part of G_0' corresponding to the Cartan subgroup of the minimal parabolic subgroups sit at the bottom.

If G_0 does not have a compact Cartan subgroup, we reduce to that case as follows. Let $H_0 = T_0 \times A_0$ be a Cartan subgroup that is as compact as possible, i.e., T_0 is a Cartan subgroup of a maximal compact subgroup $K_0 \subset G_0$. Let $P_0 = M_0A_0N_0$ be an associated cuspidal parabolic subgroup. Then just do the cascade construction for M_0, obtaining a family of Cartan subgroups $H_{M,0}\{S\} \subset M_0$ as S runs over the $W(M_0,T_0)$–conjugacy classes of strongly orthogonal sets $S \subset \Sigma(\mathfrak{m},\mathfrak{t})$ of noncompact roots of $\mathfrak{m}$. Then the $H_0\{S\} = H_{M,0}\{S\} \times A_0$ give the conjugacy classes of Cartan subgroups of G_0.

A careful examination of the character formulae (8.6), (8.7), (9.11) and (9.13) shows that the various tempered series exhaust enough of $\widehat{G_0}$ for a decomposition of $L_2(G_0)$ essentially as

(10.4) $$\sum_{H_0\in Car(G_0)} \; \sum_{\chi\otimes e^{\nu-\rho_{\mathfrak{m}}}\in\widehat{T_0}} \int_{\mathfrak{a}_0^*} H_{\pi_{\chi,\nu,\sigma}} \otimes H^*_{\pi_{\chi,\nu,\sigma}} m(H_0 : \chi : \nu : \sigma)d\sigma.$$

Here $Car(G_0)$ denotes the set of G_0–conjugacy classes of Cartan subgroups and the Borel measure $m(H_0 : \chi : \nu : \sigma)d\sigma$ is the **Plancherel measure** on $\widehat{G_0}$. In general the Plancherel density $m(H_0 : \chi : \nu : \sigma)$ has a formula that varies with the component of the regular set. This was worked out by Harish–Chandra for groups of Harish–Chandra class, and somewhat more generally by Herb and myself. Harish–Chandra's approach is based on an analysis of the structure of the Schwartz space, while Herb and I use explicit character formulae. These explicit formulae allow us to prove formula (10.4), as follows.

Start with G'_{H_0} where H_0 represents the conjugacy class of Cartan subgroups of G_0 that are as compact as possible. A look at the character formulae cited above, shows that the H_0–series representations suffice to expand functions $f \in C_0^\infty(G'_{H_0})$. That expansion formula gives us the map

(10.5a) $$C_0^\infty(G_0) \to C^\infty(G_0 \setminus G'_{H_0}) \text{ by } f \mapsto f_1$$

where r_x denotes right translation by $x \in G_0$ and

(10.5b) $$f_1(x) = f(x) - \sum_{\chi \otimes e^{\nu - \rho_m} \in \widehat{T_0}} \int_{\mathfrak{a}_0^*} \Theta_{\pi_{\chi,\nu,\sigma}}(r_x f) m(H_0 : \chi : \nu : \sigma) d\sigma.$$

Now let $\{H_0\{\alpha_1\}, \ldots, H_0\{\alpha_{m_1}\}\}$ be a set of representatives of the conjugacy classes of Cartan subgroups just below H_0. A look at the character formulae cited above, shows that the $H_0\{\alpha_i\}$–series representations suffice to expand functions $f \in C_0^\infty(G'_{H_0\{\alpha_i\}})$. Those expansions do not interact, nor do they introduce nonzero values in G'_{H_0}, so they give us a map

(10.6a) $$C^\infty(G_0 \setminus G'_{H_0}) \to C^\infty\left(G_0 \setminus (G'_{H_0} \cup \bigcup G'_{H_0\{\alpha_i\}})\right) \text{ by } f_1 \mapsto f_2$$

where

(10.6b)

$$f_2(x) - f_1(x) = \sum_{1 \leqq i \leqq m_1} \sum_{\chi \otimes e^{\nu - \rho_m} \in \widehat{T_0\{\alpha_i\}}} \int_{\mathfrak{a}_0^*\{\alpha_i\}} \Theta_{\pi_{\chi,\nu,\sigma}}(r_x f) m(H_0\{\alpha_i\} : \chi : \nu : \sigma) d\sigma.$$

Now simply proceed down one level at a time. The tricky point here is to know the character formulae completely, so that one knows f_j well enough to compute f_{j+1}. Finally, one obtains the final form

(10.7) $$f(x) = \sum_{H_0 \in Car(G_0)} \sum_{\chi \otimes e^{\nu - \rho_m} \in \widehat{T_0}} \int_{\mathfrak{a}_0^*} \Theta_{\pi_{\chi,\nu,\sigma}}(r_x f) m(H_0 : \chi : \nu : \sigma) d\sigma.$$

References for §10.

- Harish–Chandra, "The Plancherel formula for complex semisimple Lie groups", Trans. Amer. Math. Soc. **76** (1954), 485–528.
- Harish–Chandra, "Harmonic analysis on real reductive groups, I", J. Funct. Anal. **19** (1975), 104–204.
- Harish–Chandra, "Harmonic analysis on real reductive groups, II", Inventiones Math. **36** (1976), 1–55.
- Harish–Chandra, "Harmonic analysis on real reductive groups, III", Annals of Math. **104** (1976), 117–201.
- R. A. Herb, "Fourier inversion and the Plancherel formula for semisimple Lie groups", Amer. J. Math. **104** (1982), 9–58.
- R. A. Herb, "Discrete series characters and Fourier inversion on semisimple real Lie groups", Trans. Amer. Math. Soc. **277** (1983), 241–261.
- R. A. Herb and J. A. Wolf, "The Plancherel theorem for general semisimple Lie groups", Compositio Math. **57** (1986), 271–355.
- R. A. Herb and J. A. Wolf, "Rapidly decreasing functions on general semisimple Lie groups", Compositio Math. **58** (1986), 73–110.
- P. Sally and G. Warner, "The Fourier transform on semisimple Lie groups of real rank one", Acta Math. **131** (1973), 11–26.
- N. Wallach, "Real reductive groups", I, Academic Press, 1988.
- N. Wallach, "Real reductive groups", II, Academic Press, 1992.
- J. A. Wolf, "The action of a real semisimple Lie group on a complex flag manifold, II: Unitary representations on partially holomorphic cohomology spaces", Memoirs. Amer. Math. Soc., No. 138, 1974.

PART 4. GEOMETRIC REALIZATION OF THE TEMPERED SERIES.

In this part we show how the standard tempered representations occur as natural geometric objects over certain real group orbits.

§11. MEASURABLE OPEN ORBITS AND THE DISCRETE SERIES.

Fix a complex flag manifold $Z = G/P$. An open orbit $G_0(z) \subset Z$ is called **measurable** if it carries a G_0–invariant volume element. If that is the case, then the invariant volume element is the volume element of a G_0–invariant, possibly indefinite, Kaehler metric on the orbit, and the isotropy subgroup $G_0 \cap P_z$ is the centralizer in G_0 of a (compact) torus subgroup of G_0. In more detail, measurable open orbits are characterized by

11.1. Proposition. *Let $D = G_0(z)$ be an open G_0–orbit on the complex flag manifold $Z = G/P$. Then the following conditions are equivalent.*
1. *The orbit $G_0(z)$ is measurable.*
2. *$G_0 \cap P_z$ is the G_0–centralizer of a (compact) torus subgroup of G_0.*
3. *D has a G_0–invariant possibly-indefinite Kaehler metric, thus a G_0–invariant measure obtained from the volume form of that metric.*
4. *$\tau\Phi^r = \Phi^r$, and $\tau\Phi^n = -\Phi^n$ where $\mathfrak{p}_z = \mathfrak{p}_\Phi$.*

5. $\mathfrak{p}_z \cap \tau\mathfrak{p}_z$ *is reductive, i.e.* $\mathfrak{p}_z \cap \tau\mathfrak{p}_z = \mathfrak{p}_z^r \cap \tau\mathfrak{p}_z^r$.
6. $\mathfrak{p}_z \cap \tau\mathfrak{p}_z = \mathfrak{p}_z^r$.
7. $\tau\mathfrak{p}$ *is* Ad (G)*-conjugate to the parabolic subalgebra* $\mathfrak{p}^- = \mathfrak{p}^r + \mathfrak{p}^n$ *opposite to* $\mathfrak{p}$.
In particular, if one open G_0*-orbit on* Z *is measurable, then they all are measurable.*

Note that Condition 4 of Proposition 11.1 is automatic if the Cartan subalgebra $\mathfrak{h}_0$, relative to which $\mathfrak{p}_z = \mathfrak{p}_\Phi$, is the Lie algebra of a compact Cartan subgroup of G_0, for in that case $\tau\alpha = -\alpha$ for every $\alpha \in \Sigma(\mathfrak{g}, \mathfrak{h})$. In particular, if G_0 has discrete series representations, so that by a result of Harish–Chandra it has a compact Cartan subgroup, then every open G_0–orbit on Z is measurable.

Condition 4 is also automatic if P is a Borel subgroup of G, and more generally Condition 7 provides a quick test for measurability.

Now suppose that G_0 has a compact Cartan subgroup $T_0 \subset K_0$. Let $Z = G/P$ be a complex flag manifold, let $z \in Z$, set $D = G_0(z)$, and suppose that

(11.2) $\quad D$ is open in Z and G_0 has compact isotropy subgroup U_0 at z.

Passing to a conjugate, equivalently moving z within D, we may suppose $T_0 \subset U_0$.

Let $\mu \in \widehat{U_0}$, let E_μ denote the representation space, and let $\mathbb{E}_\mu \to D \cong G_0/U_0$ denote the associated holomorphic homogeneous vector bundle. Then μ is finite dimensional and is constructed as follows. First, $U_0 \cap G_0^0$ is the identity component U_0^0, and $U_0 = Z_{G_0}(G_0^0)U_0^0$. Second there are irreducible representations $[\chi] \in \widehat{Z_{G_0}(G_0^0)}$ and $[\mu^0] \in \widehat{U_0^0}$ that agree on Z_{G_0} such that $[\mu] = [\chi \otimes \mu^0]$.

Let $\beta - \rho_\mathfrak{u}$ denote the highest weight of μ^0, corresponding to infinitesimal character β, and suppose that

$$\lambda = \beta - \rho_\mathfrak{u} + \rho_\mathfrak{g} \text{ is regular.} \tag{11.3}$$

Then G_0 has a discrete series representation $\pi_{\chi,\lambda}$, whose infinitesimal character has Harish–Chandra parameter λ.

Since μ is unitary, the bundle $\mathbb{E}_\mu \to D$ has a G_0–invariant hermitian metric. Essentially as in the compact case, we have the spaces
(11.4)
$A_0^{(p,q)}(D;\mathbb{E}_\mu) : C^\infty$ compactly supported $\mathbb{E}_\mu$–valued (p,q)–forms on D,

and the Kodaira–Hodge orthocomplementation operators

$$\begin{aligned} \sharp &: A_0^{(p,q)}(D;\mathbb{E}_\mu) \to A_0^{(n-p,n-q)}(D;\mathbb{E}_\mu^*) \\ \text{and } \tilde{\sharp} &: A_0^{(n-p,n-q)}(D;\mathbb{E}_\mu^*) \to A_0^{(p,q)}(D;\mathbb{E}_\mu) \end{aligned} \tag{11.5}$$

where $n = \dim_{\mathbb{C}} D$. Thus we have a positive definite inner product on $A_0^{(p,q)}(D;\mathbb{E}_\mu)$ given by

$$(11.6) \qquad \langle F_1, F_2\rangle_D = \int_{G_0} \langle F_1, F_2\rangle_{gU_0} dg = \int_D F_1 \overline{\wedge} \sharp F_2$$

and thus

$$(11.7) \quad L_2^{(p,q)}(D;\mathbb{E}_\mu) : \text{ Hilbert space completion of } \left(A_0^{(p,q)}(D;\mathbb{E}_\mu), \langle\cdot,\cdot\rangle_D\right).$$

Let $\square$ denote the Kodaira–Hodge–Laplace operator $\overline{\partial}\,\overline{\partial}^* + \overline{\partial}^*\overline{\partial}$ of $\mathbb{E}_\mu$. Then $\square$ is a hermitian–symmetric elliptic operator on $L_2^{(0,q)}(D;\mathbb{E}_\mu)$ with domain $A_0^{(p,q)}(D;\mathbb{E}_\mu)$, and a result of Andreotti and Vesentini allows one to conclude that its closure $\widetilde{\square}$ is self–adjoint. Accordingly, we have the Hilbert spaces

$$(11.8) \qquad \mathcal{H}^{(p,q)}(D;\mathbb{E}_\mu) = \{\omega \in \text{ Domain}\,(\widetilde{\square}) \mid \widetilde{\square}(\omega) = 0\}$$

of square integrable harmonic $\mathbb{E}_\mu$–valued $(0,q)$–forms on D. The natural actions of G_0 on those spaces are unitary representations.

We write $\mathcal{H}^q(D;\mathbb{E}_\mu)$ for $\mathcal{H}^{(0,q)}(D;\mathbb{E}_\mu)$ and we write π_μ^q for the unitary representation of G_0 on $\mathcal{H}^q(D;\mathbb{E}_\mu)$.

11.9. Theorem. *Let $[\mu] = [\chi \otimes \mu^0] \in \widehat{U_0}$ where μ^0 has highest weight $\beta - \rho_{\mathfrak{u}}$ and thus has infinitesimal character β. If $\lambda = \beta - \rho_{\mathfrak{u}} + \rho_{\mathfrak{g}}$* (compare (11.3)) *is $\Sigma(\mathfrak{g},\mathfrak{t})$–singular then every $\mathcal{H}^q(D;\mathbb{E}_\mu) = 0$. Now suppose that $\lambda = \beta - \rho_{\mathfrak{u}} + \rho_{\mathfrak{g}}$ is $\Sigma(\mathfrak{g},\mathfrak{t})$–regular and define*

$$(11.10) \qquad \begin{aligned} q_{\mathfrak{u}}(\lambda) = &\,|\{\alpha \in \Sigma^+(\mathfrak{k},\mathfrak{t}) \setminus \Sigma^+(\mathfrak{u},\mathfrak{t}) \mid \langle\lambda,\alpha\rangle < 0\}| \\ &+ |\{\beta \in \Sigma^+(\mathfrak{g},\mathfrak{t}) \setminus \Sigma^+(\mathfrak{k},\mathfrak{t}) \mid \langle\lambda,\beta\rangle > 0\}|. \end{aligned}$$

Then $\mathcal{H}^q(D;\mathbb{E}_\mu) = 0$ for $q \neq q_{\mathfrak{u}}(\lambda)$, and G_0 acts irreducibly on $\mathcal{H}^{q_{\mathfrak{u}}(\lambda)}(D;\mathbb{E}_\mu)$ by the discrete series representation $\pi_{\chi,\lambda}$ of infinitesimal character λ.

An interesting variation on this result realizes the discrete series on spaces of L_2 bundle-valued harmonic spinors.

Indication of Proof. The proof of Theorem 11.9 has three major components. The first is the alternating sum formula

$$(11.11a) \qquad \sum_{q \geqq 0} (-1)^q \Theta_{{}^0\pi_\mu^q} = (-1)^{|\Sigma^+| + q_{\mathfrak{u}}(\lambda)} \Theta_{\pi_{\chi,\lambda}}$$

where ${}^0\pi_\mu^q$ is the discrete series component of the natural unitary representation π_μ^q of G_0 on $\mathcal{H}^q(D;\mathbb{E}_\mu)$, and $\Theta_{{}^0\pi_\mu^q}$ is its distribution character. It is implicit here that $\Theta_{{}^0\pi^q_\mu}$ exists. The second major component of the proof is the consequence

(11.11b)
$$\pi^q_\mu = {}^0\pi^q_\mu$$

of the Plancherel formula (10.7). The third major component of the proof is the vanishing theorem

(11.11c)
$$\mathcal{H}^q(D;\mathbb{E}_\mu) = 0 \text{ for } q \neq q_u(\lambda).$$

To simplify the argument one should carry out three reductions. First, one may assume that $G_0 = G_0^\dagger$, for the discrete series representations of G_0 are induced from those of $G_0^\dagger$ and one has the character relation (8.7). Second, one may assume that G_0 is connected, $G_0 = G_0^0$, for the discrete series characters of $G_0^\dagger$ are just products $\Theta_{\pi_{\chi,\lambda}}(zx) = \operatorname{trace}\chi(z)\Theta_{\pi^0_\lambda}(x)$, as in the second equation of (8.6a). Third, one may assume that P is a Borel subgroup of G, so $U_0 = T_0$, by using the Borel–Weil Theorem 7.15 on the fibres of $G_0/T_0 \to G_0/U_0$ to make the Leray spectral sequence explicit.

We will assume that G_0 is connected and $U_0 = T_0$ for the discussion of formulae (11.11).

We indicate the argument for the alternating sum formula (11.11a). Use the Plancherel formula to express

(11.12)
$$L_2^{(0,q)}(D;\mathbb{E}_\mu) = \int_{\widehat{G_0}} H_\pi \widehat{\otimes} \left(H_\pi^* \otimes \Lambda^q \mathfrak{n}^* \otimes E_\mu\right)^{U_0} dm(\pi)$$

where m is Plancherel measure on $\widehat{G_0}$. Here $\mathfrak{n} = \sum_{\alpha\in\Sigma^+} \mathfrak{g}_{-\alpha}$ is the nilradical $\mathfrak{p}^{-n} = \mathfrak{b}^{-n}$ as in (1.1), so $\mathfrak{n}$ represents the antiholomorphic tangent space and $\mathfrak{n}^*$ represents the fibre for the bundle of $(0,1)$–forms. Also, $\widehat{\otimes}$ denotes projective tensor product and the integral is a direct integral of Hilbert spaces. One now writes out the formulae for $\overline{\partial}$ and $\overline{\partial}^*$ and pushes them inside the integral of (11.12). They commute with the left action of G_0 and so act on the second projective tensor product factor $\left(H_\pi^* \otimes \Lambda^q \mathfrak{n}^* \otimes E_\mu\right)^{U_0}$. There they ignore the E_μ factor and act on $H_\pi^* \otimes \Lambda^q \mathfrak{n}^*$. The action on $H_\pi^* \otimes \Lambda^q \mathfrak{n}^*$ produces a certain finite dimensional Lie algebra cohomology $H^q(\pi)$ as follows.

Let H_π^0 denote the space of K_0–finite vectors in H_π. Then $\pi \otimes \mathrm{ad}^*$ gives a representation of $\mathfrak{t} = \mathfrak{p}^r$ on $H_\pi^0 \otimes \Lambda^\bullet \mathfrak{n}^*$. If $\{y_i\}$ is a basis of $\mathfrak{n}$ and $\{\omega^i\}$ is the dual basis of $\mathfrak{n}^*$ then the coboundary, for Lie algebra cohomology of $\mathfrak{t}$ relative to its representation on H_π^0, is

(11.13a)
$$\delta = \sum \left(d\pi(y_i) \otimes e(\omega^i) + \tfrac{1}{2} \otimes e(\omega^i)\, \mathrm{ad}^*(y_i)\right) : H_\pi^0 \otimes \Lambda^q \mathfrak{n}^* \to H_\pi^0 \otimes \Lambda^{q+1} \mathfrak{n}^*$$

where $e(\cdot)$ denotes exterior product. Let $i(\cdot)$ denote the dual operation, interior product. Then δ has adjoint

(11.13b)
$$\delta^* = \sum \left(-d\pi(\tau(y_i)) \otimes i(\omega^i) + \tfrac{1}{2} \otimes \mathrm{ad}^*(y_i)^* i(\omega^i)\right) : H_\pi^0 \otimes \Lambda^{q+1} \mathfrak{n}^* \to H_\pi^0 \otimes \Lambda^q \mathfrak{n}^* .$$

Then $\delta+\delta^*$ is essentially self–adjoint on $H^0_\pi \otimes \wedge^\bullet \mathfrak{n}^*$ and has finite dimensional kernel $H^q(\pi)$ on $H^0_\pi \otimes \wedge^q \mathfrak{n}^*$. One now combines (11.12) and (11.13) to obtain

$$\mathcal{H}^q_2(D;\mathbb{E}_\mu) = \int_{\widehat{G_0}} H_\pi \widehat{\otimes} \big(H^q(\pi^*) \otimes E_\mu\big)^{T_0} dm(\pi). \tag{11.14a}$$

In particular, the discrete series part ${}^0\pi^q_\mu$ of π^q_μ is given by

$${}^0\pi^q_\mu = \sum_{\pi \in \widehat{G_{0,d}}} \dim \big(H^q(\pi^*) \otimes E_\mu\big)^{T_0} \pi. \tag{11.14b}$$

If $f \in C^\infty(K_0)$ then $\pi|_{K_0}(f) = \int_{K_0} f(k)\pi(k)dk$ is a trace class operator on H_π, $f \mapsto T_\pi(f) = \text{trace}\, \pi|_{K_0}(f)$ is a distribution on K_0, and $T_\pi|_{K_0 \cap G'_0} = \Theta_\pi|_{K_0 \cap G'_0}$. These are delicate results of Harish–Chandra. The connection with (11.11a) and (11.14b) is that

$$f_\pi = \sum_{q \geqq 0} (-1)^q \big(\text{character of } T_0 = U_0 \text{ on } H^q(\pi)\big) \tag{11.15a}$$

satisfies

$$f_\pi|_{T_0 \cap G'_0} = (-1)^{|\Sigma^+|}\, \Delta_{G_0,T_0}\, e^{\rho_{\mathfrak{g}}}\, T_\pi|_{T_0 \cap G_0}\,. \tag{11.15b}$$

Now let $F_\lambda = \sum_{q \geqq 0} (-1)^q \Theta_{{}^0\pi^q_\lambda}$ and compute

$$\begin{aligned} F_\lambda &= \sum_{q \geqq 0} (-1)^q \sum_{\pi \in \widehat{G_{0,d}}} \dim\big(H^q(\pi^*) \otimes E_\mu)^{T_0}\big)\, \Theta_\pi \\ &= \sum_{\pi \in \widehat{G_{0,d}}} \left(\sum_{q \geqq 0} (-1)^q \dim\big(H^q(\pi^*) \otimes E_\mu\big)^{T_0} \right) \Theta_\pi \qquad (11.16a) \\ &= \sum_{\pi \in \widehat{G_{0,d}}} \big(\text{coefficient of } e^{-\lambda+\rho_{\mathfrak{g}}} \text{ in } f_{\pi^*}\big) \Theta_\pi\,. \end{aligned}$$

But

$$f_{\pi^*_\nu} = (-1)^{|\Sigma^+| + q_\iota(\nu)} \sum_{w \in W(G_0,T_0)} e^{-(w(\nu) - \rho_{\mathfrak{g}})}\,,$$

in which the coefficient of $e^{-\lambda}$ is equal to 0 if $\lambda \notin W(G_0,T_0)(\nu)$, is equal to $(-1)^{|\Sigma^+|+q_\iota(\lambda)}$ if $\lambda \in W(G_0,Y_0)(\nu)$. Thus we have

$$F_\lambda = (-1)^{|\Sigma^+| + q_\iota(\lambda)} \Theta_{\pi_\lambda}\,. \tag{11.16b}$$

This proves the alternating sum formula (11.11a).

The Plancherel Formula (10.7) implies that

$$\{\pi \in \widehat{G_0} \setminus \widehat{G_{0,d}} \mid T_{\pi^*}|_{K_0 \cap G'_0} = \Theta_{\pi^*}|_{K_0 \cap G'_0} \neq 0\} \tag{11.17}$$

has Plancherel measure 0. It follows from (11.14) and (11.17) that ${}^0\pi^q_\mu = \pi^q_\mu$, and (11.11b) follows.

References for §11.

- W. Schmid, "On a conjecture of Langlands", Annals of Math. **93** (1971), 1–42.
- W. Schmid, "L^2–cohomology and the discrete series", Annals of Math. **103** (1976), 375–394.
- W. Schmid and J. A. Wolf, "Geometric quantization and derived functor modules for semisimple Lie groups", J. Funct. Analysis **90** (1990), 48–112.
- J. A. Wolf, "The action of a real semisimple Lie group on a complex flag manifold, II: Unitary representations on partially holomorphic cohomology spaces", Memoirs. Amer. Math. Soc., No. 138, 1974.
- J. A. Wolf, "Partially harmonic spinors and representations of reductive Lie groups", J. Funct. Analysis **15** (1974), 117–154.
- J. A. Wolf, "Geometric realizations of discrete series representations in a nonconvex holomorphic setting", Bull. Soc. Math. de Belgique **42** (1990), 797–812.

§12. Partially Measurable Orbits and Tempered Series.

Choose a Cartan subgroup $H_0 \subset G_0$. We are going to realize the H_0–series representations of G_0 in a way analogous to the way we realized the principle series in §7, with Theorem 11.9 in place of the Bott–Borel–Weil Theorem 7.15.

Let θ be the Cartan involution of G_0 that stabilizes H_0, split $H_0 = T_0 \times A_0$ and let $Z_{G_0}(A_0) = M_0 \times A_0$ as before. Fix a positive root system $\Sigma^+ = \Sigma^+(\mathfrak{g}, \mathfrak{h})$ defined by positive root systems $\Sigma^+(\mathfrak{m}, \mathfrak{t})$ and $\Sigma^+(\mathfrak{g}_0, \mathfrak{a}_0)$ as in (7.3). Let $P_0 = M_0A_0N_0$ be the corresponding cuspidal parabolic subgroup of G_0 associated to H_0.

Following the idea of the geometric realization of the principal series, we fix a set $\Phi \subset \Psi_\mathfrak{m}$ where $\Psi_\mathfrak{m}$ is the simple root system for $\Sigma^+(\mathfrak{m}, \mathfrak{t})$. Then as in (7.5) we have

$$\begin{aligned} &\mathfrak{z}_\Phi = \{\xi \in \mathfrak{t} \mid \phi(\xi) = 0 \ \forall \phi \in \Phi\} \text{ and its real form } \mathfrak{z}_{\Phi,0} = \mathfrak{m}_0 \cap \mathfrak{r}_\Phi \mathfrak{z}_\Phi ,\\ &U_\Phi = Z_M(\mathfrak{z}_\Phi), U_{\Phi,0} = M_0 \cap U_\Phi, \text{ and Lie algebras } \mathfrak{u}_\Phi \text{ and } \mathfrak{u}_{\Phi,0} ,\\ (12.1) \quad &\mathfrak{r}_\Phi = \mathfrak{u}_\Phi + \sum_{\gamma \in \Sigma^+(\mathfrak{m},\mathfrak{t})} \mathfrak{m}_{-\gamma}, \text{ parabolic subalgebra of } \mathfrak{m} ,\\ &R_\Phi = N_M(\mathfrak{r}_\Phi), \text{ corresponding parabolic subgroup of } M , \text{ and}\\ &S_\Phi = M/R_\Phi, \text{ associated complex flag manifold.} \end{aligned}$$

Let r_Φ denote the base point, $r_\Phi = 1R_\Phi \in R_\Phi$. Since T_0 is a compact Cartan subgroup of M_0 contained in $U_{\Phi,0}$,

$$(12.2a) \qquad D_\Phi = M_0(r_\Phi) \subset S_\Phi \text{ is a measurable open } M_0\text{–orbit on } R_\Phi$$

We now assume that

$$(12.2b) \quad U_{\Phi,0} \text{ is compact, so the considerations of §11 apply to } D_\Phi \subset S_\Phi .$$

Fix $[\mu] = [\chi \otimes \mu_\beta^0] \in \widehat{U_{\Phi,0}}$ as before. Given $\sigma \in \mathfrak{a}_0^*$ we will use the Theorem 11.9 to find the H_0–series representation $\pi_{\chi\otimes\eta_\nu^0,\sigma}$ on a cohomology space related to a particular orbit in the complex flag manifold $Z_\Phi = G/P_\Phi$. Here as before, the simple root system $\Psi_{\mathfrak{m}} \subset \Psi$ by the coherence in our choice of $\Sigma^+(\mathfrak{g},\mathfrak{h})$, so $\Phi \subset \Psi$ and Φ defines a parabolic subgroup $P_\Phi \subset G$.

Let $z_\Phi = 1P_\Phi \in G/P_\Phi = Z_\Phi$. As $A_0N_0 \subset G_0 \cap P_\Phi$ we have $G_0 \cap P_\Phi = U_{\Phi,0}A_0N_0$. Thus $Y_\Phi = G_0(z_\Phi)$ is a G_0–orbit on Z_Φ, and D_Φ sits in Y_Φ as the orbit $M_0(z_\Phi)$. Here note that $P_0 = M_0A_0N_0 = \{g \in G_0 \mid gD_\Phi = D_\Phi\}$.

12.3. Lemma. *The map $Y_\Phi \to G_0/P_0$, given by $g(z_\Phi) \mapsto gP_0$, defines a G_0–equivariant fibre bundle with structure group M_0 and whose fibres gD_Φ are the maximal complex analytic submanifolds of Y_Φ.*

The data (μ,σ) defines a representation $\gamma_{\mu,\sigma}$ of $U_{\Phi,0}A_0N_0$ by

(12.4a) $$\gamma_{\mu,\sigma}(uan) = e^{(\rho_\mathfrak{g}+i\sigma)(\log a)}\mu(u) \quad \text{where } \rho_\mathfrak{g} = \tfrac{1}{2}\textstyle\sum_{\alpha\in\Sigma^+} \mathfrak{g}_\alpha .$$

This defines a G_0–homogeneous vector bundle

(12.4b) $$\mathbb{E}_{\mu,\sigma} \to G_0/U_{\Phi,0}A_0N_0 = Y_\Phi \text{ such that } \mathbb{E}_{\mu,\sigma}|_{D_\Phi} = \mathbb{E}_\mu .$$

Each $\mathbb{E}_{\mu,\sigma}|_{gD_\Phi}$ is an $\mathrm{Ad}(g)P_0$–homogeneous holomorphic vector bundle.

Since $[\mu]$ is unitary and K_0 acts transitively on G_0/P_0 we have a K_0–invariant hermitian metric on $\mathbb{E}_{\mu,\sigma}$. We will use it without explicit reference.

Consider the subbundle of the complexified tangent bundle to Y_Φ,

(12.5a) $$\begin{aligned} &\mathbb{T} \to Y_\Phi \text{ defined by:}\\ &\qquad \mathbb{T}|_{gD_\Phi} \to gD_\Phi \text{ is the holomorphic tangent bundle of } gD_\Phi . \end{aligned}$$

It defines

(12.5b) $$\begin{aligned} &\mathbb{E}^{p,q}_{\mu,\sigma} = \mathbb{E}_{\mu,\sigma} \otimes \Lambda^p(\mathbb{T}^*) \otimes \Lambda^q(\overline{\mathbb{T}}^*) \to D_\Phi ,\\ &A_0^{p,q}(Y_\Phi;\mathbb{E}_{\mu,\sigma}) : C^\infty \text{ compactly supported sections of } \mathbb{E}^{p,q}_{\mu,\sigma} \to Y_\Phi ,\\ &\mathcal{O}(\mathbb{E}_{\mu,\sigma}) : \text{ sheaf of germs of } C^\infty \text{ sections of } \mathbb{E}_{\mu,\sigma} \to Y_\Phi\\ &\qquad\qquad \text{holomorphic over every } gD_\Phi . \end{aligned}$$

$A^{p,q}(Y_\Phi;\mathbb{E}_{\mu,\sigma})$ is the space of $\mathbb{E}_{\mu,\sigma}$–valued partially (p,q)–forms on Y_Φ, and $A_0^{p,q}(Y_\Phi;\mathbb{E}_{\mu,\sigma})$ is the subspace of compactly supported forms.

The fibre E_μ of $\mathbb{E}_\mu \to D_\Phi$ has a positive definite $U_{\Phi,0}$–invariant hermitian inner product because μ is unitary; we translate this around by K_0 to obtain a K_0–invariant hermitian structure on the vector bundle $\mathbb{E}^{p,q}_{\mu,\sigma} \to Y_\Phi$. Similarly $\mathbb{T} \to Y_\Phi$ carries a K_0–invariant hermitian metric. Using these hermitian metrics we have K_0–invariant Hodge–Kodaira orthocomplementation operators

(12.6) $$\begin{aligned} &\sharp : A_0^{p,q}(Y_\Phi;\mathbb{E}_{\mu,\sigma}) \to A_0^{n-p,n-q}(Y_\Phi;\mathbb{E}^*_{\mu,\sigma})\\ &\tilde{\sharp} : A_0^{n-p,n-q}(Y_\Phi;\mathbb{E}^*_{\mu,\sigma}) \to A_0^{p,q}(Y_\Phi;\mathbb{E}_{\mu,\sigma}) \end{aligned}$$

where $n = \dim_{\mathbb{C}} D_\Phi$. The global G_0–invariant hermitian inner product on $A_0^{p,q}(Y_\Phi;\mathbb{E}_{\mu,\sigma})$ is given by taking the M_0–invariant inner product along each fibre of $Y_\Phi \to G_0/P_0$ and integrating over G_0/P_0,

$$(12.7) \qquad \langle F_1, F_2\rangle_{Y_\Phi} = \int_{K_0/(K_0\cap M_0)} \left(\int_{kD_\Phi} F_1 \bar{\wedge} \sharp F_2\right) d(k(K_0\cap M_0)).$$

where $\bar{\wedge}$ means exterior product followed by contraction of E_μ against E_μ^*.

The $\bar{\partial}$ operator of Z_Φ induces the $\bar{\partial}$ operators on each of the gD_Φ, so they fit together to give us an operator

$$(12.8a) \qquad \bar{\partial} : A_0^{p,q}(Y_\Phi;\mathbb{E}_{\mu,\sigma}) \to A_0^{p,q+1}(Y_\Phi;\mathbb{E}_{\mu,\sigma})$$

that has formal adjoint

$$(12.8b) \qquad \bar{\partial}^* : A_0^{p,q+1}(Y_\Phi;\mathbb{E}_{\mu,\sigma}) \to A_0^{p,q}(Y_\Phi;\mathbb{E}_{\mu,\sigma}) \text{ given by } \bar{\partial}^* = -\tilde{\sharp}\bar{\partial}\sharp\,.$$

This in turn defines a sub-elliptic operator, the "partial Kodaira–Hodge–Laplace operator"

$$(12.8c) \qquad \Box = \bar{\partial}\bar{\partial}^* + \bar{\partial}^*\bar{\partial} : A_0^{p,q}(Y_\Phi;\mathbb{E}_{\mu,\sigma}) \to A_0^{p,q}(Y_\Phi;\mathbb{E}_{\mu,\sigma}).$$

$A_0^{p,q}(Y_\Phi;\mathbb{E}_{\mu,\sigma})$ is a pre Hilbert space with the global inner product (12.7). Denote

$$(12.9) \qquad L_2^{p,q}(Y_\Phi;\mathbb{E}_{\mu,\sigma}) : \text{Hilbert space completion of } A_0^{p,q}(Y_\Phi;\mathbb{E}_{\mu,\sigma}).$$

Apply Andreotti–Vesentini along each gD_Φ to see that the closure of $\tilde{\Box}$ of $\Box$, as a densely defined operator on $L_2^{p,q}(Y_\Phi;\mathbb{E}_{\mu,\sigma})$ from the domain $A_0^{p,q}(Y_\Phi;\mathbb{E}_{\mu,\sigma})$, is essentially self–adjoint. Its kernel

$$(12.10) \qquad \mathcal{H}_2^{p,q}(Y_\Phi;\mathbb{E}_{\mu,\sigma}) = \{\omega \in \text{Domain}(\tilde{\Box}) \mid \tilde{\Box}\omega = 0\}$$

is the space of **square integrable partially harmonic** (p,q)–forms on Y_Φ with values in $\mathbb{E}_{\mu,\sigma}$.

The factor e^{ρ_0} in the representation $\gamma_{\mu,\sigma}$ that defines $\mathbb{E}_{\mu,\sigma}$ insures that the global inner product on $A_0^{p,q}(Y_\Phi;\mathbb{E}_{\mu,\sigma})$ is invariant under the action of G_0. The other ingredients in the construction of $\mathcal{H}_2^{p,q}(Y_\Phi;\mathbb{E}_{\mu,\sigma})$ are invariant as well, so G_0 acts naturally on $\mathcal{H}_2^{p,q}(Y_\Phi;\mathbb{E}_{\mu,\sigma})$ by isometries. This action is a unitary representation of G_0.

Essentially as before, we write $\mathcal{H}_2^{q}(Y_\Phi;\mathbb{E}_{\mu,\sigma})$ for $\mathcal{H}_2^{0,q}(Y_\Phi;\mathbb{E}_{\mu,\sigma})$, because those are the only harmonic spaces that we will use, and because $\mathcal{H}_2^{q}(Y_\Phi;\mathbb{E}_{\mu,\sigma})$ is closely related to the sheaf cohomology $H^q(Y_\Phi, \mathcal{O}(E_{\mu,\sigma}))$. The relation, which we will see later, is that they have the same underlying Harish–Chandra module.

We can now combine Theorem 11.9 with the definition ((9.2) and (9.3)) of the H_0–series, obtaining

12.11. Theorem. *Let* $[\mu] = [\chi \otimes \mu^0_\beta] \in \widehat{U_{\Phi,0}}$ *where* μ^0 *has highest weight* $\beta - \rho_{\mathfrak{u}}$ *and thus has infinitesimal character* β. *Let*

$$\nu = \beta - \rho_{\mathfrak{u}_\Phi} + \rho_{\mathfrak{m}}, \tag{12.12}$$

suppose $\sigma \in \mathfrak{a}_0^*$, *and fix an integer* $q \geqq 0$.
1. *If* $\langle \nu, \alpha \rangle = 0$ *for some* $\alpha \in \Sigma(\mathfrak{m}, \mathfrak{t})$ *then* $\mathcal{H}_2^q(Y_\Phi; \mathbb{E}_{\mu,\sigma}) = 0$.
2. *If* $\langle \nu, \alpha \rangle \neq 0$ *for all* $\alpha \in \Sigma(\mathfrak{m}, \mathfrak{t})$, *define*

$$\begin{aligned} q_{\mathfrak{u}_\Phi}(\nu) &= |\{\alpha \in \Sigma^+((\mathfrak{k} \cap \mathfrak{m}), \mathfrak{t}) \setminus \Sigma^+(\mathfrak{u}_\Phi, \mathfrak{t}) \mid \langle \nu, \alpha \rangle < 0\}| \\ &\quad + |\{\beta \in \Sigma^+(\mathfrak{m}, \mathfrak{t}) \setminus \Sigma^+((\mathfrak{k} \cap \mathfrak{m}), \mathfrak{t}) \mid \langle \nu, \beta \rangle > 0\}|. \end{aligned} \tag{12.13}$$

Then $\mathcal{H}^q(Y_\Phi; \mathbb{E}_{\mu,\sigma}) = 0$ *for* $q \neq q_{\mathfrak{u}_\Phi}(\nu)$, *and the action of* G_0 *on* $\mathcal{H}^{q_{\mathfrak{u}_\Phi}(\nu)}(Y_\Phi; \mathbb{E}_{\mu,\sigma})$ *is the* H_0*-series representation* $\pi_{\chi,\nu,\sigma}$ *of infinitesimal character* $\nu + i\sigma$.

A variation on this theorem realizes the tempered series on spaces of L_2 bundle-valued partially harmonic spinors.

References for §12.

- J. A. Wolf, "The action of a real semisimple Lie group on a complex flag manifold, II: Unitary representations on partially holomorphic cohomology spaces", Memoirs. Amer. Math. Soc., No. 138, 1974.
- J. A. Wolf, "Partially harmonic spinors and representations of reductive Lie groups", J. Funct. Analysis **15** (1974), 117–154.

PART 5. THE LINEAR CYCLE SPACE.

In this part we indicate the geometric setting for double fibration transforms, one of the current approaches to geometric construction of non-tempered representations.

§13. EXHAUSTION FUNCTIONS ON MEASURABLE OPEN ORBITS.

Bounded symmetric domains $D \subset \mathbb{C}^n$ are convex, and thus Stein, so cohomologies $H^k(D; \mathcal{F}) = 0$ for $k > 0$ whenever $\mathcal{F} \to D$ is a coherent analytic sheaf. This is a key point in dealing with holomorphic discrete series representations. More generally, for general discrete series representations and their analytic continuations, one has

13.1. Theorem. *Let* $Z = G/P$ *be a complex flag manifold,* G *semisimple and simply connected, and let* G_0 *be a real form of* G. *Let* $D = G_0(z) \subset Z = G/P$ *be a measurable open orbit. Let* $Y = K_0(z)$, *maximal compact subvariety of* D, *and let* $s = \dim_{\mathbb{C}} Y$. *Then* D *is* $(s+1)$*-complete in the sense of Andreotti–Grauert. In particular, if* $\mathcal{F} \to D$ *is a coherent analytic sheaf then* $H^k(D; \mathcal{F}) = 0$ *for* $k > s$.

Indication of Proof. Let $\mathbb{K}_Z \to Z$ and $\mathbb{K}_D = K_Z|_D \to D$ denote the canonical line bundles. Their dual bundles

$$\text{(13.2)} \qquad \mathbb{L}_Z = \mathbb{K}_Z^* \to Z \quad \text{and } \mathbb{L}_D = \mathbb{K}_D^* \to D$$

are the homogeneous holomorphic line bundles over Z associated to the character

$$\text{(13.3)} \qquad e^\lambda : P_z \to \mathbb{C} \text{ defined by } e^\lambda(p) = \det \mathrm{Ad}(p)|_{\mathfrak{p}_z^n}\,.$$

Write $D = G_0/V_0$ where V_0 is the real form $G_0 \cap P_z$ of P_z^r. Write V for the complexification P_z^r of V_0, $\rho_{G/V}$ for half the sum of the roots that occur in $\mathfrak{p}_z^n$, and $\lambda = 2\rho_{G/V}$. If $\alpha \in \Sigma(\mathfrak{g}, \mathfrak{h})$ then (i) $\langle \alpha, \lambda \rangle = 0$ and $\alpha \in \Phi^r$, or (ii) $\langle \alpha, \lambda \rangle > 0$ and $\alpha \in \Phi^n$, or (iii) $\langle \alpha, \lambda \rangle < 0$ and $\alpha \in \Phi^{-n}$. Now $\tau\lambda = -\lambda$. Decompose $\mathfrak{g}_0 = \mathfrak{k}_0 + \mathfrak{s}_0$ under the Cartan involution with fixed point set $\mathfrak{k}_0$, thus decomposing the Cartan subalgebra $\mathfrak{h}_0 \subset \mathfrak{g}_0 \cap \mathfrak{p}_z$ as $\mathfrak{h}_0 = \mathfrak{t}_0 + \mathfrak{a}_0$ with $\mathfrak{t}_0 = \mathfrak{h}_0 \cap \mathfrak{k}_0$ and $\mathfrak{a}_0 = \mathfrak{h}_0 \cap \mathfrak{s}_0$. Then $\lambda(\mathfrak{a}_0) = 0$.

View $D = G_0/V_0$ and $Z = G_u/V_0$ where G_u is the analytic subgroup of G for the compact real form $\mathfrak{g}_u = \mathfrak{k}_0 + \sqrt{-1}\,\mathfrak{s}_0$. Then e^λ is a unitary character on V_0 so

$$\text{(13.4)} \qquad \begin{array}{l} \mathbb{L}_Z \to Z = G_u/V_0 \text{ has a } G_u\text{–invariant hermitian metric } h_u\,, \\ \mathbb{L}_D \to D = G_0/V_0 \text{ has a } G_0\text{–invariant hermitian metric } h_0\,. \end{array}$$

We now have enough information to carry out a computation that results in

13.5. Lemma. *The hermitian form $\sqrt{-1}\,\partial\bar\partial h_u$ on the holomorphic tangent bundle of Z is negative definite. The hermitian form $\sqrt{-1}\,\partial\bar\partial h_0$ on the holomorphic tangent bundle of D has signature $n - 2s$ where $n = \dim_{\mathbb{C}} D$.*

13.6. Corollary. *Define $\phi : D \to \mathbb{R}$ by $\phi = \log(h_0/h_u)$. Then the Levi form $\mathcal{L}(\phi)$ has at least $n - s$ positive eigenvalues at every point of D.*

The next point is to show that ϕ is an exhaustion function for D, in other words that

$$\{z \in D \mid \phi(z) \leqq c\} \text{ is compact for every } c \in \mathbb{R}.$$

It suffices to show that $e^{-\phi}$ has a continuous extension from D to the compact manifold Z that vanishes on the topological boundary $\mathrm{bd}(D)$ of D in Z. For that, choose a G_u–invariant metric h_u^* on $\mathbb{L}_Z^* = \mathbb{K}_Z$ normalized by $h_u h_u^* = 1$ on Z, and a G_0–invariant metric h_0^* on $\mathbb{L}_D^* = \mathbb{K}_D$ normalized by $h_0 h_0^* = 1$ on D. Then $e^{-\phi} = h_0^*/h_u^*$. So it suffices to show that h_0^*/h_u^* has a continuous extension from D to Z that vanishes on $\mathrm{bd}(D)$.

The holomorphic cotangent bundle $\mathbb{T}_Z^* \to Z$ has fibre $\mathrm{Ad}(g)(\mathfrak{p}_z^n)^* = \mathrm{Ad}(g)(\mathfrak{p}_z^{-n})$ at $g(z)$. Thus its G_u–invariant hermitian metric is given on the fibre $\mathrm{Ad}(g)(\mathfrak{p}_z^{-n})$ at $g(z)$ by $F_u(\xi, \eta) = -\langle \xi, \tau\theta\eta \rangle$ where $\langle , \rangle$ is the Killing form. Similarly the G_0–invariant indefinite–hermitian metric on $\mathbb{T}_D^* \to D$ is given on the fibre $\mathrm{Ad}(g)(\mathfrak{p}_z^{-n})$ at $g(z)$ by $F_0(\xi,\eta) = -\langle\xi,\tau\eta\rangle$. But $\mathbb{K}_Z = \det \mathbb{T}_Z^*$ and $\mathbb{K}_D = \det \mathbb{T}_D^*$, so

$$h_0^*/h_u^* = c \cdot (\text{determinant of } F_0 \text{ with respect to } F_u)$$

for some nonzero real constant c. This extends from D to a C^∞ function on Z given by

$$(13.7) \qquad f(g(z)) = c \cdot (\det F_0|_{\mathrm{Ad}(g)(\mathfrak{p}_z^{-n})} \text{ relative to } \det F_u|_{\mathrm{Ad}(g)(\mathfrak{p}_z^{-n})}).$$

It remains only to show that the function f of (13.7) vanishes on $\mathrm{bd}(D)$. If $g(z) \in \mathrm{bd}(D)$ then $G_0(g(z))$ is not open in Z, so $\mathrm{Ad}(g)(\mathfrak{p}_z) + \tau\mathrm{Ad}(g)(\mathfrak{p}_z) \neq \mathfrak{g}$. Thus there exists an $\alpha \in \Sigma(\mathfrak{g}, \mathrm{Ad}(g)\mathfrak{h})$ such that $\mathfrak{g}_{-\alpha} \not\subset \mathrm{Ad}(g)(\mathfrak{p}_z) + \tau\mathrm{Ad}(g)(\mathfrak{p}_z)$. Now $\mathfrak{g}_\alpha \subset \mathrm{Ad}(g)(\mathfrak{p}_z^{-n}) \cap \tau\mathrm{Ad}(g)(\mathfrak{p}_z^{-n})$. If $\beta \in \Sigma(\mathfrak{g}, \mathrm{Ad}(g)\mathfrak{h})$ with $\mathfrak{g}_\beta \subset \mathrm{Ad}(g)(\mathfrak{p}_z^{-n})$ then $F_0(\mathfrak{g}_\alpha, \mathfrak{g}_\beta) = 0$, so $f(g(z)) = 0$. Thus ϕ is an exhaustion function for D in Z. In view of Corollary 13.6 now D is $(s+1)$–complete. Theorem 13.1 follows.

References for §13.

- A. Andreotti and H. Grauert, "Théorèmes de finitude pour la cohomologie des espaces complèxes", Bull. Soc. Math. France **90** (1962), 193–259.
- W. Schmid, "Homogeneous complex manifolds and representations of semisimple Lie groups", thesis, University of California at Berkeley, 1967.
- W. Schmid and J. A. Wolf, "A vanishing theorem for open orbits on complex flag manifolds", Proc. Amer. Math. Soc. **92** (1984), 461–464.
- J. A. Wolf, "The Stein condition for cycle spaces of open orbits on complex flag manifolds", Annals of Math. **136** (1992), 541–555.

§14. The Exhaustion Function on a General Open Orbit.

We extend Theorem 13.1 to arbitrary open orbits. The result is

14.1. Theorem. *Let $Z = G/P$ be a complex flag manifold, G semisimple and simply connected, and let G_0 be a real form of G. Let $D = G_0(z) \subset Z = G/P$ be an open orbit. Let $Y = K_0(z)$, maximal compact subvariety of D, and let $s = \dim_{\mathbb{C}} Y$. Then D is $(s+1)$–complete in the sense of Andreotti–Grauert. In particular, if $\mathcal{F} \to D$ is a coherent analytic sheaf then $H^k(D;\mathcal{F}) = 0$ for $k > s$.*

The idea of the proof is to show that the arbitrary open orbit $D = G_0(z) \subset Z$ is the base of a canonical holomorphic fibration $\pi_D : \widetilde{D} \to D$ where $\widetilde{D}$ is a measurable open G_0–orbit in a certain flag manifold W that lies over Z. We then take a close look at that fibration and its relation to the maximal compact linear subvarieties.

Fix the open orbit $D = G_0(z) \subset Z = G/P$ and consider the parabolic subalgebra $\mathfrak{p}^+ = \mathfrak{p}^r + \mathfrak{p}^n \subset \mathfrak{g}$ opposite to $\mathfrak{p}_z = \mathfrak{p} = \mathfrak{p}^r + \mathfrak{p}^{-n}$. Denote

$$\mathfrak{q} = \mathfrak{p} \cap \tau\mathfrak{p}^+. \tag{14.2}$$

As D is open, so $\mathfrak{p}^{-n} \cap \tau\mathfrak{p}^{-n} = 0$, $\mathfrak{q}$ is the sum of a nilpotent ideal $\mathfrak{q}^{-n}$ and a reductive subalgebra $\mathfrak{q}^r$ given by

(14.3)

$$\mathfrak{q}^r = \mathfrak{p}^r \cap \tau\mathfrak{p}^r \text{ and } \mathfrak{q}^{-n} = (\mathfrak{p}^r \cap \tau\mathfrak{p}^n) + (\mathfrak{p}^{-n} \cap \tau\mathfrak{p}^r) + (\mathfrak{p}^{-n} \cap \tau\mathfrak{p}^n) = (\mathfrak{p}^r \cap \tau\mathfrak{p}^n) + \mathfrak{p}^{-n}.$$

Then $\mathfrak{q}$ is a parabolic subalgebra of $\mathfrak{g}$, and $\mathfrak{q} \cap \tau\mathfrak{q} = \mathfrak{p}^r \cap \tau\mathfrak{p}^r$, which is reductive. Let Q denote the parabolic subgroup of G corresponding to $\mathfrak{q} \subset \mathfrak{g}$ and let W denote the corresponding flag manifold G/Q. Our choice of P was such that $\mathfrak{p} = \mathfrak{p}_z$ where $z \in Z$ and $D = G_0(z)$ is the open orbit under study. Note that we have implicitly made the corresponding choice on W:

14.4. Lemma. *Define $w \in W$ by $\mathfrak{q} = \mathfrak{q}_w$. Then $\widetilde{D} = G_0(w)$ is a measurable open G_0–orbit on W, and $gw \mapsto gz$ defines a surjective holomorphic projection $\pi_D : \widetilde{D} \to D$. Finally, the following are equivalent:* (i) *D is measurable,* (ii) *$\widetilde{D} = D$,* (iii) *π_D is one-to-one, and* (iv) *$Q = P$.*

The structure of the fibre of $\pi_D : \widetilde{D} \to D$ is given by

14.5. Lemma. *Let $\mathfrak{u} = (\mathfrak{p}^r \cap \tau\mathfrak{p}^{-n}) + (\mathfrak{p}^{-n} \cap \tau\mathfrak{p}^r)$, nilradical of $\mathfrak{p} \cap \tau\mathfrak{p}$, and let U be the corresponding complex analytic subgroup of G. Then U is unipotent, $\mathfrak{u}_0 = \mathfrak{g}_0 \cap \mathfrak{u}$ is a real form of $\mathfrak{u}$, $U_0 = G_0 \cap U$ is a real form of U, $U(w) = U_0(w)$, and $\pi_D : \widetilde{D} \to D$ is a holomorphic fibre bundle with structure group U and affine fibres $\pi_D^{-1}(gz) = gU_0(w)$. If $g \in G_0$ then the holomorphic tangent space to $gU_0(w)$ at $g(w)$ is represented by $\mathrm{Ad}\,(g)(\mathfrak{p} \cap \tau\mathfrak{p}^{-n})$ and the antiholomorphic tangent space is represented by $\mathrm{Ad}\,(g)(\mathfrak{p} \cap \tau\mathfrak{p}^n)$.*

Proof. Here U is the nilradical of $P \cap \tau P$ so $U_0 = G_0 \cap U$ is the nilradical of the isotropy subgroup $G_0 \cap P$ and is a real form of U. Note $\mathfrak{u} = \mathfrak{v} + \tau\mathfrak{v}$ where $\mathfrak{v} = \mathfrak{p}^r \cap \tau\mathfrak{p}^{-n} = \mathfrak{u} \cap \mathfrak{q}^n$, and where $\tau\mathfrak{v} = \mathfrak{u} \cap \mathfrak{q}^{-n}$. Both are subalgebras; $\mathfrak{v}$ represents the holomorphic tangent space of $U_0(w)$ at w and $\tau\mathfrak{v}$ represents the antiholomorphic tangent space. Note $[\mathfrak{v}, \tau\mathfrak{v}] = 0$.

Now $U(w) = V(w) = U_0(w)$ is the fibre over z of $\pi_D : \widetilde{D} \to D$, and $G_0 \cap P$ is the semidirect product of its unipotent radical U_0 and a Levy complement $G_0 \cap Q$. Thus $\pi_D : \widetilde{D} \to D$ satisfies $\pi_D^{-1}(g \cdot (G_0 \cap P)) = gU_0 \cdot (G_0 \cap Q)$; in terms of the complex groups this is the same as $gV \cdot Q$. Now we can express π_D as the quotient of $G_0/(G_0 \cap Q)$ by the action of U_0 on the right. Then the surjective holomorphic map π_D is the projection of a principle U_0–bundle. The assertions follow. □

14.6. Corollary. *Denote $\widetilde{Y} = K_0(w)$. Then $\widetilde{Y} = K(w)$, $\widetilde{Y}$ is a maximal compact complex subvariety of $\widetilde{D}$, and $\pi_D|_{\widetilde{Y}}$ is a biholomorphic diffeomorphism of $\widetilde{Y}$ onto Y.*

Now we push down the exhaustion function $\widetilde{\phi}$ of Corollary 13.6 from the measurable open orbit $\widetilde{D} = G_0(w) \subset W$ to our given open orbit $D = G_0(z) \subset$

Z. We keep the notation h_0 and h_u of §13, but applied to $\widetilde{D}$ rather than to D.

14.7. Lemma. *If $g \in G_0$ then $\sqrt{-1}\partial\overline{\partial}\log h_0|_{gU_0(w)} = 0$.*

Proof. The holomorphic tangent space $\mathfrak{u} \cap \mathfrak{q}^n = \mathfrak{r}^r \cap \tau\mathfrak{r}^{-n}$ to $U_0(w)$ at w has basis given by elements $\xi_\alpha \in \mathfrak{g}_\alpha$ as α runs over $\Gamma^n = \Phi^r \cap (-\tau\Phi^n)$. Let $\alpha, \beta \in \Gamma^n$. If $\tau\xi_\beta \in \mathfrak{g}_{-\alpha}$ then $\alpha \in \tau\Phi^r \cap \Phi^n$, so then $\alpha \in \Phi^r \cap (-\tau\Phi^n) \cap \tau\Phi^r \cap \Phi^n \subset \Gamma^r \cap \Gamma^n$, which is empty. The Lie algebra cohomology computation that leads to Lemma 13.5 shows $\sqrt{-1}\partial\overline{\partial}\log h_0(\xi_\alpha, \xi_\beta) = 0$. Take linear combinations to conclude that $\sqrt{-1}\partial\overline{\partial}\log h_0|_{U_0(w)}$ is identically zero at w. As $\sqrt{-1}\partial\overline{\partial}\log h_0$ is G_0–invariant, $\sqrt{-1}\partial\overline{\partial}\log h_0|_{gU_0(w)}$ is identically zero at gw, for every $g \in G_0$. □

14.8. Lemma. *If $g \in G_0$ then $\mathcal{L}(\widetilde{\phi})|_{gU_0(w)}$ is positive definite.*
This shows in particular that the fibres $gU_0(w)$ of $\pi_D : \widetilde{D} \to D$ are Stein manifolds. We already knew that for another reason: U is unipotent, so those fibres are affine varieties.

Proof. $\sqrt{-1}\partial\overline{\partial}\log h_0|_{gU_0(w)}$ is identically zero, by Lemma 14.7. $\sqrt{-1}\partial\overline{\partial}\log h_u$ is negative definite, so $\sqrt{-1}\partial\overline{\partial}\log h_u|_{gU_0(w)}$ is negative definite, and the difference $\mathcal{L}(\widetilde{\phi})|_{gU_0(w)} = \sqrt{-1}\partial\overline{\partial}\log h_0|_{gU_0(w)} - \sqrt{-1}\partial\overline{\partial}\log h_u$ is positive definite. □

14.9. Proposition. *If $g \in G_0$ then $\widetilde{\phi}|_{gU_0(w)}$ has a unique minimum point $m(g)$, the function $\phi : D \to \mathbb{R}$ given by*

$$\phi(g(z)) = \widetilde{\phi}(m(g)) = \min\{\widetilde{\phi}(w') \mid w' \in \pi_D^{-1}(g(z))\} \tag{14.10}$$

is well defined. Furthermore, ϕ is a real analytic exhaustion function on D.

Indication of Proof. Let $g \in G_0$. If $c > 0$ then $\widetilde{D}_c = \{w' \in \widetilde{D} \mid \widetilde{\phi}(w') \leqq c\}$ is compact because $\widetilde{\phi}$ is an exhaustion function. Thus $\widetilde{D} \cap gU_0(w)$ is compact. In particular $\widetilde{\phi}|_{gU_0(w)}$ has an absolute minimum. Let $w_1 \neq w_2$ be relative minima of $\widetilde{\phi}|_{gU_0(w)}$. Choose a smooth curve s in $gU_0(w)$ from w_1 to w_2, say $s(0) = w_1$ and $s(1) = w_2$, with $s'(t) \neq 0$ for $0 < t < 1$. Set $f(t) = d\widetilde{\phi}(s'(t)) = \frac{d}{dt}\widetilde{\phi}(s(t))$. Then f has a relative maximum at some t_0 between 0 and 1. Here we use $w_1 \neq w_2$. But Lemma 14.8 says $f''(t) > 0$ for $0 < t < 1$. Thus $w_1 = w_2$. We have proved that $\widetilde{\phi}_{gU_0(w)}$ has a unique minimum point $m(g) \in gU_0(w)$.

Now $\phi : D \to \mathbb{R}$ is well defined by (14.10). Each $\pi_D(\widetilde{D}_c) = D_c$, compact, so $\phi : D \to \mathbb{R}$ is an exhaustion function. ϕ is C^ω because $M = \{m(g) \mid g \in G_0\}$, the minimum locus just described, is a C^ω subvariety of $\widetilde{D}$. □

14.11. Remark. The first part of the argument of Proposition 14.9 shows that $m(g)$ is the unique critical point of $\widetilde{\phi}|_{gU_0(w)}$. The second part of the argument shows that the minimum locus $M = \{m(g) \mid g \in G_0\}$ is a C^ω subvariety of $\widetilde{D}$.

Define $\zeta = \phi \cdot \pi_D$, so $\zeta : \widetilde{D} \to \mathbb{R}$ by $\zeta(g(w)) = \widetilde{\phi}(m(g)) = \phi(\pi_D(g(w))$. Then the holomorphic tangent spaces of the fibres of π_D are in the kernel of the Levi form $\mathcal{L}(\zeta)$, and if $g \in G_0$ then $\mathcal{L}(\zeta)_{g(w)}$ has the same number of positive eigenvalues as $\mathcal{L}(\phi)_{g(z)}$.

Denote complex dimensions of our spaces by

(14.12) $$n = \dim_{\mathbb{C}} D\ , \quad \tilde{n} = \dim_{\mathbb{C}} \widetilde{D}\ , \qquad s = \dim_{\mathbb{C}} Y\ , \quad \tilde{s} = \dim_{\mathbb{C}} \widetilde{Y}$$

where $Y = K_0(z) \subset D$ and $\widetilde{Y} = K_0(w) \subset \widetilde{D}$ are the maximal compact subvarieties. Lemma 14.6 implies $s = \tilde{s}$.

14.13. Lemma. *Recall the minimum locus $M \subset \widetilde{D}$ of* Proposition 14.9 *and* Remark 14.11. *Let $m \in M$ and let $T_m^{(1,0)}(M)$ denote the part of the holomorphic tangent space to $\widetilde{D}$ tangent to M at m. Then $\mathcal{L}(\widetilde{\phi})|_{T_m^{(1,0)}(M)}$ has at least $n - s$ eigenvalues > 0.*

Proof. Proposition 13.6, applied to $\widetilde{D}$, says that $\mathcal{L}(\widetilde{\phi})$ has at least $\tilde{n} - \tilde{s}$ eigenvalues > 0 at m, and $\dim_{\mathbb{C}} \pi_D^{-1}\pi_D(m) = \tilde{n} - n$. So $\mathcal{L}(\widetilde{\phi})|_{T_m^{(1,0)}(M)}$ has at least $n - \tilde{s} = n - s$ eigenvalues > 0. □

14.14. Corollary. *$\mathcal{L}(\zeta)$ has at least $n - s$ eigenvalues greater than zero at every point of $\widetilde{D}$.*

Theorem 14.1 follows.

References for §14.

- J. A. Wolf, "The Stein condition for cycle spaces of open orbits on complex flag manifolds", Annals of Math. **136** (1992), 541–555.
- J. A. Wolf, "Exhaustion functions and cohomology vanishing theorems for open orbits on complex flag manifold", Math. Research Letters **2** (1995), 179–191.

§15. THE STEIN PROPERTY.

Theorem 4.4 says that

(15.1) $$Y = K_0(z) \cong K_0/(K_0 \cap P_z) \cong K/(K \cap P_z)$$

is a complex submanifold of D. Furthermore, Y is not contained in any compact complex submanifold of D of greater dimension. So Y is a maximal compact subvariety of D. We will refer to

(15.2) $$M_D = \{gY \mid g \in G \text{ and } gY \subset D\}$$

as the **linear cycle space** or the **space of maximal compact linear subvarieties** of D. Since Y is compact and D is open in Z, M_D is open in

(15.3a) $$M_Z = \{gY \mid g \in G\} \cong G/L$$

where

(15.3b) $$L = \{g \in G \mid gY = Y\}, \text{ closed complex subgroup of } G.$$

Thus M_D has a natural structure of complex manifold. Its structure is given by

15.4. Theorem. *Let D be an open G_0-orbit on a complex flag manifold $Z = G/P$. Then the linear cycle space M_D is a Stein manifold.*

The first step is Proposition 15.5 below, which gives the structure of L. Note that the kernel of the action of L on Y is $E = \bigcap_{k\in K_0} kP_z k^{-1} = \bigcap_{k\in K} kP_z k^{-1}$ and that $KE \subset L \subset KP_z$.

In general, G, P, Z, D, K and Y break up as direct products according to any decomposition of $\mathfrak{g}_0$ as a direct sum of ideals, equivalently any decomposition of G_0 as a direct product. Here we use our assumption that G is connected and simply connected. So, for purposes of determining L we may, and do, assume that G_0 is noncompact and simple, in other words that G_0/K_0 is an irreducible riemannian symmetric space of noncompact type.

As before, we say that G_0 is of **hermitian type** if the irreducible riemannian symmetric space G_0/K_0 is an hermitian symmetric space.

Let θ be the Cartan involution of G_0 with fixed point set K_0 and $\mathfrak{g} = \mathfrak{k} + \mathfrak{s}$ under θ, as usual. By the irreducibility of G_0/K_0, the adjoint action of K_0 on $\mathfrak{s}_0 = \mathfrak{g}_0 \cap \mathfrak{s}$ is irreducible. G_0 is of hermitian type if and only if this action fails to be absolutely irreducible. Then there is a positive root system ${}'\Sigma^+ = {}'\Sigma^+(\mathfrak{g}, \mathfrak{h})$ such that $\mathfrak{s} = \mathfrak{s}_+ + \mathfrak{s}_-$ where $\mathfrak{s}_+$ is a sum of ${}'\Sigma^+$–positive root spaces and represents the holomorphic tangent space of G_0/K_0, and $\mathfrak{s}_- = \overline{\mathfrak{s}_+}$ is a sum of ${}'\Sigma^+$–negative root spaces and represents the antiholomorphic tangent space. Write $S_\pm = exp(\mathfrak{s}_\pm) \subset G$. Then G_0/K_0 is an open G_0–orbit on G/KS_-.

Recall the compact real form $\mathfrak{g}_u = \mathfrak{k}_0 + \sqrt{-1}\, \mathfrak{s}_0$ of $\mathfrak{g}$. The corresponding real analytic subgroup G_u of G is a compact real form, thus is a maximal compact subgroup, and $K_0 = G_0 \cap G_u$. K_0 is its own normalizer in G_0, but its normalizer $N_{G_u}(K_0)$ in G_u can have several components.

15.5. Proposition. *Either G_0 is of hermitian type and $L = KE = KS_\pm$, connected, or*[3] *$L \subset KN_{G_u}(K)$ with identity component $L^0 = K$. In either case $G_0 \cap L = K_0$.*

The proof is a run through the structural possibilities for G_0 and $\mathfrak{p}_z$. The group $V = G \cap P_z$ is compact in Cases 1 and 2 below, is noncompact in Cases 3 and 4, and can be either compact or noncompact in Cases 5 and 6. The cases are

(1) G_0 is of hermitian type with $P_z \subset KS_-$. In this case $L = KE = KS_-$ and $G_0 \cap L = K_0$.

(2) G_0 is of hermitian type with $P_z \subset KS_+$. As in Case 1, $L = KE = KS_+$ and $G_0 \cap L = K_0$.

(3) G_0 is of hermitian type with $P_z \not\subset KS_-$, $P_z \not\subset KS_+$ and $S_- \subset P_z$. In this case $L = KE = KS_-$ and $G_0 \cap L = K_0$.

(4) G_0 is of hermitian type with $P_z \not\subset KS_-$, $P_z \not\subset KS_+$ and $S_+ \subset P_z$. Arguing as in Case 3, we conclude that $L = KE = KS_+$ and $G_0 \cap L =$

[3] This latter situation occurs both for G_0 of hermitian type and for G_0 not of hermitian type.

K_0.

(5) G_0 is of hermitian type with $P_z \not\subset KS_-$, $P_z \not\subset KS_+$, $S_- \not\subset P_z$ and $S_+ \not\subset P_z$. In this case $L^0 = K$. and $G_0 \cap L = K_0$.

(6) G_0 is not of hermitian type. In this case $L^0 = K$ and $G_0 \cap L = K_0$.

15.6. Corollary. *Either L is a parabolic subgroup $KS_\pm$ of G and $M_Z = G/L$ is a projective algebraic variety, or L is a reductive subgroup of G with identity component K and $M_Z = G/L$ is an affine algebraic variety.*

Consider the first of the two cases of Corollary 15.6. There the result is

15.7. Proposition. *Suppose that M_Z is a projective algebraic variety. Then the open orbit $D \subset Z$ is measurable and M_D is a bounded symmetric domain. In particular M_D is a Stein manifold.*

Indication of Proof. G_0 is of hermitian type and we may assume $L = KS_-$. Also, $M_Z = G/L$ is the standard complex realization of the compact hermitian symmetric space G_u/K. Now

$$(15.8) \qquad G\{D\} = \{g \in G \mid gY \subset D\}.$$

is open in G and $M_D = \{gY \mid g \in G\{D\}\}$. M_D is stable under the action of G_0 so

(15.9) $\quad G\{D\}$ is a union of double cosets $G_0 g L$ with $g \in G$.

The proof of Proposition 15.7 consists of showing that only the identity double coset occurs in $G\{D\}$. The double cosets $G_0 g L$ of (15.9) are in one-to-one correspondence with the G_0–orbits on M_Z. Those orbits are completely understood, as described in §5. Following Theorem 5.9 there is a (necessarily finite) set C of transforms $c_\Gamma c_\Delta^2$, where Γ and Δ are disjoint subsets of Ξ (see (5.8), such that (i) if $c_\Gamma c_\Delta^2, c_{\Gamma'} c_{\Delta'}^2 \in C$ with $|\Gamma| = |\Gamma'|$ and $|\Delta| = |\Delta'|$ then $\Gamma = \Gamma'$ and $\Delta = \Delta'$ and (ii) $G_{\mathbb{C}}\{D\} = \bigcup_{c \in C} GcL$. So if $c_\Gamma c_\Delta^2 \in C$ then $c_{\Gamma \cup \Delta'}^2 \in C$ for every subset $\Delta' \subset \Delta$. In particular, if $c_\Delta^2 \notin C$ whenever $\emptyset \neq \Delta \subset \Xi$ then $C = \{1\}$ and $G_{\mathbb{C}}\{D\} = GL$.

Now the proof of Proposition 15.7 is reduced to the proof that $c_\Delta^2 \notin C$ for all non–empty subsets $\Delta \subset \Xi$. That is seen by an analysis of the boundary of $G_0(1 \cdot L)$ in terms of some natural norms on $\mathfrak{g}$ and certain of its subspaces.

Next consider the second of the two cases of Corollary 15.6. There the result is

15.10. Proposition. *Suppose that M_Z is an affine algebraic variety. Then M_D is an open Stein subdomain of the Stein manifold M_Z.*

Indication of Proof. Recall the real analytic exhaustion function $\phi : D \to \mathbb{R}$ of Proposition 14.9. We use it to define $\phi_M : M_D \to \mathbb{R}^+$ by

$$(15.11) \qquad \phi_M(gY) = \sup_{y \in Y} \phi(g(y)) = \sup_{k \in K} \phi(gk(z)) .$$

$W = G\{D\} = \{g \in G \mid gY \subset D\}$ is open in G, so

$$\psi : W \times K_0 \to \mathbb{R}^+ \text{ by } \psi(g,k) = \phi(gk(z))$$

is a C^ω function on the C^ω manifold $W \times K_0$. Thus the set defined by vanishing of the differential in the K_0–variable,

$$\widetilde{U} = \{(g,k) \in W \times K_0 \mid d_{K_0}\psi(g,k) = 0\},$$

is a C^ω subvariety of $W \times K_0$. $\widetilde{U}$ is a union of C^ω subvarieties, one of which is

$$U = \{(g,k) \in W \times K_0 \mid \psi(g,k) = \sup_{k' \in K_0} \phi(gk(z))\}.$$

The map $f : U \to M_D$ given by $f(g,k) = gY$ is C^ω. If $(g,k) \in U$ then $\phi(gk(z)) = \psi(g,k) = \phi_M(gY)$. Since $f : U \to M_D$ is C^ω and surjective, and since $\psi|_U$ is C^ω, now ϕ_M is C^ω.

By construction, $\psi(g,k)$ is constant in the second variable $k \in K_0$. The Levi form $\mathcal{L}(\phi)$ has its positive eigenvalues in directions transversal to the $gY = gK_0(z)$, so the Levi form $\mathcal{L}(\phi_M)$ on M_D is positive semidefinite and ϕ_M is plurisubharmonic. Now, using the fact that ϕ is an exhaustion function on D,

15.12. Lemma. *ϕ_M is a real analytic plurisubharmonic function on M_D. If Y_∞ is a point on the boundary of M_D in M_Z and $\{Y_i\}$ is a sequence in M_D that tends to Y_∞ then* $\lim_{Y_i \to Y_\infty} \phi_M(Y_i) = \infty$.

The last step is to modify ϕ_M to obtain a strictly plurisubharmonic exhaustion function on M_D. Since M_Z an affine algebraic variety, it is Stein. Now M_Z carries a strictly plurisubharmonic exhaustion function N, and $\zeta = \phi_M + N : M_D \to \mathbb{R}$ is a strictly plurisubharmonic exhaustion function on M_D. It follows that M_D is Stein.

Proof of Theorem 15.4. Theorem 15.4 follows from Proposition 15.7 when M_Z is a projective algebraic variety, from Proposition 15.10 when M_Z is an affine algebraic variety. Proposition 15.5 says that these are the only cases.

References for §15.

- A. Andreotti and R. Narasimhan, "Oka's Heftungslemma and the Levi problem for complex spaces", Trans. Amer. Math. Soc. **111** (1964), 345–366.
- F. Docquier and H. Grauert, "Levisches Problem und Rungescher Satz für Teilgebiete Steinscher Mannigfaltigkeiten", Math. Annalen **140** (1960), 94–123.
- H. Grauert, "On Levi's problem and the imbedding of real–analytic manifolds", Annals of Math. **68** (1958), 460–472.
- J. A. Wolf, "The Stein condition for cycle spaces of open orbits on complex flag manifolds", Annals of Math. **136** (1992), 541–555.
- J. A. Wolf, "Exhaustion functions and cohomology vanishing theorems for open orbits on complex flag manifolds", Math. Research Letters **2** (1995), 179–191.

§16. Connection with Double Fibration Transforms.

In general, let $D = G_0(z)$ be an open orbit in the complex flag manifold $Z = G/P$, let Y be the maximal compact linear subvariety $K_0(z)$, and consider the linear cycle space $M_D = \{gY \mid g \in G \text{ and } gY \subset D\}$. Then we have a double fibration

$$(16.1) \qquad \begin{array}{ccc} \mathcal{Y}_D & \xrightarrow{\;p_M\;} & M_D \\ {\scriptstyle p_D}\big\downarrow & & \\ D & & \end{array}$$

where $\mathcal{Y}_D = \{(Y', y') \mid y' \in Y' \in M_D\}$, and the projections $p_M(Y', y') = Y'$ and $p_D(Y', y') = y'$.

Let $n = \dim_{\mathbb{C}} D$ and $s = \dim_{\mathbb{C}} Y$ as before. Consider a **negative** homogeneous holomorphic vector bundle $\mathbb{E} \to D$. Then we can expect nonzero cohomology only in degree s. For many purposes, for example for making estimates of one sort or another, it would be preferable to have representations of G_0 occur on spaces of functions rather than on cohomology spaces, and here we use a double fibration transform to carry $H^s(D; \mathcal{O}(\mathbb{E}))$ to a space of functions on M_D. For this, one first considers the pullback $p_D^*\mathcal{O}(\mathbb{E}) \to \mathcal{Y}_D$ and then the G_0–homogeneous s^{th} Leray direct image sheaf $\mathcal{F} = \mathcal{R}^s(p_D^*\mathcal{O}(\mathbb{E})) \to M_D$. Here $\mathcal{F}$ is locally free so it corresponds to a G_0–homogeneous holomorphic vector bundle $\mathbb{F} \to M_D$, and $\mathbb{F} \to M_D$ is holomorphically trivial because M_D is Stein. In this way one carries the G_0–module $H^s(D; \mathcal{O}(\mathbb{E}))$ to a space of sections of $\mathbb{F}$ and thus to a space of functions with values in the typical fibre $F = H^s(Y; p_D^*\mathcal{O}(\mathbb{E}))$ of $\mathbb{F}$. Of course, if M_Z is a projective algebraic variety then, by Proposition 15.7, G_0 is of hermitian type and M_D is the bounded symmetric domain G_0/K_0.

Consider the special case where $G_0 = SU(2,2)$ and Z is the complex projective space $P^3(\mathbb{C})$ and $\mathbb{E} \to D$ is a negative line bundle. Here there are two open orbits, the positive definite lines in $\mathbb{C}^{2,2}$ and the negative definite lines, and $s = 1$ for each of them. The maps of $H^s(D; \mathcal{O}(\mathbb{E}))$ to a space of F–valued functions of G_0/K_0 are the classical Penrose Transforms.

In the general case, in order to make the double fibration transform explicit one needs to know

the exact structure of M_D and the differential equations that pick out the functions $f_\sigma : M_D \to F$ that correspond to cohomologies $\sigma \in H^s(D; \mathcal{O}(\mathbb{E}))$.

The second item here is relatively straightforward. There is some recent progress on the first item by Dunne, Novak, Zierau and myself. In almost all cases Zierau and I have shown that if G_0 is of hermitian type and if M_Z is not a projective algebraic variety then $M_D \cong (G_0/K_0) \times (\overline{G_0/K_0})$.

References for §16.

- R. J. Baston and M. G. Eastwood, "The Penrose Transform: Its Interaction with Representation Theory", Clarendon Press, Oxford, 1989.

- E. G. Dunne and R. Zierau, "Twistor theory for indefinite Kaehler symmetric spaces", Contemporary Math. **154** (1993), 117–132.
- J. D. Novak, "Parametrizing Maximal Compact Subvarieties", Proc. Amer. Math. Soc. **124** (1966), 969–975.
- J. D. Novak, Ph.D. Thesis, Oklahoma State University, 1995.
- R. O. Wells, "Parametrizing the compact submanifolds of a period matrix domain by a Stein manifold", in Symposium on Several Complex Variables, Springer Lecture Notes in Mathematics **184** (1971), 121–150.
- J. A. Wolf and R. Zierau, "Riemannian exponential maps and decompositions of reductive Lie groups", in Topics in Geometry, Progress in Nonlinear Differential Equations and Applications, **20** (1996), Birkhäuser, Boston, 349–353.
- J. A. Wolf and R. Zierau, "Cayley transforms and orbit structure in complex flag manifolds", to appear.
- J. A. Wolf and R. Zierau, "Linear Cycle Spaces in Flag Domains", "Linear cycle spaces in flag domains, I: Flags corresponding to the classical groups", ESI Preprint No. 357, 1996.
- J. A. Wolf and R. Zierau, "Linear Cycle Spaces in Flag Domains, II: The general case", in preparation.

Department of Mathematics
University of California at Berkeley
Berkeley, CA 94720

Progress in Mathematics

Edited by:

Hyman Bass
Dept. of Mathematics
Columbia University
New York, NY 10010
USA

J. Oesterlé
Institut Henri Poincaré
11, rue Pierre et Marie Curie
75231 Paris Cedex 05
FRANCE

A. Weinstein
Department of Mathematics
University of California
Berkeley, CA 94720
USA

Progress in Mathematics is a series of books intended for professional mathematicians and scientists, encompassing all areas of pure mathematics. This distinguished series, which began in 1979, includes authored monographs and edited collections of papers on important research developments as well as expositions of particular subject areas.

We encourage preparation of manuscripts in some form of TeX for delivery in camera-ready copy which leads to rapid publication, or in electronic form for interfacing with laser printers or typesetters.

Proposals should be sent directly to the editors or to: Birkhäuser Boston, 675 Massachusetts Avenue, Cambridge, MA 02139, U. S. A.

123 BRYLINSKI/BRYLINSKI/GUILLEMIN/KAC. Lie Theory and Geometry: In Honor of Bertram Kostant
124 AEBISCHER/BORER/KALIN/LEUENBERGER/ REIMANN. Symplectic Geometry
125 LUBOTZKY. Discrete Groups, Expanding Graphs and Invariant Measures
126 RIESEL. Prime Numbers and Computer Methods for Factorization
127 HORMANDER. Notions of Convexity
128 SCHMIDT. Dynamical Systems of Algebraic Origin
129 DIJGRAAF/FABER/VAN DER GEER. The Moduli Space of Curves
130 DUISTERMAAT. Fourier Integral Operators
131 GINDIKIN/LEPOWSKY/WILSON (eds). Functional Analysis on the Eve of the 21st Century. In Honor of the Eightieth Birthday of I. M. Gelfand, Vol. 1
132 GINDIKIN/LEPOWSKY/WILSON (eds.) Functional Analysis on the Eve of the 21st Century. In Honor of the Eightieth Birthday of I. M. Gelfand, Vol. 2
133 HOFER/TAUBES/WEINSTEIN/ZEHNDER. The Floer Memorial Volume
134 CAMPILLO LOPEZ/NARVAEZ MACARRO (eds) Algebraic Geometry and Singularities
135 AMREIN/BOUTET DE MONVEL/GEORGESCU. C_0-Groups, Commutator Methods and Spectral Theory of N-Body Hamiltonians
136 BROTO/CASACUBERTA/MISLIN (eds). Algebraic Topology: New Trends in Localization and Periodicity
137 VIGNERAS. Représentations l-modulaires d'un groupe réductif p-adique avec $l = p$
138 BERNDT/DIAMOND/HILDEBRAND (eds).
139 BERNDT/DIAMOND/HILDEBRAND (eds). Analytic Number Theory, Vol. 2 In Honor of Heini Halberstam
140 KNAPP. Lie Groups Beyond an Introduction
141 CABANES. Finite Reductive Groups: Related Structures and Representations
142 MONK. Cardinal Invariants on Boolean Algebras
143 GONZALEZ-VEGA/RECIO. Algorithms in Algebraic Geometry and Applications
144 BELLAÏCHE/RISLER. Sub-Riemannian Geometry
145 ALBERT/BROUZET/DUFOUR (eds). Integrable Systems and Foliations Feuilletages et Systèmes Intégrables
146 JARDINE. Generalized Etale Cohomology
147 DIBIASE. Fatou Theorems
148 HUANG. Two-Dimensional Conformal Geometry and Vertex Operator Algebras
149 SOURIAU. Structure of Dynamical Systems. A Symplectic View of Physics
150 SHIOTA. Geometry of Subanalytic and Semialgebraic Sets
151 HUMMEL. Gromov's Compactness Theorem For Pseudo-holomorphic Curves
152 GROMOV. Metric Structures for Riemannian Manifolds
153 BUESCU. Exotic Attractors: From Liapunov Stability to Riddled Basins
154 BOTTCHER/KARLOVICH. Carleson Curves, Muckenhoupt Weights, and Toeplitz Operators
155 DRAGOMIR. Locally Conformal Kähler Geometry
156 GUIVARC'H/JI/TAYLOR. Compactifications of Symmetric Spaces
157 MURTY/MURTY. Non-vanishing of L-functions and Applications
158 TIRAO/VOGAN/WOLF. Geometry and Representation Theory of Real and p-adic Groups